Weak interactions of leptons and quarks

Weak interactions of leptons and quarks

EUGENE D. COMMINS
Department of Physics, University of California, Berkeley

PHILIP H. BUCKSBAUM
Bell Telephone Laboratories, Murray Hill, New Jersey

CAMBRIDGE UNIVERSITY PRESS
Cambridge
London   New York   New Rochelle
Melbourne   Sydney

Published by the Press Syndicate of the University of Cambridge
The Pitt Building, Trumpington Street, Cambridge CB2 1RP
32 East 57th Street, New York, NY 10022, USA
296 Beaconsfield Parade, Middle Park, Melbourne 3206, Australia

© Cambridge University Press 1983

First published 1983

Printed in the United States of America

Library of Congress Cataloging in Publication Data
Commins, Eugene D.
Weak interactions of leptons and quarks.
Bibliography: p.
Includes index.
1. Lepton interactions. 2. Quarks.
3. Weak interactions (Nuclear physics)
I. Bucksbaum, Philip H. II. Title.
QC794.8.W4C65 539.7'54 82-4452

ISBN 0 521 23092 6   hard covers
ISBN 0 521 27370 6   paperback

Contents

Preface ix

1 Introduction to weak interactions 1
1.1 Particles and interactions 1
1.2 Early history of weak interactions: The Fermi era 7
1.3 Parity nonconservation 10
1.4 The V–A law 12
1.5 The two-component neutrino 15
1.6 Difficulties of Fermi-type theories 16
1.7 The naive intermediate boson hypothesis 18
1.8 Charm and the GIM model 19
1.9 Classification of weak interactions 22

2 The standard model 33
2.1 Introduction 33
2.2 $U(1)$ gauge invariance and electrodynamics 37
2.3 The Yang–Mills field 41
2.4 Spontaneous symmetry breaking 44
2.5 Couplings of leptons and quarks to gauge fields 52
2.6 Lepton and quark couplings to the scalar field 58
2.7 The path integral in quantum mechanics 63
2.8 The path integral in field theory 65
2.9 Path integrals and gauge invariance in electrodynamics: The massless Yang–Mills field 68
2.10 Feynman rules for the standard model in R_ξ gauge 71
2.11 Some brief comments on renormalization 78
2.12 Left–right symmetric theories 85
2.13 The Higgs boson 86

3	**Leptonic weak interactions**	**92**
3.1	Introduction	92
3.2	Muon decay and the V–A law	92
3.3	A general muon-decay amplitude: Michel parameters	98
3.4	Experimental determination of muon parameters	101
3.5	τ leptons	108
3.6	Weak–electromagnetic interference effects in $e^+e^- \to \mu^+\mu^-$	113
3.7	Z^0 decay, $W^\pm$ decay	115
3.8	Z^0 production in e^+e^- collisions	117
3.9	$e^+e^- \to W^+W^-$; W and Z^0 production in $p\bar{p}$ collisions	118
3.10	Neutrino–electron elastic scattering and inverse muon decay	121
4	**General properties of hadronic weak currents**	**131**
4.1	Introduction	131
4.2	Unitary symmetry and the quark model	132
4.3	The six-quark model	150
4.4	Decay of charged pions	153
4.5	$K_{\mu 2}$ and K_{e2} decays	155
4.6	Other decays related to $\pi_{\mu 2}$	155
4.7	Restrictions on weak amplitudes from proper Lorentz invariance	156
4.8	Time-reversal invariance; CPT invariance	161
4.9	The conserved vector current hypothesis	165
4.10	Second-class currents and G parity	168
4.11	The partially conserved axial current hypothesis	169
4.12	Form factors for neutral currents	172
4.13	The empirical selection rules $\Delta S = \Delta Q$ and $\lvert \Delta I \rvert = \tfrac{1}{2}$	173
4.14	Current algebra	175
5	**Nuclear β-decay and muon capture**	**178**
5.1	The impulse approximation	178
5.2	Allowed β decay	180
5.3	The transition probability for allowed decays	182
5.4	β-decay coupling constants: Validity of the V–A law	191
5.5	Tests of CVC and searches for second-class currents	192
5.6	Muon capture	197
6	**Weak decays of mesons and hyperons**	**204**
6.1	Introduction	204
6.2	K_{l3} decays	204
6.3	Kaon nonleptonic decays	209
6.4	Hyperon semileptonic decays	215

6.5	Hyperon nonleptonic decays	223
6.6	Possible dynamical origins of the $|\Delta I| = \frac{1}{2}$ rule	228
6.7	Weak decays of charmed hadrons	232
6.8	Bottom mesons	241

7 Neutral K mesons and CP violation — 244

7.1	Introduction	244
7.2	Strangeness oscillations and regeneration	247
7.3	The mass and decay matrices	249
7.4	Eigenvalues and eigenvectors of $\Gamma + iM$	255
7.5	Transition amplitudes for $K_S \to \pi\pi$ and $K_L \to \pi\pi$	258
7.6	Measurement of η^{+-} and η^{00}	260
7.7	Regeneration and the $K_L - K_S$ mass difference	262
7.8	CP violation in K_{l3} decays	268
7.9	CPT conservation in K^0 decay	269
7.10	The phase of ϵ and the magnitude of Γ_{12} and M_{12}	270
7.11	The phase of ϵ'	271
7.12	Summary of CP-violation experimental data	272
7.13	Origins of CP violation	272
7.14	CP violation in heavy-quarked mesons D^0, B^0, T^0	283

8 High-energy neutrino–nucleon collisions — 289

8.1	Introduction	289
8.2	Experimental methods	290
8.3	Charged-current neutrino–nucleon elastic scattering	295
8.4	Neutral-current neutrino–nucleon elastic scattering	299
8.5	Deep inelastic charged-current neutrino–nucleon interactions	303
8.6	Charmed-particle production in neutrino beams	317
8.7	Tests of the standard model with neutral neutrino–nucleon reactions	320

9 The parity-nonconserving eN and NN interactions — 336

9.1	Introduction	336
9.2	Deep inelastic scattering of polarized electrons	339
9.3	Parity violation in atoms	343
9.4	Constraints on neutral-current parameters	352
9.5	Parity violation in nuclear forces	354

10 Lepton mixing, neutrino oscillations, and neutrino mass — 369

10.1	Introduction	369
10.2	$\Delta L = 0$ neutrino oscillations	371
10.3	Neutrino oscillations with three lepton generations	374

viii Contents

10.4	$|\Delta L| = 2$ oscillations and Majorana neutrinos	376
10.5	Neutrino refractive index	380
10.6	Neutrino oscillation experiments	382
10.7	Double β decay	385
10.8	Direct measurements of neutrino mass	390
10.9	Neutrino decay; neutrino magnetic moment	393
11	**Neutrino astrophysics**	**396**
11.1	Introduction	396
11.2	Neutrinos from the sun	397
11.3	Emission of neutrinos by hot stars	404
11.4	Supernova explosions and neutrino emission	410
11.5	Neutrino-burst detectors	412
11.6	Neutrinos of cosmological origin	413
11.7	High-energy neutrino detection	419
12	**Summary and conclusions**	**421**
	Appendix A: Notation and conventions	**426**
A.1	Relativistic notation	426
A.2	The Dirac equation	427
A.3	Representations of γ matrices	428
A.4	Dirac fields	429
A.5	Parity	430
A.6	Charge conjugation	430
A.7	Time reversal	430
	Appendix B: The S-matrix: Transition probabilities and cross sections	**432**
	Appendix C: Dimensional regularization	**436**
	Appendix D: Applications of current algebra	**439**
D.1	The Adler–Weisberger relation	439
D.2	The Ademollo–Gatto theorem	441
	Appendix E: Physical constants	**444**
	Selected bibliography	*445*
	References	*447*
	Index	*463*

Preface

Some years ago, one of us (E.D.C.) wrote a book called *Weak Interactions*, which was an attempt to describe the field in simple and concrete terms. Despite numerous shortcomings, it enjoyed modest success, and a number of students and colleagues asked when a revised edition might appear. In the meantime, however, the subject underwent a vast revolution. Renormalizable gauge theories unifying the weak and electromagnetic interactions were invented; neutral currents, τ leptons, and charmed particles were discovered; and the whole point of view regarding elementary particles changed radically. Thus, we decided to prepare this volume, which is not a revision but a new book. Our goal is still to present the subject in simple, concrete terms. For our audience, we have in mind students, experimentalists, and physicists not specializing in the field, rather than sophisticated theorists. Thus experimental results are emphasized, and very difficult or speculative theoretical topics are treated in a simplified fashion, or else omitted altogether. For example, only the most elementary notions in quantum chromodynamics are discussed here, grand unified theories receive only passing mention, and proofs of renormalizability are omitted.

The reader should find that in most places a general knowledge of elementary particle physics and the one-particle Dirac theory with the usual Feynman rules is adequate, but at several crucial points some knowledge of quantum field theory is necessary. Thus the reader should be familiar with the contents of Sakurai's excellent text: *Advanced Quantum Mechanics* (Sakurai, 1967). In the present book, many calculational details are briefly summarized or omitted, and problems are provided to help fill in some of the gaps.

We are grateful to many persons for assistance, direct or indirect. A great deal was learned from study of the excellent unpublished notes on Yang–Mills theories by L. Maiani and from useful discussions with R. Cahn, M. Chanowitz, H. Primakoff, and many others. We thank E. Segrè for wise advice, and M. Suzuki and P. Drell for reading the entire manuscript and offering very detailed and useful suggestions and criticism. Thanks also go to Claudia Madison for typing the manuscript with patience and skill.

Eugene D. Commins
Philip H. Bucksbaum[1]

1 P.H.B. contributed to this book while a post-doctoral physicist at the University of California, Berkeley.

1
Introduction to weak interactions

1.1 Particles and interactions

All matter consists of elementary particles, governed by four known interactions: the gravitational, the electromagnetic, the strong, and the weak. In this book, we study the weak interaction. Our subject was once limited to nuclear β decay, but it has grown immensely and now includes the decays of muons and tau leptons, the slow decays of mesons and baryons, muon capture by nuclei, all neutrino interactions with matter, strangeness oscillations, and many other phenomena, some only recently discovered.

Since the 1960s, very profound changes have occurred in our understanding of weak interactions. Remarkable new ideas based on the principle of *local gauge invariance* have emerged. They provide a unified account of the weak and electromagnetic interactions and show that, despite striking differences in their observed characteristics, these may be regarded as but two different manifestations of a single, more fundamental, *electroweak interaction*.

The new ideas have crystallized in a now widely accepted "standard" model of electroweak phenomena, created principally by S. Glashow (1961,80) and co-workers (70), S. Weinberg (67,72,80), A. Salam (68,80), and Salam and J. Ward (64). The principle of local gauge invariance has also been applied to strong interactions, and a promising theory, quantum chromodynamics (QCD), has emerged and is now so closely related to the electroweak theory that it is hardly possible to separate them.

In this introduction we must therefore begin by presenting the elementary particles and quanta, both observed and hypothetical, that inhabit the world of both theories. These are the leptons, the quarks, the photon, the weak intermediate bosons, the gluons, and the Higgs bosons.

The *lepton* family consists of the electron e^-, its close relatives the muon μ^-, and the tau lepton τ^- (the last discovered as recently as 1975), as well as their associated neutrinos ν_e, ν_μ, and ν_τ. Each lepton possesses spin $\frac{1}{2}$, and for each there is a corresponding antilepton: e^+, μ^+, τ^+, $\bar{\nu}_e$, $\bar{\nu}_\mu$, and $\bar{\nu}_\tau$. The leptons are divided into distinct doublets, or *generations*, as follows:

$$\begin{pmatrix} \nu_e \\ e^- \end{pmatrix}, \quad \begin{pmatrix} \nu_\mu \\ \mu^- \end{pmatrix}, \quad \begin{pmatrix} \nu_\tau \\ \tau^- \end{pmatrix} \tag{1.1}$$

in which the upper members (neutrinos) possess zero electric charge and experience only weak interactions and the lower members have charge $e = -4.8 \times 10^{-10}$ esu and suffer both electromagnetic and weak interactions. Each lepton generation also has associated with it an additive quantum number called the *lepton number* (L_e, L_μ, L_τ), which is $+1$ for each lepton in that generation, -1 for each antilepton of that generation, and 0 for all other particles. Experiment shows that each lepton number is conserved in all known processes. More lepton generations, still undiscovered, may well exist.

The strongly interacting particles, or *hadrons*, consist of the half-integral-spin *baryons* (p, n, Λ^0, etc.) and the integral-spin *mesons* (π, K, ρ, ψ, etc.). Until relatively recently, it was thought that hadrons were elementary particles, but now we know that they are composed of more elementary objects of fractional charge and spin $\frac{1}{2}$: the *quarks*. These occur in distinct varieties, or "flavors," of which five have been identified so far:

"up" u
"down" d
"charmed" c
"strange" s
"bottom" b

In addition, the existence of a sixth quark – "top," or t – is suspected. Granting this, we may arrange the quarks in distinct doublets or generations, just as we did for the leptons:

$$\begin{pmatrix} u \\ d \end{pmatrix}, \quad \begin{pmatrix} c \\ s \end{pmatrix}, \quad \begin{pmatrix} t \\ b \end{pmatrix} \tag{1.2}$$

In each quark doublet, the upper member (u, c, or t) has electric charge $Q = \frac{2}{3}|e|$ and the lower member (d, s, or b) possesses charge $Q = -\frac{1}{3}|e|$. To each quark there corresponds an antiquark – $\bar{u}$, $\bar{d}$, $\bar{c}$, $\bar{s}$,

1.1 Particles and interactions

$\bar{t}$, and $\bar{b}$ with opposite electric charge. Quarks and antiquarks experience strong, electromagnetic, and weak interactions. We may assign an additive quantum number, the *baryon number*, to quarks to be conserved in all known processes. Each quark has baryon number $B = \frac{1}{3}$, whereas for each antiquark, $B = -\frac{1}{3}$. As in the case of leptons, we cannot exclude the possibility that more quark generations exist.

One may ask if the leptons and quarks are themselves elementary! For our purposes, they will be so regarded, but many physicists believe that leptons and quarks are composites of still more fundamental objects, which various authors have named *rishons, preons,* or *haplons* (Harari 79, Shupe 79, Terazawa 80, Fritzsch and Mandelbaum 81). These ideas are very attractive, but so far they remain speculations.

Let us consider very briefly the quark structure of hadrons, saving details for later. According to the present view, a meson consists of a "valence" quark–antiquark pair $q\bar{q}$ in a bound state, together with an "ocean" of virtual quark–antiquark pairs. For example, the valence quark composition of several mesons is as follows:

$$\pi^+ = u\bar{d}$$
$$\pi^- = d\bar{u}$$
$$\phi^0 = s\bar{s}$$
$$K^+ = u\bar{s}$$
$$D^0 = c\bar{u}$$
$$F^+ = c\bar{s}$$

A baryon is composed of three valence quarks qqq and also an ocean of $q\bar{q}$ pairs. The valence quark composition of several baryons is as follows:

$$p = uud$$
$$n = ddu$$
$$\Sigma^+ = uus$$
$$\Xi^0 = uss$$
$$\Omega^- = sss$$
$$\Lambda_c^+ = udc$$

The evidence for the quark structure of hadrons is massive and convincing, but it is indirect; for there is no incontestable experimental evidence for the existence of a single free quark,[1] and indeed there are good reasons to believe that quarks must remain forever confined

---

[1] The results of one experiment by W. Fairbank and co-workers (La Rue et al. 81) appear to reveal the existence of fractional charges in matter. However, the meaning of this result is not yet clear.

within hadrons. According to quantum chromodynamics, the force binding quarks together is feeble at short distances but increases with relative separation. Thus the work required to separate a quark from its parent hadron is very large or infinite, and the expenditure of a large amount of energy in attempting this separation results in the production of quark–antiquark pairs from the vacuum. The net result is production of new hadronic states rather than liberation of the original quark.

From the fundamental fermions – the leptons and quarks – we now turn to the various intermediate bosons of electromagnetic, weak, and strong interactions, which transmit the forces between fermions. In the standard model of electroweak phenomena, there are four such bosons. They are the massless photon γ of quantum electrodynamics and three massive quanta – W^+, W^- (with electric charges $\pm|e|$, respectively), and a neutral boson Z^0. Just as the photon "mediates" processes such as electron–electron scattering in quantum electrodynamics (see Figure 1.1a), so W^+ and W^- act as intermediaries in the *charged* weak interactions such as nuclear β decay, muon decay, and so forth (see Figure 1.1b). The Z^0 plays a similar role for an entirely new class of *neutral* weak interactions (see Figure 1.1c, for example),

Figure 1.1. (a) e–e scattering by photon exchange. (b) Muon decay by W^- exchange. (c) ν_μ–e scattering by Z^0 exchange.

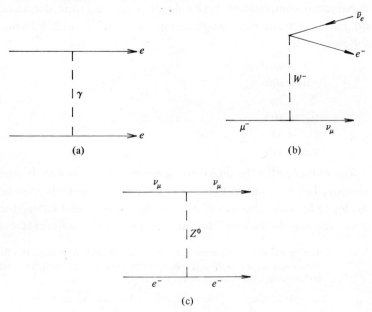

1.1 Particles and interactions

which are predicted to exist in the standard model and to possess very specific properties.

Neutral weak interactions were first observed in 1973. Subsequently, their properties have been elucidated in a number of remarkable experiments, which verify the predictions of the standard model in this domain and constitute a brilliant success for the theory. Until very recently there was no direct experimental evidence for the existence of $W^\pm$ or Z^0. Preliminary observations at CERN do confirm that W's exist with the expected mass, but the detailed properties of these important particles remain to be studied in the future.

The charged and neutral intermediate bosons of weak interactions must be massive because the weak forces have very short range. At the same time, we require of any acceptable theory of weak interactions that it be renormalizable. That is, there must be a systematic procedure for eliminating the divergences encountered in calculations of higher-order corrections to a given process, so that sensible answers for these corrections can be obtained. Quantum electrodynamics is renormalizable, but the older theories of weak interactions are not, the chief stumbling block being the necessity of having a short-range interaction. However, as was first proved by G. 't Hooft (71a,b), the new electroweak theory is renormalizable, even though it includes massive intermediate vector bosons.

In the new theory, the mechanism for imparting mass to the intermediate bosons is *spontaneous symmetry breaking,* an essential feature that, together with local gauge invariance, forms the cornerstone of the theory. An inevitable consequence of this mechanism is the appearance in the theory of certain massive scalar particles called *Higgs bosons.* A curious feature is that, although these particles are required, no definite prescription is given for their mass, which could be anywhere within the range of 7 to 1000 GeV/c^2.

The strong interaction between quarks is mediated by *gluons,* which are hypothetical spin-1 (vector) quanta of zero rest mass. The quark–quark–gluon interaction is intimately associated with the concept of *color charge,* from which quantum chromodynamics acquires its name. In addition to electric charge, quarks also possess an additive quantum number called *color,* which plays no role for leptons. A quark of a given flavor may exist in one of three color states [called, arbitrarily, red (R), blue (B), and green (G)]. An antiquark possesses *anticolor* ($\bar{R}$, $\bar{B}$, or $\bar{G}$). Hadrons do not have color, because the color of their constituent quarks and/or antiquarks cancels out. That is, hadrons exist in color-singlet states.

Gluons are similar to photons in that they have zero rest mass and spin 1. However, photons carry no charge, electrical or otherwise, whereas gluons carry color charge. In fact, it is presumed that there exist eight distinct gluon states, identified by color charge as follows:

$$R\bar{B}$$
$$B\bar{R}$$
$$R\bar{G}$$
$$G\bar{R}$$
$$G\bar{B}$$
$$B\bar{G}$$
$$\frac{1}{\sqrt{2}}(B\bar{B} - G\bar{G})$$
$$\frac{1}{\sqrt{6}}(B\bar{B} + G\bar{G} - 2R\bar{R})$$

Because of their color charge, the gluons possess nonlinear self-interactions that have no analog in electrodynamics and that account for the fact that the quark–quark interaction becomes feeble at short distances (*asymptotic freedom*). It appears that the peculiar color properties of gluons and quarks may also yield an understanding of quark confinement. There is as yet no proof that the equations of quantum chromodynamics must lead to confinement, but considerable progress has recently been made (Adler 81, Bander 81).

Even from the brief, general remarks made so far, it should be clear to the reader that an astonishing wealth of new ideas has emerged in elementary particle physics, not the least of which has occurred in weak interactions. Furthermore, in the past few years, there have been bold attempts to incorporate quantum chromodynamics and electroweak interactions in a single "grand unified theory," and even more ambitious efforts have been made to include gravity in a truly unified scheme. In this book we make use of the most elementary concepts in quantum chromodynamics in our attempts to understand certain features of the weak interactions of hadrons. However, we make only passing reference to grand unified theories, and gravity does not enter our discussions (except peripherally, in Chapter 11, on weak interactions in astrophysics).[2]

---

2 In view of the fact that gravity affects all matter and radiation and determines the large-scale structure of matter in the universe, the neglect of gravity deserves some explanation. The point is that Newton's constant, $G = 6.67 \times 10^{-8}$ g^{-1} cm^3 sec^{-2}, is very small, and therefore, gravitational effects in the microscopic domain of elementary particles at attainable energies are neg-

1.2 Early history of weak interactions: The Fermi era

Our account of the development of weak interactions begins with the discovery of radioactivity by Becquerel in 1896 and the recognition soon thereafter that, in one form of radioactivity, the decaying nucleus emits β rays (electrons).[3] In 1914, Chadwick observed that the electrons in β decay are emitted with a continuous spectrum of energies, which defies comprehension if one assumes a two-body final state. This and subsequent calorimetric measurements of β decay in the 1920s seemed to suggest an extraordinary result: that energy and linear momentum are not conserved in β decay.

In order to rescue the fundamental conservation laws, W. Pauli (31,33) proposed that a neutral particle of near-vanishing or zero rest mass and half-integral spin (later named the *neutrino* by Fermi) is emitted along with the electron in β decay and that it escapes observation because of its feeble interaction with surrounding matter. Very soon thereafter, and directly stimulated by Pauli's radical hypothesis, Enrico Fermi (34) proposed his theory of β decay. Fermi's ideas were formulated in close analogy to quantum electrodynamics, which had been created several years before by Dirac (27) and further developed by Heisenberg and Pauli (29,30).

The interaction between an electron and the radiation field is described in quantum electrodynamics by the "minimal" Lagrangian density,

$$\mathscr{L} = -e\overline{\Psi}\gamma_\mu\Psi A^\mu \qquad (1.3)$$

where A^μ is the 4-vector potential of the radiation field and $e\overline{\Psi}\gamma_\mu\Psi$ the electromagnetic current density associated with the electron field Ψ. The simplest imaginable processes in quantum electrodynamics would be the emission or absorption of a photon by a free electron.

Actually such processes cannot occur because of energy–momentum conservation. Nevertheless, the hypothetical photon emission by a free electron serves as a convenient model for constructing

---

ligible. To see this, we consider an elementary particle with a "radius" characterized by its Compton wavelength, $R = \hbar/Mc$. What must the mass of such a particle be in order that its gravitational energy be comparable in size to its rest energy? We find: $Mc^2 = MG/R = MG \cdot Mc/\hbar$. Thus, $M = (\hbar c/G)^{1/2} = 2 \times 10^{-5}$ g $\simeq 10^{19}$ GeV/c^2. This enormous "Planck mass" is equivalent to an energy of 10^{19} GeV and a Compton wavelength of $\sim 10^{-33}$ cm. Only at such high energies and short distances would gravity become of comparable importance to strong, electromagnetic, or even weak interactions.

3 For excellent reviews of the early history of β decay, see Wu (59) and Kofoed-Hansen and Christensen (62).

the Fermi theory. We consider the simplest β decay, that of a free neutron:

$$n \to pe^-\bar{\nu}_e \tag{1.4}$$

In Fermi's theory, the $e^-\bar{\nu}_e$ pair plays the role formerly assigned to the photon, whereas the transforming nucleon replaces the charged particle of electrodynamics. Thus, in a rather obvious way, the following replacements are suggested:

$$\bar{\Psi}\gamma_\mu\Psi \to \bar{\Psi}_p\gamma_\mu\Psi_n \tag{1.5}$$

$$A^\mu \to \bar{\Psi}_e\gamma^\mu\Psi_{\nu_e} \tag{1.6}$$

Also, e of the electromagnetic Lagrangian is replaced by $G_F/\sqrt{2}$, where the *Fermi coupling constant* G_F must be determined by experiment and is found to take the value

$$G_F = 1.03 \times 10^{-5} m_p^{-2} \tag{1.7}$$

in units where $\hbar = c = 1$. Thus Fermi's β decay Lagrangian was obtained as

$$\mathscr{L}_\beta = -\frac{G_F}{\sqrt{2}} \bar{\Psi}_p \gamma_\mu \Psi_n \bar{\Psi}_e \gamma^\mu \Psi_{\nu_e} \tag{1.8}$$

In 1934, the first positron β decays were observed (Curie and Joliot 34), and within a few years, orbital electron capture was discovered (Alvarez 38). These processes are readily accommodated in $\mathscr{L}_\beta$ simply by adding to (1.8) its Hermitian conjugate, yielding

$$\mathscr{L}_\beta = -\frac{G_F}{\sqrt{2}} [\bar{\Psi}_p \gamma_\alpha \Psi_n \bar{\Psi}_e \gamma^\alpha \Psi_{\nu_e} + \bar{\Psi}_n \gamma_\alpha \Psi_p \bar{\Psi}_{\nu_e} \gamma^\alpha \Psi_e] \tag{1.9}$$

Fermi's Lagrangian (1.9) gives a very good account of many observed characteristics of β decay when employed to calculate transition probabilities in first-order perturbation theory.

Constructed as it was in close analogy to quantum electrodynamics, $\mathscr{L}_\beta$ is a sum of scalar products of two (polar) vectors and is thus an invariant with respect to proper Lorentz transformations and spatial inversion (parity). However, as hinted by Fermi and discussed in detail by Gamow and Teller (36), there is a more general way to construct a β-decay Lagrangian linear in Ψ_p, Ψ_n, Ψ_e, and Ψ_{ν_e} or their conjugates and containing no derivatives of these fields. Specifically, $\mathscr{L}_\beta$ could be a linear combination of the following terms and their Hermitian conjugates, each of which is a scalar:

$$\bar{\Psi}_p \Psi_n \bar{\Psi}_e \Psi_{\nu_e} \qquad \text{scalar} \times \text{scalar} \quad (S)$$

1.2 Early history of weak interactions

$$\overline{\Psi}_p \gamma_\alpha \Psi_n \overline{\Psi}_e \gamma^\alpha \Psi_{\nu_e} \qquad \text{vector} \times \text{vector (V)}$$
$$\overline{\Psi}_p \sigma_{\alpha\beta} \Psi_n \overline{\Psi}_e \sigma^{\alpha\beta} \Psi_{\nu_e} \qquad \text{tensor} \times \text{tensor (T)}$$
$$\overline{\Psi}_p \gamma_\alpha \gamma_5 \Psi_n \overline{\Psi}_e \gamma^\alpha \gamma_5 \Psi_{\nu_e} \qquad \text{axial vector} \times \text{axial vector (A)}$$
$$\overline{\Psi}_p \gamma_5 \Psi_n \overline{\Psi}_e \gamma_5 \Psi_{\nu_e} \qquad \text{pseudoscalar} \times \text{pseudoscalar (P)}$$

The matrix element for allowed β decay could then be written as

$$\mathcal{M} = \frac{G_F}{\sqrt{2}} \sum_{\text{nucleons}} \sum_{j=\text{SVTAP}} \int C_j \overline{\psi}_p(\mathbf{x}) O_j \psi_n(\mathbf{x}) \overline{\psi}_e(\mathbf{x}) O_j \psi_{\overline{\nu}_e}(\mathbf{x}) \, d^3x \tag{1.10}$$

where the ψ's are single-particle Dirac wavefunctions,

$$O_j = 1, \gamma_\lambda, \sigma_{\lambda\nu}, \gamma_\lambda \gamma_5, \gamma_5$$

for $j = $ S, V, T, A, P, respectively, and the integration is carried out over the nuclear volume. The coupling constants C_j could only be determined by experiment. However, the following was immediately evident: the nonrelativistic limit for the nucleons is appropriate for β decay, since the initial nucleon is nonrelativistic and the momentum transferred to the leptons is typically very small (~ 1 MeV/c). In this limit, the pseudoscalar term vanishes, and it is easy to show that the following nuclear spin selection rules are obeyed for allowed β transitions:

$$\begin{aligned} &\text{For S, V:} \quad \Delta J = 0 \\ &\text{For A, T:} \quad \Delta J = 0, \pm 1, \quad \text{but } J = 0 \not\to J = 0 \end{aligned} \tag{1.11}$$

Therefore, although Fermi's vector theory and/or the scalar variant S might be appropriate for $\Delta J = 0$ transitions, some A and/or T component must be present to account for $|\Delta J| = 1$ allowed decays, which were known to occur. More than two decades would elapse, however, before the β-decay coupling constants could be determined.

With the discovery of the muon in the late 1930s (Neddermeyer and Anderson 36,37,39) and observations of muon decay and muon capture in nuclei (Conversi et al. 46,47; Wheeler 47) shortly after World War II, it became apparent that there exist other reactions in nature in addition to β decay that have the same coupling strength and similar characteristics and are merely two different manifestations of a general Fermi interaction (Wheeler and Tiomno 49). Indeed, the discovery of pions (Lattes et al. 47), K mesons (Rochester and Butler 47), and hyperons (Armenteros et al. 51), all in cosmic rays, and the elucidation of their decays by accelerator experiments in the early 1950s gave rise to the concept of a universal charged weak interaction governing all slow decays of unstable particles.

1.3 Parity nonconservation

By 1956, sufficient data on the decay systematics of K mesons had been accumulated to reveal a peculiarity known as the $\tau-\theta$ puzzle (Dalitz 53,54; Fabri 54, Orear et al. 56). Here, evidently, were two meson states, τ and θ, with the same mass, charge, spin (zero), and lifetime. However, they decayed to final states of opposite parity.

$$\theta \to \pi^+ \pi^0 \quad \text{(even parity)}$$
$$\tau \to \pi^+ \pi^+ \pi^- \quad \text{(odd parity)}$$

The phenomenon seemed inexplicable if parity was conserved in the decays, unless extremely artificial assumptions were adopted. However, while studying this problem, T. D. Lee and C. N. Yang (56) realized that invariance with respect to space inversion (parity) had always been taken for granted and had never before been adequately tested in weak interactions. They suggested that parity might actually be violated in weak interactions (which would allow τ and θ to be the same particle: K^+), and most importantly, they suggested direct experimental tests for parity violation in β decay, muon decay, and so on.

Soon thereafter and as a direct result of Lee and Yang's new ideas, parity violation was observed through detection of the fore–aft asymmetry in β emission by polarized ^{60}Co nuclei (Wu et al. 57). Following almost immediately came the observation of parity violation in the sequence of decays:

$$\pi^\pm \to \mu^\pm \nu$$
$$\hookrightarrow e^\pm \nu \nu$$

(Garwin et al. 57, Friedman and Telegdi 57). There ensued a period of intense experimental activity, which led, in a few years, to the clarification of the laws of β decay and muon decay.

Let us briefly describe how this was done for β decay, leaving the details for Chapter 5. First, once the possibility of parity violation is admitted, the β-decay Lagrangian is no longer necessarily invariant under space inversion. Instead, it may possess a pseudoscalar as well as a scalar part. This implies that, in addition to the "even" couplings $C_{S,V,T,A,P}$ already mentioned, there may also exist "odd" couplings $C'_{S,V,T,A,P}$:

$$\mathcal{M} = \frac{G_F}{\sqrt{2}} \sum_{\text{nucleons}} \sum_j \int d^3 x \, [\bar{\psi}_p O_j \psi_n][\bar{\psi}_e O_j (C_j - C'_j \gamma_5) \psi_{\bar{\nu}_e}] \quad (1.12)$$

since terms of the form $\bar{\psi}_p O_j \psi_n \bar{\psi}_e O_j \gamma_5 \psi_\nu$ are pseudoscalar. The coefficients C_j and C'_j had to be determined by experiment, and could be

1.3 Parity nonconservation

complex if time-reversal invariance was not assumed (Section 4.8). Because there are 10 coupling constants and merely 1 arbitrary overall phase, 19 real constants have to be determined. The situation is even worse if lepton conservation is not assumed [as we have tacitly done in (1.12)]. Then, there are 35 real constants to determine (Pauli 57, Pursey and Kahana 57). Even if the pseudoscalar couplings are dismissed in the nonrelativistic nucleon limit, this still leaves a very large number of constants. Nevertheless, the actual values of the constants were determined by direct appeal to a small number of key experiments, as follows.

It was observed, with the aid of Mott or Møller scattering, that electrons emitted in β decay are longitudinally polarized with a polarization, or helicity, of $P(e^-) = -v/c$, whereas positrons in β decay possess $P(e^+) = +v/c$. One thus speaks of "left-handed" e^- and "right-handed" e^+ in β decay. As we shall see in Chapter 5, this implies that $C_j = C'_j$ for V–A couplings and $C_j = -C'_j$ for S–T–P couplings. Incidentally, it is very interesting to note that, in 1928, R. T. Cox and co-workers (28) performed an experiment to search for the polarization of β rays from radium, and they found a positive effect! However, the result did not attract much attention, and further work by Cox and others tended to obscure the issue.

Next, observations of electron–neutrino angular correlations and circular polarization of γ rays following β decay revealed that the neutrino ν_e is also left-handed and the antineutrino $\bar{\nu}_e$ right-handed, the helicities being $h(\nu_e) = -1$ and $h(\bar{\nu}_e) = +1$. This implies that the S, T couplings are, in fact, zero or very small and that the β interaction is a combination of vector (V) and axial vector (A). The same conclusion is supported by data on spectral shapes.

Observations of the total decay rates for the $0^+ \to 0^+$ nuclear decays $^{10}C \to {}^{10}B$, $^{14}O \to {}^{14}N$, and so on, which are transitions in which the vector matrix element can be calculated quite precisely and the axial amplitude is zero, permitted the determination $C_V = 1$. Then, measurements of the mean life of free neutrons, where both vector and axial vector couplings enter, yielded $C_V^2 + 3C_A^2$, which leads to the result

$$\lambda \equiv |C_A/C_V| = 1.25 \tag{1.13}$$

The departure of this ratio from unity is a manifestation of strong interaction effects.

Finally, observations of the β and neutrino asymmetries in the decay of polarized neutrons, which are effects involving interference between vector and axial vector amplitudes, permitted determination of the relative phases of C_A and C_V. The final result for neutron decay was

$$\mathcal{M} = \frac{G_F}{\sqrt{2}} \int d^3x \, \bar{\psi}_p \gamma_\mu (1 - \lambda \gamma_5) \psi_n \bar{\psi}_e \gamma^\mu (1 - \gamma_5) \psi_{\bar{\nu}_e} \qquad (1.14)$$

Apart from the factor $\lambda = 1.25$, this expression signifies that the neutron and proton, as well as the electron, are left-handed in β decay and the antineutrino is right-handed. Except for these important qualifications, it turned out, after a quarter-century, that Fermi's original formulation was correct.

1.4 The V–A law

The theoretical and experimental activity stimulated by Lee and Yang's important discovery of parity violation culminated in a generalization of Fermi's theory, proposed independently by Feynman and Gell-Mann (58) and by Sudarshan and Marshak (58). According to this scheme, all charged weak processes: β decay, muon decay, charged-pion decay, and so on, are described by an effective Lagrangian density in which a "universal" charged weak current $\mathcal{J}_\lambda$ is coupled to itself, or rather to its Hermitian conjugate, at a single space-time point:

$$\mathcal{L} = -\frac{1}{2} \frac{G_F}{\sqrt{2}} (\mathcal{J}_\lambda \mathcal{J}^{\lambda\dagger} + \mathcal{J}^{\lambda\dagger} \mathcal{J}_\lambda) \qquad (1.15)$$

The current $\mathcal{J}_\lambda$ consists of a leptonic and a hadronic portion:

$$\mathcal{J}_\lambda = j_{l\lambda} + J_\lambda \qquad (1.16)$$

The leptonic part is a straightforward generalization from β decay, which we write today as

$$j_{l\lambda} = \bar{\Psi}_e \gamma_\lambda (1 - \gamma_5) \Psi_{\nu_e} + \bar{\Psi}_\mu \gamma_\lambda (1 - \gamma_5) \Psi_{\nu_\mu}$$
$$+ \bar{\Psi}_\tau \gamma_\lambda (1 - \gamma_5) \Psi_{\nu_\tau} + \cdots \qquad (1.17)$$

It describes all charged weak transformations of the following kinds:

$$\nu_l \to l^-, \quad l^+ \to \bar{\nu}_l \qquad (1.18)$$

and the Hermitian conjugate current $j_{l\lambda}{}^\dagger$ describes the reverse transformations.

In light of present knowledge, we express the charged hadronic weak current in terms of the quark model. For example, the neutron has the valence quark composition *udd;* the proton, *uud.* Neutron β decay therefore occurs because of the quark transformation $d \to u$. Thus the

1.4 The V–A law

hadronic charged weak current contains a component proportional to

$$\bar{D}\gamma_\lambda(1 - \gamma_5)U \tag{1.19}$$

where U and D are up and down-quark field operators, respectively. However, as was already known in the 1950s, there exist weak decays in which the *strangeness* of the initial and final states differ by unity, for example, $K^+ \to \mu\nu$, $\Lambda^0 \to p\pi^-$, $\Sigma^- \to ne^-\bar{\nu}$, and so forth. In such transitions, in today's language, an up quark must transform into a strange quark or vice versa. There then should exist, in addition, a component of the charged hadronic weak current proportional to

$$\bar{S}\gamma_\lambda(1 - \gamma_5)U \tag{1.20}$$

where S is the strange-quark field operator.

N. Cabibbo (63) showed that one can account for many features of the strangeness-changing ($|\Delta S| = 1$) and strangeness-conserving ($\Delta S = 0$) weak transitions involving hadrons by assuming that these components enter into the hadronic weak current in a way that we now express as

$$J_\lambda = \bar{D}_C\gamma_\lambda(1 - \gamma_5)U \tag{1.21}$$

where $D_C = \cos\theta_C\, D + \sin\theta_C\, S$, and $\theta_C \simeq 13°$ is the *Cabibbo angle*, an empirical parameter that has not yet been given a satisfactory explanation in terms of more fundamental quantities. As we shall see in subsequent chapters, Cabibbo's hypothesis accounts for the relative coupling strength of strangeness-conserving and strangeness-violating baryon semileptonic decays, for the ratio of leptonic decay rates of pions and kaons, for the ratio of vector couplings in β decay and muon decay (a ratio that is very nearly but not exactly unity), and for other important features of charged weak interactions. Cabibbo's hypothesis must be modified, as we shall see presently, to take into account weak transformations of the charmed quark and other heavy quarks.

The V–A, or $\gamma_\lambda(1 - \gamma_5)$, structure of the charged weak current in the Feynman–Gell-Mann scheme implies not only that parity is violated but also that charge conjugation invariance fails. A parity transformation, or spatial inversion, may be achieved by a mirror reflection and a rotation of 180°. Since the laws of nature are invariant under rotations, it follows that parity is violated in a natural process if the process and its mirror image occur with different probabilities. Under charge conjugation, particles are transformed to antiparticles but spins and momenta are left unchanged. Accordingly, charge conjugation, or C invariance, is violated if a process and the corresponding one in which

1 Introduction to weak interactions

particles are replaced by antiparticles do not occur with the same probability.

A simple example is afforded by the dominant mode of decay of positive and negative pions ("$\pi_{\mu 2}$"):

$$\pi^+ \to \mu^+ \nu_\mu, \qquad \pi^- \to \mu^- \bar{\nu}_\mu$$

Experimentally, it was found that μ^+ in π^+ decay has helicity $h(\mu^+) = -1$ (see Figure 1.2a). This is in accord with the fact that the pion has zero spin, with the expectation, according to the V–A law, that ν_μ is left-handed and with conservation of linear and angular momentum. A mirror reflection through the plane containing the pion and perpendicular to the lepton momenta reverses these momenta but leaves the spins unchanged. Thus the helicity of each lepton is reversed, and we obtain the situation shown in Figure 1.2b, which is not observed. Parity is therefore violated. On the other hand, a C transfor-

Figure 1.2. (a) $\pi^+ \to \mu^+ \nu_\mu$ decay. (b) The result of a mirror reflection of (a). (c) The result of a C transformation on (a). (d) The result of a CP transformation on (a).

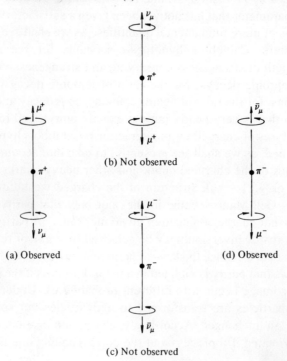

mation on the particles of Figure 1.2a yields the situation shown in Figure 1.2c, that is, the decay $\pi^- \to \mu^- \bar{\nu}_\mu$, in which $h(\mu^-) = -1$. However, experiment shows that the μ^- emitted in π^- decay actually has helicity $h(\mu^-) = +1$. Therefore, C is also violated. A combination of the two transformations yields the decay of the negative pion with the correct helicities (Figure 1.2d).

In addition to the C and P transformations, an important role in weak interactions is played by the time reversal (T) transformation, which will be discussed in detail in Section 4.8. Here spins and momenta are reversed, and initial and final states are interchanged. A combination of all three transformations (CPT) occupies a special position, because it may be shown in the CPT theorem (Pauli 55, Lüders 54) that, for any local Lorentz-invariant field theory in which the observables are represented by Hermitian operators, CPT invariance must hold. Therefore, although any two of the symmetries C, P, and T might be violated in nature, it is widely believed that CPT must be a valid symmetry. For example, in the V–A law, C and P separately fail. However, T and, therefore, the combination CP are valid symmetries, as is CPT. For T (and CP) to fail here, one would require the V and A amplitudes to be relatively complex. CPT invariance of the various interactions implies, among other things, that the masses, total lifetimes, and gyromagnetic ratios of a particle and the corresponding antiparticle are equal.

1.5 The two-component neutrino

So far we have tacitly assumed that the neutrino, like the electron, is a Dirac particle and can be described by an ordinary four-component spinor. However, if the V–A law is exact and if the neutrino mass is zero, we can employ a two-component description. Such an idea was in fact first proposed by H. Weyl (29) on purely mathematical grounds, before the neutrino's physical existence was ever imagined.

We consider the Dirac equation satisfied by a neutrino wavefunction ψ:

$$i\gamma_\mu \frac{\partial \psi}{\partial x^\mu} - m_\nu \psi = 0 \tag{1.22}$$

In the Weyl representation the 4×4 matrices take the form:

$$I = \begin{pmatrix} I & 0 \\ 0 & I \end{pmatrix}, \quad \gamma^0 = \begin{pmatrix} 0 & I \\ I & 0 \end{pmatrix}, \quad \gamma = \begin{pmatrix} 0 & -\sigma \\ \sigma & 0 \end{pmatrix}, \quad \gamma_5 = \begin{pmatrix} I & 0 \\ 0 & -I \end{pmatrix}$$

16   *1 Introduction to weak interactions*

and $\psi = \begin{pmatrix} \phi \\ \chi \end{pmatrix}$, where ϕ and χ are two-component spinors. Also, (1.22) may be written as

$$i\frac{\partial \phi}{\partial t} + i\boldsymbol{\sigma} \cdot \boldsymbol{\nabla}\phi = m_\nu \chi \tag{1.23}$$

$$i\frac{\partial \chi}{\partial t} - i\boldsymbol{\sigma} \cdot \boldsymbol{\nabla}\chi = m_\nu \phi \tag{1.24}$$

In the limit $m_\nu \to 0$, these two equations become decoupled and also for plane-wave states $i\,\partial/\partial t = E$, $-i\boldsymbol{\nabla} = \mathbf{p}$, and $E = |\mathbf{p}|$ for zero mass. Thus (Landau 57, Lee and Yang 57, Salam 57), we obtain in this limit:

$$h\phi \equiv \boldsymbol{\sigma} \cdot \mathbf{p}\phi/|\mathbf{p}| = \phi \tag{1.25}$$

$$h\chi \equiv \boldsymbol{\sigma} \cdot \mathbf{p}\chi/|\mathbf{p}| = -\chi \tag{1.26}$$

These equations reveal that the two-component spinors ϕ and χ are eigenstates of helicity h with eigenvalues ± 1, respectively. Therefore, in the limit of zero mass, the assumption that χ and ϕ represent neutrino and antineutrino, respectively, is equivalent to assumption of the V–A law. Of course, it must be emphasized that the discovery of a finite neutrino mass, no matter how small, would invalidate the two-component description.

1.6   Difficulties of Fermi-type theories

The Feynman–Gell-Mann scheme, supplemented by Cabibbo's hypothesis, provided an excellent phenomenological account of the observed charged weak interactions within the framework of first-order perturbation theory. However, it was nothing more than a generalized version of the Fermi theory, and thus it suffered from the same fundamental difficulties as the latter – difficulties that had, in fact, been recognized many years before (Heisenberg 36,38). In order to see these problems, we consider the following scattering reaction:

$$\nu_e + e^- \to e^- + \nu_e \tag{1.27}$$

which is predicted to exist in the Feynman–Gell-Mann scheme and to be described by the Lagrangian term

$$-\frac{G_F}{\sqrt{2}} j_{e\lambda} j_e^{\lambda\dagger} \tag{1.28}$$

On dimensional grounds, the cross section must be of order $G_F^2 \omega^2$, where ω is the energy of the incoming (outgoing) neutrino in the CM frame. In fact, one finds, by explicit calculation, that the differential cross section is

1.6 Difficulties of Fermi-type theories

$$\frac{d\sigma}{d\Omega} = \frac{G_F^2}{\pi^2} \omega^2 \tag{1.29}$$

In general, we may express a differential cross section as the absolute square of a scattering amplitude $f(\theta)$, and the latter may be expanded in partial waves as

$$f(\theta) = \frac{1}{\omega} \sum_{J=0}^{\infty} (J + \tfrac{1}{2}) \mathcal{M}_J P_J(\cos \theta) \tag{1.30}$$

where $\mathcal{M}_J$ is the amplitude of the Jth partial wave. In a theory of the Fermi type, however, only the $J = 0$ partial wave enters, because we are dealing with a contact interaction of zero range. Therefore,

$$\frac{d\sigma}{d\Omega} = |f(\theta)|^2 = \frac{1}{4\omega^2} |\mathcal{M}_0|^2$$

However, unitarity (conservation of probability) requires in the general case that $|\mathcal{M}_J| \leq 1$ for each J. Therefore, we must have

$$d\sigma/d\Omega \leq 1/4\omega^2 \tag{1.31}$$

It is easy to see that (1.29) and (1.31) lead to a contradiction at high energies, that is, when

$$\omega \geq \pi^{1/2}/4^{1/4} G_F^{1/2} \simeq 300 \text{ GeV}$$

Thus it is clear that our first-order calculation of the cross section fails completely at energies of order $G_F^{-1/2}$. We might naturally suppose at first that this difficulty arises from neglect of higher-order corrections and that, if these were included, the cross section would level out and remain within the bounds imposed by unitarity. However, calculation shows that the second-order diagram of Figure 1.3 yields a divergent result.

Of course, in quantum electrodynamics, one also encounters divergent integrals corresponding to higher-order diagrams. However, the divergences are removed to all orders by mass and charge renormalization. In the present case, we may attempt to eliminate the divergence of Figure 1.3 by an analogous procedure, but when we go

Figure 1.3. Second-order contribution to ν_e–e scattering.

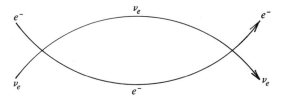

to the next higher order, a new and more severe divergence is encountered, which requires new renormalization constants. When all diagrams are taken into account, it is found that an infinite set of renormalization constants is needed; in short, the Fermi theory is not renormalizable.

Moreover, there is a serious problem with electromagnetic (radiative) corrections to weak processes as described in the Feynman–Gell-Mann scheme. To be sure, the radiative corrections in muon decay are finite to all orders in $\alpha = e^2/4\pi\hbar c$ and present no difficulty. However, in $e-\nu_e$ scattering, radiative corrections diverge in order α^2, and in nuclear β decay, the divergence occurs in order α. Therefore the generalized Fermi theory, although very successful at a certain level, contains grave difficulties and cannot possibly be correct.

1.7 The naive intermediate boson hypothesis

The notion that all interactions between fermions are transmitted by intermediate bosons has its origin in quantum electrodynamics and in Yukawa's meson theory of nuclear forces (Yukawa 35). Once the V–A law was established, it seemed reasonable, indeed compelling, to suppose that the charged weak interactions also proceed by exchange of intermediate vector bosons $W^\pm$, which would necessarily be quite massive, since the weak interaction has very short range.

Let us attempt to construct a theory of charged weak interactions in the simplest possible manner by employing such bosons. First of all, it can be shown (see Section 2.5) that the momentum-space propagator for a massive vector boson is

$$D_{\mu\nu}^W = -\frac{g_{\mu\nu} - (1/m_W^2)q_\mu q_\nu}{q^2 - m_W^2} \tag{1.32}$$

According to the Feynman–Gell-Mann scheme, the amplitude for muon decay, $\mu^- \to e^- \nu_\mu \bar{\nu}_e$, is

$$\mathcal{M} = \frac{G_F}{\sqrt{2}} \bar{u}_e \gamma_\lambda (1 - \gamma_5) v_{\nu_e} \bar{u}_{\nu_\mu} \gamma^\lambda (1 - \gamma_5) u_\mu \tag{1.33}$$

where u_e, u_μ, u_{ν_μ}, and v_{ν_e} are single-particle Dirac spinors for the electron, muon, ν_μ, and $\bar{\nu}_e$, respectively. In our W boson theory, the amplitude would instead take the form

$$\mathcal{M}' = -g\bar{u}_e\gamma_\lambda(1-\gamma_5)v_{\nu_e}\frac{g^{\lambda\sigma}-(1/m_W^2)q^\lambda q^\sigma}{q^2-m_W^2}g\bar{u}_{\nu_\mu}\gamma_\sigma(1-\gamma_5)u_\mu \tag{1.34}$$

where $g\gamma_\lambda(1 - \gamma_5)$ is a "semiweak" vertex factor and q the 4-momentum transfer (see Figure 1.4). Now in muon decay, the momentum transfer is very small: $m_W^2 \gg |q^2|$. Therefore, (1.34) reduces to

$$\mathcal{M}' = (g^2/m_W^2)\bar{u}_e\gamma_\lambda(1 - \gamma_5)v_{\nu_e}\bar{u}_{\nu_\mu}\gamma^\lambda(1 - \gamma_5)u_\mu \tag{1.35}$$

Furthermore, we know that the Feynman–Gell-Mann scheme provides a very accurate description of muon decay, and therefore (1.33) must be correct to a very good approximation. Consequently, we should equate (1.35) and (1.33), which determine the semiweak coupling constant g in terms of the W mass:

$$g = m_W G_F^{1/2} 2^{-1/4} \tag{1.36}$$

It is now possible to calculate processes that lead to difficulty in the Fermi theory, such as neutrino–electron scattering. One readily finds that, in the partial wave expansion of the scattering amplitude for this process (1.30), each partial-wave amplitude grows logarithmically. For example,

$$\mathcal{M}_0 = \frac{G_F^2 m_W^2}{2\pi} \ln\left(1 + \frac{4\omega^2}{m_W^2}\right) \tag{1.37}$$

Thus unitarity is again violated at sufficiently large energy. The unitarity limit is violated even more strongly in the cross section for $\nu + \bar{\nu} \to W^+ + W^-$.

Figure 1.4. Muon decay by W exchange.

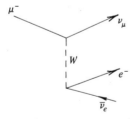

Furthermore, higher-order terms diverge strongly because of the term proportional to $q^\lambda q^\sigma/m_W^2$ in the W propagator, (1.32). Thus the naive intermediate boson theory is also nonrenormalizable, and in addition, radiative corrections to most weak processes in this model diverge in all orders of α.

1.8 Charm and the GIM model

Despite these profound difficulties, it not only seemed plausible in the 1960s that the charged weak interactions should be mediated by $W^\pm$ bosons but also that there should exist neutral weak interactions, mediated by neutral vector bosons (Bludman 58, Glashow 61, Salam and Ward 64). However, a major problem presented itself: strangeness-changing neutral weak interactions simply do not occur. Experiment revealed that the neutral decay[4] $K_L^0 \to \mu^+\mu^-$ proceeds with a branching ratio of only 9×10^{-9}, almost all of which can be accounted for by radiative effects. This result was very perplexing in view of the fact that the analogous charged decay $K^+ \to \mu^+\nu_\mu$ is fully allowed, with a branching ratio of 63 percent. According to the quark model and the intermediate boson hypothesis, these allowed and forbidden decays are represented by Figures 1.5a and b, respectively. The K^+ is a bound state of $u\bar{s}$; when it decays, the u transforms to an s by W^+ emission (or the $\bar{s}$ transforms to a $\bar{u}$) and the remaining $q\bar{q}$ pair annihilates. Evidently the analogous $d \to s$ neutral weak transformation cannot occur to generate $K_L^0 \to \mu^+\mu^-$.

Figure 1.5. (a) K^+ allowed decay. (b) K_L^0 forbidden decay.

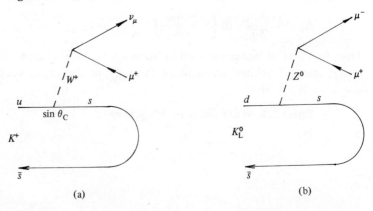

The virtual absence of the latter transition seemed even more puzzling when it was recognized that, even if, for some reason, the neutral amplitude of Figure 1.5b turns out to be zero, there should still be a second-order charged weak contribution to the decay amplitude, as shown in Figure 1.6a, where the Cabibbo angle factors are displayed at their respective vertices.

The key to the solution of this puzzle is the charmed quark (first proposed in 1964 and only observed for the first time in 1974). In 1970, Glashow, Iliopoulos, and Maiani (70) took seriously the notion of a

---

[4] K_L^0 is the long-lived component of the neutral K meson.

1.8 Charm and the GIM model

Figure 1.6. Second-order contributions to $K_L^0 \to \mu^+\mu^-$.

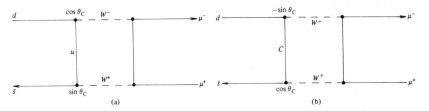

charmed quark and proposed that, just as the u quark is coupled to the combination

$$d_C = d \cos\theta_C + s \sin\theta_C \tag{1.38}$$

in the hadronic charged weak current, so also is the charmed quark coupled to the combination

$$s_C = -d \sin\theta_C + s \cos\theta_C \tag{1.39}$$

The result is that the charged hadronic weak current must now be written as

$$J_\lambda = \bar{D}_C \gamma_\lambda (1-\gamma_5) U + \bar{S}_C \gamma_\lambda (1-\gamma_5) C \tag{1.40}$$

which replaces (1.21). In the Glashow–Iliopoulos–Maiani (GIM) model, it is also assumed that the neutral component of the hadronic weak current takes the following form, where we suppress space-time indexes:

$$\bar{U}U + \bar{C}C + \bar{D}D + \bar{S}S = \bar{U}U + \bar{C}C + \bar{D}_C D_C + \bar{S}_C S_C$$

The significance of this expression for the present discussion is that it contains no cross terms $\bar{D}S$ or $\bar{S}D$. Thus, one expects no neutral weak interactions in which a d quark transforms to an s quark or vice versa, as well as no transitions in which $u \to c$ or vice versa. This readily explains why the amplitude corresponding to Figure 1.5b is zero.

As for the second-order charged contribution, according to (1.40), there should exist an additional amplitude, corresponding to Figure 1.6b, as well as the amplitude attached to Figure 1.6a. Whereas the latter is proportional to $\cos\theta_C \sin\theta_C$, the new amplitude is proportional to $-\sin\theta_C \cos\theta_C$. In the limit where u and c quarks have the same mass, these two amplitudes exactly cancel, thus eliminating the second-order contribution.

The GIM formula (1.40) for the charged hadronic weak current represents a major generalization of Cabibbo's hypothesis. It is further generalized to take into account six quarks in the Kobayashi–Maskawa model (see Section 4.3). Here, the single Cabibbo angle θ_C is replaced by three angles θ_1, θ_2, and θ_3 and a fourth parameter δ, the so-called *CP*-violating phase.

1.9 Classification of weak interactions

Before we undertake a detailed discussion of the modern theory of weak interactions ("standard model"), which incorporates the notions described in the previous section, let us classify the various weak processes. This will give us an opportunity to become acquainted with the weak phenomena in a general way.

In Table 1.1, the weak interactions are categorized as charged, neutral, or charged/neutral, according to whether they occur by W exchange, Z exchange, or a combination of the two. They are also classified as leptonic, semileptonic, or nonleptonic. Going step-by-step through the table, we begin with the purely leptonic charged weak interactions:

(1) muon decay,
(2) inverse muon decay,
(3) leptonic decays of $\tau^\pm$.

Muon decay has been studied in great detail for many years in numerous precise experiments, and the results are in very good agreement with the V–A law. However, one cannot exclude the possibility of appreciable S, T, P couplings nor the appearance of some $V + A$ component even if S, T, P couplings are excluded. In muon decay, the lepton numbers L_e and L_μ are separately conserved. The transitions $\mu^- \to e^-\gamma$, $\mu^- \to e^-e^+e^-$, on the contrary, violate L_e and L_μ conservation and do not occur, even though they are allowed energetically and conserve electric charge. One finds

$$\Gamma(\mu \to e\gamma)/\Gamma(\mu \to e\nu\nu) < 1.9 \times 10^{-10}$$

and

$$\Gamma(\mu \to eee)/\Gamma(\mu \to e\nu\nu) < 1.9 \times 10^{-9}$$

The results of *inverse muon decay* experiments, which employ high-energy ν_μ beams, corroborate the V–A law and also help to rule out an alternative description of lepton conservation that seemed possible for a number of years. Here, an ordinary lepton number L was defined such that $L = +1$ for all leptons and $L = -1$ for all antileptons, and in addition, a multiplicative quantum number, called *muon parity*, was invoked, having the value -1 for all leptons and antileptons in the second (muon) generation and $+1$ for all other particles.

Incidentally, the question of lepton conservation is intimately con-

Table 1.1. *Classification of weak interactions*

Interaction	Charged		Neutral		Charged/neutral	
Leptonic	(1)	$\mu \to e\nu\nu$	(4)	$\nu_\mu e \to \nu_\mu e$	(7)	$\nu_e e \to \nu_e e$
	(2)	$\nu_\mu e \to \mu \nu_e$	(5)	$\bar{\nu}_\mu e \to \bar{\nu}_\mu e$	(8)	$\bar{\nu}_e e \to \bar{\nu}_e e$
	(3)	$\tau \to l\nu\nu$	(6)	$e^+ e^- \to l^+ l^-$	(9)	$e^+ e^- \to \nu_e \bar{\nu}_e$
Semileptonic						
Meson	(10)	$\pi^+ \to \mu\nu, e\nu$				
		$K^+ \to \mu\nu, e\nu$				
		$\to F^+ \to \tau^+ \nu$ *[now called D_s^+]*				
	(11)	$\pi^+ \to \pi^0 e\nu$				
	(12)	$K^+ \to \pi^0 l\nu$				
		$K_L^0 \to \pi^\pm l\nu$				
	(13)	$D \to \begin{pmatrix} \pi \\ K \\ K^* \end{pmatrix} l\nu$				
Baryon	(14)	$\mu^- B \to B' \nu$	(17)	$\nu N \to \nu N, \nu N\pi, \nu X$		
	(15)	$B \to B' l\nu$	(18)	$\bar{\nu}_e + D \to n + p + \bar{\nu}_e$		
	(16)	$\nu B \to B' l$	(19)	$eN \to eN, eX$		
Nonleptonic						
Meson	(20)	$K \to \pi\pi$				
	(21)	$K \to 3\pi$				
	(22)	$D \to KK, K\pi, K2\pi, K3\pi$				
	(23)	$B^{0,\pm} \to D\pi, DK$				
Baryon	(24)	$\Lambda \to N\pi$			(26)	$NN \to NN$
		$\Sigma \to N\pi$				
		$\Xi \to N\pi$				
	(25)	$\Lambda_c^- \to pK^- \pi^+$				

nected with the topics of double β decay, neutrino mass, and neutrino oscillations, all to be considered in detail in Chapter 10. According to the present view, lepton and baryon conservation are not truly fundamental and exact principles, like charge conservation, and at some level, they are expected to break down, giving rise to phenomena such as proton decay.

The τ leptons ($\tau^\pm$) are produced in e^+e^- collisions, and the *τ-leptonic decays* have by now been observed in considerable detail. The experimental results are in accord with the V–A law and signify that the ν_τ and τ are "sequential" leptons, that is, essentially similar to the more familiar leptons of the first two generations.

It is important to emphasize that the weak processes described so far, and indeed all weak processes observed up to 1981, are "low-energy" phenomena for which $|q^2| \ll m_{W,Z}^2$. Crudely speaking, they all possess amplitudes of order

$$G_F \simeq g^2/m_{W,Z}^2$$

where g is a semiweak coupling constant defined here very loosely and to be made precise in Chapter 2 [see also (1.33) and (1.35)]. According to the standard model, g is the same in order of magnitude as the electron–electron–photon vertex factor in quantum electrodynamics, namely, $(4\pi\alpha)^{1/2}$. Thus, low-energy weak interactions are *not* weak because of the smallness of g but rather because the intermediate boson mass is so large. In the standard model, one predicts $m_W \simeq 80$ GeV/c^2 and $m_Z \simeq 90$ GeV/c^2.

Amplitudes for weak processes at large q^2 should thus attain considerable size. In the coming decade, the semiweak coupling may be detected directly in the decay $W \to l\nu$ (Figure 1.7).

We pass now to the *neutral leptonic reactions*. The processes

(4) $\nu_\mu e^- \to \nu_\mu e^-$

(5) $\bar{\nu}_\mu e^- \to \bar{\nu}_\mu e^-$

Figure 1.7. Decay $W \to l\nu$.

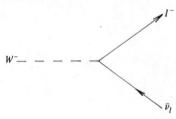

1.9 Classification of weak interactions

(Figures 1.8a and b) are induced with high-energy ν_μ, $\bar{\nu}_\mu$ beams and provide important tests of the standard model.

Reactions

(6) $e^+e^- \to e^+e^-$; $\mu^+\mu^-$; $\tau^+\tau^-$

(Figures 1.9a and b) can occur not only by photon exchange but also by Z^0 exchange. The coexistence of weak and electromagnetic amplitudes leads to characteristic interference effects, for example, an expected charge asymmetry in the angular distribution of muons with respect to the axis of e^+ and e^- in a colliding-beam machine, in $e^+e^- \to \mu^+\mu^-$.

At energies where $|q^2| \simeq m_Z^2$, the weak amplitude is expected to dominate greatly over the electromagnetic, and it should then be possible, within the coming decade, to generate Z^0s in copious numbers by e^+e^- semiweak collisions at the highest-energy colliding-beam accelerators. Presumably, Z^0 can decay semiweakly in a variety of modes:

$$Z^0 \to \nu_e \bar{\nu}_e; e^+e^-; \nu_\mu \bar{\nu}_\mu; \mu^+\mu^-; \ldots ; q\bar{q}$$

The numbers and types of these decay channels determine the mean life of Z^0 and thus the width of the Z^0 production resonance. This may provide a means to determine the number of distinct neutrino flavors, including those as yet undiscovered.

Figure 1.8. Neutral leptonic reactions: (a) ν_μ; (b) $\bar{\nu}_\mu$.

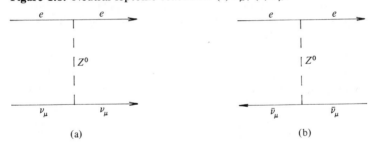

Figure 1.9. Neutral leptonic reactions: (a) γ exchange; (b) Z^0 exchange.

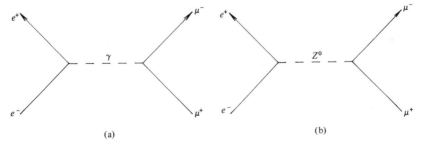

The *charged/neutral* leptonic weak processes in which W or Z^0 may be exchanged include:

(7) $\nu_e e^- \to \nu_e e^-$
(8) $\bar{\nu}_e e^- \to \bar{\nu}_e e^-$
(9) $e^+ e^- \to \nu_e \bar{\nu}_e$

Reaction (8) (Figures 1.10a and b) has been observed with intense low-energy $\bar{\nu}_e$ fluxes from a nuclear reactor, and the results help to fix parameters of the standard model. Reaction (9) is thought to be of crucial importance as an energy-loss mechanism in the late stages of evolution of hot dense stars.

Figure 1.10. Charged/neutral leptonic reactions: (a) Z^0 exchange; (b) W exchange.

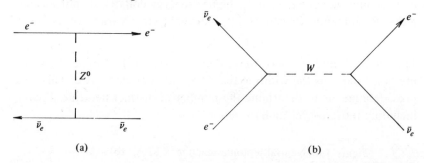

Next we come to weak processes involving hadrons. In general, these are much more difficult to analyze than the purely leptonic interactions because of the complications of strong interactions, and this is particularly true of the nonleptonic (purely hadronic) transitions. Among the semileptonic transitions, we first discuss decays of the form *meson → vacuum + leptons* [items (10) in Table 1.1]. The dominant mode of charged-pion decay is $\pi_{\mu 2}$, as we have already noted. This is illustrated schematically in Figure 1.11a. The analogous transition, $K^+ \to \mu^+ \nu_\mu$, shown in Figure 1.5a, has a much smaller amplitude because $\sin \theta_C$ is a small number. Perhaps, in the coming decade, it will be possible to observe the decay $F^+ \to \tau^+ \nu_\tau$ (Figure 1.16). In each of these transitions, it is the *axial* part of the charged-hadronic weak current that is responsible for the meson → vacuum transformation. One can demonstrate that the contribution of a pseudoscalar coupling, which is another possibility a priori, is in fact very small or zero by considering the very small branching ratio $\pi_{e2}/\pi_{\mu 2}$.

Items (11)–(13) of Table 1.1 are transitions of the form *meson → meson + leptons*. Pion β decay is illustrated in Figure 1.12. This very

1.9 Classification of weak interactions

rare mode of pion decay has been observed, and the results are in good agreement with a calculation based on the conserved vector current (CVC) hypothesis. The K_{l3} decays are important for a variety of reasons, including tests of various symmetries and the possibility of placing stringent limits on scalar and tensor couplings.

The *baryon semileptonic transitions* include nuclear β decay, orbital electron capture, muon capture by a proton

(14) $\mu^- + p \rightarrow n + \nu_\mu$

or by a complex nucleus, and the semileptonic decays of the metastable hyperons:

(15) $\Lambda^0 \rightarrow pe^-\bar{\nu}_e, \quad \Sigma^- \rightarrow ne^-\bar{\nu}_e, \quad \Sigma^+ \rightarrow \Lambda^0 e^+ \nu_e, \ldots$

The latter transitions provide very important tests of Cabibbo's hypothesis.

Charged semileptonic transitions may also be induced by collisions

Figure 1.11. Leptonic decays M_{l2} of (a) π^+; (b) F^+.

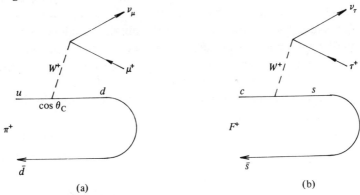

Figure 1.12. Pion β decay.

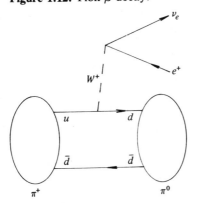

of neutrinos with nucleons through W exchange. In this category (16), we have inverse β decay

$$\bar{\nu}_e + p \to n + e^+$$

and inverse muon capture

$$\nu_\mu + n \to p + \mu^-$$

We must also include the high-energy, deep-inelastic collisions

$$\nu_\mu + N \to \mu^- + X$$
$$\bar{\nu}_\mu + N \to \mu^+ + X$$

in which X stands for highly excited hadronic matter. In such transitions, the W boson exchanged between neutrino and nucleon is regarded as a probe to study hadronic structure, just as a photon is similarly employed in electron–nucleon deep-inelastic collisions. The momentum transfer is very large, and to a good first approximation, one can treat the nucleon as an assembly of free, pointlike "partons" with parallel momenta – transverse momentum and binding forces being neglected. The partons consist of the quarks in valence states and the quarks and antiquarks of the ocean. This treatment leads to "scaling" (see Chapter 8), a phenomenon shown by experimental results to be approximately true. Violations of scaling are due in part to the residual interactions between partons, which can be treated by QCD on the basis of perturbation theory.

In some high-energy neutrino–nucleon collisions, one observes "dilepton" (e.g., $\mu^- e^+$) events associated with strange-particle production:

$$\nu_\mu + N \to \mu^- e^+ + (K^0 \text{ or } \Lambda^0) + \text{anything}$$

Such transitions can be understood as follows: a nucleon d quark is converted to a charmed quark c, thus producing a charmed hadron, which decays to a lepton pair and a strange hadron:

$$\nu_\mu + d \to \mu^- + c + \text{anything}$$
$$\hookrightarrow s(\text{strange hadron}) + l^+ \nu_l$$

In the coming decade, we anticipate detailed observations of semileptonic transitions involving the b quark and, possibly, semileptonic decays of hadrons containing the t quark.

We pass now to the *neutral semileptonic transitions*. Here, neutrino–nucleon collisions result in the following reactions:

1.9 Classification of weak interactions

(17) $\nu_\mu(\bar{\nu}_\mu) + N \to \nu_\mu(\bar{\nu}_\mu) + N$ (elastic)
$\nu_\mu(\bar{\nu}_\mu) + N \to \nu_\mu(\bar{\nu}_\mu) + N + (\pi \text{ or } K \text{ or } \ldots)$ (semiinclusive)
$\nu_\mu(\bar{\nu}_\mu) + N \to \nu_\mu(\bar{\nu}_\mu) + X$ (deep-inelastic, inclusive)

Such reactions occur by Z^0 exchange. They have been observed with high-energy neutrino beams and provide important results in support of the standard model. The reaction

(18) $\bar{\nu}_e + D \to \bar{\nu}_e + n + p$

has been observed with the aid of low-energy $\bar{\nu}_e$ and also yields results in agreement with the standard model. The electron–nucleon interaction

(19) $e^- + N \to e^- + N, \quad e^- + N \to e^- + X$

occurs not only by photon exchange but also by Z^0 exchange. Thus, once again, we have weak–electromagnetic interference, which here gives rise to observable parity-violating effects in high-energy, deep-inelastic scattering of polarized electrons on nucleons and in low-energy atomic spectroscopy. The experimental results in these two cases yield complementary information on the neutral weak-coupling parameters. The results are, again, in very good agreement with the standard model.

We come finally to the nonleptonic weak transitions. The decays of the neutral K mesons (semileptonic and nonleptonic) are of interest because of the peculiar particle-mixture properties of the K^0–$\bar{K}^0$ system, first pointed out by Gell-Mann (53),[5] whereas the transitions $K^0 \to \pi\pi$ are of special importance because of CP violation (see Chapter 7). Mesons K^0 and $\bar{K}^0$ are charge conjugates of one another and possess definite strangeness eigenvalues ($+1$ and -1, respectively). Thus it is appropriate to think in terms of the states $|K^0\rangle$ and $|\bar{K}^0\rangle$, which describe these particles at rest, when discussing the strong and electromagnetic interactions, which conserve strangeness. Yet, $|K^0\rangle$ and $|\bar{K}^0\rangle$ do not possess definite lifetimes for weak decay, because the weak interactions do not conserve strangeness. We may imagine a fictitious world in which the weak interactions are "turned off" and $|K^0\rangle$ and $|\bar{K}^0\rangle$ are degenerate eigenstates of the strong plus electromagnetic Hamiltonian. Then, any independent linear combinations of $|K^0\rangle$ and $|\bar{K}^0\rangle$ are also eigenstates of this zero-order Hamiltonian. When the weak interaction is "turned on," the appropriate linear combinations (called $|K_L^0\rangle$ and

---

5 See also Nakano and Nishijima (53).

$|K_S^0\rangle$ for long and short, respectively) will be those states with definite lifetimes, but they will no longer have definite strangeness nor exactly the same energies (masses). This results in the phenomenon of strangeness oscillations.

The short-lived K_S^0 decays in only two significant modes: $\pi^+\pi^-$ and $\pi^0\pi^0$, and each of these final states has CP eigenvalue $+1$. On the other hand, there are many known modes of decay for the long-lived component K_L^0, including the fully allowed decay to $\pi^+\pi^-\pi^0$ in which the final state is predominantly an eigenstate of CP with eigenvalue -1. In 1957, Landau (57) suggested that although C and P are separately violated in weak interactions, CP is a valid symmetry. Thus it was believed for a number of years that CP invariance is valid for all interactions, and in particular, that $|K_S^0\rangle$ and $|K_L^0\rangle$ are eigenstates of CP with eigenvalues $+1$ and -1, respectively. Thus, because

$$CP|K^0\rangle = -|\bar{K}^0\rangle$$
$$CP|\bar{K}^0\rangle = -|K^0\rangle$$

(the negative signs are used because K^0 and $\bar{K}^0$ are particles with negative intrinsic parity), it followed that

$$|K_S^0\rangle = \frac{1}{\sqrt{2}}(|K^0\rangle - |\bar{K}^0\rangle), \qquad CP = +1 \qquad (1.41)$$

$$|K_L^0\rangle = \frac{1}{\sqrt{2}}(|K^0\rangle + |\bar{K}^0\rangle), \qquad CP = -1 \qquad (1.42)$$

According to this scheme, the decays $K_L^0 \to \pi^+\pi^-$ and $K_L^0 \to \pi^0\pi^0$ are strictly forbidden.

However, in 1964, Christenson, Cronin, Fitch, and Turlay (64) discovered that the decay $K_L^0 \to \pi^+\pi^-$ actually occurs with small but finite probability, and the transition $K_L^0 \to \pi^0\pi^0$ was also soon observed. Thus violation of CP invariance was demonstrated, and the description given by (1.41) and (1.42) had to be modified. Although the explanation for the phenomenon of CP violation is not yet clear, there is reason to believe that it is associated with the "Cabibbo rotations" in the six-quark model (see Section 4.3 and Chapter 7).

With the nonleptonic decays of K mesons and of hyperons, we encounter the approximate selection rule $|\Delta I| = \frac{1}{2}$; that is, the initial and final isospins differ by one-half. It appears that this important rule has its origin in the suppression of certain amplitudes and the enhancement of others by QCD (gluon-radiative) corrections to the effective weak Hamiltonian.

Such effects are also believed to be of importance in hadronic decays of the charmed mesons D and F and bottom mesons B. These decays are complex and involved, and not yet well understood (see Chapter 6).

Finally, there exist weak contributions to the nuclear force, which must coexist alongside the much larger electromagnetic and strong contributions. The weak component can be isolated experimentally from the others because it violates parity. Here, in general, both W and Z^0 exchanges can occur. The situation is very complicated, and the experimental evidence somewhat inconclusive, but it does appear to confirm predictions of the standard model.

Thus we conclude our preliminary, superficial survey of weak interactions. We must now set about constructing the modern theory of electroweak interactions and then apply it to more detailed discussions of these phenomena.

Problems

1.1 Consider the matrix element for allowed nuclear β decay according to the original Fermi theory as given by (1.10). Verify that, in the nonrelativistic limit for the nucleons, the selection rules (1.11) hold and the contribution of the pseudoscalar term is negligible.

1.2 By taking into account conservation of angular momentum and the intrinsic parity of pions, show that the final state parities in the decays

"θ": $K^+ \to \pi^+ \pi^0$
"τ": $K^+ \to \pi^+ \pi^+ \pi^-$

are even and odd, respectively.

1.3 Consider e–ν_e scattering by W-boson exchange, according to the naive intermediate vector boson hypothesis. Show that in lowest order and in the limit of zero electron mass, the differential cross section is

$$\frac{d\sigma}{d\Omega} = \frac{G_F^2}{\pi^2} \frac{m_W^4 \omega^2}{[m_W^2 + 2\omega^2(1 - \cos\theta)]^2}.$$

where ω and θ are the neutrino energy and scattering angle, respectively, in the CM frame. Thus, show that the total cross section is

$$\sigma = \frac{4G^2}{\pi} \frac{\omega^2}{1 + \omega^2/m_W^2}$$

Utilize the partial wave expansion (1.30) to show that the $J = 0$ partial wave amplitude $\mathcal{M}_0$ is given by (1.37).

1.4 The Pauli–Dirac representation for Dirac matrices γ_μ and spinors ψ is commonly used (see Appendix A).

(a) Find the unitary transformation W, which transforms the Pauli–Dirac γ_μ into the corresponding matrices γ'_μ in the Weyl representation

$$\gamma'_\mu = W\gamma_\mu W\dagger$$

(b) Let $\gamma''_\mu = M\gamma_\mu M\dagger$, where $M = \gamma^0(1 + \gamma^2)/\sqrt{2}$. Show that γ''_0 is imaginary and the γ'' are real (Majorana representation).

2
The standard model

2.1 Introduction

The modern theory of weak and electromagnetic interactions is based on an intricate combination of complex and subtle ideas that we shall explore in this chapter. Before embarking on detailed discussions, however, let us take an overview, so that we may gain some perspective on the problems to be faced. The single most important principle to be assumed in the construction of the theory is that it must be invariant under certain local gauge transformations. We thus begin the chapter by discussing global and local $U(1)$ gauge invariance in a familiar theory, electrodynamics. It is found that global gauge invariance is equivalent to charge conservation, whereas the more restrictive condition of local gauge invariance gives rise to the familiar minimal coupling between current j_μ and electromagnetic field A_μ. The gauge quanta that appear here are the massless photons.

We then consider an important generalization, first proposed by Yang and Mills (54), to local $SU(2)$ gauge transformations on an isodoublet field. This theory gives rise to massless vector quanta with electric charges $+1$, -1, and 0. Attractive possibilities are thus suggested for the weak interactions: Might the Yang–Mills quanta be identified as charged and neutral intermediate bosons? The difficulty is that the real weak bosons, if they exist, must be massive, whereas the Yang–Mills quanta are massless. We cannot repair this defect simply by adding a mass term to the Yang–Mills Lagrangian; if we do this, gauge invariance is violated and we are back to the naive vector boson theory with all its grave defects. Another way must be found to impart mass to the Yang–Mills quanta; moreover we seek a related mechanism that might generate the lepton and quark masses.

2 The standard model

At this point, we turn to another idea, which at first seems quite unrelated to the problems just discussed but ultimately plays a direct role in their solution. That is the phenomenon of spontaneous symmetry breaking, with particular reference to a complex isodoublet scalar field. Spontaneous symmetry breaking refers to the situation in which the Lagrangian of a field theory possesses a certain symmetry not shared by the ground or vacuum state of the system.

For example, we might consider an ensemble of identical particles located at the corners of an infinite cubic lattice, each possessing a spin and magnetic moment. The Lagrangian of this system is clearly rotationally invariant. If the temperature T is sufficiently high, the spins are randomly oriented, but below a critical value $T = T_c$, the spins begin to line up in one direction, and this alignment becomes complete at $T = 0$. Since one direction has been singled out, the rotational symmetry of the system itself is broken. In the case of the complex isodoublet scalar field, which has four independent real components, we shall see that the state of affairs depends on a continuous parameter μ^2 in the Lagrangian. For $\mu^2 > 0$, the ground state of the system possesses the full symmetry of the Lagrangian and the quanta that appear in the theory are four real scalar bosons of identical positive mass μ. However, for $\mu^2 < 0$, the symmetry is broken in the ground state and we obtain one scalar quantum with positive mass (the so-called Higgs boson) and three massless scalar bosons. The latter are called Goldstone bosons, after the author who first pointed out that, for theories of this type, a massless scalar field always appears for each degree of freedom in which the symmetry is spontaneously broken (Goldstone 61).

Now, there is no experimental evidence for such Goldstone bosons, and it would thus appear that we have introduced a new problem with no physical relevance, instead of solving our old ones. However, if we now modify the Lagrangian of the complex isodoublet scalar field so that it is invariant under local $SU(2) \times U(1)$ gauge transformations, the situation is changed radically. In addition to the four scalar quanta, we now have four vector gauge quanta [corresponding to three from $SU(2)$ and one from $U(1)$]. Moreover, when μ^2 is set less than zero to give spontaneous symmetry breaking, the Goldstone bosons do not appear; instead the three degrees of freedom to which they correspond appear as additional (longitudinal polarization) degrees of freedom, one for each of three of the gauge quanta.[1] The appearance of longitudinal

---

[1] That gauge fields evade the Goldstone theorem was first established by Higgs (64a,b,66); see also Englert and Brout (64), Guralnik et al. (64), and Kibble (67).

2.1 Introduction

polarization in a vector field is equivalent to mass, so we obtain three massive gauge quanta with charges ±1 and 0. These are to be identified as the charged and neutral weak vector bosons. The fourth vector quantum remains massless and is identified as the photon. Of the four original scalar bosons, only the single massive Higgs boson survives. The four gauge fields and the Higgs field possess unique and specific nonlinear interactions with one another that have interesting and important physical implications and are an inevitable consequence of this type of theory, which is called non-Abelian.

The next step is to introduce the leptons and quarks and couple them to the gauge fields. The choices here are very much constrained by the requirement that the new theory reproduce the known and valid results of electrodynamics and of the V−A Feynman−Gell-Mann scheme for charged weak interactions. According to the model proposed independently by S. Weinberg (67) and A. Salam (68), who were the first to recognize the crucial significance of spontaneous symmetry breaking for a theory of weak interactions, one assumes that the fermions appear in weak "left-handed isodoublets" and "right-handed isosinglets." Thus the V−A law is inserted "by hand" into the new theory.

One might argue that this is somewhat unsatisfactory, since a fundamental theory of weak interactions ought somehow to explain the striking fact of parity violation. On the other hand, the fermion−gauge field couplings are now determined not only for electrodynamics and the charged weak interactions but also for the neutral weak interactions. In fact, the specific properties of these latter interactions were predicted by the new model before they were actually observed for the first time in 1973. Since then, detailed observations of neutral weak couplings in neutrino−nucleon, neutrino−electron, and electron−nucleon scattering and in low-energy atomic physics have verified the predictions of the Weinberg−Salam model in brilliant detail.

There remains the problem of the fermion masses. This is dealt with in the new theory by introduction of a gauge invariant fermion−fermion−Higgs coupling of the Yukawa type, which leads to the possibility of neutrino masses and neutrino oscillations. The quark−quark−Higgs coupling is also related to the Cabibbo angle(s) and to the parameter describing *CP* violation.

The new theory, described by a lengthy and complicated Lagrangian, must be quantized and a set of Feynman rules obtained before specific calculations of physical processes can be performed. In "ordinary" field theory, one can simply read the Feynman rules by inspection, directly from the Lagrangian. However, for a non-Abelian gauge field, this is not the case. On the contrary, formidable difficulties are encountered because a gauge field is not defined uniquely in a given physical situation but can be specified only up to a gauge transformation.

To surmount these obstacles, it is necessary to adopt a rather indirect approach to quantization, namely, the path integral method, which was first developed by R. P. Feynman for quantum electrodynamics in 1948. This method has the advantage that it demonstrates quite clearly how the gauge symmetry of the classical Lagrangian affects the quantum-mechanical properties of the system. However, it has the disadvantage of relative unfamiliarity, since it has been to a large extent ignored in standard quantum mechanics courses.[2]

Our treatment of this subject will, of necessity, be abbreviated, simple, and therefore quite superficial. We shall introduce path integrals by considering them in connection with nonrelativistic quantum mechanics in one spatial dimension. We shall then indicate how the essential ideas can be generalized without much difficulty to ordinary scalar field theory, where one can obtain the Feynman–Dyson expansion and Wick's theorem by the path integral technique (Abers and Lee 73).

The special problem associated with gauge fields must then be encountered. As first shown by L. D. Fadeev and V. N. Popov (67), this is achieved by the introduction of certain "ghost" fields, which are complex scalar fields obeying Fermi–Dirac statistics.[3] In electrodynamics, the ghost field appears but may be ignored, whereas in the Yang–Mills theory, an additional type of Feynman vertex occurs in which the ghost and gauge fields are coupled together. This emerges in addition to the diagrams that could have been read directly from the original Lagrangian. It turns out that the ghost–ghost–gauge boson vertex plays an essential role in analysis of the renormalizability of the theory, and without it, the theory is inconsistent.

The general set of Feynman rules for the Yang–Mills field with spontaneous symmetry breaking, including fermions, is most conveniently stated in the "generalized renormalizable (R_ξ) gauge." One finds that, although the rules depend on the particular choice of gauge, all physi-

---

2 An introduction to the path integral method is given in the text *Quantum Mechanics and Path Integrals* by R. P. Feynman and A. R. Hibbs (65).
3 See also Feynman (63), de Witt (64,67), and Mandelstam (68a,b).

cal results (S-matrix elements) are gauge-invariant. In particular, the ghost fields, though they must be encountered at intermediate stages, play no role in the final results (Fujikawa et al. 72).

Ultimately, one comes to the question of renormalization. If this is a complex and difficult problem in quantum electrodynamics, it is much more so here because of the appearance of nonlinear self-interactions and ghost fields. In 1971, 't Hooft (71a,b) succeeded in proving that the massless Yang–Mills theory is renormalizable and, shortly afterward, that this is also true for the complete theory with spontaneous symmetry breaking. This result, which was independently obtained by B. Lee (72) and Lee and R. Zinn-Justin, (Lee, 72a–c) must be regarded as a truly remarkable achievement in theoretical physics. The method of dimensional regularization introduced by 't Hooft and Veltman (72a,b) is an extremely useful tool, not only for these theoretical developments, but also for practical perturbation calculations.

Our aims in this chapter are quite modest: to obtain the classical Lagrangian of the standard model, to sketch briefly the path integral method and indicate how the ghost fields arise, and to state the Feynman rules in the R_ξ gauge, with an indication of how it is related to other gauges. With regard to renormalization, we shall attempt no more than the most elementary discussion of "naive power counting" and dimensional regularization. Thus, beyond supplying the reader with practical rules for carrying out specific calculations in the chapters to follow, we can provide only a glimpse of the remarkable, elaborate, and difficult edifice constructed by many theoretical physicists over the past two decades in gauge field theory.[4]

2.2 $U(1)$ gauge invariance and electrodynamics

Let us now consider a field $\Psi(x)$ governed by the Lagrangian density

$$\mathscr{L} = \mathscr{L}(\Psi, \partial_\mu \Psi) \tag{2.1}$$

We are interested in phase transformations of the following type:

$$\Psi \to e^{i\alpha}\Psi = \Psi' \tag{2.2}$$

where α is real. These are called $U(1)$ field-phase or gauge transformations [$U(1)$ because $e^{i\alpha}$ is a 1×1 unitary matrix]. The quantity α may be a constant, in which case the transformation is said to be *global*, or

---

[4] The reader who is primarily interested in applications may omit Secs. 2.7–2.11.

may vary from one point to another in space-time, $\alpha = \alpha(x_\mu)$, in which case we have a *local* gauge transformation.

For many purposes, it is sufficient to consider infinitesimal gauge transformations. In this case,

$$\delta\Psi = \Psi' - \Psi \simeq +i\alpha\Psi \tag{2.3}$$

There may be several field components Ψ rather than just one Ψ. For example, in the Dirac electron theory, we may consider the electron field Ψ and its conjugate $\overline{\Psi}$ to be independent components. Then, whereas Ψ satisfies (2.3), we have

$$\delta\overline{\Psi} = -i\alpha\overline{\Psi} \tag{2.4}$$

Let us investigate the change in $\mathscr{L}$ when Ψ varies. Writing out just the terms in one field component, we have

$$\delta\mathscr{L} = \left[\frac{\partial\mathscr{L}}{\partial\Psi}\delta\Psi + \frac{\partial\mathscr{L}}{\partial(\partial_\mu\Psi)}\delta(\partial_\mu\Psi)\right] \tag{2.5}$$

Since $\delta(\partial_\mu\Psi) = \partial_\mu(\delta\Psi)$, (2.5) becomes

$$\delta\mathscr{L} = \left[\frac{\partial\mathscr{L}}{\partial\Psi}\delta\Psi + \partial_\mu\left(\frac{\partial\mathscr{L}}{\partial(\partial_\mu\Psi)}\delta\Psi\right) - \partial_\mu\left(\frac{\partial\mathscr{L}}{\partial(\partial_\mu\Psi)}\right)\delta\Psi\right] \tag{2.6}$$

However $\mathscr{L}$ satisfies the Euler–Lagrange equation

$$\frac{\partial\mathscr{L}}{\partial\Psi} - \partial_\mu\frac{\partial\mathscr{L}}{\partial(\partial_\mu\Psi)} = 0 \tag{2.7}$$

Therefore, using (2.3), (2.6) becomes

$$\delta\mathscr{L} = \partial_\mu\left[\frac{\partial\mathscr{L}}{\partial(\partial_\mu\Psi)}\cdot i\alpha\Psi\right] \tag{2.8}$$

Let us now define the current density J^μ as

$$J^\mu = -i\frac{\partial\mathscr{L}}{\partial(\partial_\mu\Psi)}\cdot\Psi$$

Then

$$\delta\mathscr{L} = -\alpha\,\partial_\mu J^\mu - (\partial_\mu\alpha)J^\mu \tag{2.9}$$

For constant α (global transformation), we then have $\delta\mathscr{L} = -\alpha\,\partial_\mu J^\mu$ and "current conservation" is equivalent to global $U(1)$ gauge invariance of $\mathscr{L}$.

For example, let us consider the Lagrangian for the Dirac electron field:

$$\mathscr{L}_e = \tfrac{1}{2}\overline{\Psi}[i\gamma^\mu\,\partial_\mu - m]\Psi - \tfrac{1}{2}[i\,\partial_\mu\overline{\Psi}\gamma^\mu + m\overline{\Psi}]\Psi \tag{2.10}$$

2.2 U(1) gauge invariance and electrodynamics

The Euler–Lagrange equation for $\bar{\Psi}$, namely,

$$\frac{\partial \mathscr{L}_e}{\partial \bar{\Psi}} - \partial_\mu \frac{\partial \mathscr{L}_e}{\partial(\partial_\mu \bar{\Psi})} = 0$$

is just the Dirac equation

$$i\gamma^\mu \partial_\mu \Psi - m\Psi = 0 \tag{2.11}$$

and the Euler–Lagrange equation for Ψ is just the Dirac conjugate equation

$$i\,\partial_\mu \bar{\Psi} \gamma^\mu + m\bar{\Psi} = 0 \tag{2.12}$$

The current is

$$J^\mu = -i\left[\frac{\partial \mathscr{L}_e}{\partial(\partial_\mu \Psi)}\Psi - \bar{\Psi}\frac{\partial \mathscr{L}_e}{\partial(\partial_\mu \bar{\Psi})}\right] = \bar{\Psi}\gamma^\mu \Psi \tag{2.13}$$

and $\partial_\mu J^\mu = 0$ from (2.11) and (2.12). Thus $\mathscr{L}_e$ is invariant under global $U(1)$ gauge transformations.

As another example, consider a complex scalar field ϕ with Lagrangian

$$\mathscr{L} = \frac{1}{2}\left[\frac{\partial \phi^*}{\partial x_\mu}\frac{\partial \phi}{\partial x^\mu} - \mu^2 \phi^* \phi\right] \tag{2.14}$$

where μ is a mass. The field equations are

$$\Box \phi^* + \mu^2 \phi^* = 0$$
$$\Box \phi + \mu^2 \phi = 0 \tag{2.15}$$

and the conserved current is

$$J^\mu = \frac{1}{2}\left[\phi^* \frac{\partial \phi}{\partial x_\mu} - \frac{\partial \phi^*}{\partial x_\mu}\phi\right] \tag{2.16}$$

The Lagrangian of (2.14) is also invariant under global $U(1)$ gauge transformations.

We now consider *local* $U(1)$ gauge transformations, where $\alpha = \alpha(x_\mu)$. For the Dirac Lagrangian $\mathscr{L}_e$, since $\partial_\mu J^\mu = 0$, (2.9) becomes

$$\delta \mathscr{L}_e = -(\partial_\mu \alpha) J^\mu \tag{2.17}$$

which shows that $\mathscr{L}_e$ is not invariant under local gauge transformations. However, that invariance may be restored if we add to $\mathscr{L}_e$ a term $\mathscr{L}_1 = -eJ_\mu A^\mu$, where e is a constant and A^μ a vector field satisfying the condition that when $\Psi \to \Psi' = e^{i\alpha}\Psi$ and $\bar{\Psi} \to \bar{\Psi}' = e^{-i\alpha}\bar{\Psi}$, we have

$$A^\mu \to A^{\mu\prime} = A^\mu - (1/e)\partial^\mu \alpha \tag{2.18}$$

With (2.18), we obtain
$$\mathcal{L}_I' = -eJ_\mu A^{\mu\prime} = -eJ_\mu A^\mu + J_\mu \partial^\mu \alpha \tag{2.19}$$
Combining (2.19) and (2.17), we see that
$$\delta(\mathcal{L}_e + \mathcal{L}_I) = 0 \tag{2.20}$$
so that $\mathcal{L}_e + \mathcal{L}_I$ is invariant. Of course, $\mathcal{L}_I$ is the familiar minimal interaction of electrodynamics and (2.18) is the well-known gauge transformation condition for the vector potential.

To obtain the full Lagrangian of electrodynamics, we have only to add a gauge-invariant term describing the radiation itself:
$$\mathcal{L}_R = -\tfrac{1}{4} F_{\mu\nu} F^{\mu\nu} \tag{2.21}$$
where
$$F_{\mu\nu} = \partial_\mu A_\nu - \partial_\nu A_\mu \tag{2.22}$$
is the antisymmetric electromagnetic field tensor and we employ Heaviside–Lorentz units. Note that (2.21) describes a field whose quanta (photons) possess zero mass. If the photon mass m_γ were greater than zero, we would be obliged to add a gauge-noninvariant mass term
$$\mathcal{L}_R \to \mathcal{L}_R' = -\tfrac{1}{4} F_{\nu\mu} F^{\nu\mu} + \tfrac{1}{2} m_\gamma^2 A_\mu A^\mu \tag{2.23}$$
which would defeat the purpose for which A was introduced.

The preceding discussion for $\mathcal{L}_e + \mathcal{L}_I$ may be reexpressed in a convenient notation. We define the *covariant derivative* of Ψ and $\overline{\Psi}$ by
$$D_\mu \Psi = (\partial_\mu + ieA_\mu)\Psi \tag{2.24}$$
$$D_\mu \overline{\Psi} = (\partial_\mu - ieA_\mu)\overline{\Psi} \tag{2.25}$$
where $D_\mu \Psi$ has the useful property that it transforms in the same way as Ψ under the gauge transformation. That is,
$$\begin{aligned}\delta(D_\mu\Psi) &= \delta(\partial_\mu + ieA_\mu)\Psi \\ &= \partial_\mu(\delta\Psi) + ie\,\delta A_\mu\,\Psi + ieA_\mu\,\delta\Psi \\ &= i(\partial_\mu\alpha)\Psi + i(\alpha\,\partial_\mu\Psi) + ie\,\delta A_\mu\,\Psi - eA_\mu\alpha\Psi\end{aligned} \tag{2.26}$$
However $\delta A_\mu = -\partial_\mu \alpha / e$. Therefore (2.26) becomes
$$\begin{aligned}\delta(D_\mu\Psi) &= i(\alpha\,\partial_\mu\Psi) - eA_\mu\alpha\Psi \\ &= i\alpha(\partial_\mu + ieA_\mu)\Psi \\ &= i\alpha(D_\mu\Psi)\end{aligned} \tag{2.27}$$
In place of $\mathcal{L}_e + \mathcal{L}_I + \mathcal{L}_R$, we may write the gauge-invariant expression
$$\mathcal{L} = \tfrac{1}{2}\overline{\Psi}[i\gamma^\mu D_\mu - m]\Psi - \tfrac{1}{2}[iD_\mu\overline{\Psi}\gamma^\mu + m\overline{\Psi}]\Psi - \tfrac{1}{4}F_{\mu\nu}F^{\mu\nu} \tag{2.28}$$

Frequently we write this in the shorter form

$$\mathcal{L}_{QED} = \overline{\Psi}[i\gamma^\mu_\mu D_\mu - m]\Psi - \tfrac{1}{4}F_{\mu\nu}F^{\mu\nu} \tag{2.29}$$

To summarize, we found that the Dirac Lagrangian $\mathcal{L}_e$ is invariant under global $U(1)$ gauge transformations, which is equivalent to the condition of current conservation: $\partial_\mu J^\mu = 0$. However, $\mathcal{L}_e$ is not invariant under local $U(1)$ gauge transformations, though that invariance is restored if we add the minimal interaction term, provided that $A'_\mu = A_\mu - \partial_\mu \alpha/e$. That is achieved by replacing ∂_μ by a covariant derivative D_μ in $\mathcal{L}_e$.

The essential point is that electromagnetism enters the Lagrangian $\mathcal{L}$, not in an ad hoc way, but as the natural consequence of the symmetry of $\mathcal{L}$ with respect to local $U(1)$ gauge transformations, just as conservation of momentum follows from translational invariance or charge conservation comes from global gauge invariance.

2.3 The Yang–Mills field

We now extend the ideas just considered to the case of an isodoublet Dirac field $\Psi = \begin{pmatrix} \Psi_1 \\ \Psi_2 \end{pmatrix}$. We shall be interested in local $SU(2)$ gauge transformations of the following form:[5]

$$\begin{aligned} \Psi &\to \Psi' = \exp(i\boldsymbol{\epsilon}\cdot\mathbf{t})\Psi \\ \overline{\Psi} &\to \overline{\Psi}' = \overline{\Psi}\exp(-i\boldsymbol{\epsilon}\cdot\mathbf{t}) \end{aligned} \tag{2.30}$$

where $\boldsymbol{\epsilon}\cdot\mathbf{t} = \epsilon_1 t_1 + \epsilon_2 t_2 + \epsilon_2 t_3$, the ϵ_i are three real parameters depending on x_μ, and the t_1, t_2, and t_3 are isospin-$\tfrac{1}{2}$ matrices:

$$t_i = \tfrac{1}{2}\tau_i$$

where

$$\tau_1 = \begin{pmatrix} 0 & 1 \\ 1 & 0 \end{pmatrix}, \quad \tau_2 = \begin{pmatrix} 0 & -i \\ i & 0 \end{pmatrix}, \quad \tau_3 = \begin{pmatrix} 1 & 0 \\ 0 & -1 \end{pmatrix}$$

Then the t_i satisfy the $SU(2)$ commutation relations

$$[t_i, t_j] = i\epsilon_{ijk} t_k \tag{2.31}$$

with ϵ_{ijk} the completely antisymmetric unit tensor. For infinitesimal ϵ_i, we have

$$\delta\Psi = i\boldsymbol{\epsilon}\cdot\mathbf{t}\Psi, \qquad \delta\overline{\Psi} = -i\overline{\Psi}\boldsymbol{\epsilon}\cdot\mathbf{t} \tag{2.32}$$

Transformations of the type (2.30) or (2.32) are different from the $U(1)$ transformations previously discussed because they mix components

---

[5] For a detailed discussion of groups $SU(n)$, see Section 4.1².

Ψ_1 and Ψ_2 with one another or multiply Ψ_1 and Ψ_2 by opposite, rather than equal, phase factors.

Let us calculate the change in a Lagrangian $\mathscr{L}$ resulting from a transformation $\delta\Psi = i\boldsymbol{\epsilon} \cdot \mathbf{t}\Psi$. Repeating the steps that led to (2.8) and writing out only the terms in one field component, we find

$$\delta\mathscr{L} = \partial_\mu \left(\frac{\partial \mathscr{L}}{\partial(\partial_\mu \Psi)} i\boldsymbol{\epsilon} \cdot \mathbf{t}\Psi \right) \tag{2.33}$$

Defining the current as

$$\mathbf{J}_\mu{}^v = -i \frac{\partial \mathscr{L}}{\partial(\partial_\mu \Psi)} \mathbf{t}\Psi \tag{2.34}$$

(2.33) becomes

$$\delta\mathscr{L} = -\boldsymbol{\epsilon} \cdot \partial_\mu \mathbf{J}^{v\mu} - \partial_\mu \boldsymbol{\epsilon} \cdot \mathbf{J}^{v\mu}$$

which is analogous to (2.9). For constant $\boldsymbol{\epsilon}$ (global $SU(2)$ gauge transformations), $\delta\mathscr{L} = 0$ if $\partial_\mu \mathbf{J}^{v\mu} = 0$. Such would be the case for an isodoublet, each component of which satisfied the Dirac equation, with

$$\mathbf{J}^{v\mu} = \bar{\Psi}\gamma^\mu \mathbf{t}\Psi \tag{2.35}$$

However, if we have a local gauge transformation $\boldsymbol{\epsilon} = \boldsymbol{\epsilon}(x_\mu)$, then (2.34) reveals that $\mathscr{L}$ is not invariant.

Once more we try to restore gauge invariance by introducing a compensating minimal interaction. For this purpose, we now employ a triplet of gauge fields $A_\mu{}^1$, $A_\mu{}^2$, and $A_\mu{}^3$, one for each isospin matrix t_i. We also define the covariant derivative

$$D_\mu = \partial_\mu + ig\mathbf{A}_\mu \cdot \mathbf{t} \tag{2.36}$$

by analogy with (2.24). What properties should $\mathbf{A}_\mu$ have in order that $D_\mu\Psi$ transform in the same way as Ψ? We require that

$$\delta(D_\mu\Psi) = i\boldsymbol{\epsilon} \cdot \mathbf{t}(D_\mu\Psi) \tag{2.37}$$

Now

$$\begin{aligned}
\delta(D_\mu\Psi) &= \delta(\partial_\mu\Psi) + ig\,\delta(\mathbf{A}_\mu \cdot \mathbf{t}\Psi) \\
&= \partial_\mu(\delta\Psi) + ig\,\delta\mathbf{A}_\mu \cdot \mathbf{t}\Psi + ig\mathbf{A}_\mu \cdot \mathbf{t}\,\delta\Psi \\
&= i(\partial_\mu\boldsymbol{\epsilon}) \cdot \mathbf{t}\Psi + i\boldsymbol{\epsilon} \cdot \mathbf{t}\,\partial_\mu\Psi + ig\,\delta\mathbf{A}_\mu \cdot \mathbf{t}\Psi - g\mathbf{A}_\mu \cdot \mathbf{t}\boldsymbol{\epsilon} \cdot \mathbf{t}\Psi
\end{aligned}$$

Equating this to (2.37), we find

$$\delta\mathbf{A}_\mu \cdot \mathbf{t} = -\frac{1}{g}\partial_\mu\boldsymbol{\epsilon} \cdot \mathbf{t} + i(\boldsymbol{\epsilon} \cdot \mathbf{t}\,\mathbf{A}_\mu \cdot \mathbf{t} - \mathbf{A}_\mu \cdot \mathbf{t}\,\boldsymbol{\epsilon} \cdot \mathbf{t})$$

Now, using (2.31), we obtain

$$\delta\mathbf{A}_\mu = -\frac{1}{g}\partial_\mu\boldsymbol{\epsilon} - (\boldsymbol{\epsilon} \times \mathbf{A}_\mu) \tag{2.38}$$

2.3 The Yang–Mills field

which defines the gauge transformation property of $\mathbf{A}_\mu$ under the infinitesimal transformation $\delta\Psi = i\boldsymbol{\epsilon}\cdot\mathbf{t}\Psi$. In (2.38), the first term on the right-hand side is analogous to that obtained in the $U(1)$ case (2.18). However, the second term is entirely new and arises because $\boldsymbol{\epsilon}\cdot\mathbf{t}$ and $\mathbf{A}_\mu\cdot\mathbf{t}$ are 2×2 matrices that do not commute. One refers to this essential new feature by stating that $\mathbf{A}_\mu$ is a non-Abelian gauge field.

If quantities ϵ_i are finite and $\Psi' = \exp(i\boldsymbol{\epsilon}\cdot\mathbf{t})\Psi$ and $\bar{\Psi}' = \bar{\Psi}\exp(-i\boldsymbol{\epsilon}\cdot\mathbf{t})$ then the gauge transformation property of $\mathbf{A}_\mu$ is

$$\mathbf{A}'_\mu\cdot\mathbf{t} = \exp(i\boldsymbol{\epsilon}\cdot\mathbf{t})\left(\mathbf{A}_\mu\cdot\mathbf{t} - \frac{i}{g}\partial_\mu\right)\exp(-i\boldsymbol{\epsilon}\cdot\mathbf{t}) \tag{2.39}$$

The next step is to formulate a gauge invariant "radiation" term analogous to $\mathscr{L}_\mathrm{R} = -\tfrac{1}{4}F_{\mu\nu}F^{\mu\nu}$ in electrodynamics. We shall now demonstrate that

$$\mathscr{L}_\mathrm{YM} = -\tfrac{1}{4}\mathbf{E}_{\mu\nu}\cdot\mathbf{E}^{\mu\nu} \tag{2.40}$$

where YM signifies "Yang–Mills" and where

$$\mathbf{E}_{\mu\nu} = \partial_\mu\mathbf{A}_\nu - \partial_\nu\mathbf{A}_\mu - g(\mathbf{A}_\mu\times\mathbf{A}_\nu) \tag{2.41}$$

has the required properties of local $SU(2)$ gauge invariance. From (2.38), we obtain

$$\delta(\partial_\mu\mathbf{A}_\nu - \partial_\nu\mathbf{A}_\mu) = \partial_\mu\,\delta\mathbf{A}_\nu - \partial_\nu\,\delta\mathbf{A}_\mu$$
$$= -\frac{1}{g}\partial_\mu\,\partial_\nu\boldsymbol{\epsilon} - \partial_\mu(\boldsymbol{\epsilon}\times\mathbf{A}_\nu) + \frac{1}{g}\partial_\nu\,\partial_\mu\boldsymbol{\epsilon} + \partial_\nu(\boldsymbol{\epsilon}\times\mathbf{A}_\mu)$$
$$= (\partial_\nu\boldsymbol{\epsilon})\times\mathbf{A}_\mu - (\partial_\mu\boldsymbol{\epsilon})\times\mathbf{A}_\nu - \boldsymbol{\epsilon}\times(\partial_\mu\mathbf{A}_\nu - \partial_\nu\mathbf{A}_\mu)$$

Also

$$-\delta[g(\mathbf{A}_\mu\times\mathbf{A}_\nu)] = -g(\delta\mathbf{A}_\mu)\times\mathbf{A}_\nu - g\mathbf{A}_\mu\times\delta\mathbf{A}_\nu$$
$$= +(\partial_\mu\boldsymbol{\epsilon})\times\mathbf{A}_\nu - (\partial_\nu\boldsymbol{\epsilon})\times\mathbf{A}_\mu + g(\boldsymbol{\epsilon}\times\mathbf{A}_\mu)\times\mathbf{A}_\nu$$
$$+ g\mathbf{A}_\mu\times(\boldsymbol{\epsilon}\times\mathbf{A}_\nu)$$

Combining these quantities and utilizing the identity

$$(\boldsymbol{\epsilon}\times\mathbf{A}_\mu)\times\mathbf{A}_\nu + \mathbf{A}_\mu\times(\boldsymbol{\epsilon}\times\mathbf{A}_\nu) = \boldsymbol{\epsilon}\times(\mathbf{A}_\mu\times\mathbf{A}_\nu)$$

we obtain

$$\delta\mathbf{E}_{\mu\nu} = -(\boldsymbol{\epsilon}\times\mathbf{E}_{\mu\nu})$$

Therefore,

$$\delta\mathscr{L}_\mathrm{YM} = -\tfrac{1}{4}(\delta\mathbf{E}_{\mu\nu}\cdot\mathbf{E}^{\mu\nu} + \mathbf{E}_{\mu\nu}\cdot\delta\mathbf{E}^{\mu\nu})$$
$$= +\tfrac{1}{4}[(\boldsymbol{\epsilon}\times\mathbf{E}_{\mu\nu})\cdot\mathbf{E}^{\mu\nu} + (\boldsymbol{\epsilon}\times\mathbf{E}^{\mu\nu})\cdot\mathbf{E}_{\mu\nu}]$$
$$= 0 \tag{2.42}$$

Thus $\mathscr{L}_\mathrm{YM}$, as defined in (2.41), is indeed gauge-invariant. The existence of the additional non-Abelian term $-g(\mathbf{A}_\mu\times\mathbf{A}_\nu)$ in $\mathbf{E}_{\mu\nu}$ implies that in $\mathscr{L}_{YM}$ we have not only "zeroth-order" terms similar to those in electrodynamics

$$-\tfrac{1}{4}(\partial_\mu \mathbf{A}_\nu - \partial_\nu \mathbf{A}_\mu) \cdot (\partial^\mu \mathbf{A}^\nu - \partial^\nu \mathbf{A}^\mu) \tag{2.43}$$

but also terms of order g

$$+\tfrac{1}{4}g(\partial_\mu \mathbf{A}_\nu - \partial_\nu \mathbf{A}_\mu) \cdot (\mathbf{A}^\mu \times \mathbf{A}^\nu) + \tfrac{1}{4}g(\mathbf{A}_\mu \times \mathbf{A}_\nu) \cdot (\partial^\mu \mathbf{A}^\nu - \partial^\nu \mathbf{A}^\mu) \tag{2.44}$$

and terms of order g^2

$$-\tfrac{1}{4}g^2(\mathbf{A}_\mu \times \mathbf{A}_\nu) \cdot (\mathbf{A}^\mu \times \mathbf{A}^\nu) \tag{2.45}$$

The nonlinear self-interaction terms of order g and g^2 (called, respectively, A^3 and A^4 terms) have no analog in quantum electrodynamics. They arise from the non-Abelian character of $\mathbf{A}_\mu$.

In order to obtain a preliminary understanding of the nature of the quanta of the Yang–Mills field, we take the limit $g \to 0$ in $\mathscr{L}_{YM}$:

$$\lim_{g \to 0} \mathscr{L}_{YM} = -\tfrac{1}{4}(\partial_\mu \mathbf{A}_\nu - \partial_\nu \mathbf{A}_\mu)(\partial^\mu \mathbf{A}^\nu - \partial^\nu \mathbf{A}^\mu) \tag{2.46}$$

It is clear from the form of this expression and by analogy to quantum electrodynamics that the Yang–Mills quanta must be zero-mass vector bosons. Furthermore, since the components $(A_\mu^1 + iA_\mu^2)/\sqrt{2}$, A_μ^3 and $(A_\mu^1 - iA_\mu^2)/\sqrt{2}$ form an isotriplet vector, the Yang–Mills quanta must have charges $+1$, -1, and 0.

The Yang–Mills theory just set forth for local $SU(2)$ gauge transformations of an isodoublet field may be generalized to some other group of local gauge transformations on a field with n components. In particular, we may consider $SU(3)$ gauge transformations on a field with three color components. In place of the $n^2 - 1 = 3$ generators $\tfrac{1}{2}\tau_i$ of $SU(2)$, we now have the $n^2 - 1 = 8$ generators $\tfrac{1}{2}\lambda_i$ of $SU(3)$; the corresponding 8-gauge field quanta are then identified as the gluons of quantum chromodynamics.

2.4 Spontaneous symmetry breaking

After local gauge invariance, spontaneous symmetry breaking is the next fundamental principle of the new theory. Let us introduce it by means of an example from mechanics involving elastic instability. (This problem was first studied by the great mathematician L. Euler.) Consider a thin metal rod of length L and circular cross section of radius R. The rod is clamped at both ends and subjected to a compressional force F directed along the rod axis (z axis) (see Figure 2.1a). Let the deflections of the rod in the x and y directions as a function of z be

2.4 Spontaneous symmetry breaking

$X(z)$ and $Y(z)$, respectively. Then it may be shown from the theory of elasticity that the conditions for equilibrium are

$$IE\frac{d^4X}{dz^4} + F\frac{d^2X}{dz^2} = 0 \qquad (2.47)$$

$$IE\frac{d^4Y}{dz^4} + F\frac{d^2Y}{dz^2} = 0 \qquad (2.48)$$

where E is Young's modulus and $I = \frac{1}{4}\pi R^4$. The boundary conditions for the clamped rod are

$$X(0) = X(L) = 0 \qquad (2.49)$$
$$X'(0) = X'(L) = 0, \qquad X' = dX/dz \qquad (2.50)$$

with similar conditions on Y.

Quite obviously, if we start with a very small force F, the equilibrium condition will be that the rod is straight; that is, we will have the solutions $X(z) = Y(z) = 0$, and the rod will be stable against small perturbations from straightness. However, if F is increased until it reaches a certain critical value F_c, then the rod will become unstable against small perturbations from straightness and will bend (see Figure 2.1b). Although in general the deflection will be large and (2.47) and (2.48)

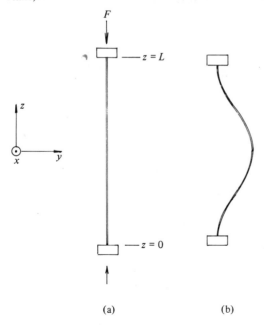

Figure 2.1. A problem of elastic instability (see text discussion for details).

(valid for small deflections) will not apply to the situation of Figure 2.1b, these equations can still be used to find F_c, which is the value of F for neutral stability. Thus, we seek a solution of (2.47) with $X(z) \neq 0$ subject to boundary conditions (2.49) and (2.50). It is not difficult to show that such a solution is

$$X(z) = \text{const} \cdot \sin^2 \pi z/L \qquad (2.51)$$

which corresponds to $F_c = 4\pi^2 IE/L^2$.

We may now consider small transverse oscillations of the rod about its equilibrium position. When the rod is straight, it may be shown that the eigenfrequencies for small oscillations ω are given by

$$\cos kL \cosh kL = 1 \qquad (2.52)$$

where $k = \omega^{1/2}(\mu_0/EI)^{1/4}$ and μ_0 is the mass per unit length. Obviously the frequencies of oscillation are identical for small oscillations in any direction, that is, in any plane containing the z axis.

When the rod is bent (Figure 2.1b), we may also have small oscillations. In general, however, the characteristic frequencies will be different; and in particular, the frequencies for oscillation in the plane of bending will not be the same as in the perpendicular direction.

The Lagrangian of this system is axially symmetric. It contains a continuous parameter, the compressional force F. For small values of F the lowest energy state of the system, which we may call the *ground state*, also has axial symmetry (the rod is straight). For $F \geq F_c$, the system, when perturbed infinitesimally, jumps to a new ground state in which the symmetry is broken (the bent rod). This new state is degenerate; that is, it is but one of an infinite number of possible states, since the rod can be bent in any plane containing the z axis. Small oscillations about the original axially symmetric ground state themselves have axial symmetry, whereas this is no longer true for the bent rod.

We can now discuss spontaneous symmetry breaking in field theory. In place of the rod deflection as a function of z, we consider a quantum field $\phi = \phi(x_\mu)$ with a Lagrangian possessing a certain symmetry and containing a continuous parameter, the analog of the compressional force F. For certain values of the parameter, the ground, or vacuum, state of the field also possesses the symmetry. If the parameter exceeds a critical value, however, the symmetry is spontaneously broken in the ground state, which now becomes degenerate. Small oscillations are analogous in quantum field theory to the appearance of particles, the relationship between their energy and momentum being expressed in terms of mass. In general, the mass of the particles will differ for the

2.4 Spontaneous symmetry breaking

symmetric ground state and for the ground state in which the symmetry is spontaneously broken. Although the problem we shall consider is initially merely of "academic" interest because it does not correspond directly to a real physical situation, it possesses features of general importance.

In particular, let us consider a complex isodoublet scalar field

$$\phi = \frac{1}{\sqrt{2}} \begin{pmatrix} \phi_1 + i\phi_2 \\ \phi_3 + i\phi_4 \end{pmatrix}$$

where $\phi_1, \ldots, \phi_4$ are real scalar fields. We take for the Lagrangian

$$\mathscr{L} = (\partial_\mu \phi^\dagger)(\partial^\mu \phi) - \mu^2 \phi^\dagger \phi - \lambda(\phi^\dagger \phi)^2 \qquad (2.53)$$

In this expression, the first term on the right-hand side (RHS) corresponds to kinetic energy and the second, for $\mu^2 > 0$, corresponds to mass. If we ignore the third term, the field equations are just the Klein–Gordon equations (2.15). The third term in (2.53) represents a "ϕ^4" self-interaction. Such a Lagrangian is known to be renormalizable. Note that $\mathscr{L}$ is invariant under global $SU(2) \times U(1)$ gauge transformations.

The corresponding Hamiltonian density is obtained from the defining equation

$$\mathscr{H} = \dot{\phi}^\dagger \frac{\partial \mathscr{L}}{\partial \dot{\phi}^\dagger} + \frac{\partial \mathscr{L}}{\partial \dot{\phi}} \dot{\phi} - \mathscr{L}$$

to be

$$\mathscr{H} = \partial_0 \phi^\dagger \, \partial_0 \phi + \nabla \phi^\dagger \cdot \nabla \phi + \mu^2 \phi^\dagger \phi + \lambda(\phi^\dagger \phi)^2 \qquad (2.54)$$

For a classical field, this is just the energy density, and since we require it to be positive for arbitrary ϕ, we must have $\lambda \geq 0$. Let us seek a minimum of $\mathscr{H}$ with respect to ϕ. All the derivative terms are positive unless ϕ is stationary with respect to x_μ, therefore we seek a constant value of ϕ to minimize

$$V = \mu^2 \phi^\dagger \phi + \lambda(\phi^\dagger \phi)^2 \qquad (2.55)$$

Let $y = \phi^\dagger \phi > 0$; then if $\mu^2 > 0$, V reaches a minimum for $y = 0$. The vacuum state is therefore $\phi = 0$, which obviously has the full symmetry of the Lagrangian. If we make small perturbations about this vacuum state, we find that the Lagrangian can be expressed in terms of these perturbations and that we obtain four real scalar fields, all with mass μ.

However, for $\mu^2 < 0$, the minimum value of V is attained where

$$\partial V / \partial y = \mu^2 + 2\lambda y = 0$$

that is, where
$$\phi^\dagger \phi = y = -\mu^2/2\lambda \equiv \eta^2 > 0 \qquad (2.56)$$
Clearly, many different values of ϕ satisfy (2.56), since any arbitrary global $SU(2) \times U(1)$ transformation on ϕ leaves y invariant. For definiteness, let us choose
$$\phi_0 = \begin{pmatrix} 0 \\ \eta \end{pmatrix} \qquad (2.57)$$
as our vacuum state in some frame F with a particular orientation of isospin axes. Note that ϕ_0 spontaneously breaks the global $SU(2) \times U(1)$ symmetry, since it selects a particular direction in charge space. Now we introduce a small perturbation on ϕ_0, which depends on $x = x_\mu$. In some frame F' whose axes vary with respect to those of F from point to point, the field will now be
$$\phi_{F'}(x) = \begin{pmatrix} 0 \\ \eta + \sigma(x)/\sqrt{2} \end{pmatrix} \qquad (2.58)$$
where $\sigma(x)$ is a small quantity. However, it is necessary for us to find ϕ in the original frame F, since we want to compare its effect in $\mathscr{L}$ with that of ϕ_0, and the Lagrangian is not invariant with respect to local $SU(2) \times U(1)$ gauge transformations. To find ϕ in frame F, we must perform an infinitesimal isospin rotation on $\phi_{F'}(x)$:
$$\phi_F(x) = U\phi_{F'}(x) = e^{i\bar{\theta}(x)\cdot t}\phi_{F'} \simeq (I + i\bar{\theta}\cdot t)\begin{pmatrix} 0 \\ \eta + \sigma(x)/\sqrt{2} \end{pmatrix}$$
Since the parameters $\bar{\theta}_i$ and σ are both small, we obtain to first order
$$\Phi(x) \equiv \phi_F - \phi_0 \simeq \begin{pmatrix} \frac{1}{2}(i\bar{\theta}_1 + \bar{\theta}_2)\eta \\ \sigma/\sqrt{2} - \frac{1}{2}i\bar{\theta}_3\eta \end{pmatrix} \qquad (2.59)$$
We may now compute the Lagrangian (2.53) in terms of the variables $\bar{\theta}$ and σ. Taking into account that $\eta^2 = \mu^2/2\lambda > 0$ and ignoring a constant term, we find, after some simple algebra, that
$$\mathscr{L} = \tfrac{1}{4}\eta^2(\partial^\mu\bar{\theta}\cdot\partial_\mu\bar{\theta}) + \tfrac{1}{2}\partial_\mu\sigma\cdot\partial^\mu\sigma + \mu^2\sigma^2 + O(\sigma^3) + O(\sigma^4) \qquad (2.60)$$
Now, making the change of variables
$$\theta_i = \frac{1}{\sqrt{2}}\eta\bar{\theta}_i \qquad (2.61)$$
we finally obtain
$$\mathscr{L} = \tfrac{1}{2}[\partial^\mu\theta\cdot\partial_\mu\theta + \partial_\mu\sigma\cdot\partial^\mu\sigma + 2\mu^2\sigma^2 + O(\sigma^3)] \qquad (2.62)$$
This expression describes three scalar fields θ_1, θ_2, and θ_3 that are massless (since there is no term in $\mathscr{L}$ corresponding to the θ_i^2) and a

2.4 Spontaneous symmetry breaking

scalar field σ with mass $m_\sigma = (-2\mu^2)^{1/2} > 0$. There are also nonlinear self-interaction terms for the latter field (terms of order σ^3, etc.). The three massless scalar quanta are just the Goldstone bosons mentioned in the introduction (Section 2.1).

At this point we must remind ourselves that a cornerstone of our new theory is the assumption of local gauge invariance. Therefore, although the problem we just discussed has many interesting features, it cannot correspond to physical reality because the Lagrangian of (2.53) is not locally gauge-invariant. At this stage, then, it seems reasonable for us to consider a modified problem in which there appears once again a complex isodoublet scalar field with nonlinear self-interaction term $-\lambda(\phi^\dagger\phi)^2$, but where the Lagrangian *is* invariant under local $SU(2) \times U(1)$ transformations.

For this purpose, we replace the partial derivatives in (2.53) by covariant derivatives

$$D_\mu\phi = \partial_\mu\phi + ig\mathbf{A}_\mu \cdot \mathbf{t}\phi + \tfrac{1}{2}ig'B_\mu\phi \tag{2.63}$$

$$(D_\mu\phi)^\dagger = \partial_\mu\phi^\dagger - ig\phi^\dagger\mathbf{A}_\mu \cdot \mathbf{t} - \tfrac{1}{2}ig'B_\mu\phi^\dagger \tag{2.64}$$

where g and g' are two independent coupling constants and the $\mathbf{A}_\mu$ form an isotriplet of gauge fields and B_μ an isosinglet gauge field, corresponding to $SU(2)$ and $U(1)$ gauge transformations, respectively. Also, we include in the Lagrangian the appropriate contributions from gauge fields alone:

$$\mathcal{L} = (D_\mu\phi)^\dagger(D^\mu\phi) - \mu^2\phi^\dagger\phi - \lambda(\phi^\dagger\phi)^2 - \tfrac{1}{4}\mathbf{E}_{\mu\nu} \cdot \mathbf{E}^{\mu\nu} - \tfrac{1}{4}f_{\mu\nu}f^{\mu\nu} \tag{2.65}$$

where $\mathbf{E}_{\mu\nu} = \partial_\mu\mathbf{A}_\nu - \partial_\nu\mathbf{A}_\mu - g(\mathbf{A}_\mu \times \mathbf{A}_\nu)$ and $f_{\mu\nu} = \partial_\mu B_\nu - \partial_\nu B_\mu$.

We may now retrace the steps taken earlier. In the new Hamiltonian density, we eliminate the positive terms containing derivatives by requiring the fields to be constant. Thus for the ground state, we have some constant field ϕ_0 and, a priori, constant fields $\mathbf{A}_0^\mu$ and B_0^μ. However, if the latter are nonzero, a particular direction or set of directions are chosen in space-time, which would violate the notion that the vacuum state ought to exhibit space-time isotropy. Therefore, we assume that $\mathbf{A}_0^\mu$ and B_0^μ are zero. For $\mu^2 < 0$, the vacuum state once again consists of a constant field $\phi_0 = \begin{pmatrix} 0 \\ \eta \end{pmatrix}$. As before, we introduce a small perturbation $\sigma(x)/\sqrt{2}$:

$$\phi_{F'}(x) = \begin{pmatrix} 0 \\ \eta + \sigma(x)/\sqrt{2} \end{pmatrix} \tag{2.66}$$

2 The standard model

Once more we might ask: What is ϕ in the original frame F? Evidently $\phi_F = U(x)\phi_{F'}$, where $U(x)$ is some unitary transformation. However, our Lagrangian is locally gauge-invariant, which means that we can rotate ϕ_F by any amount and shift its phase by any amount at any point in space-time and the Lagrangian will be unaffected. This implies that we can fix the gauge by writing

$$\phi(x) = U^{-1}\phi_F = U^{-1}U\phi_{F'} = \phi_{F'} = \begin{pmatrix} 0 \\ \eta + \sigma(x)/\sqrt{2} \end{pmatrix} \qquad (2.67)$$

In this step, since no rotation is applied, no Goldstone bosons appear.[6] We emphasize that this is of crucial importance.

We may now proceed to calculate the Lagrangian in terms of the quantities η and $\sigma/\sqrt{2}$ as follows:

$$D_\mu \phi = \begin{pmatrix} \partial_\mu + \tfrac{1}{2}ig'B_\mu + \tfrac{1}{2}igA_{\mu 3} & \tfrac{1}{2}ig(A_{\mu 1} - iA_{\mu 2}) \\ \tfrac{1}{2}ig(A_{\mu 1} + iA_{\mu 2}) & \partial_\mu + \tfrac{1}{2}ig'B_\mu - \tfrac{1}{2}igA_{\mu 3} \end{pmatrix} \begin{pmatrix} 0 \\ \eta + \sigma/\sqrt{2} \end{pmatrix}$$

$$= \begin{pmatrix} \tfrac{1}{2}ig(A_{\mu 1} - iA_{\mu 2})(\eta + \sigma/\sqrt{2}) \\ (\partial_\mu + \tfrac{1}{2}ig'B_\mu - \tfrac{1}{2}igA_{\mu 3})(\eta + \sigma/\sqrt{2}) \end{pmatrix}$$

Therefore,

$$D_\mu \phi^\dagger D^\mu \phi = \tfrac{1}{2}\partial_\mu \sigma \partial^\mu \sigma + [\tfrac{1}{4}g^2(A_{\mu 1}A_1^\mu + A_{\mu 2}A_2^\mu) + \tfrac{1}{4}g^2(A_{\mu 3} - (g'/g)B_\mu)^2](\eta + \sigma/\sqrt{2})^2 \qquad (2.68)$$

When the theory is quantized, we will want gauge fields to correspond to particles of definite charge. In preparation for this, we now define

$$W_\mu^+ = \frac{1}{\sqrt{2}}(A_{\mu 1} - iA_{\mu 2}) \qquad (2.69)$$

$$W_\mu^- = \frac{1}{\sqrt{2}}(A_{\mu 1} + iA_{\mu 2}) \qquad (2.70)$$

The two remaining fields, $A_{\mu 3}$ and B_μ, are both neutral, so we may choose new linear combinations

$$Z_\mu = \cos\theta_W A_{\mu 3} - \sin\theta_W B_\mu \qquad (2.71)$$

$$A_\mu = \sin\theta_W A_{\mu 3} + \cos\theta_W B_\mu \qquad (2.72)$$

where θ_W (the Weinberg angle) is an adjustable parameter. Comparing these two equations to (2.68), we see that

$$\tan\theta_W \equiv g'/g$$

Also, we have the following, with $\eta^2 = -\mu^2/2\lambda$:

---

[6] The Goldstone bosons disappear because we made a fortunate choice of gauge: the so-called *unitary* gauge.

2.4 Spontaneous symmetry breaking

$$V = -\mu^2 \phi^\dagger \phi - \lambda(\phi^\dagger \phi)^2$$
$$= -\mu^2(\eta + \sigma/\sqrt{2})^2 - \lambda(\eta + \sigma/\sqrt{2})^4$$
$$= -\left(\frac{\mu^2}{2}\eta^2 - \mu^2\sigma^2 - \frac{\mu^2\sigma^3}{\sqrt{2}\eta} - \frac{\mu^2\sigma^4}{8\eta^2}\right) \quad (2.73)$$

Collecting the various terms and ignoring the constant $\frac{1}{2}\mu^2\eta^2$, the Lagrangian (2.65) becomes

$$\mathcal{L} = \frac{1}{2}\partial_\mu \sigma \, \partial^\mu \sigma + \frac{g^2\eta^2}{4}(W_\mu^{+\dagger}W^{+\mu} + W_\mu^{-\dagger}W^{-\mu}) + \frac{g^2\eta^2}{4\cos^2\theta_W}Z_\mu Z^\mu$$
$$+ \mu^2\sigma^2 + 2^{-1/2}\eta^{-1}\mu^2\sigma^3 + 2^{-3/2}\eta^{-2}\mu^2\sigma^4$$
$$+ \left(\frac{g^2}{2\sqrt{2}}\eta\sigma + \frac{g^2\sigma^2}{8}\right)\left(W_\mu^{+\dagger}W^{+\mu} + W_\mu^{-\dagger}W^{-\mu} + \frac{1}{\cos^2\theta_W}Z_\mu Z^\mu\right)$$
$$- \tfrac{1}{4}\mathbf{E}_{\mu\nu} \cdot \mathbf{E}^{\mu\nu} - \tfrac{1}{4}f_{\mu\nu}f^{\mu\nu} \quad (2.74)$$

Before we can discuss the physical meaning of this Lagrangian, we must clarify the last two terms. Writing them out, we have

$$-\tfrac{1}{4}\mathbf{E}_{\mu\nu} \cdot \mathbf{E}^{\mu\nu} - \tfrac{1}{4}f_{\mu\nu}f^{\mu\nu} = -\tfrac{1}{4}[\partial_\mu \mathbf{A}_\nu - \partial_\nu \mathbf{A}_\mu - g(\mathbf{A}_\mu \times \mathbf{A}_\nu)]$$
$$\times [\partial^\mu \mathbf{A}^\nu - \partial^\nu \mathbf{A}^\mu - g(\mathbf{A}^\mu \times \mathbf{A}^\nu)]$$
$$- \tfrac{1}{4}(\partial_\mu B_\nu - \partial_\nu B_\mu)(\partial^\mu B^\nu - \partial^\nu B^\mu)$$

Now, using (2.71) and (2.72) to transform this expression and inserting in the Lagrangian (2.74), we finally obtain, after some algebra,

$$\mathcal{L}_0 = \frac{1}{2}\partial_\mu \sigma \, \partial^\mu \sigma + \mu^2\sigma^2 - \frac{1}{4}A_{\mu\nu}A^{\mu\nu}$$
$$- \frac{1}{4}(W_{\mu\nu}^{+\dagger}W^{\mu\nu+} + W_{\mu\nu}^{-\dagger}W^{\mu\nu-}) + \frac{g^2\eta^2}{4}(W_\mu^{+\dagger}W^{+\mu} + W_\mu^{-\dagger}W^{-\mu})$$
$$- \frac{1}{4}(Z_{\mu\nu}Z^{\mu\nu}) + \frac{g^2\eta^2}{4\cos^2\theta_W}Z_\mu Z^\mu$$
$$+ \frac{g}{4}[(\mathbf{A}_\mu \times \mathbf{A}_\nu)(\partial^\mu \mathbf{A}^\nu - \partial^\mu \mathbf{A}^\nu) + (\partial_\mu \mathbf{A}_\nu - \partial_\nu \mathbf{A}_\mu)(\mathbf{A}^\mu \times \mathbf{A}^\nu)]$$
$$- \frac{g^2}{4}(\mathbf{A}_\mu \times \mathbf{A}_\nu) \cdot (\mathbf{A}^\mu \times \mathbf{A}^\nu)$$
$$+ \left(\frac{g^2}{2\sqrt{2}}\eta\sigma + \frac{g^2\sigma^2}{8}\right)\left(W_\mu^{+\dagger}W^{+\mu} + W_\mu^{-\dagger}W^{-\mu} + \frac{1}{\cos^2\theta_W}Z_\mu Z^\mu\right)$$
$$+ \frac{\mu^2}{\eta\sqrt{2}}\sigma^3 + \frac{\mu^2}{8\eta^2}\sigma^4 \quad (2.75)$$

where $A_{\mu\nu} = \partial_\mu A_\nu - \partial_\nu A_\mu$, $Z_{\mu\nu} = \partial_\mu Z_\nu - \partial_\nu Z_\mu$, and so on. In the first line of (2.75), we have terms corresponding to the kinetic energy and mass of the Higgs boson, which already appeared at an earlier stage in the discussion. In the second line appears a field tensor $A_{\mu\nu} = \partial_\mu A_\nu - \partial_\nu A_\mu$, where A_μ is a linear combination of isoscalar and isovector [see (2.72)]. We identify A_μ as the electromagnetic field; its quanta are photons. (Note that $A_{\mu\nu}$ is usually written as $F_{\mu\nu}$.) In the covariant derivative $D_\mu = \partial_\mu + ig\mathbf{A}_\mu \cdot \mathbf{t} + \tfrac{1}{2} ig' B_\mu I$, the quantity $i(gA_{\mu 3} t_3 + \tfrac{1}{2} g' B_\mu I)$ can be written as

$$\tfrac{1}{2} ig A_{\mu 3} \tau_3 + \tfrac{1}{2} ig' B_\mu I = \tfrac{1}{2} ig(\cos\theta_W Z_\mu \tau_3 + \sin\theta_W A_\mu \tau_3) \\ + \tfrac{1}{2} ig'(-\sin\theta_W Z_\mu + \cos\theta_W A_\mu) I$$

Taking into account that $g'/g = \tan\theta$, the terms in A_μ on the RHS give: $iA_\mu g \sin\theta[(I + \tau_3)/2]$. Now $(I + \tau_3)/2$ is just the charge operator. Therefore, since we identify A_μ as the electromagnetic field, it is clear that we must have

$$g \sin\theta_W = |e| \qquad (2.76)$$

The third line of (2.75) describes two vector fields, W_μ^+ and W_μ^-, with mass $m_W = g\eta/\sqrt{2}$. The quanta are identified as the charged intermediate vector bosons of weak interactions. Similarly, the fourth line represents a neutral massive vector field with quanta of mass $m_Z = g\eta/(\sqrt{2} \cos\theta_W)$. These are identified as mediators of neutral weak interactions. The masses of the $W^\pm$ and Z bosons arose from the spontaneous symmetry-breaking procedure. In the fifth and sixth lines, we have the characteristic Yang–Mills nonlinear self-interaction terms of the vector field $\mathbf{A}_\mu$. The seventh line describes higher-order couplings between the Higgs field and the vector fields $W^\pm$ and A, and the last line describes higher-order self-interactions of the Higgs field.

2.5 Coupling of leptons and quarks to gauge fields

We now introduce the material particles – leptons and quarks – into the theory. We begin by recalling that the leptons all fall naturally into generations or doublets:

$$E = \begin{pmatrix} \nu_e \\ e^- \end{pmatrix}, \quad M = \begin{pmatrix} \nu_\mu \\ \mu^- \end{pmatrix}, \quad T = \begin{pmatrix} \nu_\tau \\ \tau^- \end{pmatrix} \qquad (2.77)$$

Of course, the mass of each neutrino is zero, or at least much less than that of its charged counterpart. But at this initial stage, we can assume all fermion masses to be zero, and we shall later invoke a suitable mechanism to generate the masses.

As noted in Chapter 1, all experimental evidence is consistent with the assumption that charged weak currents are of the V–A form. That is, they involve only the left-handed lepton components. If we are to

2.5 Coupling of leptons and quarks to gauge fields

build a theory on the principle of local gauge invariance, it follows that we shall have to be concerned with $SU(2)$ rotations of the *left-handed* components:

$$E_L = \tfrac{1}{2}(1 - \gamma_5)E, \quad M_L = \tfrac{1}{2}(1 - \gamma_5)M, \quad T_L = \tfrac{1}{2}(1 - \gamma_5)T \quad (2.78)$$

This can be done by considering, in addition to ordinary $SU(2)$ transformations

$$\delta E = i\boldsymbol{\epsilon} \cdot \mathbf{t} E \quad (2.79)$$

also the *chiral* $SU(2)$ transformations

$$\delta_5 E = i\gamma_5 \boldsymbol{\eta} \cdot \mathbf{t} E \quad (2.80)$$

Here the quantities η_1, η_2, and η_3 are three real infinitesimal parameters. The generators of all these transformations form an $SU(2) \times SU(2)$ group:

$$\begin{aligned}[\mathbf{t}_i, \mathbf{t}_j] &= i\epsilon_{ijk}\mathbf{t}_k \\ [\mathbf{t}_i, \mathbf{t}_j\gamma_5] &= i\epsilon_{ijk}\mathbf{t}_k\gamma_5 \\ [\mathbf{t}_i\gamma_5, \mathbf{t}_j\gamma_5] &= i\epsilon_{ijk}\mathbf{t}_k \end{aligned} \quad (2.81)$$

Now let us define $L_i = \tfrac{1}{2}(1 - \gamma_5)\mathbf{t}_i$ and $R_i = \tfrac{1}{2}(1 + \gamma_5)\mathbf{t}_i$. Then (2.81) can be rewritten as

$$\begin{aligned}[L_i, L_j] &= i\epsilon_{ijk}L_k \\ [R_i, R_j] &= i\epsilon_{ijk}R_k \\ [L_i, R_j] &= 0 \end{aligned} \quad (2.82)$$

and we can say that E, M, and T form weak isodoublets under transformations generated by the L_i. However, experiment shows that the right-handed states $\tfrac{1}{2}(1 + \gamma_5)\nu_e, \tfrac{1}{2}(1 + \gamma_5)e^-, \ldots$, do not participate in the charged weak interaction. We therefore assume that all right-handed lepton states form isosinglets, i.e., are invariant under isospin rotations. Thus we are assuming that leptons form weak left-handed isodoublets and weak right-handed isosinglets, and in this way, we incorporate the known facts about parity violation in charged weak interactions into our theory without explaining them in terms of anything more fundamental.

Incidentally, a Dirac Lagrangian is not invariant under chiral isospin transformations if $m \neq 0$. Consider the *axial* current defined by

$$\mathbf{J}_\mu^A = -i \frac{\partial \mathcal{L}}{\partial(\partial_\mu \Psi)} \gamma_5 \mathbf{t}\Psi$$

which is analogous to (2.34). We find for the divergence of $\mathbf{J}_\mu^A$:

$$\partial^\mu \mathbf{J}_\mu^A = \partial^\mu(\overline{\Psi} \partial_\mu \gamma_5 \mathbf{t}\Psi) = 2im\overline{\Psi}\gamma_\mu\gamma_5\mathbf{t}\Psi \neq 0$$

Having previously described the effect of $SU(2)$ transformations on the lepton fields $E_L, M_L, \ldots$, we turn our attention to the $U(1)$ trans-

2 The standard model

formations. This can be understood quite easily in the following way. First, form a column vector involving all lepton fields:

$$\chi = \begin{bmatrix} \nu_{eL} \\ e_L \\ \nu_{eR} \\ e_R \\ \nu_{\mu L} \\ \mu_L \\ \cdot \\ \cdot \\ \cdot \end{bmatrix} \quad (2.83)$$

This vector has as many components n as there are left- and right-handed lepton states. Next, we consider a $U(1)$ transformation on χ. This can be written as

$$\delta\chi = \tfrac{1}{2} i\epsilon Y \chi \quad (2.84)$$

where Y is an $n \times n$ diagonal matrix and ϵ a real infinitesimal parameter. When gauge fields are introduced to preserve local $SU(2) \times U(1)$ invariance, they yield minimal couplings of the form

$$\bar{\chi}\gamma^\mu(gA_\mu^1 L^1 + gA_\mu^2 L^2 + gA_\mu^3 L^3 + \tfrac{1}{2}g'B_\mu Y)\chi \quad (2.85)$$

In particular, let us consider the following terms in this last expression:

$$\begin{aligned} gA_\mu^3 L^3 + \tfrac{1}{2}g'B_\mu Y &= g(\cos\theta_W Z_\mu + \sin\theta_W A_\mu)L^3 \\ &\quad + \tfrac{1}{2}g'(-\sin\theta_W Z_\mu + \cos\theta_W A_\mu)Y \\ &= g\sin\theta_W(L^3 + \tfrac{1}{2}Y)A_\mu \\ &\quad + g\cos\theta_W(L^3 - \tfrac{1}{2}\tan^2\theta_W Y)Z_\mu \end{aligned} \quad (2.86)$$

The term in (2.86) involving the electromagnetic field A_μ is

$$g\sin\theta_W(L^3 + \tfrac{1}{2}Y)A_\mu = |e|A_\mu(L^3 + \tfrac{1}{2}Y) \quad (2.87)$$

Thus $(L^3 + \tfrac{1}{2}Y)$ must be the electric charge operator Q, or

$$Y = 2(Q - L_3) \quad (2.88)$$

We now present a tabulation of values of "weak hypercharge."

	ν_{eL}	e_L	ν_{eR}	e_R	$\nu_{\mu L}$	μ_L	$\nu_{\mu R}$	μ_R	$\cdots$
Q	0	-1	0	-1	0	-1	0	-1	
L^3	$\tfrac{1}{2}$	$-\tfrac{1}{2}$	0	0	$\tfrac{1}{2}$	$-\tfrac{1}{2}$	0	0	
Y	-1	-1	0	-2	-1	-1	0	-2	

From this tabulation, it is clear how to construct the appropriate covariant derivatives of the lepton fields:

2.5 Coupling of leptons and quarks to gauge fields

$$D_\mu E_L = (\partial_\mu + ig\mathbf{A}_\mu \cdot \mathbf{t} - \tfrac{1}{2}ig'B_\mu)E_L$$
$$D_\mu e_R = (\partial_\mu - ig'B_\mu)e_R$$
$$D_\mu \nu_{eR} = \partial_\mu \nu_{eR} \tag{2.89}$$

⋮

The Lagrangian for the lepton fields may now be written as

$$\mathscr{L}_l = i\bar{E}_L\gamma^\mu D_\mu E_L + i\bar{M}_L\gamma^\mu D_\mu M_L + \cdots ,$$
$$+ i(\bar{e}_R\gamma^\mu D_\mu e_R + \cdots) + i(\bar{\nu}_{rR}\gamma^\mu D_\mu \nu_{eR} + \cdots) \tag{2.90}$$

and this must be added to the Lagrangian that describes the Higgs field and the gauge fields (2.75).

Weak and electromagnetic interactions of leptons arise from the interaction terms in (2.90). Employing (2.69)–(2.72) once more, these terms become, after some algebra,

$$\mathscr{L}_l^{INT} = -\frac{g}{2\sqrt{2}}\bar{\nu}_e\gamma^\mu(1-\gamma^5)eW_\mu^+ - \frac{g}{2\sqrt{2}}\bar{e}\gamma^\mu(1-\gamma^5)\nu_e W_\mu^-$$
$$- \frac{g}{4\cos\theta_W}[\bar{\nu}_e\gamma^\mu(1-\gamma^5)\nu_e - \bar{e}\gamma^\mu(1-4\sin^2\theta_W - \gamma^5)e]Z_\mu$$
$$- e\bar{e}\gamma^\mu eA_\mu + (\text{muon, tau, } \ldots \text{ terms}) \tag{2.91}$$

In (2.91), the first line describes the coupling of the charged leptonic weak current to the charged intermediate vector bosons $W^\pm$. We recall that the $W^\pm$ fields are described by the terms in (2.75) as

$$\mathscr{L}_{W^\pm} = -\tfrac{1}{4}(W_{\mu\nu}^{+\dagger}W^{+\mu\nu} + W_{\mu\nu}^{-\dagger}W^{-\mu\nu})$$
$$+ \tfrac{1}{4}g^2\eta^2(W_\mu^{+\dagger}W^{+\mu} + W_\mu^{-\dagger}W^{-\mu}) \tag{2.92}$$

Let us digress for a moment on the properties of such a vector field. For this purpose, we consider a Lagrangian of the same general form as the first line of (2.91) plus (2.92), namely,

$$\mathscr{L}_V = -\tfrac{1}{4}f_{\mu\nu}f^{\mu\nu} - \tfrac{1}{2}m^2\phi_\mu\phi^\mu - \phi_\mu J^\mu$$

where ϕ_μ is the vector field and $f_{\mu\nu} = \partial_\mu\phi_\nu - \partial_\nu\phi_\mu$. We also assume that ϕ_μ is coupled to some vector current J^μ.

The field equations are obtained from the Euler–Lagrange equations

$$\frac{\partial\mathscr{L}}{\partial\phi_\mu} - \partial_\nu\frac{\partial\mathscr{L}}{\partial(\partial_\nu\phi_\mu)} = 0$$

and they are

$$\partial_\nu\partial^\nu\phi^\mu - \partial_\nu\partial^\mu\phi^\nu + m^2\phi^\mu = J^\mu \tag{2.93}$$

Differentiating both sides with respect to x^μ and noting that

2 The standard model

$\partial_\mu \partial_\nu \partial^\nu \phi^\mu = \partial_\mu \partial^\mu \partial_\nu \phi^\nu$, we obtain $\partial_\mu \phi^\mu = 1/m^2 \, \partial_\mu J^\mu$. Thus (2.93) becomes

$$\Box \phi^\mu + m^2 \phi^\mu = \left(g^{\mu\nu} + \frac{1}{m^2} \partial^\mu \partial^\nu\right) J_\nu \tag{2.94}$$

Now let us Fourier-analyze both sides of (2.94). Writing

$$\phi^\mu = \frac{1}{(2\pi)^4} \int f^\mu e^{ikx} \, d^4k$$

and

$$J_\nu = \frac{1}{(2\pi)^4} \int j_\nu e^{ikx} \, d^4k$$

taking derivatives, and equating the resulting integrands, we obtain

$$f^\mu = -\left(\frac{g^{\mu\nu} - k^\mu k^\nu/m^2}{k^2 - m^2}\right) j_\nu$$

The propagator for the free vector field is then

$$D^{\mu\nu} = -i \left(\frac{g^{\mu\nu} - k^\mu k^\nu/m^2}{k^2 - m^2}\right) \tag{2.95}$$

In the present case of the $W^\pm$ fields, the mass, from (2.92), is $m_W = g\eta/\sqrt{2}$.

Now let us recall our previous discussion of the naive intermediate vector boson theory (Section 1.7). It can be seen that our present theory [first line of (2.91)] and the naive theory yield identical forms for the W charged lepton coupling. As noted in Section 1.7, the known strength of charged weak processes at low energy (muon decay), which are correctly described by the Feynman–Gell-Mann scheme, fix the $l\nu_l W^\pm$ coupling strength. We need only replace (1.36) by the relation

$$g/2\sqrt{2} = m_W G_F^{1/2} 2^{-1/4}$$

Now, since $m_W = g\eta/\sqrt{2}$ and $g \sin \theta_W = |e| = (4\pi\alpha)^{1/2}$, we obtain

$$m_W^2 = \left(\frac{\pi\alpha}{\sqrt{2} G_F}\right) \frac{1}{\sin^2 \theta_W} = \frac{(37.5 \text{ GeV}/c^2)^2}{\sin^2 \theta_W} \tag{2.96}$$

As we shall see, a variety of experimental results fixes the value

$$\sin^2 \theta_{W,\text{expt}} = 0.23 \tag{2.97}$$

which yields the prediction

$$m_W = 78 \text{ GeV}/c^2 \tag{2.98}$$

The mass of Z is given by the formula $m_Z = m_W/\cos \theta_W$. This leads to the prediction

$$m_Z = 89 \text{ GeV}/c^2 \tag{2.99}$$

2.5 Coupling of leptons and quarks to gauge fields

To be sure, the actual W and Z^0 masses are expected to deviate from these zeroth-order values by several percent because of radiative corrections.

The second line of (2.91) describes the coupling of neutral weak leptonic currents to the neutral intermediate boson Z. We note that the neutrino current contains the familiar $1 - \gamma_5$ factor; however, the electronic current is *not* pure V–A; in fact, since (as experiment reveals) $\sin^2 \theta_W = 0.23$, the vector portion nearly vanishes. Also, the neutrino and electron axial neutral currents enter with opposite signs.

The coupling of quark weak currents to charged and neutral intermediate bosons is constructed in an analogous way to the lepton case. We start with the Glashow–Iliopoulos–Maiani (70) scheme, according to which weak couplings of the first two generations of quarks are determined by the following multiplets:

$$\begin{pmatrix} u \\ d_C \end{pmatrix}_L, \quad u_R, \quad d_{CR}; \quad \begin{pmatrix} c \\ s_C \end{pmatrix}_L, \quad c_R, \quad s_{CR} \tag{2.100}$$

Here we ignore, for the present, the third quark generation and we assume that, as in the lepton case, we have weak left-handed isodoublets and weak right-handed isosinglets. Once again we compute the weak hypercharge Y and present the results in a table, with $Y_q = 2(Q_p - L_{3q})$.

	u_L	d_{CL}	u_R	d_{CR}	c_L	s_{CL}	c_R	s_{CR}
Q_q	$\tfrac{2}{3}$	$-\tfrac{1}{3}$	$\tfrac{2}{3}$	$-\tfrac{1}{3}$	$\tfrac{2}{3}$	$-\tfrac{1}{3}$	$\tfrac{2}{3}$	$-\tfrac{1}{3}$
L_{3q}	$\tfrac{1}{2}$	$-\tfrac{1}{2}$	0	0	$\tfrac{1}{2}$	$-\tfrac{1}{2}$	0	0
Y_q	$\tfrac{1}{3}$	$\tfrac{1}{3}$	$\tfrac{4}{3}$	$-\tfrac{2}{3}$	$\tfrac{1}{3}$	$\tfrac{1}{3}$	$\tfrac{4}{3}$	$-\tfrac{2}{3}$

It is now a simple matter to write out the covariant derivatives:

$$D_\mu \begin{pmatrix} u_L \\ d_{CL} \end{pmatrix} = (\partial_\mu + ig\mathbf{A}_\mu \cdot \mathbf{t} + \tfrac{1}{6} ig' B_\mu) \begin{pmatrix} u_L \\ d_{CL} \end{pmatrix}$$
$$D_\mu u_R = (\partial_\mu + \tfrac{2}{3} ig' B_\mu) u_R \tag{2.101}$$
$$D_\mu d_{CR} = (\partial_\mu - \tfrac{1}{3} ig' B_\mu) d_{CR}$$

We may then write the quark portion of the Lagrangian as

$$\mathscr{L}_Q = i(\bar{u}_L, \bar{d}_{CL}) \gamma^\mu D_\mu \begin{pmatrix} u_L \\ d_{CL} \end{pmatrix} + i\bar{u}_R \gamma^\mu D_\mu u_R + i\bar{d}_{CR} \gamma^\mu D_\mu u_{CR} + \cdots \tag{2.102}$$

and add it to the leptonic Lagrangian (2.91) and the Higgs–gauge field Lagrangian (2.75). As in the leptonic case, weak and electromagnetic interactions of quarks arise from the interaction terms of (2.102). Employing the relations $A_{\mu 3} = \cos\theta_W Z_\mu + \sin\theta_W A_\mu$ and $B_\mu = \cos\theta_W A_\mu - \sin\theta_W Z_\mu$, and some straightforward algebra, we arrive at the result

$$\begin{aligned}\mathcal{L}_Q^{\text{INT}} = &-(g/2\sqrt{2})\bar{u}\gamma^\mu(1-\gamma_5)d_c W_\mu^+ - (g/2\sqrt{2})\bar{d}_c\gamma^\mu(1-\gamma_5)u W_\mu^-\\&-\frac{g}{4\cos\theta_W}[\bar{u}\gamma^\mu(1-\tfrac{8}{3}\sin^2\theta_W - \gamma_5)u\\&\quad - \bar{d}\gamma^\mu(1-\tfrac{4}{3}\sin^2\theta_W - \gamma_5)d]Z_\mu + \tfrac{2}{3}e\bar{u}\gamma^\mu u A_\mu\\&-\tfrac{1}{3}e\bar{d}\gamma^\mu d A_\mu + \text{second-generation terms} + \cdots\end{aligned} \quad (2.103)$$

In this equation, the first line describes coupling of the charged quark weak current to the $W^\pm$ bosons, the second line gives the neutral quark weak coupling to the Z boson, and the third line gives the electromagnetic interaction of quarks.

It is important to emphasize the relationship between the various currents entering expression (2.103). The electromagnetic current $J_{\text{EM}} = \bar{\chi}_q \gamma^\mu Q_q \chi_q$, with $Q_q = \tfrac{1}{2}Y_q + t_3$, contains an isoscalar Y and an isovector portion τ_3. The charged weak current may be written as

$$J_-^\mu = (J_-^{V\mu} - J_-^{A\mu}) = \bar{\chi}_q \gamma^\mu(1-\gamma_5)t_- \chi_q$$

The neutral weak current may be expressed as

$$J_Z^\mu = \tfrac{1}{2}J_3^\mu - 2\sin^2\theta_W J_{\text{EM}} \quad (2.104)$$

where $J_3^\mu = \bar{\chi}_q \gamma^\mu(1-\gamma_5)t_3 \chi_q$.

Clearly $J_{1,2,3}^{V\mu}$ is just the vector isospin current of (2.34) and is conserved because of global $SU(2)$ gauge invariance. Thus we have the *conserved vector current* (CVC) hypothesis, originally proposed by Feynman and Gell-Mann in 1958, that states that the vector hadronic weak current, its Hermitian conjugate, and the isovector portion of the EM current form a single isospin triplet of conserved currents. As we shall see, CVC has important consequences for hadronic weak interactions (Section 4.9).

2.6 Lepton and quark couplings to the scalar field

The construction of a theoretical framework for electroweak interactions must include a renormalizable, gauge-invariant prescription for imparting masses to the leptons and quarks. For this purpose, let us begin with the leptons and consider a term in the Lagrangian of the general form

$$\mathcal{L}_S = g_S \phi \bar{\psi}\psi \quad (2.105)$$

where g_S is a constant, $\phi = \begin{pmatrix} 0 \\ \eta + \sigma/\sqrt{2} \end{pmatrix}$ the scalar field, and ψ a

2.6 Lepton and quark couplings to the scalar field

lepton field. Before we examine the detailed properties of this coupling, we shall consider the situation qualitatively.

Clearly there are two contributions to $\mathscr{L}_S$:

$$\mathscr{L}_S = \mathscr{L}_{S1} + \mathscr{L}_{S2} \tag{2.106}$$

where

$$\mathscr{L}_{S1} = g_S \eta \bar{\psi}\psi \tag{2.107a}$$

and

$$\mathscr{L}_{S2} = g_S [\sigma(x)/\sqrt{2}\,]\bar{\psi}\psi \tag{2.107b}$$

Since η is a constant, $\mathscr{L}_{S1}$ is an effective mass term and $\mathscr{L}_{S2}$ is a Yukawa interaction term. Now,

$$g_S \eta \simeq m_{\text{lepton}}$$

and

$$g\eta \simeq m_W$$

Therefore,

$$g_S/g \simeq m_l/m_W$$

Consider next a Yukawa interaction in which the scalar boson σ is exchanged between two leptons (see Figure 2.2). Each vertex contributes

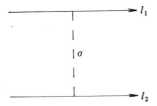

Figure 2.2. Higgs exchange.

a factor $g_S/\sqrt{2}$, and the σ propagator gives a factor

$$-1/(q^2 - m_\sigma^2) \simeq 1/m_\sigma^2$$

at low energies. Thus the amplitude is

$$\text{amplitude} \simeq \frac{g_S^2}{m_\sigma^2} \simeq \frac{g^2}{m_\sigma^2}\frac{m_l^2}{m_W^2} \simeq G_F \frac{m_l^2}{m_\sigma^2}$$

Compared with a typical lepton–lepton weak amplitude (as in muon decay), the Higgs exchange amplitude just written is of relative order m_l^2/m_σ^2. For $m_\sigma \gg m_l$, this quantity is quite negligible, and we ignore it for the present, concentrating our attention on $\mathscr{L}_{S1}$.

Leaving the question of renormalizability aside, our immediate problem is to formulate $\mathcal{L}_{S1}$ so that it is gauge-invariant. Now, ϕ is an isodoublet under $SU(2)$ gauge transformations. Therefore $\bar{E}_L\phi$ and $\bar{M}_L\phi$ are $SU(2)$ scalars, as are $\phi^\dagger E_L$ and $\phi^\dagger M_L$. (For now, we consider only two lepton generations.) However, these quantities are not invariant under $U(1)$ because, since $Y(E_L) = -1$,

$$\delta E_L = -\tfrac{1}{2}i\epsilon E_L, \qquad \delta \bar{E}_L = \tfrac{1}{2}i\epsilon \bar{E}_L$$

whereas

$$\delta\phi = \tfrac{1}{2}i\epsilon\phi$$

Therefore

$$\delta(\bar{E}_L\phi) = \delta(\bar{E}_L)\phi + \bar{E}_L\delta\phi = i\epsilon \bar{E}_L\phi$$

Thus to form a $U(1)$ invariant, we must combine $\bar{E}_L\phi$ with an isoscalar for which $Y = -2$. For example, we can have $\bar{E}_L\phi e_R$ or $\bar{E}_L\phi\mu_R$. If we also require $\mathcal{L}_S$ to be Hermitian, then a possible Lagrangian is

$$\begin{aligned}\mathcal{L}_{S1} = {}& g_{11}(\bar{E}_L\phi e_R + \bar{e}_R\phi^\dagger E_L) + g_{12}(\bar{E}_L\phi\mu_R + \bar{\mu}_R\phi^\dagger E_L) \\ &+ g_{21}(\bar{M}_L\phi e_R + \bar{e}_R\phi^\dagger M_L) + g_{22}(\bar{M}_L\phi\mu_R + \bar{\mu}_R\phi^\dagger M_L)\end{aligned} \quad (2.108)$$

where the g's are arbitrary coefficients. However, (2.108) is still not the most general expression because we can construct a new isodoublet whose components are linear combinations of the components of $\phi^\dagger$ but that transform like ϕ under $SU(2)$. To see this, write the following:

$$\phi = \begin{pmatrix}\phi_1 \\ \phi_2\end{pmatrix} \tag{2.109}$$

and

$$\tilde{\phi} \equiv (-i\phi^\dagger \cdot \tau_2)_t \tag{2.110}$$

where the subscript t means transpose. Now,

$$\tilde{\phi} = \left[-i(\phi_1^\dagger, \phi_2^\dagger)\begin{pmatrix}0 & -i \\ i & 0\end{pmatrix}\right]_t = (\phi_2^\dagger, -\phi_1^\dagger)_t = \begin{pmatrix}\phi_2^\dagger \\ -\phi_1^\dagger\end{pmatrix}$$

Also, under $SU(2)$,

$$\begin{aligned}\delta\tilde{\phi} &= -i[-i\phi^\dagger\boldsymbol{\epsilon}\cdot\mathbf{t}\tau_2]_t \\ &= -(\tau_2)_t(\boldsymbol{\epsilon}\cdot\mathbf{t})_t(\tau_2)_t(\tau_2)_t\phi_1^\dagger \\ &= -i\left[\begin{pmatrix}0 & -i \\ i & 0\end{pmatrix}\begin{pmatrix}\tfrac{1}{2}\epsilon_3 & \tfrac{1}{2}(\epsilon_1 - i\epsilon_2) \\ \tfrac{1}{2}(\epsilon_1 + i\epsilon_2) & -\tfrac{1}{2}\epsilon_3\end{pmatrix}\begin{pmatrix}0 & -i \\ i & 0\end{pmatrix}\right]_t \tilde{\phi}\end{aligned}$$

Therefore,

$$\delta\tilde{\phi} = i\boldsymbol{\epsilon}\cdot\mathbf{t}\tilde{\phi} \tag{2.111}$$

As a result, $\bar{E}_L\tilde{\phi}$ is an $SU(2)$ singlet. Furthermore, $Y(\bar{E}_L) = +1$ and $Y(\tilde{\phi}) = -1$, so $\bar{E}_L\tilde{\phi}$ is also a $U(1)$ singlet. To form a Lorentz scalar, we

2.6 Lepton and quark couplings to the scalar field

need to include ν_{eR} or $\nu_{\mu R}$, each of which are $SU(2)$ singlets with $Y = 0$. Thus, a generalized version of (2.108) is

$$\begin{aligned}\mathcal{L}_S = &\; g_{11}(\bar{E}_L \phi e_R + \bar{e}_R \phi^\dagger E_L) + g_{12}(\bar{E}_L \phi \mu_R + \bar{\mu}_R \phi^\dagger E_L) \\ &+ g_{21}(\bar{M}_L \phi e_R + \bar{e}_R \phi^\dagger M_L) + g_{22}(\bar{M}_L \phi \mu_R + \bar{\mu}_R \phi^\dagger E_L) \\ &+ h_{11}(\bar{E}_L \tilde{\phi} \nu_{eR} + \bar{\nu}_{eR} \tilde{\phi}^\dagger E_L) + h_{12}(\bar{E}_L \tilde{\phi} \nu_{\mu R} + \bar{\nu}_{\mu R} \tilde{\phi}^\dagger E_L) \\ &+ h_{21}(\bar{M}_L \tilde{\phi} \nu_{eR} + \bar{\nu}_{eR} \tilde{\phi}^\dagger M_L) + h_{22}(\bar{M}_L \tilde{\phi} \nu_{\mu R} + \bar{\nu}_{\mu R} \tilde{\phi}^\dagger M_L)\end{aligned} \quad (2.112)$$

where the g's and h's are numerical coefficients.

Let us define three "vectors":

$$V_1 = \begin{pmatrix} E_L \\ M_L \end{pmatrix}, \quad V_2 = \begin{pmatrix} e_R \\ \mu_R \end{pmatrix}, \quad V_3 = \begin{pmatrix} \nu_{eR} \\ \nu_{\mu R} \end{pmatrix} \quad (2.113)$$

Ignoring space-time indexes, the leptonic Lagrangian (2.90) may be written in terms of these vectors as

$$\mathcal{L}_l = i[\bar{V}_1 V_1 + \bar{V}_2 V_2 + \bar{V}_3 V_3]$$

Now let us perform three independent "rotations" on the three vectors V_1, V_2, and V_3:

$$V_1' = O_1 V_1, \quad V_2' = O_2 V_2, \quad V_3' = O_3 V_3$$

The quantities O_1, O_2, O_3 are 2×2 real orthogonal matrices, and each is characterized by a single real parameter (angle). Obviously $\mathcal{L}_l$ is invariant under these rotations. Since we have three real parameters at our disposal, we may choose them so that in (2.112) the following conditions are satisfied:

$$g_{12} = g_{21}, \quad h_{12} = h_{21}$$

and

$$g_{12} = 0 \quad \text{or} \quad h_{12} = 0$$

The choice $g_{12} = g_{21} = 0$ is equivalent to the assumption that electron and muon mass eigenstates are the same as the electron and muon states that participate in the weak interaction. Making this choice and using the facts that

$$\phi = \begin{pmatrix} 0 \\ \eta + \sigma/\sqrt{2} \end{pmatrix} \simeq \begin{pmatrix} 0 \\ \eta \end{pmatrix}$$

and

$$\bar{e}_L e_R + \bar{e}_R e_L = \bar{e}e, \ldots$$

we finally arrive at the expression

$$\mathcal{L}_S = \eta[g_{11} \bar{e}e + g_{22} \bar{\mu}\mu + h_{11} \bar{\nu}_e \nu_e + h_{22} \bar{\nu}_\mu \nu_\mu + h_{12}(\bar{\nu}_e \nu_\mu + \bar{\nu}_\mu \nu_e)] \quad (2.114)$$

In this formula, the constants ηg_{11}, ηg_{22}, ηh_{11}, ηh_{22}, refer to e, μ, ν_e, and ν_μ masses, respectively. The quantity $h_{12} = h_{21}$ allows for the possibility of "neutrino oscillations," as we shall discuss in Chapter 10, in which the present discussion will be generalized to include three lepton doublets.

The quarks may be treated in a similar manner. We restrict ourselves for the present to the first two-quark generations within the framework of the GIM scheme and note the correspondence:

$$\begin{pmatrix} u \\ d_C \end{pmatrix}_L \leftrightarrow E_L, \qquad \begin{pmatrix} c \\ s_C \end{pmatrix}_L \leftrightarrow M_L$$

$$u_R \leftrightarrow \nu_{eR}, \qquad c_R \leftrightarrow \nu_{\mu R}$$
$$d_{CR} \leftrightarrow e_R, \qquad s_{CR} \leftrightarrow \mu_R$$

Step-by-step repetition of the previous analysis generates coefficients $g'_{11}, g'_{12}, \ldots, h'_{11}, \ldots$ with the supplementary conditions:

$$g'_{12} = g'_{21}, \qquad h'_{12} = h'_{21} = 0$$

or

$$g'_{12} = g'_{21} = 0, \qquad h'_{12} = h'_{21}$$

Now the quark Lagrangian $\mathcal{L}_Q$ is diagonal in the Cabibbo-rotated states, but these are not the states of definite mass; instead the latter are the states $u, c, s,$ and d. In order to account for this condition, we must therefore make the choice $h'_{12} = h'_{21} = 0$, which leads to

$$\mathcal{L}'_S = \eta[g'_{11}\bar{d}_C d_C + g'_{22}\bar{s}_C s_C + g'_{12}(\bar{d}_C s_C + \bar{s}_C d_C) + h'_{11}\bar{u}u + h'_{22}\bar{c}c] \tag{2.115}$$

with the additional condition that this expression must be diagonal in states d and s; that is, it must contain no cross terms of the form $\bar{d}s$ or $\bar{s}d$. It is easy to show that this requirement is equivalent to the condition

$$\tan 2\theta_C = 2g'_{12}/(g'_{22} - g'_{11}) \tag{2.116}$$

We have now formulated the Lagrangian of weak and electromagnetic interactions in the so-called unitary gauge. Let us summarize its properties. The Lagrangian includes:

(i) $\mathcal{L}_0$ in (2.75) for the gauge fields and Higgs scalar field;
(ii) $\mathcal{L}_l$ in (2.90), which contains the lepton fields and their interactions with the gauge fields;
(iii) $\mathcal{L}_Q$ in (2.103), which describes the quark fields and their coupling to the gauge fields;
(iv) $\mathcal{L}_S$ in (2.114), which contains the lepton–scalar field couplings and therefore describes lepton masses and neutrino oscillations;

(v) $\mathscr{L}'_S$ in (2.115), which contains quark–scalar field couplings and thus yields quark masses.

Thus far our Lagrangian is a classical entity, and we must now consider how it is to be quantized.

2.7 The path integral in quantum mechanics[7]

In ordinary quantum mechanics, we frequently deal with the scattering of a spinless particle (which we may take to have unit mass) from a potential $V(x)$. The wavefunction ψ of the particle is usually specified for each x (we may work in one spatial dimension at present) at some initial time t_0, and we seek to find ψ at each x and at some later time t'. The usual approach is to solve the Schrödinger equation:

$$i\hbar\, \partial\psi/\partial t = -\tfrac{1}{2}\hbar^2 \nabla^2 \psi + V(x)\psi \qquad (2.117)$$

subject to the initial condition $\psi = \psi(x, t_0)$. An equivalent approach makes use of the Green's function or kernel $K(x', t'; x, t_0)$, which relates $\psi(x', t')$ to $\psi(x, t_0)$ as follows:

$$\psi(x', t') = \int K(x', t'; x, t_0)\psi(x, t_0)\, dx \qquad (2.118)$$

It is very useful to express K as a path integral, the properties of which we now outline without proof. Let us first divide the time interval $t'-t_0$ into $n + 1$ small segments of magnitude ϵ as shown in Figure 2.3. To each value of t we associate a value of the coordinate x:

$$t = t_0 : x_0$$
$$t_1 : x_1$$
$$t_2 : x_2$$
$$\cdot$$
$$\cdot$$
$$\cdot$$
$$t_n : x_n$$
$$t' = t_{n+1} : x_{n+1} = x'$$

The values x_0 and x' are fixed and determine the endpoints of a "path." The other values $x_1, \ldots, x_n$ determine the rest of the path. The latter quantities are arbitrary; once they are chosen, a path is determined, but another set of values would determine an equally valid path.

Next we calculate the action S for each path:

$$S = \int_{t_0}^{t'} L\, dt$$

---

[7] Sections 2.7–2.11 contain difficult theoretical material, which, for reasons of space, is admittedly treated in an abbreviated and superficial manner. The practical-minded reader may wish to omit them.

where $L = \frac{1}{2}p^2 - V(x)$ is the classical Lagrangian. Then we multiply the action by $i/\hbar$, exponentiate, and sum over all possible paths. The result, when multiplied by a suitable normalizing factor, is the desired Green's function:

$$K(x', t'; x, t_0) = \lim_{n\to\infty, \epsilon\to 0} \left(\frac{1}{2\pi i\epsilon}\right)^{(n+1)/2} \int \prod_{i=1}^{n} dx_i \exp\left(\frac{i}{\hbar}\int L\, dt\right)$$
(2.119)

which is frequently written shortened to

$$K = \int [dx] \exp(iS/\hbar)$$
(2.120)

In fact, (2.119) is not completely general, being inadequate if the potential is velocity-dependent. However, that is a detail that we ignore for the present.

Equation (2.119) is a mathematical expression of Huygens' principle, and with it the distinction between classical and quantum mechanics, or equivalently, ray and wave optics, can be made clearly. In classical mechanics, action S is generally very large compared with $\hbar$. Hence $\exp(iS/\hbar)$ oscillates very rapidly when we go from one path to another, and in the sum over paths, there is essentially complete cancellation

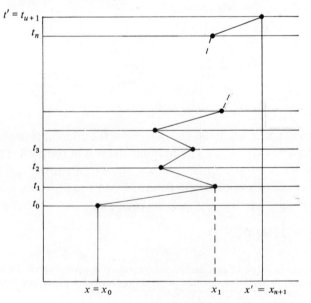

Figure 2.3. Construction of path between points x_0 and x' (see text discussion for details).

except in the immediate vicinity of that path for which S reaches an extreme value (in most cases a minimum). Hence in classical mechanics, the only path of importance (the "classical" path) is the path of least action. In quantum mechanics, however, S may be comparable to or smaller than $\hbar$, in which case many different paths contribute more or less equally.

A special example of significance is the case of a free particle, where $L = L_0 = \tfrac{1}{2}p^2$. The kernel is

$$K_0 = \left(\frac{1}{2\pi i\epsilon}\right)^{(n+1)/2} \int_{-\infty}^{\infty} dx_1 \cdots \int_{-\infty}^{\infty} dx_n$$
$$\times \exp[\tfrac{1}{2}i\epsilon(x_1 - x_0)^2]\exp[\tfrac{1}{2}i\epsilon(x_2 - x_1)^2] \cdots$$
$$\times \exp[\tfrac{1}{2}i\epsilon(x_{n+1} - x_n)^2] \qquad (2.121)$$

By employing the well-known formula

$$I_0 = \int_{-\infty}^{\infty} du \, \exp(iau^2) = (i\pi/a)^{1/2} \qquad (2.122)$$

and also mathematical induction, it may easily be shown that (2.121) becomes

$$K_0(x', t'; x, t_0) = \left[\frac{1}{2\pi i(t' - t_0)}\right]^{1/2} \exp\left[\frac{i}{2(t' - t_0)}(x' - x)^2\right]$$
$$\times \theta(t' - t_0) \qquad (2.123)$$

Here we have also included a factor $\theta(t' - t_0)$, which is unity for $t' > t_0$ and 0 otherwise, in order to account for the fact that in the Schrödinger theory waves propagate forward in time. Equation (2.123) is familiar from many texts in elementary quantum mechanics.

2.8 The path integral in field theory

Formula (2.120) may be generalized to field theory quite readily. Instead of dividing the time interval into small elements ϵ, we now divide space-time into four-dimensional cells of volume ϵ^4. Also, where we previously defined a path by associating a value of x_i to each time instant t_i, we now define a "path" by specifying the value of a field ϕ (which may have many components) in each cell. The path integral now consists of summing over all possible values of the field in each cell. Setting $\hbar = 1$, one finds a kernel

$$K = \int [d\phi] \exp(i \int \mathcal{L}(x) \, d^4x) \qquad (2.124)$$

where $\mathcal{L}$ is the Lagrangian density. The potential $V(x)$ in the one-particle theory is analogous to an interaction term in $\mathcal{L}$ that couples a field to itself or couples various fields together. In particular, the free-particle Lagrangian $L_0 = \frac{1}{2}p^2$ of the one-particle theory, which leads to the kernel K_0 of (2.123), is analogous to a free-field Lagrangian $\mathscr{L}_0$ in field theory.

A given physical process in the latter theory is specified by the number of initial and final quanta in each state of momentum and spin. It proves convenient to build a mechanism into the formalism that automatically generates the initial and final particle states from the vacuum, quite apart from any consideration of interactions. This is done by arbitrary source terms introduced into the Lagrangian density. We define a generating functional $Z[J]$ by the path integral:

$$Z[J] = \int [d\phi] \exp\{i \int [\mathscr{L}(x) + J(x)\phi(x)] d^4x\} \quad (2.125)$$

which is the same as (2.124) except that it contains an additional term in the exponent with integrand $J(x)\phi(x)$, where $J(x)$ is an arbitrary source function. $Z[J]$ plays a central role in the development of quantum field theory according to the method of path integrals, for, with the aid of this generating functional, all the standard results of field theory can be obtained. For example, it can be shown that

$$\left.\frac{\delta^n Z[J]}{\delta J(x_1)\,\delta J(x_2) \cdots \delta J(x_n)}\right|_{J=0} = i^n \langle 0|T[\phi(x_1)\phi(x_2) \cdots \phi(x_n)]|0\rangle$$

$$= i^n G(x_1, \ldots, x_n) \quad (2.126)$$

where $\langle 0|T[\phi(x_1) \cdots \phi(x_n)]|0\rangle$ is the vacuum expectation value of the time-ordered product of n fields ϕ and $G(x_1, \ldots, x_n)$ the n-point Green's function. It may also be shown that the *connected* n-point graphs are described by

$$G_C(x_1, \ldots, x_n) = (-i)^{n-1} \frac{\delta^n W[J]}{\delta J(x_1) \cdots \delta J(x_n)} \quad (2.127)$$

where $Z[J] = \exp[iW(J)]$.

This may be illustrated by the example of a real scalar field ϕ, with Lagrangian $\mathscr{L} = \mathscr{L}_0 + \mathscr{L}_1$, where $\mathscr{L}_0 = \frac{1}{2}(\partial_\mu \phi\, \partial^\mu \phi - \mu^2 \phi^2)$ and $\mathscr{L}_1 = \mathscr{L}_1(\phi)$ is an interaction term. Ignoring $\mathscr{L}_1$ for the moment, we have, after a partial integration,

$$Z_0[J] = \int [d\phi] \exp\{i \int d^4x [-\frac{1}{2}\phi(\Box + \mu^2 - i\epsilon)\phi + J\phi]\} \quad (2.128)$$

The integral in the exponent is the limit of a sum over four-dimensional cells. Labeling the field in cell α as ϕ_α, that in adjoining cell β as ϕ_β,

2.8 The path integral in field theory

and so on, we may regard $\frac{1}{2}(\Box + \mu^2 - i\epsilon)$ as the limit of a symmetric matrix $A_{\alpha\beta}$ that connects neighboring cells. Now it can be shown that if A is symmetric,

$$\int_{-\infty}^{\infty} \prod_{i=1}^{N} dx_i \exp(-x_i A_{ij} x_j + 2 S_k x_k) = (\pi/\det A)^{1/2} \exp(S_i A_{ij}^{-1} S_j) \quad (2.129)$$

Employing this result with $S_i = J(x)/2$ and dropping an (infinite) multiplicative factor that is independent of J, it may be shown that

$$Z_0[J] = \exp\{-\tfrac{1}{2} i \int d^4x \, d^4y \, J(x)[\Box + \mu^2 - i\epsilon]^{-1} J(y)\} \quad (2.130)$$

The quantity $\Delta_F(x - y) = (\Box + \mu^2 - i\epsilon)^{-1}$ can be expressed as a Fourier integral

$$\Delta_F(x) = \int \frac{d^4k}{(2\pi)^4} e^{-ik\cdot x} \left(\frac{1}{k^2 - \mu^2 + i\epsilon}\right) \quad (2.131)$$

The factor $-i(k^2 - \mu^2 + i\epsilon)^{-1}$ is, of course, the Feynman propagator for a scalar meson of mass μ.

The generating functional $Z_0[J]$ of (2.130) has a straightforward physical interpretation: It is analogous to K_0 of (2.123). The significance of the exponent on the right-hand side of (2.130) is as follows: $J(y)$ creates a free scalar meson that propagates from y to x and is destroyed by $J(x)$. Expanding the exponential gives us a series of terms corresponding to 0, 1, 2, ... free scalar mesons.

Let us now include the interaction terms $\mathscr{L}_I(\phi)$, which we assume can be expressed as a power series in ϕ. It can readily be seen that

$$Z[J] = \int [d\phi] \exp\{i \int d^4x [\mathscr{L}_0 + \mathscr{L}_I(\phi) + J\phi]\}$$

$$= \exp\left[i \int d^4x \, \mathscr{L}_I\left(\frac{1}{i}\frac{\delta}{\delta J(x)}\right)\right] \int [d\phi] \exp\left\{i \int d^4x [\mathscr{L}_0 + J\phi]\right\}$$

$$\sim \exp\left[i \int d^4x \, \mathscr{L}_I\left(\frac{1}{i}\frac{\delta}{\delta J(x)}\right)\right]$$

$$\times \exp\left[\frac{-i}{2} \int d^4y \, J(x) \Delta_F(x-y) J(y)\right] \quad (2.132)$$

When (2.132) is employed with (2.126), one obtains the usual Feynman–Dyson expansion and it is a straightforward matter to recover the conventional rules of perturbation theory for this case. For a detailed analysis, the reader is referred to the review by Abers and Lee (73).

2.9 Path integrals and gauge invariance in electrodynamics: The massless Yang–Mills field

Let us now attempt to follow similar reasoning for electrodynamics. We start with the field equation

$$\partial_\nu F^{\mu\nu} = \partial_\nu \partial^\mu A^\nu - \partial_\nu \partial^\nu A^\mu = j^\mu \qquad (2.133)$$

The Green's function for this equation [analogous to $\Delta_F(x-y) = (\Box + \mu^2 - i\epsilon)^{-1}$ of the previous section] is called $\Delta_{\mu\nu}(x-y)$. It must satisfy

$$(\partial^\nu \partial^\mu - g^{\mu\nu}\Box)\Delta_{\mu\sigma}(x-y) = g_\sigma{}^\nu \delta^4(x-y) \qquad (2.134)$$

However, there is a serious difficulty, which can be seen if we differentiate both sides by applying ∂_ν. The left-hand side of (2.134) becomes zero, but the right-hand side becomes $\partial_\sigma \delta^4(x-y)$, which requires $\Delta_{\mu\sigma}$ to be infinite. The problem arises because the operator $O^{\mu\nu} = \partial^\nu \partial^\mu - g^{\mu\nu}\Box$ has no inverse. This is so because A_ν is a gauge field such that $F_{\mu\nu}$ is invariant under an arbitrary gauge transformation: $A_\nu \to A_\nu + \partial_\nu \Lambda$. Thus, $O^{\mu\nu} \partial_\nu \Lambda = 0$; that is, $O^{\mu\nu}$ possesses zero eigenvalues and is therefore singular.

A related fact is that the path integral $\int [dA] e^{iS(A)}$ is infinite, for, when integrating over all possible vector potential values in each cell ϵ^4, one must include, for each distinct A, all values of A equivalent up to a gauge transformation; that is, one must integrate over all values of Λ. However, for these distinct values of Λ, $\mathscr{L}$ and, therefore, the action S are invariant. We thus have a constant integrand integrated over an infinite number of paths. Evidently what is desired is the selection of just one path for each gauge-inequivalent A. That this can be done consistently was first demonstrated for non-Abelian gauge fields by Fadeev and Popov (67). We shall present here a very simple and heuristic argument to show how the solution to the problem results in the introduction of "ghost" fields.

Let us divide the A_μ of electrodynamics into a set such that none of the members is related to any of the other members by a gauge transformation. For each member $\bar{A}_\mu$, we consider all possible gauge transformations Λ. Then

$$Z = \int [dA] \exp(i \int \mathscr{L}) = \int [d\bar{A}] \exp(i \int \mathscr{L}) \int [d\Lambda] \qquad (2.135)$$

where the right-hand side can be written as indicated precisely because $\mathscr{L}$ is invariant under gauge transformations.

To deal with the infinite constant $\int [d\Lambda]$, we multiply by a factor $\exp(-i \int C^2/2)$, where C is a gauge-noninvariant function of the A_μ

2.9 The massless Yang–Mills field

and we render the resulting integral independent of the choice of C by multiplying by the Jacobian $\det(\partial C/\partial \Lambda)$. Thus,

$$\int [d\Lambda] \to \int [d\Lambda] \det(\partial C/\partial \Lambda) \exp(-i \int \tfrac{1}{2} C^2)$$
$$= \int dC \exp(-i \int \tfrac{1}{2} C^2) \quad (2.136)$$

This is equivalent to multiplying Z by an overall constant, which is always permissible because in the end it only affects normalization factors. Thus,

$$Z = \int [d\bar{A}] \int [d\Lambda] \det\left(\frac{\partial C}{\partial \Lambda}\right) \exp\left[i \int (\mathscr{L} - \tfrac{1}{2} C^2)\right] \quad (2.137)$$

Now suppose that when we perform a gauge transformation

$$A_\mu \to A_\mu + \partial_\mu \Lambda \quad (2.138)$$

the function C is altered as follows:

$$C \to C + M\Lambda \quad (2.139)$$

where M is some operator that may include derivatives. Then, in terms of space-time cells,

$$C_\alpha \to C_\alpha + M_{\alpha\beta} \Lambda_\beta \quad (2.140)$$

and also $\det(\partial C/\partial \Lambda) = \det M$. We now employ a useful technique for evaluating $\det M$. For any Hermitian matrix A, it can be shown that

$$(\det A)^{-1} = \pi^{-n} i^{-n} \int dz_1 \int dz_2 \cdots \int dz_n \exp(i\langle z|A|z\rangle) \quad (2.141)$$

where $\langle z|A|z\rangle = z_1^* A z_1 + z_2^* A z_2 + \cdots + z_n^* A z_n$. For, we know that

$$\int_{-\infty}^{\infty} dx \int_{-\infty}^{\infty} dy \exp[-a(x^2 + y^2)] = \pi/a$$

Let $z = x + iy$ and let $dx\, dy$ be written as dz. Then

$$\int dz \exp(-az^*z) = \pi/a$$

$$\prod_{i=1}^{n} \int dz_i \exp(-a_i z_i^* z_i) = \pi^n/a_1 \cdots a_n \quad (2.142)$$

where the a_i are real.

Now suppose that A is a diagonal $n \times n$ matrix:

$$A = \begin{pmatrix} a_1 & & \\ & \ddots & \\ & & a_n \end{pmatrix}$$

Then $\Sigma\, a_n z_n^* z_n = \langle z|A|z\rangle$, and replacing each a_j by $-ia_j$, we find

$$\int dz_1 \int dz_2 \cdots \int dz_n \exp(i\langle z|A|z\rangle) = i^n \pi^n / a_i \cdots a_n$$
$$= i^n \pi^n / \det(A)$$

Finally, we make a unitary transformation

$$z' = Uz, \quad A' = UAU^{-1}$$

Then $dz'_1 \cdots dz'_n = dz_1 \cdots dz_n$ and (2.141) is obtained.

Apart from an irrelevant constant factor, (2.141) is the cell equivalent of the equation

$$(\det M)^{-1} = \int [d\psi] \exp(i \int \psi^* M \psi) \tag{2.143}$$

For example, suppose we choose $C = \partial_\mu A^\mu$. Then, if $A^\mu \to A^\mu + \partial^\mu \Lambda$, $M = \Box$ and, from (2.143), ψ is a complex scalar field not connected to any source. However, we really want to know det M, whereas (2.143) is a formula for $(\det M)^{-1}$. Let us suppose that it is possible to represent det M by an equation similar to (2.143):

$$\det M = \int [d'\phi] \exp(i \int \phi^* M \phi) \tag{2.144}$$

where ϕ is also a complex scalar field and the symbol $[d'\phi]$ is clarified later. Then,

$$\det M (\det M)^{-1} = 1 = \int [d\psi] \exp(i \int \psi^* M \psi) \int [d'\phi] \\ \times \exp(i \int \phi^* M \phi) \tag{2.145}$$

Now, because ψ is a complex scalar field not connected to any sources and $\psi^* M \psi$ appears in every term in the Green's function expansion, the ψ lines must appear in closed loops in all Feynman diagrams. The ϕ lines appear in closed loops for exactly the same reason. However, since the left-hand side of (2.145) is unity, the contributions of the ψ and ϕ loops must cancel order by order in the perturbation expansion. This can only be achieved if we associate a minus sign with each ϕ loop to cancel the plus sign associated with each ψ loop. This is the significance of the notation $[d'\phi]$. We thus arrive at the result that (2.144) may be interpreted as a path integral over the complex scalar ϕ field if we associate a factor of (-1) with each closed loop. This implies Fermi–Dirac statistics; that is, the "wrong" statistics for the complex scalar ghost field.

The net result of this admittedly sketchy discussion is that an infinite constant factor associated with gauge invariance has been separated from $Z[J]$. A specific choice of gauge fixes the factor C, which implies the ghost factor $\exp(i \int \phi^* M \phi)$, where M depends on the choice of C. As we have emphasized, in electrodynamics, the ghost field plays only a trivial role and may be ignored.

However, let us now consider the $SU(2)$ Yang–Mills field $\mathbf{A}_\mu$. As we recall, the Lagrangian for the gauge field is $\mathscr{L}_{YM} = -\frac{1}{4} \mathbf{E}_{\mu\nu} \cdot \mathbf{E}^{\mu\nu}$, where $\mathbf{E}_{\mu\nu} = \partial_\mu \mathbf{A}_\nu - \partial_\nu \mathbf{A}_\mu - g(\mathbf{A}_\mu \times \mathbf{A}_\nu)$ and $\mathscr{L}_{YM}$ is invariant under local gauge transformations of the form

$$\delta \Psi = i\boldsymbol{\epsilon} \cdot \mathbf{t} \Psi, \qquad \delta \overline{\Psi} = -\overline{\Psi} i\boldsymbol{\epsilon} \cdot \mathbf{t}$$

The gauge transformation property of $\mathbf{A}_\mu$ is

$$\delta \mathbf{A}_\mu = -(1/g)\, \partial_\mu \boldsymbol{\epsilon} - (\boldsymbol{\epsilon} \times \mathbf{A}_\mu)$$

and the covariant derivative is

$$D_\mu = \partial_\mu + ig \mathbf{A}_\mu \cdot \mathbf{t}$$

By analogy with the preceding discussion for electrodynamics, it is suggested that we define a function corresponding to the Landau gauge condition $\partial^\mu \mathbf{A}_\mu = 0$:

$$\mathbf{C} = \partial^\mu \mathbf{A}_\mu \tag{2.146}$$

Then since

$$\begin{aligned}\partial^\mu \mathbf{A}'_\mu &= \partial^\mu \mathbf{A}_\mu - (1/g)\, \partial^\mu(\partial_\mu \boldsymbol{\epsilon}) - \partial^\mu(\boldsymbol{\epsilon} \times \mathbf{A}_\mu)\\ &= \partial^\mu \mathbf{A}_\mu - (1/g)\, \partial^\mu D_\mu \boldsymbol{\epsilon}\end{aligned}$$

we find

$$\mathbf{C}' = \mathbf{C} - (1/g)\, \partial^\mu(D_\mu \boldsymbol{\epsilon}) \tag{2.147}$$

and therefore that

$$M = -(1/g)\, \partial^\mu(D_\mu)$$

As before, we have a complex scalar ghost field ϕ with an additional factor:

$$\int [d'\phi]\, \exp(i \int \phi^* \, \partial^\mu D_\mu \phi) \tag{2.148}$$

where $[d'\phi]$ signifies that a factor of -1 is associated with each closed loop. The new feature rising in the massless Yang–Mills case is the covariant derivative in the exponent, which implies a coupling of the ghost field to the field $\mathbf{A}_\mu$. As a result, in addition to the "A^3" and "A^4" self-interaction terms, which give trilinear and quadrilinear vertices, respectively, there is also a ghost–ghost–vector boson vertex, the existence of which would not have been suspected from the original Lagrangian.[8] This vertex must be taken into account, otherwise the theory is inconsistent and not renormalizable.

2.10 Feynman rules for the standard model in R_ξ gauge

Let us briefly recall the essential ingredients of the spontaneously broken Yang–Mills theory, presented earlier in this chapter. Initially, one introduces a complex isodoublet scalar field, the Lagrangian of which contains a continuous parameter μ^2. For $\mu^2 < 0$, three of the

---

[8] The crude heuristic argument we presented here can be replaced by a more rigorous derivation leading to the same result; see Fadeev and Popov (67) and Abers and Lee (73).

four real field components are identified as zero-mass Goldstone bosons, whereas the fourth is associated with a massive scalar Higgs field. If the condition of invariance with respect to local $SU(2) \times U(1)$ gauge transformations is now imposed on the Lagrangian, four vector gauge fields appear, and by means of a suitable unitary transformation (choice of "unitary" gauge) the three degrees of freedom associated with the Goldstone bosons are absorbed as longitudinal polarization degrees of freedom, one for each of three gauge fields. The three corresponding gauge quanta thus acquire mass, whereas the Goldstone bosons disappear and the fourth gauge quantum (photon) remains massless. Here use was made of a field with a classical potential V; the lowest energy of the classical field is found by minimizing the potential. In the quantized version of the theory, with a general multicomponent field ϕ_i and corresponding set of sources J_i, one may start with $Z[J] = \exp\{iW[J]\}$ and define a set of quantities

$$\Phi_i(x) = \delta W[J]/\delta J_i(x)$$

It may be shown that $\Phi_i(x)$ is the vacuum expectation value of ϕ_i in the presence of $J_i(x)$. For $J_i(x) = 0$, $\Phi_i(x)|_{J_i=0} = v_i$ is the vacuum expectation value of ϕ_i in the absence of sources; and *this is not necessarily zero*. Now, defining a Legendre transformation Γ by

$$\Gamma[\Phi] = W[J] - \int d^4x \, \mathbf{J}(x) \cdot \mathbf{\Phi}(x)$$

and a "superpotential" $\mathcal{V}$ by

$$\Gamma[\Phi = 0] = -(2\pi)^4 \, \delta^4(0) \mathcal{V}(\phi)$$

it is possible to demonstrate that $\mathcal{V}$ has properties strictly analogous to V. In short, the simple classical analysis of V that led to the properties of the Goldstone bosons and the Higgs phenomenon (Section 2.4) remains valid in the quantum field theory (Abers and Lee 73).

Although the classical theory was presented in unitary, or "U," gauge, the quantum theory may be formulated in a variety of gauges. Those most frequently discussed in addition to U gauge are the R gauge and the 't Hooft–Feynman gauge. Each is a special case of the so-called generalized renormalizable, or "R_ξ," gauge, characterized by continuous real parameters α, ξ, and η. The Green's functions of the theory depend on these parameters but the physical results (S-matrix elements) do not (Fujikawa et al. 72).

Let us identify the various particles that appear in R_ξ gauge. There are, of course, the usual fermions, as well as $W^\pm$, Z^0, and γ. In addition, one has the Higgs scalar σ and three unphysical scalar bosons $s^\pm$ and χ, as well as an isotriplet of scalar–fermion ghosts. In the limit ξ,

2.10 Feynman rules for the standard model in R_ξ gauge

$\eta \to 0$ (unitary gauge), the unphysical scalar bosons disappear, whereas for $\eta, \xi \to \infty$, they are the Goldstone bosons.

We now present the propagators for vector mesons, Higgs, and unphysical scalars and ghost fields in R_ξ gauge.

(i) Vector mesons:

$$D_{\mu\nu}^{W^\pm}(k) = -i\left[g_{\mu\nu} - \frac{k_\mu k_\nu}{k^2 - m_W^2/\xi}(1 - \xi^{-1})\right]\frac{1}{k^2 - m_W^2 + i\epsilon} \quad (2.149)$$

$$D_{\mu\nu}^{Z}(k) = -i\left[g_{\mu\nu} - \frac{k_\mu k_\nu}{k^2 - m_Z^2/\eta}(1 - \eta^{-1})\right]\frac{1}{k^2 - m_Z^2 + i\epsilon} \quad (2.150)$$

$$D_{\mu\nu}^{\gamma}(k) = -i\left[g_{\mu\nu} - \frac{k_\mu k_\nu}{k^2}(1 - \alpha)\right]\frac{1}{k^2 + i\epsilon} \quad (2.151)$$

(ii) Higgs boson (mass m_σ):

$$\sigma: \quad D^\sigma(k) = \frac{i}{k^2 - m_\sigma^2 + i\epsilon} \quad (2.152)$$

(iii) Unphysical scalar bosons:

$$s^\pm: \quad D^{s^\pm}(k) = \frac{i}{k^2 - m_W^2/\xi + i\epsilon} \quad (2.153)$$

$$\chi: \quad D^\chi(k) = \frac{i}{k^2 - m_Z^2/\eta + i\epsilon} \quad (2.154)$$

(iv) Fermion scalar ghosts:

$$D_\pm^g(k) = \frac{i}{k^2 - m_W^2/\xi} \quad (2.155)$$

$$D_0^g(k) = \frac{i}{k^2 - m_Z^2/\eta} \quad (2.156)$$

Although the photon propagator (2.151) contains the arbitrary parameter α, this has no effect on S-matrix elements, since $D_{\mu\nu}^\gamma(k)$ is always sandwiched between conserved currents, so that the term $(k_\mu k_\nu/k^2)(1 - \alpha)$ always yields zero in S-matrix elements. It can be seen that all the other propagators vary as k^{-2} for large k^2 when $\xi, \eta \neq 0$. This suggests renormalizability by "naive power counting" (see Section 2.11). We also note that (2.149) can be rewritten as

$$D_{\mu\nu}^{W^\pm}(k) = \frac{-i[g_{\mu\nu} - (1/m_W^2)k_\mu k_\nu]}{k^2 - m_W^2} - \frac{i}{m_W^2}\frac{k_\mu k_\nu}{k^2 - m_W^2/\xi} \quad (2.157)$$

The second term on the right-hand side of (2.157) has a pole that is canceled in the S-matrix by the pole in the $s^\pm$ propagator (2.153).

In the limit $\xi, \eta \to \infty$, we obtain the "R" gauge. Here

$$\lim_{\xi \to \infty} D_{\mu\nu}{}^{W^\pm}(k) = -i\left(g_{\mu\nu} - \frac{k_\mu k_\nu}{k^2}\right)\frac{1}{k^2 - m_W{}^2} \qquad (2.158)$$

and similarly for the Z^0 boson when $\eta \to \infty$. For $\xi = 1$ and $\eta = 1$, we obtain the 't Hooft–Feynman gauge

$$D_{\mu\nu}{}^{W^\pm}(k) = -\frac{ig_{\mu\nu}}{k^2 - m_W{}^2} \qquad (2.159)$$

$$D^{s^\pm}(k) = \frac{i}{k^2 - m_W{}^2} = D_\pm{}^g(k) \qquad (2.160)$$

Finally, we take the limit as $\xi, \eta \to 0$. In fact, it can be shown that this limit must be taken with care (i.e., after S-matrix elements are computed), otherwise ambiguities arise. We obtain, in U gauge,

$$\lim_{\xi \to 0} D_{\mu\nu}{}^{W^\pm}(k) = \frac{-i(g_{\mu\nu} - k_\mu k_\nu/m_W{}^2)}{k^2 - m_W{}^2} \qquad (2.161)$$

which is the familiar form of the vector meson propagator already introduced in (2.95). This gauge has the advantage that ghost and unphysical boson propagators are zero. However, for large k^2, $D_{\mu\nu}{}^{W^\pm}(k^2) = O(1)$, and renormalizability does not appear evident by naive power counting. On the other hand, since the S-matrix is independent of ξ and η, it has become possible to prove renormalization for $\xi, \eta \neq 0$ and then to take the limit $\xi, \eta \to 0$.

Next we consider some of the vertex factors associated with the Weinberg–Salam Lagrangian.

The lepton–gauge boson vertices are presented in Figure 2.4 and (2.162)–(2.165).

Next we consider the trilinear gauge boson term (2.75):

$$\text{``}A^3\text{''} = \tfrac{1}{2}g[\mathbf{A}_\mu \times \mathbf{A}_\nu \cdot (\partial^\mu A^\nu - \partial^\nu A^\mu)] \qquad (2.166)$$

Employing the substitutions (2.69)–(2.72) with $g'/g = \tan \theta_W$, we find, after some algebra, that (2.166) becomes

$$\begin{aligned}\text{``}A^3\text{''} = ig[&W_\mu{}^- W_\nu{}^+(\cos\theta_W Z^{\mu\nu} + \sin\theta_W A^{\mu\nu}) \\ &+ W_\mu{}^+ W^{-\mu\nu}(\cos\theta_W Z_\nu + \sin\theta_W A_\nu) \\ &- W^{+\mu\nu} W_\mu{}^-(\cos\theta_W Z_\nu + \sin\theta_W A_\nu)]\end{aligned} \qquad (2.167)$$

Figure 2.4. Lepton–gauge boson vertices.

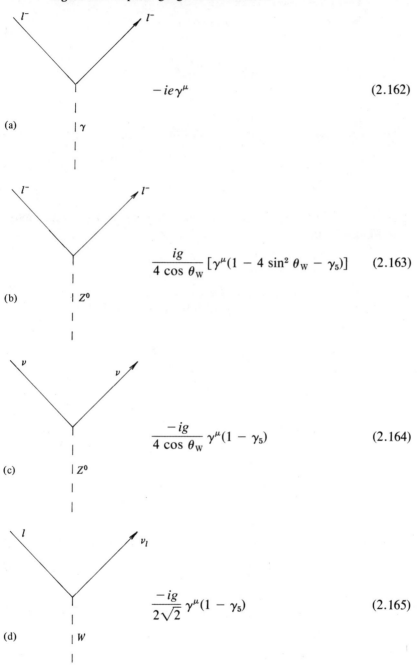

(a) $\quad -ie\gamma^\mu \quad$ (2.162)

(b) $\quad \dfrac{ig}{4\cos\theta_W}[\gamma^\mu(1 - 4\sin^2\theta_W - \gamma_5)] \quad$ (2.163)

(c) $\quad \dfrac{-ig}{4\cos\theta_W}\gamma^\mu(1 - \gamma_5) \quad$ (2.164)

(d) $\quad \dfrac{-ig}{2\sqrt{2}}\gamma^\mu(1 - \gamma_5) \quad$ (2.165)

Figure 2.5. $WW\gamma$ vertex.

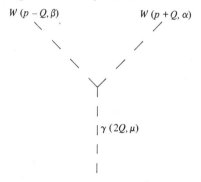

This may easily be shown to yield the following $WW\gamma$ vertex factor (see Figure 2.5):

$$\begin{aligned}V_{\alpha\beta\mu} &= ie[i(p+Q)_\mu g_{\alpha\beta} - i(p+Q)_\beta g_{\alpha\mu} \\ &\quad - i(p-Q)_\alpha g_{\beta\mu} + i(p-Q)_\mu g_{\alpha\beta} \\ &\quad - 2iQ_\beta g_{\alpha\mu} + 2iQ_\alpha g_{\beta\mu}] \\ &= -e[2p_\mu g_{\alpha\beta} - (p-3Q)_\alpha g_{\beta\mu} - (p+3Q)_\beta g_{\alpha\mu}]\end{aligned} \quad (2.168)$$

The "A^4" term is as follows:

$$\text{``}A^4\text{''} = -\tfrac{1}{4}g^2(\mathbf{A}_\mu \times \mathbf{A}_\nu)\cdot(\mathbf{A}^\mu \times \mathbf{A}^\nu)$$

which becomes

$$\text{``}A^4\text{''} = -\tfrac{1}{4}g^2[2W_\mu^+ W^{-\mu}W_\nu^+ W^{-\nu} - 2W_\mu^+ W^{+\mu}W_\nu^- W^{-\nu} \\ + 4W_\mu^+ W^{-\mu}A_{3\nu}A_3^\nu - 4W_\mu^+ A_3^\mu W_\nu^- A_3^\nu] \quad (2.169)$$

If we concern ourselves with only those terms in "A^3" and "A^4" involving coupling of W to the electromagnetic field, we find

$$\text{``}A^3\text{''} + \text{``}A^4\text{''} = -ie(W_\mu^- W_\nu^+ A^{\mu\nu} + W_\nu^+ W^{-\mu\nu}A_\mu - W_\nu^- A_\mu W^{+\mu\nu}) \\ - e^2(W^+ W^- AA - W^+ AW^- A) \quad (2.170)$$

It is interesting to compare (2.170) with the result obtained when the minimal substitution $\partial_\mu W_\nu^\pm \to (\partial_\mu \mp ieA_\mu)W_\nu^\pm$ is made in the Lagrangian for the old-fashioned intermediate boson theory. It is found that in the new theory an additional term $\tfrac{1}{2}ieW_\mu^- W_\nu^+ A^{\mu\nu}$ arises, which implies an additional contribution of $e\hbar/2m_W c$ to the magnetic moment of W^-, over and above the "normal" magnetic moment of $\mu_W^0 = e\hbar/2m_W c$. The total W^- magnetic moment is thus expected to be

$$\mu_W = 2\mu_W^0 \quad (2.171)$$

The vertex factors for coupling of the unphysical bosons in R_ξ gauge are shown in Figure 2.6.

Figure 2.6. Unphysical-boson vertices.

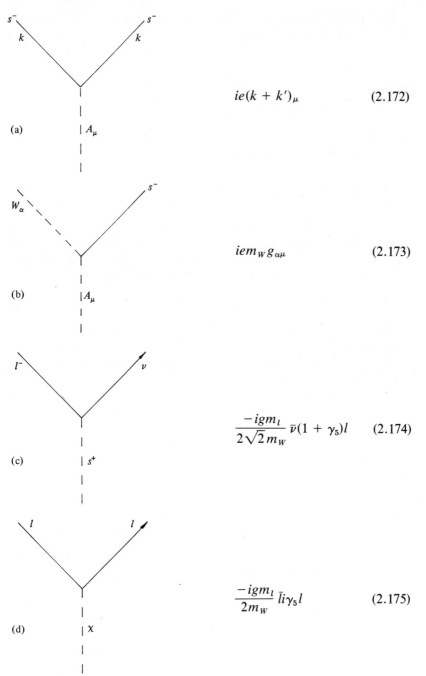

(a) $\quad ie(k + k')_\mu \quad$ (2.172)

(b) $\quad iem_W g_{\alpha\mu} \quad$ (2.173)

(c) $\quad \dfrac{-igm_l}{2\sqrt{2}\, m_W} \bar{\nu}(1 + \gamma_5)l \quad$ (2.174)

(d) $\quad \dfrac{-igm_l}{2m_W} \bar{l}i\gamma_5 l \quad$ (2.175)

78   2 The standard model

In order to illustrate the application of the Feynman rules in various gauges, let us summarize the results of calculations of the very small corrections to the g-factor anomaly $a = (g - 2)/2$ in the magnetic moment of the muon that arise from weak interaction and unphysical-boson diagrams. The relevant graphs are shown in Figure 2.7 and the results presented in Table 2.1. It is very interesting to note that for some graphs a different contribution arises for each distinct gauge. However, the total contribution of all six lowest-order graphs is gauge-invariant, as of course it should be.

2.11 Some brief comments on renormalization

Although the problem of renormalization is a very complicated one and beyond the scope of this book, it may be useful to consider the question in some of its elementary aspects. The problem arises in connection with the ultraviolet divergence of integrals over internal momenta in a Feynman diagram. It is useful to define the "degree of divergence" $D = N_n - N_d$, where N_n and N_d are the number of momentum factors in the numerator and denominator, respectively. In general (although this must be qualified carefully), if $D < 0$, one has convergence; if $D = 0$, there is a logarithmic divergence; and if $D > 0$, there is a divergence. For a given Feynman diagram, it is useful to define the following quantities:

F_e number of external fermion lines
F_i number of internal fermion lines
B_e^k number of external boson lines of type k
B_i^k number of internal boson lines of type k
C^j number of vertices of type j

We note that since a fermion line does not end, F_e must be even. Also, depending on the theory, there can be boson lines of more than one type k. For example, in electrodynamics, one has only photons. However, in the Yang–Mills theory, there are gauge bosons, Higgs bosons, unphysical bosons (and scalar ghosts, which obey Fermi statistics but have bosonlike propagators). In electrodynamics there is also only one type of vertex, whereas in the present theory there are a number of different types.

Let us now consider those factors that are significant in determining the degree of divergence D. A given process is specified by the external lines, and the order of the process is determined by the internal lines. The initial discussion will be given for a single boson type and single vertex type.

Figure 2.7. Lowest-order graphs contributing to the g-factor anomaly of μ^-, from W-boson, Z-boson, and unphysical-boson effects. See Table 2.1.

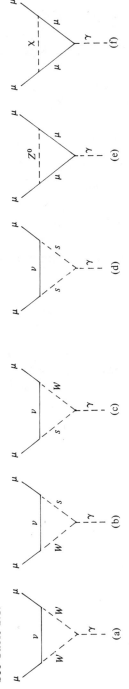

Table 2.1. Contributions of Figure 2.7 to the anomaly $a = (g-2)/2$ in units of $G_F m_\mu^2/8\pi^2\sqrt{2}$

Gauge	Figure part				Figure part		
	(a)	(b) + (c)	(d)	Subtotal	(e)	(f)	Subtotal
Unitary $\xi \to 0$	$\frac{10}{3}$	0	0	$\frac{10}{3}$	$\frac{1}{3}[(3 - 4\cos^2\theta_W)^2 - 5]$	0	$\frac{1}{3}[(3 - 4\cos^2\theta_W)^2 - 5]$
't Hooft–Feynman $\xi = 1$ $\eta = 1$	$\frac{7}{3}$	1	0	$\frac{10}{3}$	$\frac{1}{3}[(3 - 4\cos^2\theta_W)^2 - 5]$	0	$\frac{1}{3}[(3 - 4\cos^2\theta_W)^2 - 5]$
$R(\xi, \eta \to \infty)$	$\frac{4}{3}$	1	1	$\frac{10}{3}$	$\frac{1}{3}[(3 - 4\cos^2\theta_W)^2 - 5] + 1$	-1	$\frac{1}{3}[(3 - 4\cos^2\theta_W)^2 - 5]$

80   2 The standard model

  (i) Each internal line contributes $+4$ to D because of d^4p:
      $+4(F_i + B_i)$
  (ii) Each vertex corresponds to a four-dimensional delta function. This is equivalent to four factors of momentum in the denominator, but one delta function is required for energy and momentum conservation. Thus we have:
      $-4(C - 1)$
  (iii) Each internal fermion line corresponds to a propagator $1/(\not{p} - m)$. Thus we have a factor:
      $-F_i$
  (iv) Each internal scalar boson line, photon line, or vector boson line in R_ξ gauge with $\xi \neq 0$ corresponds to a propagator that varies as k^{-2} for large k^2 and contributes a factor:
      $-2B_i$
  (v) If the coupling has a derivative at the vertex, there is an additional factor:
      $+C$

Ignoring derivative couplings for the moment, we thus have

$$D = 4 - 4C + 3F_i + 2B_i \tag{2.176}$$

However, the numbers of fermion and boson lines are related to the number of vertices. For example, in electrodynamics, two fermion lines and one boson line appear at each vertex. Thus the number of endings of fermion and boson lines is $2C$ and C, respectively. Also, each internal line has two endings and each external line has one. Therefore,

$$2C = 2F_i + F_e \tag{2.177}$$

and

$$C = 2B_i + B_e \tag{2.178}$$

Substituting (2.177) and (2.178) in (2.176), one obtains

$$D = 4 - B_e - 3F_e/2 \quad \text{(electrodynamics)} \tag{2.179}$$

This analysis is called "naive power counting," or "superficial" analysis of divergence. It can be shown that it is valid for n external lines if and only if the result of carrying out any arbitrarily chosen $4(n - 1)$ subintegrations is finite; in other words, if the divergence comes about only because of the last 4-momentum integration. Graphs satisfying this criterion are called *primitive divergents,* and it can be shown that a diagram is primitively divergent whenever there is convergence if one of the internal lines is cut and replaced by two external

2.11 Comments on renormalization

lines. An enumeration of the primitive divergents corresponds to a list of the fundamental divergents from which all other divergents can be obtained by appropriate insertions.

Generally speaking, a theory is renormalizable if
  (i) There are a finite number of primitive divergents.
  (ii) Each can be eliminated (or "subtracted") by an appropriate counter-term inserted in the Lagrangian, such that the counter-term takes the same form as a term already present in the Lagrangian and contributes only to a rescaling of a coupling constant or a mass. The mathematical procedure of subtraction is called *regularization*.
  (iii) When higher-order divergences are built up from the primitive divergents by insertions, no further counterterms are required.

Let us briefly summarize the situation in electrodynamics, a theory that satisfies these requirements and is renormalizable. We enumerate the primitive divergents in Figure 2.8 in light of (2.179). As can be seen from Figure 2.8, the two outstanding divergences are compensated by rescaling (renormalizing) the electron mass and charge. Since there is no way to determine the bare mass and charge of an electron, writing the Lagrangian in terms of the physical mass and charge means that the counter-terms are already included.

Before we proceed to the Yang–Mills case, let us give an example of a theory that is easily seen from (2.176) to be nonrenormalizable. This is the Fermi theory, where there are no bosons and four fermion lines meet at each vertex. Thus we have $F_e + 2F_i = 4C$, and

$$D = 4 + 2C - 3F_e/2 \quad \text{(Fermi's theory)} \tag{2.180}$$

Here, D increases as the number of vertices increases, so there are an infinite number of primitive divergents, and an infinite number of counter-terms would be required.

In the case of the massless Yang–Mills theory, the situation is much more complicated than for electrodynamics. There exist three types of vertices (C_1, C_2, C_3) corresponding to the "A^3," "A^4," and ghost–ghost–A_μ vertices, respectively. Vertex C_1 has three gauge boson (B) lines, C_2 has four B lines, and C_3 has one B line and two ghost (G) lines. Also C_1 and C_3 involve derivative couplings. One thus finds $D = 4 - B_e - G_e$. The primitive divergents are shown in Figure 2.9.

One can show that each set of graphs in Figure 2.9 is canceled by counterterms having the same form as the original terms in the Lagrangian, and this can be extended to the spontaneously broken

82   2 The standard model

symmetry case with fermions included. The general proof of renormalizability (including higher-order divergents) is very subtle. Among other things, one must overcome a difficulty caused by the coupling of gauge fields to the axial portions of fermion currents, which gives rise to the so-called Adler–Bell–Jackiw anomaly.

It is the so-called triangle diagrams that are troublesome here (Adler 69, Bell and Jackiw 69). For example, let us consider electron–photon scattering (Compton effect). In quantum electrodynamics, there is no

Figure 2.8. (a) $F_e = 0$; $B_e = 0$: This is the "vacuum self-energy" diagram; $D = +4$. However, this graph has no external lines and is disconnected from all graphs of real processes; therefore, it need not be considered. (b) $F_e = 2$: Fermion self-energy; $D = +1$. However, it can be shown that if D is odd, only the next lowest divergence actually occurs. Thus, in effect, $D = 0$, and we have a logarithmic divergence here. It corresponds to a counterterm correcting the electron mass. (c) $F_e = 0$; $B_e = 2$: Boson self-energy; $D = +2$, a quadratic divergence. However, in QED, this vacuum-polarization diagram is reduced to a logarithmic divergence by gauge invariance. This is compensated for by a rescaling of the electron charge. (d) $F_e = 2$; $B_e = 1$: Vertex correction; $D = 0$, a logarithmic divergence. However, in QED, this divergence is canceled by that in two-fermion self-energy diagrams. (e) $F_e = 0$; $B_e = 3$: Triangle diagram; $D = 1$. However, charge-conjugation symmetry, expressed in Furry's theorem, shows that there is no contribution from diagrams of this type. (f) $F_e = 0$; $B_e = 4$: Light by light scattering; $D = 0$. Gauge invariance reduces this to $D = -4$, convergent.

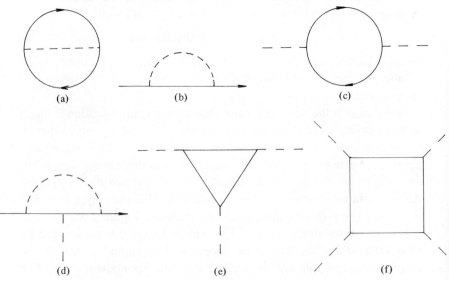

2.11 Comments on renormalization

difficulty in making the usual higher-order corrections; unitarity is obeyed and the theory is renormalizable. However, let us also include in the higher-order corrections diagrams containing the massive intermediate vector bosons, which can couple to the *axial vector* as well as vector portions of the fermion currents. In particular, diagrams of the type shown in Figure 2.10 arise.

It can be shown that the scattering amplitude associated with this diagram grows beyond the bounds associated with unitarity at sufficiently high energy. Renormalizability would thus be spoiled, except that in each case of this type the net anomaly is proportional to the weak hypercharge Y summed over all weak isodoublets. We recall that $Y = -1$ for each leptonic isodoublet component (ν_{eL}, e_L, etc.) and $Y = +\frac{1}{3}$ for each quark isodoublet component. However, there exist three quark colors, and thus the anomaly is zero because of a most fortunate cancellation (see, for example, Bouchiat et al. 72).

Figure 2.9. Primitive divergents for the massless Yang–Mills case: ----, ghost; ———, gauge boson. Graphs adding together to result in a given counterterm are grouped together in (a)–(d).

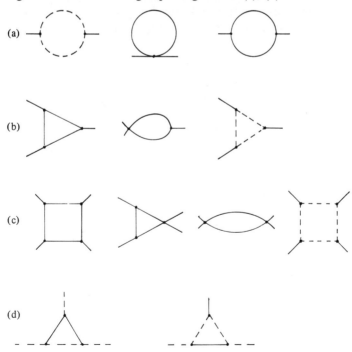

Figure 2.10. An example of the triangle anomaly in Compton scattering.

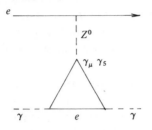

The method of dimensional regularization proposed by 't Hooft and Veltman (72a,b) proves to be very useful in practical perturbation calculations, as well as in formal development of the theory of renormalization. Here we shall attempt to explain it very briefly by means of an example, leaving some details for Appendix C. Consider the electric charge of the neutrino, which is known to be zero from experiment to extremely high accuracy. (From charge conservation, we can deduce from neutron β decay that the neutrino charge is less than $10^{-19}e$.) On the other hand, it appears that there ought to be several contributions to the charge in higher-order perturbation theory, even if it is zero in zeroth order. For example, consider Figures 2.11a and b, which yield a net contribution to the neutrino charge in the limit of zero photon 4-momentum, unless the amplitudes associated with these two diagrams should happen to cancel. (Note that there is a trilinear $WW\gamma$ vertex in Figure 2.11b).

Unfortunately, before regularization, each loop integral is infinite, and it is not at all clear how the amplitudes in Figures 2.11a and b should be compared. On the other hand, one can express each integral formally as a continuous function of the dimensionality of space-time

Figure 2.11. Contributions to neutrino charge in the limit of zero photon 4-momentum.

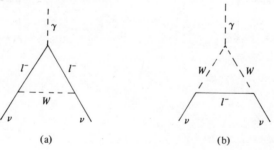

(in reality, of course, $n = 4$). Then it turns out that each integral is finite for $1 < n < 4$, and the difference between the two amplitudes may be taken before one goes to the limit $n = 4$. In this way it may be shown that the amplitudes corresponding to Figures 2.11a and b do in fact cancel (Bardeen et al. 72). The procedure is covariant, satisfies unitarity in the limit $n \to 4$, and is suitable for renormalization because the form of the integrand is retained. However, there are subtle difficulties associated with the definition of Dirac matrices for continuous n, and in particular, there exists an ambiguity in the definition of γ_5 that may be traced to a corresponding ambiguity in the triangle anomaly (see, for example, Chanowitz et al. 79).

2.12 Left–right symmetric theories

It is possible to consider variants on the standard model that yield identical predictions for all low-energy electroweak phenomena. In particular, there is considerable interest in "left–right symmetric" models, (see, for example, Pati and Salam 74, Fritzsch and Minkowski 76, Mohapatra and Sidhu 77). One starts with a Lagrangian containing equivalent massless left- and right-handed vector bosons that therefore exhibits no parity violation initially. Spontaneous symmetry breaking then causes the vector bosons to acquire different masses, with $m_R \gg m_L$. The result is that, at low q^2, the right-handed (V + A) couplings are much weaker than the left-handed ones. The charged weak currents might plausibly enter into the Lagrangian as follows (Beg et al. 77, Holstein and Treiman 77):

$$\mathscr{L} = -\frac{g}{2\sqrt{2}}[(V - A)W_L^{(0)} + (V + A)W_R^{(0)}] + \text{h.c.} \quad (2.181)$$

where $W_L^{(0)}$ and $W_R^{(0)}$ are related to the mass eigenstates W_L and W_R (with $m_R \gg m_L$) by

$$\begin{aligned} W_L &= W_L^{(0)} \cos \zeta + W_R^{(0)} \sin \zeta \\ W_R &= W_R^{(0)} \cos \zeta - W_L^{(0)} \sin \zeta \end{aligned} \quad (2.182)$$

which is analogous to the Cabibbo mixing of quark states d and s in the GIM scheme. Thus at low q^2, the effective Fermi interaction would be described by the formula

$$\mathscr{L}_{\text{eff}} = \frac{G_F}{2\sqrt{2}}[V^\dagger V + \eta_{AA} A^\dagger A + \eta_{AV}(V^\dagger A + A^\dagger V)] \quad (2.183)$$

where

$$\eta_{AA} = (\epsilon^2 m_R^2 + m_L^2)/(\epsilon^2 m_L^2 + m_R^2) \quad (2.184)$$

$$\eta_{AV} = -\epsilon(m_R^2 - m_L^2)/(\epsilon^2 m_L^2 + m_R^2) \tag{2.185}$$

and

$$\epsilon = (1 + \tan \zeta)/(1 - \tan \zeta) \tag{2.186}$$

In the limit of pure V–A, $m_R \to \infty$ and $\epsilon \to 1$. Experimental data may be used to limit the mixing angle ζ and the ratio m_L^2/m_R^2. Relevant experimental quantities for this purpose are the ρ parameter and the polarization parameter ξP in muon decay (see Sections 3.3 and 3.4), the electron polarization in β decay (see Section 5.3), and the β-decay asymmetry parameter $A(^{19}\text{Ne})$ (see Section 5.3). One finds that the present lower limit on m_R is only about a factor of three larger than the value of m_L expected in the standard model:

$$m_R \geq 240 \quad \text{GeV}/c^2 \tag{2.187}$$

Evidently then, left–right symmetric models are very much a possibility.

2.13 The Higgs boson

In this chapter we have attempted to outline the main features of the standard model of electroweak phenomena. Certainly it represents a major advance in theoretical physics, and since, as we shall see, its predictions of low-energy neutral weak phenomena are very well verified by experiment, many features of the new theory surely seem here to stay. There is, however, one fundamental aspect of the new theory that appears artificial and obscure. That is the Higgs mechanism and, in particular, the Higgs boson.

To begin, we do not know how many Higgs bosons exist. In the simplest version of the theory, expounded earlier in this chapter, we start with a single complex isodoublet field. After spontaneous symmetry breaking, we are left with one massive neutral Higgs scalar. Yet one could retain the $SU(2)_L \times U(1)$ content of the theory and have more than one isodoublet. The result after spontaneous symmetry breaking would be a retinue of Higgs bosons, some charged and some neutral. Similarly, this would happen in more complicated gauge theories. For example, the left–right-symmetric models mentioned briefly in the previous section have a complex Higgs structure, with charged and neutral scalars.

Next, we do not know what the mass of the Higgs boson(s) should be. Confining ourselves to the simplest case of one neutral scalar, the most we can do at present of a definite nature is to set upper and lower limits on m_σ. If the theory is to be well behaved in the usual sense, an

2.13 The Higgs boson

upper limit of about 1 TeV/c^2 can be arrived at in the following way (Lee et al. 77, Veltman 77). We consider the hypothetical scattering process $W^+W^- \rightarrow W^+W^-$, which can occur by γ or Z exchange and also by Higgs exchange. Now, it can be shown that the amplitude for γ or Z exchange varies as $G_F s$, where $\sqrt{s}$ is the total CM energy. This is largely canceled by the Higgs exchange diagrams in lowest order, leaving a residual amplitude that achieves the constant value $-(4G_F/\sqrt{2})m_\sigma^2$ as $s \rightarrow \infty$. Now, this residual amplitude can be expanded in partial waves J, and each partial-wave amplitude must satisfy the unitarity condition $|M_J| \leq 1$. One easily finds that this cannot be satisfied for the $J = 0$ amplitude unless $m_\sigma^2 \lesssim 4\pi\sqrt{2}/G_F$. Thus perturbation theory breaks down for the standard model if $m_\sigma \geq 1$ TeV/c^2, and the weak interaction becomes "strong" above that energy threshold.

One might expect that such a large mass and the resulting strong interactions would be manifested in some observable way in low-energy weak processes, perhaps in radiative corrections. For example, although the ratio m_W/m_Z is given by $\cos \theta_W$ in the zeroth-order theory, one might expect substantial corrections to this ratio if m_σ were very large. Yet, it has been shown by Veltman (68,70) that such modifications depend only logarithmically on the Higgs mass and are, for all practical purposes, quite negligible. In other words, the standard model might account very well for low-energy weak phenomena, yet lead to difficulties at very high energies that are profound and not at all understood.

A lower limit to m_σ of the order of $\alpha G_F^{-1/2}$ may be arrived at by consideration of radiative corrections involving loops of gauge bosons, which lead to induced self-couplings of the scalar field and therefore to mass contributions (Coleman and Weinberg 73, Linde 76, Weinberg 76a). It can be shown in this manner that $m_\sigma \geq 6.6$ GeV/c^2 for $\sin^2 \theta_W = 0.23$. Thus, we can say little more about m_σ than that it is expected to be in the range $\alpha G_F^{-1/2}$ to $G_F^{-1/2}$.

Given this large range of possible values for mass, it is very difficult to design experiments to search for Higgs bosons. Analyses of various possibilities have been carried out by a number of authors (e.g., Ellis et al. 76a, Gaemers and Goumaris 78, Donoghue and Li 79). Two particular mechanisms for Higgs production have received special consideration. Figure 2.12 describes one of them in which e^+e^- collide to produce a real Z^0, which decays to a virtual Z^0 (hence to a lepton pair) and a Higgs boson. Detailed numerical estimates have been carried out by

Figure 2.12. Production of Higgs boson in e^+e^- annihilation.

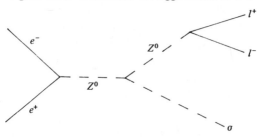

Bjorken (76). Another possibility is the process $\Upsilon(b\bar{b}) \to \gamma + \sigma$ (see, e.g., Wilczek 77), which is estimated to have a branching ratio of about 1 percent. The analogous decay in "toponium" ($t\bar{t}$) may have a much larger branching ratio, however.

Since the Higgs–fermion–fermion coupling depends on the fermion mass, the Higgs boson should decay predominantly into the heaviest fermion pair possible, according to the formula

$$\Gamma(\sigma \to f\bar{f}) = c \frac{G_F m_f^2 m_\sigma}{\sqrt{2}\,\pi} \left(1 - \frac{4m_f^2}{m_\sigma^2}\right)^{3/2}$$

where c is a color factor: 1 for leptons and 3 for quarks.

Finally we mention that the Higgs boson may be introduced into the theory of weak interactions in a rather intuitive way, without any explicit mention of spontaneous symmetry breaking (Cornwall et al. 73, Llewellyn-Smith 73). Let us require a theory of weak interactions to include coupling of leptonic currents to charged and neutral intermediate bosons in order to explain the experimental results. Moreover, let us require that the perturbation series for any given process (e.g., $\nu\bar{\nu} \to W^+W^-$) be well behaved in the sense that divergences cancel order by order and unitarity is obeyed. This is possible of course only for suitable choice of the couplings for W and Z; to cure the worst divergences, these couplings must be of the Yang–Mills type. However, after this is done, there remain residual divergences that can be eliminated only by introducing couplings of the fermions to a new scalar field. This turns out to be precisely the Higgs–fermion–fermion coupling.

Given the various difficulties associated with the Higgs boson or bosons, it seems quite possible that this portion of the theory will undergo major changes in the future. In the chapters to follow, however, we shall ignore these difficulties and concentrate for the most part on the more well-established parts of the theory and its comparisons with experiment.

Problems

Problems

2.1 In the text, an example of spontaneous symmetry breaking is taken from classical mechanics in which a long thin rod is bent when subject to a sufficiently large axial force. Think of some other examples of spontaneous symmetry breaking in classical physics.

2.2 Consider the Lagrangian for a massive vector field ϕ_μ coupled to a vector current J^μ:
$$\mathscr{L} = -\tfrac{1}{4} f_{\mu\nu} f^{\mu\nu} - \tfrac{1}{2} m^2 \phi_\mu \phi^\mu - \phi_\mu J^\mu$$
where $f_{\mu\nu} = \partial_\mu \phi_\nu - \partial_\nu \phi_\mu$. Derive the field equations (2.93) and thus obtain (2.94). Carry out the Fourier analysis described in the text to obtain (2.95).

2.3 Consider the modifications of Maxwell's equations that would result if the electromagnetic Lagrangian were altered to accommodate finite photon mass, as in (2.23). (This problem is related to the previous one.)

(a) Show that one can define electric and magnetic fields **E** and **B** in terms of the scalar and vector potentials ϕ and **A** in the usual way:
$$\mathbf{E} = -\nabla \phi - \frac{1}{c} \frac{\partial \mathbf{A}}{\partial t}, \qquad \mathbf{B} = \nabla \times \mathbf{A}$$
but that Maxwell's equations become (in cgs units):
$$\nabla \cdot \mathbf{E} = 4\pi \rho - \kappa^2 \phi$$
$$\nabla \times \mathbf{B} = 4\pi \frac{\mathbf{j}}{c} + \frac{1}{c} \frac{\partial \mathbf{E}}{\partial t} - \kappa^2 \mathbf{A}$$
$$\nabla \cdot \mathbf{B} = 0$$
$$\nabla \times \mathbf{E} = -\frac{1}{c} \frac{\partial \mathbf{B}}{\partial t}$$
where the additional terms in the constant κ^2 arise from finite photon mass m_γ. How are κ^2 and m_γ related?

(b) How is the equation of continuity modified and what special condition on the potentials must be satisfied if
$$\nabla \cdot \mathbf{j} + \partial \rho / \partial t = 0$$

(c) Show that the "Coulomb" potential of a point charge e is $(e/r)e^{-\kappa r}$.

(d) In plane-wave solutions of the form $\mathbf{E} = \mathbf{E}_0 e^{i(\mathbf{k} \cdot \mathbf{r} - \omega t)}$, what is the relationship between **k** and ω? Show that plane waves are no longer strictly transverse in this modified electrodynamics.

(e) How is the experimental limit $m_\gamma \leq 10^{-16}$ eV/c^2 arrived at?

2.4 Consider the quantum-mechanical description of a particle in one spatial dimension q. If the particle is in the state $|\Psi(t)\rangle$, the wavefunction is

$$\psi(q, t) = \langle q|\Psi(t)\rangle = \langle q(t)|\Psi(0)\rangle$$

with $\langle q(t)| \equiv \langle q|e^{-iHt}$, where H is the Hamiltonian. In this problem we obtain an expression for the kernel $K(q', t'; q, t)$, which relates $\psi(q', t')$ to $\psi(q, t)$, $t' > t$:

$$\psi(q', t') = \int K(q', t'; q, t)\psi(q, t)\, dq$$

(a) Suppose we divide the time interval $t'-t$ into $n+1$ small intervals ϵ as follows:

$$t = t_0$$
$$t_1 = t_0 + \epsilon$$
$$t_2 = t_0 + 2\epsilon$$
$$\cdot$$
$$\cdot$$
$$\cdot$$
$$t_n = t_0 + n\epsilon$$
$$t_{n+1} = t_0 + (n+1)\epsilon = t'$$

By means of completeness, show that

$$\langle q't'|qt\rangle = \int_{-\infty}^{\infty} dq_1 \cdots \int_{-\infty}^{\infty} dq_n \, \langle q't'|q_n t_n\rangle\langle q_n t_n|q_{n-1} t_{n-1}\rangle$$
$$\times \cdots \langle q_1 t_1|qt\rangle$$
$$= \int dq_1 \cdots \int dq_n$$
$$\times \langle q_{n+1}|e^{-iH\epsilon}|q_n\rangle\langle q_n|e^{-iH\epsilon}|q_{n-1}\rangle \cdots \langle q_1|e^{-iH\epsilon}|q\rangle$$

(b) Let P and Q be the operators for momentum and position, respectively: $P|p\rangle = p|p\rangle$; $Q|q\rangle = q|q\rangle$. Assuming that $e^{-iH\epsilon}$ can be expressed as a power series in P and Q and using the results:

$$\langle q_{i+1}|p\rangle = (2\pi)^{-1/2} \exp(ipq_{i+1})$$
$$\langle p|q_i\rangle = (2\pi)^{-1/2} \exp(-ipq_i)$$

show that

Problems

$$\langle q't'|qt\rangle = \prod_{i=1}^{n} \int dq_i \prod_{i=1}^{n+1} \int \frac{dp_i}{2\pi}$$

$$\times \exp\left\{i\sum_{i=1}^{n} \epsilon \left[p_i \left(\frac{q_i - q_{i-1}}{\epsilon}\right) - H\right]\right\}$$

(c) Consider the special case $H(p, q) = \frac{1}{2}p^2 + V(q)$. By means of the formula

$$I_0 = \int_{-\infty}^{\infty} du\, e^{iau^2} = (i\pi/a)^{1/2}$$

show that $\langle q't'|qt\rangle = \int [dq]\, \exp(i \int L\, dt)$, where $L = p\dot{q} - H$ and

$$\int [dq] \equiv \frac{1}{(2\pi i\epsilon)^{n+1/2}} \prod_{i=1}^{n} \int dq_i$$

(d) Use mathematical induction to derive (2.123) from (2.121).

2.5 Consider (2.170) and compare it with the result obtained when the minimal substitution $\partial_\mu W^\pm \to (\partial_\mu \mp ieA_\mu)W^\pm$ is made in the Lagrangian for the naive intermediate vector boson theory. Show that an additional term arises in the case of (2.170), which implies that the W^- possesses an anomalous magnetic moment $\mu_W = 2e\hbar/2m_W c$.

3

Leptonic weak interactions

3.1 Introduction

We shall now consider various leptonic processes, such as muon decay, production and decay of τ leptons, neutrino–electron elastic scattering, and so on. These are of fundamental importance for the study of weak interactions, for in lowest order they are free of the complications of strong interactions, and the cross sections and transition probabilities can be calculated precisely. We shall present many of the calculations in some detail here because of their intrinsic interest and because the calculational methods find many applications in later chapters.

Until quite recently, the only observed leptonic weak process was muon decay. However, in the last few years, remarkable experiments have been carried out with τ leptons, with neutrino–electron elastic scattering, and with inverse muon decay. It will be seen that all experimental results obtained so far for the leptonic weak interactions are in very good agreement with predictions based on the standard model. Many important experiments remain to be done, however, and perhaps none is more important than the attempt to observe production and decay of the charged and neutral intermediate bosons. Prospects for such observations seem quite favorable in the coming decade.

3.2 Muon decay and the V–A law

Let us begin by calculating the transition probability for muon decay. It is described in lowest order by Figure 3.1, and from (2.161) and (2.165), the invariant amplitude is

3.2 Muon decay and the V–A law

$$\mathcal{M} = \left[\frac{-ig}{2\sqrt{2}} \bar{u}_2 \gamma_\lambda (1 - \gamma_5) u_\mu\right] \left(\frac{g^{\lambda\sigma} - q^\lambda q^\sigma / m_W^2}{q^2 - m_W^2}\right)$$
$$\times \left[\frac{-ig}{2\sqrt{2}} \bar{u}_e \gamma_\sigma (1 - \gamma_5) v_1\right] \quad (3.1)$$

where u_2 and v_1 are ν_μ and $\bar{\nu}_e$ spinors, respectively. Of course, in muon decay, $|q^2| \ll m_W^2$, and thus, to a very good approximation,

$$\frac{g^{\lambda\sigma} - q^\lambda q^\sigma / m_W^2}{q^2 - m_W^2} \to -\frac{g^{\lambda\sigma}}{m_W^2}$$

Therefore, taking into account that $g^2/8m_W^2 = G_F/\sqrt{2}$, we find that $\mathcal{M}$ becomes:

$$\mathcal{M} = \frac{G_F}{\sqrt{2}} \bar{u}_2 \gamma_\lambda (1 - \gamma_5) u_\mu \bar{u}_e \gamma^\lambda (1 - \gamma_5) v_1 \quad (3.2)$$

which is, of course, the same amplitude for muon decay that appears in the older V–A theory of Feynman and Gell-Mann.

Figure 3.1. Muon decay.

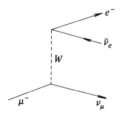

The differential transition probability per unit time is calculated in first-order perturbation theory by means of the "golden rule." We assume that the reader is familiar with this general method, summarized in Appendix B, and we go directly to the differential transition probability for muon decay:

$$d\Gamma = (2\pi)^4 \, \delta^4(p_e + p_1 + p_2 - p_\mu)$$
$$\times \frac{m_e}{E_e} \frac{m_\mu}{E_\mu} \frac{m_{\nu_e}}{E_1} \frac{m_{\nu_\mu}}{E_2} \frac{d^3\mathbf{p}_e}{(2\pi)^3} \frac{d^3\mathbf{p}_1}{(2\pi)^3} \frac{d^3\mathbf{p}_2}{(2\pi)^3} |\mathfrak{M}|^2 \quad (3.3)$$

(In this formula, the factors m_{ν_e}/E_1 and m_{ν_μ}/E_2 cause no difficulty even if the neutrino masses are zero because $m_{\nu_e} m_{\nu_\mu}$ is canceled by a factor $m_{\nu_e} m_{\nu_\mu}$ in the denominator of $|\mathcal{M}|^2$, as we shall see).

Let us now evaluate $|\mathcal{M}|^2$. Taking the complex conjugate of (3.2), we obtain

$$\mathcal{M}^* = \frac{G_F}{\sqrt{2}} \bar{u}_\mu \gamma_\sigma (1 - \gamma_5) u_2 \bar{v}_1 \gamma^\sigma (1 - \gamma_5) u_e$$

Therefore,[1]
$$|\mathcal{M}|^2 = \tfrac{1}{2} G_F^2 T_1 T_2 \tag{3.4}$$
where
$$T_1 = \mathrm{tr}[u_2 \bar{u}_2 \gamma_\lambda (1 - \gamma_5) u_\mu \bar{u}_\mu \gamma_\sigma (1 - \gamma_5)] \tag{3.5}$$
and
$$T_2 = \mathrm{tr}[u_e \bar{u}_e \gamma^\lambda (1 - \gamma_5) v_1 \bar{v}_1 \gamma^\sigma (1 - \gamma_5)] \tag{3.6}$$
Now
$$u_e \bar{u}_e = \frac{1}{2} \left(\frac{\not{p}_e + m_e}{2 m_e} \right) (1 + \gamma_5 \not{s}_e) \tag{3.7}$$
and
$$u_\mu \bar{u}_\mu = \frac{1}{2} \left(\frac{\not{p}_\mu + m_\mu}{2 m_\mu} \right) (1 + \gamma_5 \not{s}_\mu) \tag{3.8}$$
where s_e and s_μ are 4-polarizations for e^- and μ^-, respectively. Also, for very small or zero neutrino masses, we may write
$$u_2 \bar{u}_2 = \not{p}_2 / 2 m_{\nu_\mu} \tag{3.9}$$
$$v_1 \bar{v}_1 = \not{p}_1 / 2 m_{\nu_e} \tag{3.10}$$
[where, as mentioned, the factors m_{ν_μ} and m_{ν_e} in the denominators are canceled by the kinematic factors in (3.3)]. Thus
$$T_1 = \frac{1}{8 m_{\nu_\mu} m_\mu} \mathrm{tr}[\not{p}_2 \gamma_\lambda (1 - \gamma_5)(\not{p}_\mu + m_\mu)(1 + \gamma_5 \not{s}_\mu) \gamma_\sigma (1 - \gamma_5)] \tag{3.11}$$
$$T_2 = \frac{1}{8 m_e m_{\nu_e}} \mathrm{tr}[(\not{p}_e + m_e)(1 + \gamma_5 \not{s}_e) \gamma^\lambda (1 - \gamma_5) \not{p}_1 \gamma^\sigma (1 - \gamma_5)] \tag{3.12}$$
Now we use the following results to simplify T_1 and T_2:
  (i) The trace of a product of an odd number of γ matrices is zero.
  (ii) $\gamma_5^2 = 1$, $\gamma_5(1 - \gamma_5) = -(1 - \gamma_5)$, and $(1 - \gamma_5)^2 = 2(1 - \gamma_5)$.
  (iii) $\gamma_5 \gamma_\beta = -\gamma_\beta \gamma_5$ for any $\beta = 0, 1, 2, 3$.
Thus we find that
$$T_1 = \frac{1}{4 m_{\nu_\mu} m_\mu} \mathrm{tr}[\not{p}_2 \gamma_\lambda (\not{p}_\mu - m_\mu \not{s}_\mu) \gamma_\sigma (1 - \gamma_5)] \tag{3.13}$$
and

---

[1] We assume that the reader is familiar with the standard trace techniques for evaluating squares of matrix elements. These are summarized in Appendix B. The Dirac equation and Dirac algebra are summarized in Appendix A.

3.2 Muon decay and the V–A law

$$T_2 = \frac{1}{4m_e m_{\nu_e}} \text{tr}[(\not{p}_e - m_e \not{s}_e)\gamma^\lambda \not{p}_1 \gamma^\sigma (1 - \gamma_5)] \quad (3.14)$$

In order to evaluate the product $T_1 T_2$ we make use of Table 3.1, which summarizes some properties of γ matrices used repeatedly in this and succeeding chapters. We employ formula (e) of Table 3.1 to arrive at the result:

$$|\mathcal{M}|^2 = \tfrac{1}{2} G_F^2 T_1 T_2 = 2 G_F^2 (p_e - m_e s_e) \cdot p_2 (p_\mu - m_\mu s_\mu) \cdot p_1 \quad (3.15)$$

The neutrinos are undetected in all experiments to date in muon decay. Thus we must integrate $d\Gamma$ over the neutrino momenta in order to obtain an expression that can be compared with experiment. For this purpose, we insert (3.15) in the differential transition rate $d\Gamma$ (3.3) and integrate over $d^3\mathbf{p}_1 \, d^3\mathbf{p}_2$:

$$d\Gamma = \frac{2 G_F^2}{(2\pi)^5} \frac{d^3 \mathbf{p}_e}{E_e E_\mu} (p_e - m_e s_e)^\alpha (p_\mu - m_\mu s_\mu)^\beta I_{\alpha\beta} \quad (3.16)$$

where

$$I_{\alpha\beta} = \iint \delta^4(p_1 + p_2 - k) p_{2\alpha} p_{1\beta} \, d^3\mathbf{p}_1 \, d^3\mathbf{p}_2 / E_1 E_2 \quad (3.17)$$

and

$$k = p_\mu - p_e = p_1 + p_2$$

The second-rank tensor $I_{\alpha\beta}$ may be expressed quite generally as

$$I_{\alpha\beta} = A k^2 g_{\alpha\beta} + B k_\alpha k_\beta \quad (3.18)$$

with numerical coefficients A and B. To find the latter, we multiply both sides of (3.18) by $g^{\alpha\beta}$ and $k^\alpha k^\beta$ to obtain

$$g^{\alpha\beta} I_{\alpha\beta} = 4 A k^2 + B k^2 \quad (3.19)$$

$$k^\alpha k^\beta I_{\alpha\beta} = A k^4 + B k^4 \quad (3.20)$$

From (3.17) it is also evident that

Table 3.1. *Useful properties of γ matrices*

(a) $\text{tr } \gamma_\mu \gamma_\nu = 4 g_{\mu\nu}$
(b) $\text{tr } \gamma_\alpha \gamma_\beta \gamma_\rho \gamma_\sigma = 4(g_{\alpha\beta} g_{\rho\sigma} - g_{\alpha\rho} g_{\beta\sigma} + g_{\alpha\sigma} g_{\beta\rho})$
(c) $\text{tr } \gamma_\alpha \gamma_\beta \gamma_\rho \gamma_\sigma \gamma_5 = -4i \epsilon_{\alpha\beta\rho\sigma}$
(d) $\text{tr}[\gamma_\alpha \gamma_\beta \gamma_\rho \gamma_\sigma (c_1 - c_2 \gamma_5)] \, \text{tr}[\gamma^\theta \gamma^\beta \gamma^\phi \gamma^\sigma (c_3 - c_4 \gamma_5)] = 32(c_1 c_3)(g^\theta_\alpha g^\phi_\rho + g^\phi_\alpha g^\theta_\rho) + 32 c_2 c_4 (g^\theta_\alpha g^\phi_\rho - g^\phi_\alpha g^\theta_\rho)$
where c_1, c_2, c_3, and c_4 are constants.
(e) For $c_1 = c_2 = c_3 = c_4 = 1$, $\text{tr}[\gamma_\alpha \gamma_\beta \gamma_\rho \gamma_\sigma (1 - \gamma_5)] \, \text{tr}[\gamma^\theta \gamma^\beta \gamma^\phi \gamma^\sigma (1 - \gamma_5)] = 64 g^\theta_\alpha g^\phi_\rho$

$$g^{\alpha\beta}I_{\alpha\beta} = \iint \frac{\delta^4(p_1 + p_2 - k)p_1 \cdot p_2}{E_1 E_2} d^3\mathbf{p}_1 d^3\mathbf{p}_2 \tag{3.21}$$

$$k^\alpha k^\beta I_{\alpha\beta} = \iint \frac{\delta^4(p_1 + p_2 - k)(p_1 \cdot p_2)^2}{E_1 E_2} d^3\mathbf{p}_1 d^3\mathbf{p}_2 \tag{3.22}$$

Since $g^{\alpha\beta}I_{\alpha\beta}$ and $k^\alpha k^\beta I_{\alpha\beta}$ are invariants, we are at liberty to choose any coordinate system in which to evaluate integrals (3.21) and (3.22). For convenience, we choose that frame for which $\mathbf{p}_1 = -\mathbf{p}_2$ [thus for which $k = (k_0, 0)$]. Now

$$\int \delta^3(\mathbf{p}_1 + \mathbf{p}_2 - \mathbf{k}) d^3\mathbf{p}_2 = 1$$

Thus (3.21) and (3.22) become, respectively,

$$g^{\alpha\beta}I_{\alpha\beta} = 8\pi \int E_1^2 \, dE_1 \, \delta(2E_1 - k_0) = \pi k^2 \tag{3.23}$$

and

$$k^\alpha k^\beta I_{\alpha\beta} = 16\pi \int E_1^4 \, dE_1 \, \delta(2E_1 - k_0) = \tfrac{1}{2}\pi k^4 \tag{3.24}$$

Now, inserting (3.23) and (3.24) into (3.19) and (3.20), we find

$$g^{\alpha\beta}I_{\alpha\beta} = \pi k^2 = 4Ak^2 + Bk^2 \tag{3.25}$$

$$k^\alpha k^\beta I_{\alpha\beta} = \tfrac{1}{2}\pi k^4 = Ak^4 + Bk^4 \tag{3.26}$$

which yield $A = \pi/6$, $B = \pi/3$. Thus finally we have

$$I_{\alpha\beta} = \tfrac{1}{6}\pi(k^2 g_{\alpha\beta} + 2k_\alpha k_\beta) \tag{3.27}$$

Now, inserting this into (3.16), we obtain

$$d\Gamma = \frac{\pi}{3}\frac{G_F^2}{(2\pi)^5}\frac{d^3\mathbf{p}_e}{E_e E_\mu}(p_e - |m_e s_e)^\alpha(p_\mu - m_\mu s_\mu)^\beta (k^2 g_{\alpha\beta} + 2k_\alpha k_\beta) \tag{3.28}$$

In order to complete the calculation, we now go to the muon rest frame, where

$$k = (k_0, \mathbf{k}), \quad k_0 = m_\mu - E_e, \quad \mathbf{k} = -\mathbf{p}_e$$

Also, $s_\mu = (0, \hat{\mathbf{s}})$ and $s_{e0} = \mathbf{p} \cdot \hat{\mathbf{s}}_e/m_e$, and $\mathbf{s} = \hat{\mathbf{s}}_e + [(\mathbf{p} \cdot \hat{\mathbf{s}}_e)\mathbf{p}/m_e(E_e + m_e)]$, where $\hat{\mathbf{s}}_e$ is the electron polarization in the electron rest frame. With these substitutions, (3.28) becomes

$$d\Gamma = \frac{\pi G_F^2}{3(2\pi)^5 m_\mu} d\Omega_e |\mathbf{p}_e|\, dE_e \left\{(m_\mu^2 + E_e^2 - 2m_\mu E_e - \mathbf{p}_e^2)\right.$$

$$\times \left[(E_e - \mathbf{p}_e \cdot \hat{\mathbf{s}}_e)m_\mu + m_\mu\left(\mathbf{p}_e - m_e \hat{\mathbf{s}}_e - \frac{\mathbf{p}_e \cdot \hat{\mathbf{s}}_e}{E_e + m_e}\mathbf{p}_e\right) \cdot \hat{\mathbf{s}}_\mu\right]$$

$$+ 2\left[(E_e - \mathbf{p}_e \cdot \hat{\mathbf{s}}_e)(m_\mu - E_e)\right.$$

$$+ \left.\left(\mathbf{p}_e - m_e \hat{\mathbf{s}}_e - \frac{\mathbf{p}_e \cdot \hat{\mathbf{s}}_e}{E_e + m_e}\mathbf{p}_e\right) \cdot \mathbf{p}_e\right]$$

$$\left.\times [m^2 - m_\mu E_e - m_\mu \hat{\mathbf{s}}_\mu \cdot \mathbf{p}_e]\right\} \tag{3.29}$$

3.2 Muon decay and the V–A law

Since $m_\mu = 206 m_e$, we may safely neglect the electron rest mass in first approximation. Thus we have

$$\mathbf{p}_e = E_e \hat{n}$$

where $\hat{n}$ is a unit vector in the direction of electron motion in the muon rest frame. Therefore $d\Gamma$ becomes

$$d\Gamma = \frac{\pi G_F^2}{3(2\pi)^5 m_\mu} d\Omega_e |\mathbf{p}_e| dE_e (1 - \hat{n} \cdot \hat{s}_e) E_e m_\mu^2 [(3m_\mu - 4E_e) + (m_\mu - 4E_e)\hat{n} \cdot \hat{s}_\mu] \quad (3.30)$$

The maximum electron energy and momentum occur when the two neutrinos are emitted in one direction and the electron is emitted in the opposite direction. In this case we obtain

$$E_{e,\max} = (m_\mu^2 + m_e^2)/2m_\mu \quad \text{and} \quad |\mathbf{p}_{e,\max}| = (m_\mu^2 - m_e^2)/2m_\mu$$

Neglecting m_e, we obtain

$$E_{e,\max} = m_\mu/2$$

We write $\epsilon = E_e/E_{e,\max}$, substitute $\epsilon m_\mu/2$ for E_e, and choose the muon spin axis along the z direction in polar coordinates. Thus we obtain for μ^- decay:

$$d\Gamma = \frac{G_F^2 m_\mu^5}{3 \cdot 2^6 \pi^3} [2\epsilon^2(3 - 2\epsilon)] \left[1 + \left(\frac{1 - 2\epsilon}{3 - 2\epsilon}\right) \cos\theta\right]$$

$$\times \left[\frac{1 - \hat{n} \cdot \hat{s}_e}{2}\right] d\epsilon \frac{\sin\theta\, d\theta\, d\phi}{4\pi} \quad (3.31)$$

On the right-hand side of (3.31), the first factor in brackets is the normalized electron energy spectrum $n(\epsilon) = 2\epsilon^2(3 - 2\epsilon)$. The second factor in brackets describes the asymmetry in electron emission with respect to the muon polarization axis. For μ^+ decay, it is easy to show that the asymmetry factor is $1 - (1 - 2\epsilon)\cos\theta/(3 - 2\epsilon)$ instead. The third factor in brackets describes the electron helicity h; we see that $h(e^-) = -1$ independent of electron energy, in accord with the V–A law and the approximation $m_e = 0$. (In μ^+ decay, of course, we have $h(e^+) = +1$ in the same approximation.) Taking into account that $h(e^-) = h(\nu_\mu) = -1$ and $h(\bar{\nu}_e) = +1$, it is easy to see why the asymmetry coefficient is $\alpha = -1$ for maximum electron energy $\epsilon = 1$ (Figure 3.2).

We obtain the total transition rate for unpolarized muons by averaging over muon spin, summing over electron spin, integrating over electron energy, and integrating over solid angle. In these operations, each bracketed factor in (3.31) contributes a factor of unity, so

98   3 *Leptonic weak interactions*

Figure 3.2. Schematic diagram illustrating μ^- decay for $\epsilon = 1$. Neutrinos have equal momentum and opposite spin. Therefore, the electron must have spin in the same direction as that of the muon and must therefore be emitted in the opposite direction.

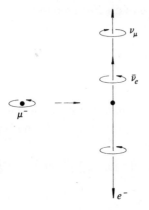

the final result for the total decay rate of unpolarized muons, still unmodified by radiative corrections, is

$$\Gamma = G_F^2 m_\mu^5 / 192 \pi^3 \tag{3.32}$$

This formula is of general importance and will be employed repeatedly in discussions of other three-body decays.

3.3   A general muon-decay amplitude: Michel parameters

Before we proceed to apply radiative corrections to result (3.31) and compare it with experiments, it is appropriate to write a very general decay amplitude and ask to what extent experimental results restrict it to the V–A form. In constructing this general amplitude, we shall continue to assume that it is invariant under proper Lorentz transformations, is linear in the four Dirac spinors (or their conjugates), and contains no derivatives of these spinors. Also, we shall assume lepton conservation; otherwise our discussion would be considerably more complicated (Pauli 57). Then, as in β decay (see Sections 1.3 and 5.4), the most general amplitude is

$$\mathcal{M} = \frac{G_F}{\sqrt{2}} \sum_i \bar{u}_e (A_i + A_i' \gamma_5) O_i v_1 \bar{u}_2 O_i \bar{u}_\mu \tag{3.33}$$

where the O_i are the usual Dirac matrices and the index i runs through S, V, T, A, and P. The coupling constants A_i and A_i' must be determined by experiment. Since parity is not conserved, both "even" (A_i)

3.3 A general muon-decay amplitude

and "odd" (A_i') coefficients appear, and these can be complex if time reversal invariance is not assumed a priori (see Section 4.8). Therefore, taking into account that an overall phase is arbitrary, we conclude that 19 real parameters must be determined by experiment.

It would in fact appear at first that the amplitude of (3.33) is not unique, because we can still satisfy the stated assumptions with a new amplitude in which two of the spinors are permuted, for example,

$$\mathcal{M} = \frac{G_F}{\sqrt{2}} \sum \bar{u}_e O_i u_\mu \bar{u}_2 O_i (C_i + C_i' \gamma_5) v_1 \qquad (3.34)$$

However, it was shown by Fierz (37) that coefficients C_i and C_i' may be expressed as linear combinations of A_i and A_i', so that the two forms are equivalent. In what follows we shall employ form (3.34) (the so-called charge-retention ordering), since it leads to relatively convenient expressions.

We now define certain useful combinations of the coupling constants called *Michel parameters* (Michel 50, Bouchiat and Michel 57) as follows:

$$\rho = (3g_A^2 + 3g_V^2 + 6g_T^2)/D \qquad (3.35)$$

$$\eta = (g_S^2 - g_P^2 + 2g_A^2 - 2g_V^2)/D \qquad (3.36)$$

$$\xi = (6g_S g_P \cos \phi_{SP} - 8g_A g_V \cos \phi_{AV} + 14g_T^2 \cos \phi_{TT})/D \qquad (3.37)$$

$$\delta = (-6g_A g_V \cos \phi_{AV} + 6g_T^2 \cos \phi_{TT})/D\xi \qquad (3.38)$$

$$h = (2g_S g_P \cos \phi_{SP} - 8g_A g_V \cos \phi_{AV} - 6g_T^2 \cos \phi_{TT})/D \quad (3.39)$$

where

$$D = g_S^2 + g_P^2 + 4g_V^2 + 6g_T^2 + 4g_A^2 \qquad (3.40)$$

$$g_i^2 = |C_i|^2 + |C_i'|^2 \qquad (3.41)$$

and

$$\cos \phi_{ij} = \text{Re}(C_i^* C_j' + C_i' C_j^*) \qquad (3.42)$$

It is then convenient to separate the ensuing discussion into two parts, each referring to a different class of experiments. In the first class, the electron helicity is observed in the decay of unpolarized muons. Here, the electron polarization is detected over the entire energy spectrum, corrections for finite electron mass are unimportant because of the limited accuracy of the measurements, and radiative corrections are also negligible. In the second class, the electron spectrum, including the asymmetry, is observed in the decay of polarized or unpolarized muons. Measurements of the electron spectrum, at least at the low-energy end, must take into account the finite electron mass, and radiative corrections are important.

First, we present an expression derived from (3.34) that is useful for discussing the electron helicity. Assuming $m_e = 0$, one may show that the transition probability per unit time averaged over muon spin and integrated over electron energies is

$$d\Gamma = \frac{G_F^2 m_\mu^5}{3 \times 2^{10} \pi^3} \frac{\sin\theta \, d\theta \, d\phi}{4\pi} D[1 + h\hat{\mathbf{n}} \cdot \hat{\mathbf{s}}_e] \tag{3.43}$$

Next we write a formula for the transition rate for *polarized* muons, summed over final e^- spins, in which we take into account finite electron mass and radiative corrections. We find (Michel 50, Bouchiat and Michel 57) that

$$d\Gamma = \frac{D}{16} \frac{G_F^2 m_\mu^5}{192\pi^3} \frac{\sin\theta \, d\theta \, d\phi}{4\pi}$$

$$\times \left\{ \frac{1 + h(x)}{1 + 4\eta(m_e/m_\mu)} \left[12(1-x) + \frac{4}{3}\rho(8x - 6) \right. \right.$$

$$\left. + 24 \frac{m_e}{m_\mu} \frac{(1-x)}{x} \eta \right]$$

$$\left. + \xi \cos\theta \left[4(1-x) + \frac{4}{3}\delta(8x - 6) + \frac{\alpha}{2\pi} \frac{g(x)}{x^2} \right] \right\} x^2 \, dx \tag{3.44}$$

Here $x = |\mathbf{p}_e|/|\mathbf{p}_{e,\text{max}}|$ and $h(x)$ and $g(x)$ are radiative corrections.[2,3] The quantities ρ and δ characterize the shape of the isotropic and anisotropic spectra at the high energy end, ξ characterizes the asymmetry, and η characterizes the low-energy end of the isotropic spectrum.

For $C_S = C_S' = C_T = C_T' = C_P = C_P' = 0$ (in other words, for an amplitude with only V and A components), one finds

$$D = 4(g_V^2 + g_A^2) \tag{3.45}$$

$$\rho = \tfrac{3}{4} \tag{3.46}$$

$$\eta = \frac{1}{2} \frac{|C_V|^2 - |C_A|^2}{|C_V|^2 + |C_A|^2} \tag{3.47}$$

$$\xi = -\frac{2g_A g_V \cos\phi_{AV}}{g_A^2 + g_V^2} \tag{3.48}$$

$$\delta = \tfrac{3}{4} \tag{3.49}$$

---

2 The radiative correction $h(x)$ is not to be confused with h, the helicity.
3 See Behrends et al. 56, Kinoshita and Sirlin 57a–d, Kinoshita 59, Berman 58, Kuznetsov 60,61, Berman and Sirlin 62, Sirlin 78.

Figure 3.3. Lowest-order radiative corrections to muon decay: (a)–(c) virtual photon diagrams; (d) and (e) inner *bremsstrahlung* diagrams.

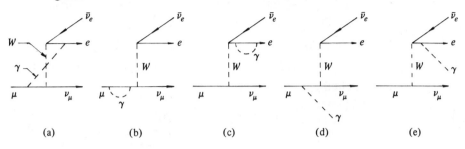

If, in addition, we impose $C_i = -C'_i$ (which is a result of the electron helicity measurements), then

$$\xi = \frac{C_V C_A^* + C_A C_V^*}{|C_V|^2 + |C_A|^2} \tag{3.50}$$

Finally, if the V–A law holds ($C_A = -C_V$), we have

$$D = 16|C_V|^2, \quad \rho = \delta = \tfrac{3}{4}, \quad \eta = 0, \quad \xi = -1 \tag{3.51}$$

In the limit of zero mass and negligible radiative corrections, (3.44) with values (3.51) and $C_V = 1$ reduces to our original V–A transition probability, (3.31). The lowest-order radiative corrections, unambiguous only for V and A amplitudes, are illustrated in Figure 3.3. Although inner *bremsstrahlung* is very important for the low-energy part of the electron spectrum, the virtual photon diagrams are relatively more important at higher energies. Including these effects and taking finite electron mass into account, one obtains the corrected V–A total decay rate:

$$\tau^{-1} = \Gamma^{\text{corr}} = \frac{G_F m_\mu^5}{192\pi^3}\left[1 - \frac{\alpha}{2\pi}\left(\pi^2 - \frac{25}{4}\right)\right] f\left(\frac{m_e^2}{m_\mu^2}\right) \tag{3.52}$$

where $f(x) = 1 - 8x + 8x^3 - x^4 + 12x^2 \ln(1/x)$. It can be shown by explicit computation that all divergences cancel in higher-order corrections to decay, when Z^0 effects are also included according to the standard model (Rajasekaran 72).

3.4 Experimental determination of muon parameters

3.4.1 *Muon magnetic moment, mass, and lifetime*

We begin our discussion of muon experiments with consideration of the muon magnetic moment and g value. These quantities are important because they are related to the mass and lifetime and, thus, to a determination of G_F. According to Dirac's theory of relativistic quantum mechanics, the magnetic moment of an electron or muon should be given by the formula

$$\mu_l = ge\hbar/2m_lc \tag{3.53}$$

where $g = 2$. Actually, g is not exactly 2 because of radiative effects; instead, defining $a = \frac{1}{2}(g - 2)$, we have

$$a_{\mu,\text{theor}} = \frac{\alpha}{2\pi} + 0.76578 \frac{\alpha^2}{\pi^2} + O(\alpha^3) \tag{3.54}$$

$$a_{e,\text{theor}} = \frac{\alpha}{2\pi} - 0.328478 \frac{\alpha^2}{\pi^2} + O(\alpha^3) \tag{3.55}$$

The anomalies a_μ and a_e (which are not equal because of differing vacuum polarization contributions in order α^2, α^3, . . .) are of the most fundamental interest in quantum electrodynamics. Anomaly a_e has been determined by extraordinarily refined experiments (Van Dyck et al. 77) in which a single electron is confined for hours in a trapping electromagnetic field. One finds:

$$a_{e,\text{expt}} = (1{,}159{,}652{,}410 \pm 200) \times 10^{-12} \tag{3.56}$$

in good agreement with the theoretical value. Anomaly a_μ is determined for μ^+ and μ^- in the elegant CERN storage ring experiment (Bailey et al. 77,79; Farley and Picasso 79). Here an external pion beam is deflected into a circular region with a magnetic field (see Figure 3.4). Pions revolve in the ring and decay in flight, giving rise to muons that acquire stable orbits and are stored for the duration of their lifetimes. Although the muons are longitudinally polarized at first, the spin precesses faster than the momentum by a factor $1 + (e\bar{B}/m_\mu c)a_\mu$ in the magnetic field, which has average value $\bar{B}$. Thus one can determine a_μ by measuring the precession difference frequency from observations of the asymmetry in muon decay. An experimental curve is shown in Figure 3.5. The results of this beautiful experiment are

$$a(\mu^-) = (11{,}659.37 \pm 0.12) \times 10^{-7} \tag{3.57}$$

$$a(\mu^+) = (11{,}659.11 \pm 0.11) \times 10^{-7} \tag{3.58}$$

which are in good agreement with the theoretical formula (3.54) and thus provide a remarkable confirmation of quantum electrodynamics. The equality of $a(\mu^+)$ and $a(\mu^-)$ (within two standard deviations) is also a test of *CPT* invariance, according to which the g values of particle and antiparticle must be equal.

Figure 3.4. Muon $g - 2$ experiment: (a) plan view of muon storage ring; (b) a cross section through 1 of the 40 magnets. For most of the data, 22 electron detectors were used. Diameter = 14 m, muon momentum = 3.094 GeV/c, time dilation factor = 29.3. (From Farley and Picasso 79. Reprinted with permission.)

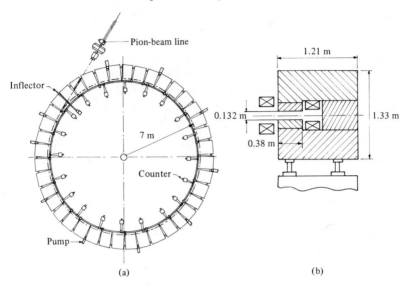

Figure 3.5. Muon $g - 2$ experiment. Decay electron counts versus time (in microseconds) after injection. Range of time for each line is shown on the right (in microseconds). (From Farley and Picasso 79. Reprinted with permission.)

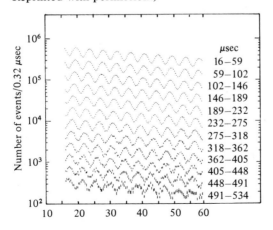

The same experiment also yields a precise test of the time-dilation effect of special relativity. Alternatively, if special relativity is assumed to be valid, one obtains the most precise values of the muon mean lifetime:

$$\tau(\mu^-) = (2.1948 \pm 0.0010) \times 10^{-6} \text{ sec} \quad (3.59)$$

$$\tau(\mu^+) = (2.1966 \pm 0.0020) \times 10^{-6} \text{ sec} \quad (3.60)$$

The magnetic moment of the muon is determined in experiments of the Garwin–Lederman type. Here muons stop in a nondepolarizing target (e.g., bromine) immersed in a magnetic field perpendicular to the muon spin. The muons then precess in the plane normal to B at the Larmor frequency. If a fixed counter is located in this plane and gated on by the arrival of the muon, then the distribution of electron events from muon decay is

$$e^{-t/\tau}[1 - A \cos(g_\mu eB/2m_\mu c)t]$$

where A is the experimental asymmetry. The magnetic field is determined by proton resonance, and the result is given as the muon/proton magnetic moment ratio (Camani et al. 78):

$$\mu(\mu^+)/\mu(p) = 3.1833448(29) \quad (\pm 0.9 \, ppm) \quad (3.61)$$

Alternatively, one can observe Zeeman transitions in the ground state of muonium (μ^+e^-) at strong magnetic field to measure $\mu(\mu^-)/\mu(p)$ (Casperson et al. 77). The determination of $g - 2$, with measurements of $\mu(\mu^+)$, gives the most accurate values of muon mass, which we express in terms of electron mass as follows:

$$m_\mu/m_e = 206.76859(29) \quad (3.62)$$

or directly as

$$m_\mu = 105.65945 \pm 0.00033 \text{ MeV}/c^2 \quad (3.63)$$

Results (3.63) and (3.59) or (3.60) may be combined in (3.52) to yield a precise value of Fermi's coupling constant (Shrock and Wang 78):

$$\begin{aligned} G_F &= (1.16632 \pm 0.00004) \times 10^{-5} \text{ GeV}^{-2} c^4 \\ &= (1.43582 \pm 0.00004) \times 10^{-49} \text{ erg cm}^3 \end{aligned} \quad (3.64)$$

3.4.2 e± helicities in muon decay

The quantity $h(e^-)$ has been determined by observation of Møller (e^-e^-) scattering of e^- incident on polarized target electrons in a magnetized iron foil (Schwartz 67). The cross sections for this process and the related Bhabha (e^+e^-) scattering are spin-dependent (Bincer 57). The helicity $h(e^+)$ has been observed not only by Bhabha

3.4 Experimental determination of muon parameters

scattering, which gives precise results (Duclos et al. 64, Egger et al. 80), but also by e^+e^- annihilation in a magnetized iron target (Buhler et al. 63) and by circular polarization of *bremsstrahlung* from polarized e^+ (Bloom et al. 64). The results, from Schwartz (67) and Egger et al. (80), respectively, in agreement with the V–A law, are:

$$h(e^-) = -0.89 \pm 0.28 \tag{3.65}$$

$$h(e^+) = +0.94 \pm 0.08 \tag{3.66}$$

Since opposite helicities are expected, these results are consistent and may be combined to yield

$$|h(e^\pm)| = 0.94 \pm 0.08 \tag{3.67}$$

3.4.3 Momentum spectrum of electrons in decay of unpolarized muons: Measurement of Michel parameters

Many experiments have been carried out to measure ρ (Plano 60, Bardon et al. 65, Sherwood 67, Fryberger 68). A very precise result was obtained with the apparatus shown in Figure 3.6 (Bardon et al. 65), yielding the results shown in Figure 3.7. The best value of ρ, based on this and other observations, is

Figure 3.6. Apparatus used to determine ρ. The entire setup is in a homogeneous magnetic field. The two circular trajectories have been drawn for the minimum and maximum positron momenta accepted at a field of 6.62 kG. The two 0.003-in. Mylar windows for each chamber, which are not shown, are the only other material in the path of the positrons. (From Bardon et al. 65. Reprinted with permission.)

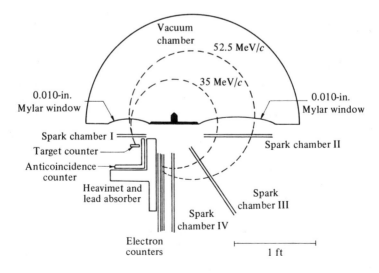

3 Leptonic weak interactions

$$\rho = 0.7518 \pm 0.0026 \tag{3.68}$$

The quantity η, which should be zero according to the V–A law, was measured by stopping π^+ and μ^+ in a bubble chamber in a strong magnetic field. Over 400,000 exposures were taken, including 2,070,000 μ^+ decays. Momentum measurements calibrated with internal conversion e^- tracks of known momentum from a radioactive source inside the bubble chamber yielded (Derenzo 69):

$$\eta = -0.12 \pm 0.21 \tag{3.69}$$

3.4.4 Asymmetry in the decay of polarized muons

The parameter ξ should take the value -1 according to the V–A law, and experimental measurement of ξ is important because observation of departure of ξ from -1 would signify the existence of right-handed (V + A) currents, as expected in left–right symmetric models (see Section 2.12). The experimental problem is very difficult because one actually observes the product $P\xi$, where P is the polariza-

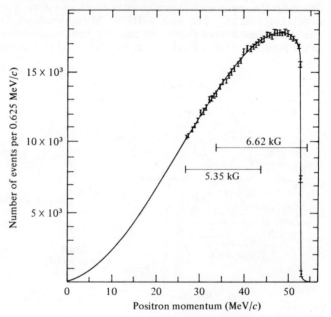

Figure 3.7. Results of experiments to determine ρ. Experimental points are plotted, together with a theoretical curve for $\rho = \frac{3}{4}$ corrected for radiative effects and ionization loss. (From Bardon et al. 65. Reprinted with permission.)

tion of the muons. This quantity is inevitably somewhat less than unity because of miscellaneous muon depolarization mechanisms, not all of which can easily be accounted for in a given experiment. The integrated asymmetry $a = \xi/3$ was determined by observing e^+ from μ^+ stopping in photoemulsion in a magnetic field of 140,000 G, parallel to the muon spin. The result (Gurevich et al. 64):

$$\xi = -0.972 \pm 0.013 \tag{3.70}$$

was found, in accord with the V–A law, but with a precision insufficient to constrain left–right symmetric models very severely (see Section 2.12). In a measurement of δ, the μ^+ with polarization P are stopped in the magnetic field of a wire spark-chamber spectrometer, which is used for both momentum analysis and precession. The parameters δ and P are determined by fitting predicted asymmetry functions to experimental results. The best value of δ from this experiment (Fryberger 68) and others is

$$\delta = 0.7551 \pm 0.0085 \tag{3.71}$$

3.4.5 Summary

Table 3.2 presents the various muon parameters and compares them with predictions of the V–A theory. Obviously, the agreement is

Table 3.2. *Properties of the muon and decay parameters*

Parameter	Value	
Muon mass m_μ	105.65945 ± 0.00033 MeV/c^2	
m_μ/m_e	206.76859(29)	
ν_μ mass m_{ν_μ}	≤ 0.57 MeV/c^2 *	
Maximum electron energy in muon decay assuming $m_{\nu e} = m_{\nu \mu} = 0$	52.827 MeV	
Muon mean life	$(2.1948 \pm 0.0010) \times 10^{-6}$ sec	
	Results	
Michel parameters	Experimental	V–A
ρ	0.7518 ± 0.0026	$\tfrac{3}{4}$
η	-0.12 ± 0.21	0
ξ	-0.972 ± 0.0013	-1
δ	0.7551 ± 0.0085	$\tfrac{3}{4}$
$h(e^-)$	$(-)0.94 \pm 0.08$	-1

* At 90% confidence limit (from Daum et al. 79).

very satisfactory, but there is considerable latitude in the choice of coupling constants for the more general amplitude (3.34). One finds that the following range of values is consistent with experiment (Derenzo 69):

$$g_S/g_V \leq 0.33, \quad g_T/g_V \leq 0.28, \quad g_P/g_S \leq 0.33$$
$$0.76 \leq g_A/g_V \leq 1.20, \quad \phi_{AV} = 180° \pm 15° \quad (3.72)$$

Thus, we see that the muon-decay experiments alone cannot exclude appreciable S, T, and P couplings and there is considerable uncertainty in the ratio g_A/g_V. This is not so surprising in view of the fact that the amplitude (3.34) contains 19 independent real parameters a priori, whereas there are only 6 independent determinations if the outgoing neutrino(s) are not observed. On the other hand, the outgoing ν_μ in μ^- decay can be replaced by an incoming ν_μ in inverse muon decay, observation of which is important in confirming the V–A law (see Section 3.10).

3.5 τ leptons

τ leptons are produced in e^+e^- collisions via one-photon exchange (Figure 3.8) and were first observed in 1975 at the Stanford Positron–Electron Asymmetric Ring (SPEAR) (Perl et al. 75). Since then, observations of τ leptons have also been carried out at other e^+e^- colliding-beam facilities. The results, though not as precise as for muon decay, are consistent with the standard description of ν_τ and τ as "sequential leptons," which form a third lepton generation similar to the first two (electron and muon) doublets (see the review by Perl 80).

Let us first calculate the cross section for $e^+e^- \to \tau^+\tau^-$ by single-photon exchange. Although this is an electromagnetic, not a weak, process, this calculation and its result play an important role in our subsequent discussions. The differential cross section is given by the "golden rule" formula (see Appendix B):

Figure 3.8. Production of τ leptons by one-photon exchange.

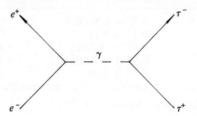

3.5 τ leptons

$$d\sigma = \frac{(2\pi)^4 \,\delta^4(p'_+ + p'_- - p_+ - p_-)}{[(p_+ \cdot p_-)^2 - m_e^4]^{1/2}} \frac{m_e^2}{E'_+} \frac{m_\tau^2}{E'_-} \frac{d^3\mathbf{p}'_+}{(2\pi)^3} \frac{d^3\mathbf{p}'_-}{(2\pi)^3} |\mathcal{M}|^2 \quad (3.73)$$

where p_+, p_-, p'_+, and p'_- are the 4-momenta of e^+, e^-, τ^+, and τ^- described by spinors v, u, v', and u', respectively and we assume that τ leptons possess spin $\frac{1}{2}$. From (2.162), the invariant amplitude is

$$\mathcal{M} = (-ie\bar{u}'\gamma_\lambda v') \frac{g^{\lambda\sigma}}{q^2} (-ie\bar{v}\gamma_\sigma u) \quad (3.74)$$

where

$$q = p_- + p_+ = p'_- + p'_+$$

In the center-of-momentum (CM) frame, which we employ from now on, $q^2 = 4E^2 = s$, where $E = E_- = E_+ = E'_- = E'_+$ and $\sqrt{s} = 2E$ is the total energy. Also, $p_+ \cdot p_- = 2E^2$. Thus,

$$\mathcal{M} = -\frac{4\pi\alpha}{s} \bar{u}'\gamma_\lambda v' \bar{v}\gamma^\lambda u \quad (3.75)$$

We shall be interested in $|\mathcal{M}|^2$ averaged over e^+ and e^- spins and summed over τ^+ and τ^- spins. This we call $|\bar{\mathcal{M}}|^2$:

$$|\bar{\mathcal{M}}|^2 = \frac{(4\pi\alpha)^2}{s^2} T_1 T_2 \quad (3.76)$$

where

$$T_1 = \frac{1}{4m_\tau^2} \text{tr}[(\not{p}'_- - m_\tau)\gamma_\lambda(\not{p}'_+ - m_\tau)\gamma_\sigma] \quad (3.77)$$

and

$$T_2 = \frac{1}{16m_e^2} \text{tr}[(\not{p}_+ - m_e)\gamma^\lambda(\not{p}_- + m_e)\gamma^\sigma] \simeq \frac{1}{16m_e^2} \text{tr}[\not{p}_+\gamma^\lambda\not{p}_-\gamma^\sigma] \quad (3.78)$$

Employing formulas (a) and (d) of Table 3.1 we obtain

$$|\bar{\mathcal{M}}|^2 = \frac{(4\pi\alpha)^2}{s^2} \frac{1}{64m_\tau^2 m_e^2} 32[p'_- \cdot p_+ p'_+ \cdot p_- \\ + p'_- \cdot p_- p'_+ \cdot p_+ + m_\tau^2 p_+ \cdot p_-] \quad (3.79)$$

Let θ be the angle between $\mathbf{p}_-$ and $\mathbf{p}'_-$. Then, for $E \gg m_e$,

$$p'_- \cdot p_+ = p'_+ \cdot p_- = E^2(1 + \beta \cos\theta) \quad (3.80)$$
$$p'_- \cdot p_- = p'_+ \cdot p_+ = E^2(1 - \beta \cos\theta) \quad (3.81)$$

where β is the speed of τ^+ or τ^-. Making use of these formulas and setting $[(p_+ \cdot p_-)^2 - m_e^4]^{1/2} \simeq 2E^2$, we obtain

$$d\sigma = \frac{\alpha^2}{8E^6}[m_\tau^2 + E^2(1 + \beta^2 \cos^2 \theta)]\delta^4(p'_+ + p'_- - p_+ - p_-)$$
$$\times d^3\mathbf{p}'_+ \, d^3\mathbf{p}'_- \tag{3.82}$$

Now, $m_\tau^2 + E^2(1 + \beta^2 \cos^2 \theta) = E^2(2 - \beta^2 \sin^2 \theta)$. Then, integrating over $d^3\mathbf{p}'_+$ and using $\int \delta^3(\mathbf{p}'_+ + \mathbf{p}'_- - \mathbf{p}_+ - \mathbf{p}_-) \, d^3\mathbf{p}'_+ = 1$ and $d^3\mathbf{p}'_- = \beta E'^2_- \, dE'_- \, d\Omega'_-$ we arrive at

$$d\sigma = \frac{\alpha^2}{8E^4}\beta(2 - \beta^2 \sin^2 \theta)E'^2_- \, dE'_- \, \delta(2E'_- - 2E_-) \, d\Omega'_- \tag{3.83}$$

When we integrate over E'_-, this becomes

$$d\sigma = \frac{\alpha^2}{16E^2}\beta(2 - \beta^2 \sin^2 \theta) \, d\Omega'_- \tag{3.84}$$

Integration over the solid angle yields the total cross section:

$$\sigma = \frac{4\pi\alpha^2}{3s}\beta\left(\frac{3 - \beta^2}{2}\right) \tag{3.85}$$

In the limit as $\beta \to 1$, this is

$$\lim_{\beta \to 1} \sigma = 4\pi\alpha^2/3s \tag{3.86}$$

which is also the cross section for $e^+e^- \to \mu^+\mu^-$ when the energy is much larger than the muon mass. We may therefore define the useful ratio:

$$R_\tau = \sigma/\sigma_{ee \to \mu\mu} = \beta(3 - \beta^2)/2 \tag{3.87}$$

Figure 3.9 shows the experimental values of R_τ obtained near threshold for τ production, together with the curve of (3.87) and theoretical R_τ curves calculated for other possible τ spins. The data show clearly that τ possesses spin $\frac{1}{2}$. Measurements of the threshold serve to determine the mass of τ:

$$m_\tau = 1{,}782 \begin{Bmatrix} +3 \\ -4 \end{Bmatrix} \text{ MeV}/c^2 \tag{3.88}$$

The yield of τ^+ and τ^- versus $s^{1/2}$ for large s may be compared with the theoretical prediction (3.86) as a test of quantum electrodynamics (although Z^0 exchange effects, to be discussed in the next section, must also be considered as a small correction for the range of s considered). The results, shown in Figure 3.10, are in good agreement with quantum electrodynamics (Brandelik et al. 80).

Let us now consider the weak decays of τ leptons. They can decay via leptonic modes:

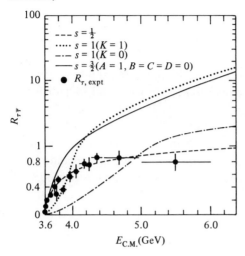

Figure 3.9. Comparison of $R_{\tau,\text{expt}}$ (Kirkby 79) with theoretical curves for pointlike particles with various spins. The constants A, B, C, and D are vertex parameters (Tsai 78). (From Perl 80. Reprinted with permission.)

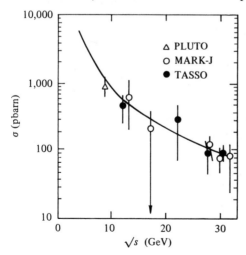

Figure 3.10. The cross section for $e^+e^- \to \tau^+\tau^-$ vs. $\sqrt{s}$ for large s. The curve is the QED prediction. The data have been corrected for radiative effects. (From Brandelik et al. 80. Reprinted with permission.)

$$\tau^- \to e^- \bar{\nu}_e \nu_\tau$$
$$\tau^- \to \mu^- \bar{\nu}_\mu \nu_\tau$$

or via hadronic modes:

$$\tau^- \to (\bar{u}d)\nu_\tau$$
$$\phantom{\tau^- \to (\bar{u}d)}\!\!\!\downarrow\!\text{hadron(s)}$$

(Note that τ^- does not have sufficient mass to decay to hadronic states that contain $\bar{c}$, s or $\bar{t}$, b quarks). Since quarks have three colors, the branching ratios for decay to $e\nu_e$, to $\mu\nu_\mu$, and to hadrons should then be approximately $B_e = \frac{1}{5}$, $B_\mu = \frac{1}{5}$, $B_h = \frac{3}{5}$, respectively. More refined estimates (Thacker and Sakurai 71, Tsai 71, Gilman and Miller 78, Kawamoto and Sanda 78, Pham et al. 78) take into account final-particle masses as well as τ masses, as well as specific features of the hadronic weak currents, a discussion of which is postponed to Chapter 4. One thus finds $B_{e,\text{theor}} = 16.4\text{--}18.0\%$, $B_{\mu,\text{theor}} = 16.0\text{--}17.5\%$, in good agreement with $B_{e,\text{expt}} = (16.5 \pm 1.5)\%$ and $B_{\mu,\text{expt}} = (18.6 \pm 1.9)\%$. The mean life of the τ can be calculated from B_e and a formula similar to (3.32):

$$\Gamma(\tau^- \to e^- \bar{\nu}_e \nu_\tau) = G_F^2 m_\tau^5 / 192\pi^3 \tag{3.89}$$

One thus obtains

$$T_{\tau,\text{theor}} = (2.8 \pm 0.2) \times 10^{-13} \text{ sec} \tag{3.90a}$$

The τ lifetime has been measured in an experiment carried out at PEP with the Mark-II detector (Feldman et al. 82). Three-pronged τ-decay events were analyzed and the τ flight distances (0.7 mm on average) were determined. The result is

$$T_{\tau,\text{expt}} = (4.6 \pm 1.9) \times 10^{-13} \text{ sec} \tag{3.90b}$$

in agreement with (3.90a). One might argue that decays of the form $\tau \to l\nu\bar{\nu}$ (where $l = \mu$ or e) could occur by means of a hypothetical $\tau\nu_l$ current coupled to an $l\nu_l$ current. However, the strengths of such couplings are constrained by experimental limits on τ production with ν_μ and ν_e beams (Ushida et al. 81). These results establish that in order to account for the observed τ lifetime (3.90b) one must assume that a third neutrino ν_τ exists.

The momentum spectrum of electrons or muons in τ decay may be used to test the V–A hypothesis, if one assumes $m(\nu_\tau) = 0$. Experiment (Bacino et al. 79) shows that the Michel parameter ρ is given by

$$\rho = 0.72 \pm 0.10 \tag{3.91}$$

in agreement with the V–A law. Alternatively, one may assume the

3.6 Weak–EM interference effects in $e^+e^- \to \mu^+\mu^-$

V–A law and use the data to place a limit on the mass of ν_τ. One then finds that

$$m_{\nu_\tau} < 250 \text{ MeV}/c^2 \tag{3.92}$$

Finally, as in muon decay, certain transitions are forbidden by lepton conservation. One finds the following experimental upper limits on branching ratios:

$$\begin{aligned} \tau^- &\to e^-\gamma, & B &< 2.6\% \\ &\to \mu^-\gamma, & B &< 1.3\% \\ &\to (3l)^-, & B &< 0.6\% \end{aligned}$$

These are consistent with lepton conservation but, of course, are much too crude to provide useful tests of that conservation law.

3.6 Weak–electromagnetic interference effects in $e^+e^- \to \mu^+\mu^-$

In the previous section, we considered the reactions $e^+e^- \to \tau^+\tau^-$ and $e^+e^- \to \mu^+\mu^-$ by single-photon exchange. At sufficiently large s, however, Z^0 exchange must also be considered. Let us therefore take both these possibilities into account in discussing $e^+e^- \to \mu^+\mu^-$. The Feynman diagrams are as shown in Figure 3.11.

The invariant amplitude is

$$\begin{aligned} \mathcal{M} = -\bigg[&\frac{4\pi\alpha}{s} \bar{u}'\gamma_\lambda v' \bar{v}\gamma^\lambda u + \frac{g^2}{16 \cos^2\theta_W} \frac{1}{(s - m_Z^2)} \\ &\times \bar{u}'\gamma_\lambda(1 - 4\sin^2\theta_W - \gamma_5)v'(g^{\lambda\sigma} - q^\lambda q^\sigma/m_Z^2) \\ &\times \bar{v}\gamma_\sigma(1 - 4\sin^2\theta_W - \gamma_5)u \bigg] \end{aligned} \tag{3.93}$$

Figure 3.11. (a) Single-photon exchange; (b) Z^0 exchange.

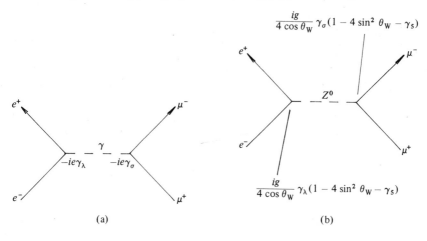

114   3 Leptonic weak interactions

This expression simplifies somewhat because

$$\frac{q^\lambda q^\sigma}{m_Z^2} \bar{v}\gamma_\sigma(1 - 4\sin^2\theta_W - \gamma_5)u$$

$$= \frac{q^\lambda}{m_Z^2} \bar{v}(\slashed{p}_+ + \slashed{p}_-)(1 - 4\sin^2\theta_W - \gamma_5)u$$

$$= \frac{q^\lambda}{m_Z^2} [\bar{v}\slashed{p}_+(1 - 4\sin^2\theta_W - \gamma_5)u$$

$$+ \bar{v}(1 - 4\sin^2\theta_W + \gamma_5)\slashed{p}_- u]$$

Neglecting electron mass, we have $\bar{v}\slashed{p}_+ = \slashed{p}_- u = 0$ from the Dirac equation; therefore, the term $q^\lambda q^\sigma/m_Z^2$ is negligible.

Now, making the following substitutions:

$$g^2/16\cos^2\theta_W = G_F m_Z^2/2\sqrt{2}, \qquad g_V = 2\sin^2\theta_W - \tfrac{1}{2},$$
$$g_A = -\tfrac{1}{2} \qquad (3.94)$$

we find

$$|\mathcal{M}|^2 = A + B + C \qquad (3.95)$$

where

$$A = \frac{e^4}{64\, s^2 m_e^2 m_\mu^2} \operatorname{tr}[\slashed{p}'_-\gamma_\mu \slashed{p}'_+\gamma_\sigma]\operatorname{tr}[\slashed{p}_+\gamma^\mu \slashed{p}_-\gamma^\sigma] \qquad (3.96\text{a})$$

$$B = \frac{e^2}{s}\frac{G_F m_Z^2}{\sqrt{2}(s - m_Z^2)} \frac{1}{32 m_e^2 m_\mu^2}$$
$$\times \{\operatorname{tr}[\slashed{p}'_-\gamma_\mu \slashed{p}'_+\gamma_\sigma(g_V - g_A\gamma_5)]\operatorname{tr}[\slashed{p}_+\gamma^\mu \slashed{p}_-\gamma^\sigma(g_V - g_A\gamma_5)]$$
$$+ \operatorname{tr}[\slashed{p}'_-\gamma_\mu(g_V - g_A\gamma_5)\slashed{p}'_+\gamma_\sigma]\operatorname{tr}[\slashed{p}_+\gamma^\mu(g_V - g_A\gamma_5)\slashed{p}_-\gamma^\sigma]\} \qquad (3.96\text{b})$$

$$C = \frac{G_F^2 m_Z^4}{s - m_Z^2} \frac{1}{32 m_e^2 m_\mu^2} \operatorname{tr}[\slashed{p}'_-\gamma_\mu(g_V - g_A\gamma_5)\slashed{p}'_+\gamma_\sigma(g_V - g_A\gamma_5)]$$
$$\times \operatorname{tr}[\slashed{p}_+\gamma^\mu(g_V - g_A\gamma_5)\slashed{p}_-\gamma^\sigma(g_V - g_A\gamma_5)] \qquad (3.96\text{c})$$

Here A is the pure electromagnetic term analogous to (3.76) (but where we now neglect m_μ compared with $\slashed{p}'_-$ or $\slashed{p}'_+$), C is a pure neutral weak term, and B represents weak–electromagnetic interference. Employing (3.80), (3.81), and Table 3.1 to reduce these terms, we arrive at

$$A = \frac{e^4}{16 m_\mu^2 m_e^2}(1 + \cos^2\theta) \qquad (3.97)$$

$$B = \frac{e^2}{4\sqrt{2}\, m_\mu^2 m_e^2}\frac{G_F m_Z^2 s}{s - m_Z^2}[g_V^2(1 + \cos^2\theta) + 2g_A^2\cos\theta]$$
$$(3.98)$$

3.7 Z^0 decay, $W^\pm$ decay

$$C = \frac{G_F^2 m_Z^4}{8m_\mu^2 m_e^2} \frac{s^2}{(s-m_Z^2)^2}$$
$$\times [(g_V^2 + g_A^2)^2(1 + \cos^2\theta) + 16 g_V^2 g_A^2 \cos\theta] \quad (3.99)$$

For $s \ll m_Z^2$, it is clear that $A \gg B \gg C$. Therefore, ignoring C for the moment, we find, for $s \ll m_Z^2$, that

$$\frac{d\sigma}{d\Omega_-'} = \frac{\alpha^2}{16E^2}(1+\cos^2\theta)\left[\left(1 - \frac{G_F g_V^2}{\pi\alpha\sqrt{2}}s\right) - \frac{G_F}{2\pi\alpha\sqrt{2}}\frac{\cos\theta}{1+\cos^2\theta}s\right] \quad (3.100)$$

It is interesting to compare this expression with that obtained by taking the limit $G_F = 0$ (pure electrodynamics). The first term on the right-hand side of (3.100) reveals that the total cross section for $\mu^+\mu^-$ production is less than that for pure electrodynamics by the factor $1 - (G_F g_V^2/\pi\alpha\sqrt{2})s$. The second term on the right-hand side of (3.100) describes the interesting "charge-asymmetry" effect, in which the probabilities for μ^- and for μ^+ emission at a given angle θ are unequal. It is easy to see that this asymmetry is *not* a parity-violating effect. Preliminary experimental observations of $\mu^+\mu^-$ production at large s yield results in agreement with (3.100) (Adeva et al. 82).

Let us now consider the regime $s \simeq m_Z^2$, which will be investigated with future e^+e^- colliding-beam accelerators. Here, the term C (3.99) becomes dominant, and we are dealing with the Z^0 production resonance. It is necessary to take into account the fact that once Z^0 is produced, it quickly decays. Therefore, the resonance will possess a natural width characterized by the replacement

$$m_Z \to m_Z - \tfrac{1}{2}i\Gamma_T \quad (3.101)$$

where Γ_T is the total decay rate of Z^0.

3.7 Z^0 decay, $W^\pm$ decay

We proceed to calculate Γ_T by considering first the transition $Z^0 \to \nu\bar\nu$. In the Z^0 rest frame, the differential transition probability per unit time is

$$d\Gamma_{\nu\bar\nu} = (2\pi)^4\,\delta^4(p_{\bar\nu}+p_\nu-p_Z)\frac{1}{2m_Z}\frac{m_\nu^2}{E_\nu E_{\bar\nu}}\frac{d^3\mathbf{p}_\nu}{(2\pi)^3}\frac{d^3\mathbf{p}_{\bar\nu}}{(2\pi)^3}|\mathcal{M}|^2 \quad (3.102)$$

where

$$-i\mathcal{M} = \frac{-ig}{4\cos\theta_W}\bar u \slashed{\epsilon}(1-\gamma_5)v \quad (3.103)$$

u and v are Dirac spinors for ν and $\bar\nu$, respectively, and ϵ_μ is the 4-polarization of Z^0. We have

$$|\mathcal{M}|^2 = \frac{g^2}{16m_\nu^2 \cos^2\theta_W} [\not{p}_\nu \not{\epsilon}(1-\gamma_5)\not{p}_{\bar\nu}\not{\epsilon}(1-\gamma_5)] \tag{3.104}$$

$$= \frac{g^2}{8\cos^2\theta_W m_\nu^2} \cdot [2p_{\bar\nu}\cdot\epsilon p_\nu\cdot\epsilon - p_\nu\cdot p_{\bar\nu}\epsilon\cdot\epsilon]$$

In the Z^0 rest frame, $\epsilon\cdot\epsilon = 1$, $p_\nu\cdot p_{\bar\nu} = 2E^2$, and $p_\nu\cdot\epsilon p_{\bar\nu}\cdot\epsilon = -E^2\cos^2\theta_0$, where $E = E_\nu = E_{\bar\nu}$, and θ_0 is the angle between $\mathbf{p}_\nu$ and $\hat{\epsilon}$. Thus,

$$|\mathcal{M}|^2 = \frac{g^2 E^2}{4\cos^2\theta_W m_\nu^2}(1-\cos^2\theta_0)$$

Inserting into (3.102) and integrating over $d^3p_{\bar\nu}$ with $\int \delta^3(\mathbf{p}_\nu + \mathbf{p}_\nu)\, d^3p_{\bar\nu} = 1$, we find

$$d\Gamma_{\nu\bar\nu} = \frac{g^2}{32\pi^2 \cos^2\theta_W m_Z} E^2\, dE\, \delta(2E - m_Z)(1-\cos^2\theta_0)\, d\Omega_\nu$$

Then integrating over E and Ω_ν, we finally obtain

$$\Gamma_{\nu\bar\nu} = G_F m_Z^3/12\sqrt{2}\,\pi \tag{3.105}$$

To calculate $Z \to e^+e^-$, $\mu^+\mu^-$, $\tau^+\tau^-$, we need only change the vertex factor from $(-ig/4\cos\theta_W)\not{\epsilon}(1-\gamma_5)$, appropriate for $\nu\bar\nu$, to $(ig/4\cos\theta_W)\not{\epsilon}(1 - 4\sin^2\theta_W - \gamma_5)$ and carry through the same steps. Similar remarks apply for $Z \to q\bar q$. We then find the following result:

$$\Gamma = \frac{G_F}{24\pi\sqrt{2}} m_Z^3 (a^2 + b^2) \tag{3.106}$$

where a and b are given by the following tabulation ($\sin^2\theta_W = 0.23$):

	a	b	$a^2 + b^2$	
$\nu\bar\nu$	1	1	2	$= 2$
e^+e^-	$1 - 4\sin^2\theta_W$	1	$1 + (1 - 4\sin^2\theta_W)^2$	$= 1.0064$
$u\bar u$	$1 - \frac{8}{3}\sin^2\theta_W$	1	$1 + (1 - \frac{8}{3}\sin^2\theta_W)^2$	$= 1.150$
$d\bar d$	$1 - \frac{4}{3}\sin^2\theta_W$	1	$1 + (1 - \frac{4}{3}\sin^2\theta_W)^2$	$= 1.480$

The total transition rate is given by

$$\Gamma_T = 3\Gamma_{\nu\bar\nu} + 3\Gamma_{e^+e^-} + 9\Gamma_{u\bar u} + 9\Gamma_{d\bar d} \tag{3.107}$$

Here, a factor of 3 occurs for three lepton or quark flavors and an additional factor of 3 enters for the three quark colors. (If more flavors exist, this result must of course be modified.) Employing (3.107) and the preceding tabulation, we find

$$\Gamma_T = 2.5 \quad \text{GeV} \tag{3.108}$$

A very similar calculation yields the transition rate for decay of a W to a lepton pair:

$$\Gamma(W^- \to l^-\bar{\nu}) = \frac{G_F}{6\pi\sqrt{2}} m_W^3$$

The total width of W^- is then

$$\Gamma_{\text{TOTAL}}(W^-) = \frac{G_F}{6\pi\sqrt{2}} m_W^3 (3+9) = \frac{\sqrt{2}}{\pi} G_F m_W^3 = 2.5 \quad \text{GeV}$$

3.8 Z^0 production in e^+e^- collisions

We now return to the question of Z^0 production in e^+e^- collisions. We see from (3.108) that $\Gamma_T \ll m_Z \simeq 90$ GeV. Therefore, to a good approximation, we can replace $(s - m_Z^2)^{-2}$ in (3.99) by $[(s - m_Z^2)^2 + \Gamma_T^2 m_Z^2]^{-1}$ from (3.101). It is then easy to show that (3.99) implies

$$\sigma(e^+e^- \to Z^0 \to F) = \frac{12\pi s}{m_Z^2} \frac{\Gamma(Z^0 \to e^+e^-)\Gamma(Z^0 \to F)}{(s - m_Z^2)^2 + \Gamma_T^2 m_Z^2} \tag{3.109}$$

where F is a particular final state. The total cross section for Z^0 production is then

$$\sigma_{\text{TOTAL}} = \sum_F \sigma(e^+e^- \to Z^0 \to F) = \frac{12\pi s}{m_Z^2} \frac{\Gamma_{e^+e^-}\Gamma_T}{(s - m_Z^2)^2 + \Gamma_T^2 m_Z^2} \tag{3.110}$$

On resonance (i.e., when $s = m_Z^2$), this becomes

$$\sigma_{\text{TOTAL}}(s = m_Z^2) = \frac{12\pi}{m_Z^2} \frac{\Gamma_{e^+e^-}}{\Gamma_T} \tag{3.111}$$

This formula may be compared with the usual one-photon cross section for $e^+e^- \to \mu^+\mu^-$:

$$R_T = \frac{\sigma_{\text{TOTAL}}(s = m_Z^2)}{4\pi\alpha^2/3m_Z^2} = \frac{9}{\alpha^2} \frac{\Gamma_{e^+e^-}}{\Gamma_T} \tag{3.112}$$

Subtracting the contribution from leptons, we obtain

$$R_h(s = m_Z^2) = \frac{9}{\alpha^2} B(Z \to e^+e^-) \left(1 - \sum_i B_l\right) \tag{3.113}$$

R_h is plotted versus $\sqrt{s}$ in Figure 3.12. Even though the result we obtained is reduced somewhat by radiative corrections (see, for example, Greco et al. 80), there is still an extremely strong resonance. For example, for an e^+e^- colliding-beam machine with a luminosity $L \sim 10^{31}$

cm^{-2}/sec at the Z^0 resonance, one expects approximately 20,000 Z^0 events per day. It thus appears possible to study in great detail the properties of Z^0 in the forthcoming decade, with the aid of e^+e^- colliding beams.

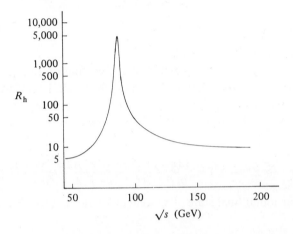

Figure 3.12. The ratio R_h is plotted vs. $\sqrt{s}$. There is a very strong resonance at the Z^0 mass. Note the logarithmic ordinate scale. Here $\sin^2 \theta_W = 0.25$ is assumed.

3.9 $e^+e^- \to W^+W^-$; W and Z^0 production in $p\bar{p}$ collisions

The process $e^+e^- \to W^+W^-$, which may also be observed at future e^+e^- colliding-beam facilities, is an interesting example of a transition in which several diagrams contribute (Figure 3.13).

The vertices γW^+W^- and $Z^0W^+W^-$ in Figures 3.13b and 3.13c, respectively, involve the trilinear couplings characteristic of the Yang–Mills theory [see (2.168)]. The actual calculation of the total cross section $\sigma(e^+e^- \to W^+W^-)$ is quite complicated, and therefore we merely present the results, referring to the original literature for details (Alles et al. 77). Let us define the variables:

$$\beta = (1 - 4m_W^2/s)^{1/2}, \qquad \sin^2 \theta_W = x, \qquad L = \ln \left| \frac{1 + \beta}{1 - \beta} \right|$$

Also, for the cross section, we write

$$\sigma = \frac{\pi \alpha^2}{8x^2} \beta \frac{1}{s} \sum_{ij} \bar{\sigma}_{ij} = \sum_{ij} \sigma_{ij} \qquad (3.114)$$

where subscripts ij refer to Figures 3.13a–c. For example, $\sigma_{\nu\nu}$ is the term arising from the square of the amplitude associated with Figure

3.13a, $\bar{\sigma}_{\nu Z}$ is the interference term arising from Figures 3.13a and c, and so forth. One finds:

$$\bar{\sigma}_{\nu\nu} = \sigma_1 \tag{3.115}$$

$$\bar{\sigma}_{\gamma\gamma} = x^2\sigma_2 \tag{3.116}$$

$$\bar{\sigma}_{ZZ} = (x^2 - \tfrac{1}{2}x + \tfrac{1}{8})[s^2/(s - m_Z^2)^2]\sigma_2 \tag{3.117}$$

$$\bar{\sigma}_{Z\gamma} = 2(\tfrac{1}{4} - x)x[s/(s - m_Z^2)]\sigma_2 \tag{3.118}$$

$$\bar{\sigma}_{\nu Z} = (x - \tfrac{1}{2})[s/(s - m_Z^2)]\sigma_3 \tag{3.119}$$

$$\bar{\sigma}_{\gamma\nu} = -x\sigma_3 \tag{3.120}$$

where

$$\sigma_1 = 2(s/m_W^2) + \tfrac{1}{12}(s/m_W^2)^2\beta^2 + 4[(1 - 2m_W^2/s)(L/\beta) - 1] \tag{3.121}$$

$$\sigma_2 = 16(s/m_W^2)\beta^2 + \tfrac{2}{3}\beta^2[(s/m_W^2)^2 - 4(s/m_W^2) + 12] \tag{3.122}$$

$$\sigma_3 = 16 - 32(m_W^2/s)(L/\beta) + 8\beta^2(s/m_W^2) + \tfrac{1}{3}\beta^2(s/m_W^2)^2(1 - 2m_W^2/s)$$
$$+ 4(1 - 2m_W^2/s) - 16(m_W^2/s)^2(L/\beta) \tag{3.123}$$

Figure 3.13. Lowest-order diagrams for $e^+e^- \to W^+W^-$.

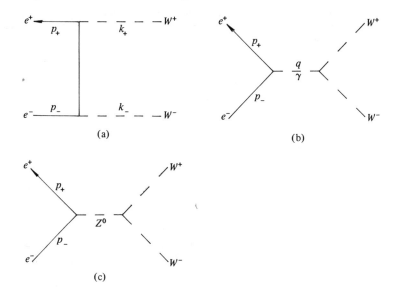

Although each diagram (Figures 3.13a–c) separately yields a cross section that increases with s and violates unitarity, the interference terms produce cancellations that cause the total cross section to reach a maximum at finite s and then decrease to zero, within the bounds imposed by unitarity (see Figure 3.14).

The intermediate bosons $W^\pm$ and Z^0 may also be produced in proton–antiproton ($p\bar{p}$) collisions at sufficiently high energies. In terms of the proton valence quarks u, u, and d and antiproton valence antiquarks $\bar{u}$, $\bar{u}$, and $\bar{d}$, the chief production mechanisms are

$$u + \bar{d} \to W^+ + X$$
$$d + \bar{u} \to W^- + X$$
$$u + \bar{u} \to Z^0 + X$$
$$d + \bar{d} \to Z^0 + X$$

In addition, at the energies contemplated at actual accelerators, the following reactions occur with somewhat lower probability:

$$u + \bar{d} \to W^+ + \gamma + X$$
$$d + \bar{u} \to W^- + \gamma + X$$
$$u + \bar{u} \to Z^0 + \gamma + X$$
$$u + \bar{u} \to W^+ + W^- + X$$

$\cdot$
$\cdot$
$\cdot$

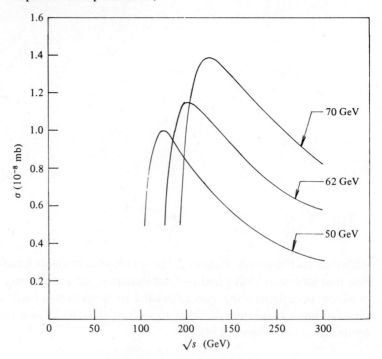

Figure 3.14. $\sigma(e^+e^- \to W^-W^+)$ in the Weinberg–Salam model as a function of $\sqrt{s}$ for $m_W = 50$, 62, and 70 GeV. (From Alles et al. 77. Reprinted with permission.)

3.10 ν-e elastic scattering and inverse muon decay

The CERN Super Proton Synchrotron (SPS) has actually been converted into a $p\bar{p}$ colliding-beam accelerator for this purpose (see, for example, Cline et al. 82). Each beam (p, $\bar{p}$) has 270-GeV energy, resulting in a CM energy of 540 GeV, equivalent to 1.5×10^5 GeV antiprotons on a fixed proton target. At such an energy, the total cross section for $p\bar{p}$ collisions is

$$\sigma_{\text{TOTAL}}(p\bar{p}) = 5 \times 10^{-26} \text{ cm}^2$$

It is expected that the SPS collider will achieve a luminosity of 10^{29} cm^{-2} sec^{-1}, which is equivalent to a $p\bar{p}$ collision rate of about 100 sec^{-1} at each of the two-beam intersections. This should yield roughly 10 W's and 1 Z^0 per day. Substantially higher production rates should be achieved at the Fermilab Tevatron $p\bar{p}$ collider later in the decade.

One hopes to detect W's by observing single muons or electrons with very high transverse momenta, which arise in the decay $W \rightarrow l\nu$. This is extremely difficult because of several features of the final states in $p\bar{p}$ collisions, now recognized from preliminary observations at CERN. One finds that the mean multiplicity of charged hadrons resulting from $p\bar{p}$ at 540 GeV is large: 27.4 ± 2, and the distribution of multiplicities about the mean is very broad. In addition, there are many final-state particles with extremely high transverse momentum – the number being much larger than that expected on the basis of the naive quark–parton model (see, for example, Section 8.5). Thus in experiments to detect W production, background problems are very severe. In spite of this, preliminary observations at CERN reveal the existence of W's with the expected mass (~ 80 GeV/c^2).

3.10 Neutrino–electron elastic scattering and inverse muon decay

We now consider the scattering processes:

$$\nu_\mu e^- \rightarrow \nu_\mu e^- \quad (3.124)$$

$$\bar{\nu}_\mu e^- \rightarrow \bar{\nu}_\mu e^- \quad (3.125)$$

$$\bar{\nu}_e e^- \rightarrow \bar{\nu}_e e^- \quad (3.126)$$

$$\nu_e e^- \rightarrow \nu_e e^- \quad (3.127)$$

$$\nu_\mu e^- \rightarrow \mu^- \nu_e \quad (3.128)$$

Reactions (3.124), (3.125), and (3.128) are generated with ν_μ and $\bar{\nu}_\mu$ beams from high-energy proton accelerators,[4] and (3.126) is observed with $\bar{\nu}_e$ from a fission reactor.

---

[4] Chapter 8 contains a discussion of experimental aspects of neutrino beams.

3 Leptonic weak interactions

According to the Feynman rules, the invariant amplitude for $\nu_\mu e \to \nu_\mu e$ is

$$\mathcal{M}(\nu_\mu e) = \left[\frac{-ig}{4\cos\theta_W} \bar{u}_{\nu 2}\gamma_\lambda(1-\gamma_5)u_{\nu 1}\right] \frac{g^{\lambda\sigma} - q^\lambda q^\sigma/m_Z^2}{q^2 - m_Z^2}$$

$$\times \left[\frac{-ig}{2\cos\theta_W} \bar{u}_{e2}\gamma_\sigma(g_V - g_A\gamma_5)u_{e1}\right] \quad (3.129)$$

where, in the standard model, $g_V = 2\sin^2\theta_W - \tfrac{1}{2}$ and $g_A = -\tfrac{1}{2}$. For $|q^2| \ll m_Z^2$, this becomes

$$\mathcal{M}(\nu_\mu e) = (G_F/\sqrt{2})\bar{u}_{\nu 2}\gamma_\lambda(1-\gamma_5)u_{\nu 1} \cdot \bar{u}_{e2}\gamma^\lambda(g_V - g_A\gamma_5)u_{e1} \quad (3.130)$$

Let the 4-momenta of initial and final neutrinos be k_1 and k_2 and of the initial and final electrons be p_1 and p_2, respectively. Averaging over initial and summing over final electron spins, we obtain for $|\mathcal{M}|^2$:

$$\overline{|\mathcal{M}(\nu_\mu e)|^2} = (G_F^2/64 m_\nu^2 m_e^2) T_1 \cdot T_2 \quad (3.131)$$

where

$$T_1 = \mathrm{tr}[\slashed{k}_2\gamma_\lambda(1-\gamma_5)\slashed{k}_1\gamma_\sigma(1-\gamma_5)] = 2\,\mathrm{tr}[\slashed{k}_2\gamma_\lambda\slashed{k}_1\gamma_\sigma(1-\gamma_5)] \quad (3.132)$$

and

$$T_2 = \mathrm{tr}[\slashed{p}_2 + m_e)\gamma^\lambda(g_V - g_A\gamma_5)(\slashed{p}_1 + m_e)\gamma^\sigma(g_V - g_A\gamma_5)]$$
$$= \mathrm{tr}[\slashed{p}_2\gamma^\lambda\slashed{p}_1\gamma^\sigma(g_V^2 + g_A^2 - 2g_V g_A\gamma_5)]$$
$$+ m_e^2\,\mathrm{tr}[(\gamma^\lambda\gamma^\sigma)\cdot(g_V^2 - g_A^2)] \quad (3.133)$$

Multiplying T_1 and T_2 and making use of Table 3.1, we find:

$$\overline{|\mathcal{M}(\nu_\mu e)|^2} = (G_F^2/64 m_\nu^2 m_e^2)[k_2\cdot p_2 k_1\cdot p_1(g_V + g_A)^2$$
$$+ k_2\cdot p_1 k_1\cdot p_2(g_V - g_A)^2$$
$$+ (g_A^2 - g_V^2)m_e^2 k_1\cdot k_2] \quad (3.134)$$

It is interesting to note that the first two terms on the right-hand side of (3.134) are invariant under the replacement $g_A \leftrightarrow g_V$. The third term is not, but of course, it is negligible compared with the first two for high-energy neutrinos.

In the case $\bar{\nu}_\mu e \to \bar{\nu}_\mu e$, it is easy to see that we obtain $\overline{|\mathcal{M}(\bar{\nu}_\mu e)|^2}$ from (3.134) by making the replacement $k_1 \leftrightarrow k_2$:

$$\overline{|\mathcal{M}(\bar{\nu}_\mu e)|^2} = (G_F^2/64 m_\nu^2 m_e^2)[k_2\cdot p_2 k_1\cdot p_1(g_V - g_A)^2$$
$$+ k_2\cdot p_1 k_1\cdot p_2(g_V + g_A)^2$$
$$+ (g_A^2 - g_V^2)m_e^2 k_1\cdot k_2] \quad (3.135)$$

Reactions $\nu_e e \to \nu_e e$ and $\bar{\nu}_e e \to \bar{\nu}_e e$ involve both charged and neutral currents, and the relative sign of the two amplitudes is important. In order to determine it correctly, we consider as an example the two diagrams that contribute to $\bar{\nu}_e e \to \bar{\nu}_e e$ scattering (Figure 3.15a and b).

Figure 3.15. $\bar{\nu}_e e \to \bar{\nu}_e e$ scattering.

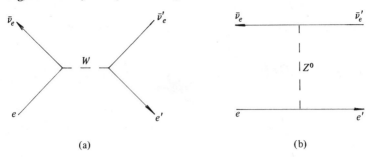

The matrix element corresponding to each of these figures involves certain products of fermion field operators, and since these satisfy anticommutation relations, we must pay attention to the order in which they appear. Ignoring space-time indexes and other irrelevant factors for the moment, we find that Figure 3.15a corresponds to the combination

$$\bar{\Psi}_e \Psi_\nu \bar{\Psi}_\nu \Psi_e \to b_e'^\dagger d_\nu'^\dagger d_\nu b_e$$

and Figure 3.15b corresponds to

$$\bar{\Psi}_\nu \Psi_\nu \bar{\Psi}_e \Psi_e \to d_\nu d_\nu'^\dagger b_e'^\dagger b_e = -d_\nu'^\dagger d_\nu b_e'^\dagger b_e = -b_e'^\dagger d_\nu'^\dagger d_\nu b_e$$

Therefore the relative sign of the two amplitudes is -1:

$$\mathcal{M}_Z(\bar{\nu}_e e) = (G_F/\sqrt{2})\bar{v}_{\nu 1}\gamma_\lambda(1-\gamma_5)v_{\nu 2}\bar{u}_{e2}\gamma^\lambda(g_V - g_A\gamma_5)u_{e1} \quad (3.136)$$

$$\mathcal{M}_W(\bar{\nu}_e e) = -(G_F/\sqrt{2})\bar{v}_{\nu 1}\gamma_\lambda(1-\gamma_5)u_{e1}\bar{u}_{e2}\gamma^\lambda(1-\gamma_5)v_{\nu 2} \quad (3.137)$$

It is convenient to perform a rearrangement of the last expression (Fierz transformation) in which interchange of u_{e1} and $v_{\nu 2}$ is accompanied by a mere sign change:

$$\mathcal{M}_W(\bar{\nu}_e e) = +(G_F/\sqrt{2})\bar{v}_{\nu 1}\gamma_\lambda(1-\gamma_5)v_{\nu 2}\bar{u}_{e2}\gamma^\lambda(1-\gamma_5)u_{e1} \quad (3.138)$$

Combining (3.136) and (3.138), we obtain

$$|\mathcal{M}(\bar{\nu}_e e)|^2 = (G_F^2/m_e^2 m_\nu^2)[k_2 \cdot p_2 k_1 \cdot p_1(g_V' - g_A')^2 \\ + k_2 \cdot p_1 k_1 \cdot p_2(g_V' + g_A')^2 \\ + m_3^2(g_A'^2 - g_V'^2)k_1 \cdot k_2] \quad (3.139)$$

where $g_V' = g_V + 1$ and $g_A' = g_A + 1$. The third term in (3.139) is *not* negligible compared with the first two for low-energy $\bar{\nu}_e$.

We proceed to calculate cross sections in the laboratory frame, using the following kinematic relations, which are valid for all of the elastic neutrino–electron reactions:

$$k_1 \cdot p_1 = k_2 \cdot p_2 = m_e E_{\nu 1}$$
$$k_1 \cdot p_2 = k_2 \cdot p_1 = m_e E_{\nu 2} = m_e(m_e + E_{\nu 1} - E_{e2}) = m_e(E_{\nu 1} - T_e)$$
$$= m_e E_{\nu 1}(1 - y)$$

where $y = T_e/E_{\nu 1}$ and $k_1 \cdot k_2 = m_e E_{\nu 1} y$. Then, each differential cross section takes the form

$$d\sigma = \frac{G_F^2}{4\pi^2} \frac{A_0 m_e^2 E_{\nu 1}^2 + B_0 m_e^2 E_{\nu 1}^2 (1-y)^2 + C_0 m_e^3 E_{\nu 1} y}{m_e E_{\nu 1} E_{e2} E_{\nu 2}} d^3p_{e2}\, d^3p_{\nu 2}$$
$$\times \delta^4(p_2 + k_2 - p_1 - k_1) \qquad (3.140)$$

where the constants A_0, B_0, and C_0 are given by the following tabulation:

Transition	A_0	B_0	C_0
$\nu_\mu e \to \nu_\mu e$	$(g_V + g_A)^2$	$(g_V - g_A)^2$	$g_A^2 - g_V^2$
$\bar\nu_\mu e \to \bar\nu_\mu e$	$(g_V - g_A)^2$	$(g_V + g_A)^2$	$g_A^2 - g_V^2$
$\nu_e e \to \nu_e e$	$(g_V' + g_A')^2$	$(g_V' - g_A')^2$	$g_A'^2 - g_V'^2$
$\bar\nu_e e \to \bar\nu_e e$	$(g_V' - g_A')^2$	$(g_V' + g_A')^2$	$g_A'^2 - g_V'^2$

After integrating (3.140) over $\mathbf{p}_{e2}$ with $\int d^3p_{e2}\, \delta^3(\mathbf{p}_2 + \mathbf{k}_2 - \mathbf{k}_1) = 1$, we next integrate with respect to $d\Omega_{\nu 2}$ over the remaining δ function:

$$\delta(E_{e2} + E_{\nu 2} - m_e - E_{\nu 1})$$
$$= \delta[E_{\nu 2} - m_e - E_{\nu 1} + (m_e^2 + E_{\nu 1}^2 + E_{\nu 2}^2 - 2E_{\nu 1}E_{\nu 2}\cos\theta)^{1/2}]$$

where θ is the angle between $\mathbf{k}_1$ and $\mathbf{k}_2$. The result is

$$\frac{d\sigma}{dy} = (G_F^2 m_e E_{\nu 1}/2\pi)[A_0 + B_0(1-y)^2 + C_0 m_e y/E_{\nu 1}] \quad (3.141)$$

and the total cross section is obtained by integrating over y from 0 to 1:

$$\sigma_{\text{TOTAL}} = (G_F^2 m_e E_{\nu 1}/2\pi)(A_0 + \tfrac{1}{3}B_0 + C_0 m_e/2E_{\nu 1}) \quad (3.142)$$

Of course, as we already noted, the term in C_0 is negligible for $E \gg m_e$, that is, for accelerator neutrinos.

The y dependence of the B_0 term in (3.141) may be understood by considering $\nu_e e$ and $\bar\nu_e e$ scattering according to the old charged-current theory, where $g_V' = g_A' = 1$. For this purpose, we rewrite (3.141) in terms of the scattering angle θ of the neutrino in the CM frame ($y = \cos\theta$) and of the neutrino energy ω in the CM frame. It is then easy to see that we obtain the following:

3.10 ν-e elastic scattering and inverse muon decay

$$d\sigma/d\Omega = \begin{cases} G_F^2\omega^2/\pi^2 & \text{for } \nu_e e \text{ scattering} \\ (G_F^2\omega^2/\pi^2)(1 - \cos\theta)^2 & \text{for } \bar{\nu}_e e \text{ scattering} \end{cases}$$

In $\nu_e e$ scattering in the CM frame, both initial particles are left-handed; thus scattering into any angle is possible and is consistent with conservation of angular momentum (Figure 3.16a). However, for $\bar{\nu}_e e$ scattering, it is impossible for the $\bar{\nu}_e$ to be scattered backward ($\theta = \pi$); otherwise angular momentum would not be conserved (Figure 3.16b).

Figure 3.16. (a) $\nu_e e \to \nu_e e$; (b) $\bar{\nu}_e e \to \bar{\nu}_e e$.

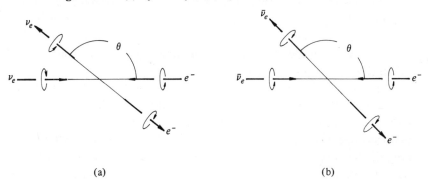

Let us now turn our attention to inverse muon decay (3.128). If we assume that the amplitude contains only V and A components, then we can write, as in (3.34),

$$\mathcal{M} = -\frac{G_F}{\sqrt{2}} \bar{u}_\mu \gamma_\lambda (1 - \gamma_5) u_e \bar{u}_{\nu_e} \gamma^\lambda (c_V - c_A \gamma_5) u_{\nu_e} \quad (3.143)$$

where $c_V = c_A = +1$ in the standard theory. It can then be shown that in the center-of-mass frame, the differential cross section is

$$\frac{d\sigma}{d\cos\theta_0} = \frac{G_F^2(s - m_\mu^2)^2}{32\pi s^3}(|c_V|^2 + |c_A|^2)[(a + b) - \lambda(a - b)] \quad (3.144)$$

where θ_0 is the angle between ν_μ and μ^-, $s = 2E_\nu m_e$, $a = [(s + m_e^2) - (s - m_e^2)\cos\theta_0][(s + m_\mu^2) - (s - m_\mu^2)\cos\theta_0]$, $b = 4s^2$, and $\lambda = 2\text{Re}(c_V^* c_A)/(|c_V|^2 + |c_A|^2)$ ($\lambda = 1$ in the standard theory). Note that in the laboratory frame, there is a finite threshold for inverse muon decay:

$$E_{\nu 1} \geq m_\mu^2/2m_e$$

Let us now consider some aspects of neutrino–electron scattering experiments. In $\nu_\mu e$ or $\bar{\nu}_\mu e$ scattering at accelerator energies, the angle of scattering θ_e of the electron with respect to the neutrino beam direction in the laboratory frame satisfies

$$\sin^2 \theta_e = \frac{2m_e}{T_e + 2m_e}\left[1 + \frac{T_e}{E_{\nu 1}} - \frac{m_e T_e}{2E_{\nu 1}^2}\right] \quad (3.145)$$

where T_e is the electron kinetic energy. The electron momentum is thus strongly peaked in the forward direction and $\theta_e^2 \leq 2m_e/E_e$. The chief problem in a typical experiment is to separate the recoil electron event from background events. For example, π^0 mesons may be produced by neutrino–nucleon collisions, and these give rise to energetic γ rays, and thus to e^+e^- pairs. However, these have a rather flat $E_e\theta_e^2$ distribution and may thus be distinguished from the true electron recoil events. The residual backgrounds are then quite small in number. For example, we consider the $\nu_\mu e \to \nu_\mu e$ experiment carried out at Fermilab (Heisterberg et al. 80).

Some 9.5×10^{18} protons of 350 GeV were delivered to a target, yielding pions, which were focused and then decayed in flight in a pion-beam decay pipe, giving rise to ν_μ. The resulting "wideband" neutrino beam had an average energy of 20 GeV and a maximum energy of 100 GeV. The $\bar{\nu}_\mu$ contamination was about 11 percent, whereas that of ν_e and $\bar{\nu}_e$ was less than 0.5 percent. The detection apparatus, located about 500 m downstream from the pion decay pipe, consisted of an electromagnetic shower electron detector with high angular resolution. It was constructed with 49 basic modules, each consisting of an aluminum plate within which the shower was generated, a multiple-wire proportional chamber, and plastic scintillators. The detector recorded 249,000 neutrino-induced interactions. Of these, 46 events were observed with the expected small angular distribution (see Figure 3.17).

Of the 46 events, 34 were attributed to genuine events $\nu_\mu e \to \nu_\mu e$, and the remainder (12) to background: 6 from $\nu_e + n \to e^- + p$ and 6 from neutral current reactions. The number of genuine events and the estimated neutrino flux may be combined to yield an experimental total cross section:

$$\sigma_{\text{expt}}(\nu_\mu e \to \nu_\mu e) = (1.40 \pm 0.30) \times 10^{-42} E_\nu \text{ cm}^2 \quad (3.146)$$

It is easy to see from formula (3.141) or (3.142) that a determination of σ_{TOTAL} for any one of the processes $\nu_\mu e \to \nu_\mu e$, $\bar{\nu}_\mu e \to \bar{\nu}_\mu e$, $\nu_e e \to \nu_e e$, or $\bar{\nu}_e e \to \bar{\nu}_e e$ constrains the permissible values of g_A and g_V to lie within an annular elliptical ring in the g_A-g_V plane. The results of various neutrino–electron scattering experiments are thus sum-

3.10 ν-e elastic scattering and inverse muon decay

marized in Figure 3.18. In addition, one may obtain a further constraint on g_A and g_V from $e^+e^- \to e^+e^-$, $\mu^+\mu^-$, $\tau^+\tau^-$ observed cross sections and limits obtained to date on the $e^+e^- \to \mu^+\mu^-$ charge asymmetry. This is also shown in Figure 3.18. It is clear from this figure that the neutrino data alone are consistent with two distinct solutions, indicated by the black patches. However, when the $e^+e^- \to l^+l^-$ data are also included, only one solution survives and the other is ruled out with more than 95 percent confidence. The final result is in very good agreement with the standard model for $\sin^2 \theta_W = 0.23$ (see Blietschau et al. 76, Reines et al. 76, Faissner et al. 78, Reithler et al. 79, Heisterberg et al. 80, Barber et al. 81, Bartel et al. 81).

Inverse muon decay has been observed with a detector exposed to the CERN SPS horn focused wideband neutrino beam. A total of 171 ± 29 events was found and the observed rate agrees well with that predicted by the standard (V−A) theory for two-component neutrinos (Jonker et al. 80a).

Figure 3.17. Angular distribution of measured events per unit solid angle satisfying either $\theta_x \leq 50$ mrad or $\theta_y \leq 50$ mrad. (From Heisterberg et al. 80. Reprinted with permission.)

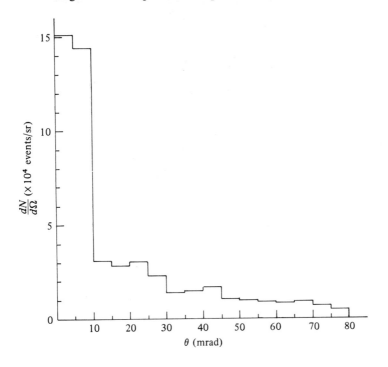

$$R_{\text{expt}}/R_{\text{V-A}} = 0.98 \pm 0.18 \tag{3.147}$$

Incidentally, in the same experiment, a search was carried out for $\bar{\nu}_\mu e^- \to \mu^- \bar{\nu}_e$, which is allowed by multiplicative lepton conservation but forbidden by the conventional lepton conservation law. The following experimental limit is obtained with 90 percent confidence:

$$\sigma(\bar{\nu}_\mu e^- \to \mu^- \bar{\nu}_e)/\sigma(\nu_\mu e^- \to \mu^- \nu_e) < 0.09 \tag{3.148}$$

In conclusion, we may briefly summarize the contents of this chapter as follows: Muon decay has been studied experimentally in great detail and the results are in very good agreement with the V–A law. However, these results cannot rule out appreciable S, T, and P couplings, and furthermore, there is considerable uncertainty in the ratio $|g_A/g_V|$ and in

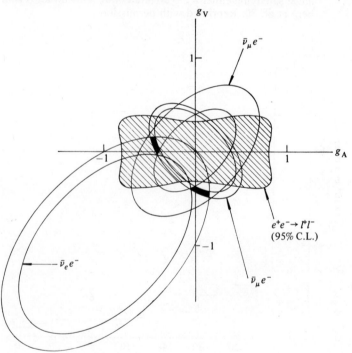

Figure 3.18. Results obtained from neutrino experiments and e^+e^- data (Mark J) in terms of limits on g_V and g_A. The region between the concentric ellipses corresponds to 68% confidence limits from neutrino data. $\nu_\mu e^-$: Heisterberg et al. 80; $\bar{\nu}_\mu e^-$: Blietschau et al. 76, Faissner et al. 78; $\bar{\nu}_e e^-$: Reines et al. 76. The shaded area represents 95% confidence-level contours from the $e^+e^- \to l^+l^-$ Mark-J experiments (Barber et al. 81, Bartel et al. 81).

the relative phase $\phi_{A,V}$. In principle, observations of inverse muon decay offer the possibility of constraining the couplings further, and the results for this reaction are in agreement with the V–A law for two-component neutrinos. However, they are not yet sufficiently accurate to improve the situation very much. Observations of τ lepton production and decay in e^+e^- colliding-beam experiments are in good agreement with the hypothesis that ν_τ and τ are sequential leptons as described by the standard model. Furthermore, neutrino–electron scattering experiments give results in good agreement with the standard model. However, by late 1982, only preliminary results on weak–electromagnetic interference in $e^+e^- \to l^+l^-$ reactions had yet been obtained, and the most important experiments of all, namely, production of Z^0 and W^+W^- in e^+e^- collisions and study of subsequent decays, remained hopes for the future.

Problems

3.1 Prove that if u_1, u_2, w_1, and w_2 are Dirac spinors, then

$$\bar{u}_1 \gamma_\lambda (1 - \gamma_5) u_2 \bar{w}_1 \gamma^\lambda (1 - \gamma_5) w_2$$
$$= -\bar{u}_1 \gamma_\lambda (1 - \gamma_5) w_2 \bar{w}_1 \gamma^\lambda (1 - \gamma_5) u_2$$

This is a special case of the Fierz transformation (Fierz 37).

3.2 Consider the energy spectrum of electrons in muon decay. Neglecting electron mass and radiative corrections, one has

$$n(\epsilon) \propto \epsilon^2 [12(1 - \epsilon) + \tfrac{4}{3}\rho(8\epsilon - 6)] \tag{1}$$

where

$$\rho = \frac{3g_A^2 + 3g_V^2 + 6g_T^2}{g_S^2 + g_P^2 + 4g_V^2 + 4g_A^2 + 6g_T^2} \tag{2}$$

and

$$g_i^2 = [C_i|^2 + |C_i'|^2 \tag{3}$$

In the text, we derived $n(\epsilon)$ for the V–A law where $\rho = \tfrac{3}{4}$. Carry out a similar derivation assuming an interaction of the form

$$\overline{\Psi}_{\nu_\mu}(1 - \gamma_5)\Psi_\mu \overline{\Psi}_e(1 - \gamma_5)\Psi_{\nu_e}$$

and show that one obtains $n(\epsilon)$ with $\rho = 0$, as expected from expression (2).

3.3 The muonium atom consists of a positive muon and an electron, which form a hydrogenic system. Muonium lasts only a short time because the muon decays. Another possibility is the capture reaction: $\mu^+ e^- \to \nu_e \bar{\nu}_\mu$. Show that in the $1\,^2S_{1/2}$ state of muonium, the ratio of rates for these two processes is the very small number:

$$R = \frac{\Gamma(\mu^+ e^- \to \nu_e \bar{\nu}_\mu)}{\Gamma(\mu^+ \to e^+ \nu_e \bar{\nu}_\mu)} = 48 \frac{m_e^3}{m_\mu^3} \alpha^3$$

3.4 Derive (3.97)–(3.99) from (3.96a–c) to obtain (3.100). Verify that the "charge-asymmetry" effect in $e^+ e^- \to \mu^+ \mu^-$ violates neither P nor C invariance. What charge asymmetry should one expect to observe at $s = m_z^2$?

3.5 Derive (3.109) from (3.99) using the conditions stated in the text.

3.6 Consider inverse muon decay:

$\nu_\mu + e \to \mu^- + \nu_e$

Derive (3.144) from (3.143) with the conditions stated in the text and show that the threshold for the reaction in the laboratory frame is $E_{\nu_1} \geq m_\mu^2 / 2m_e$.

4

General properties of hadronic weak currents

4.1 Introduction

In this chapter we begin to study hadronic weak interactions. A proper understanding of this subject can develop only from detailed knowledge of the structure of hadrons and the dynamics of their constituent quarks and gluons. It is widely believed that, in principle, one may ultimately obtain this knowledge from the equations of quantum chromodynamics (QCD). However, hadrons are such complex objects that, even if this is the correct approach, it will be necessary to resort to perturbation theory to make practical calculations of hadronic properties. Now, according to QCD, the coupling of one quark to another is weak at very short distances or large q^2 ("asymptotic freedom") and strong at very large distances or small q^2 ("infrared slavery"). Therefore, perturbation theory is applicable in only the large-q^2, or "hard-gluon," limit. On the other hand, when one considers the internal structure of an ordinary hadron such as the proton or pion, low q^2, or "soft-gluon," exchanges are very important, and the methods of QCD are not applicable because we cannot use perturbation theory.

In order to make progress in the study of hadronic weak interactions despite this obstacle, we must rely on those tools available. One very useful, if admittedly semiempirical, tool to be employed is the constituent quark model and the approximate flavor–unitary symmetries $SU(3)_f$, $SU(4)_f$, and so on, associated with it. A fundamental understanding of the basis for the semiempirical quark model does not exist, yet, it explains a great many facts about hadrons extremely well and certainly contains a great deal of truth.

We shall also make use of various exact and approximate symmetry principles, such as proper Lorentz invariance, time-reversal and *CPT* invariance, and the conserved vector current hypothesis. In this chapter our main task is to assemble all these tools, and we shall provide only a small number of important but relatively simple examples. Most of the detailed applications of the general methods to be developed here are reserved for later chapters.

4.2 Unitary symmetry and the quark model

Throughout any discussion of the quark model, a very important role is played by the group $SU(n)$. This is the name given to the group of unitary transformations U on an n-dimensional complex vector space, for which

$$\det U = +1 \tag{4.1}$$

Let us outline the principal algebraic properties of $SU(n)$ that are important for an understanding of the quark model.

We begin by writing operator U in terms of a Hermitian operator H as follows:

$$U = e^{iH} \tag{4.2}$$

Now H can be expressed in terms of certain standard Hermitian operators F_i, called generators of the group $SU(n)$:

$$H = \alpha^i F_i \tag{4.3}$$

where the numerical parameters α^i characterize H and thus U, and we employ the repeated index summation convention. Since $U = \exp(i\alpha^k F_k)$, (4.1) implies:

$$\text{tr}(F_k) = 0 \tag{4.4}$$

In general there are n^2 independent $n \times n$ matrices. With the imposition of (4.4), the number of linearly independent traceless Hermitian generators F_k is $n^2 - 1$. One simple (nonunique) prescription for finding them is to start by forming the $n - 1$ diagonal matrices:

$$\begin{pmatrix} 1 & & & \\ & -1 & & \\ & & 0 & \\ & & & \ddots \\ & & & & 0 \end{pmatrix}, \quad 1/\sqrt{3} \begin{pmatrix} 1 & & & \\ & 1 & & \\ & & -2 & \\ & & & 0 \\ & & & & \ddots \\ & & & & & 0 \end{pmatrix}, \quad \ldots,$$

$$\sqrt{2/n(n-1)} \begin{pmatrix} 1 & & & \\ & 1 & & \\ & & 1 & \\ & & & \ddots \\ & & & & -(n-1) \end{pmatrix}$$

4.2 Unitary symmetry and the quark model

Then we form the $(n^2 - n)/2$ off-diagonal matrices with 1 in a given off-diagonal position, 1 in the transposed position, and zeros elsewhere. Also, we form the $(n^2 - n)/2$ off-diagonal matrices with $-i$ in a given off-diagonal position, $+i$ in the transposed position, and zeros elsewhere.

This set of $n^2 - 1$ independent $n \times n$ Hermitian matrices will be called λ_k ($k = 1, \ldots, n^2 - 1$) with the following conditions obviously satisfied:

$$\text{tr } \lambda_k = 0 \tag{4.5}$$

$$\text{tr } \lambda_i \lambda_k = 2\delta_{ik} \tag{4.6}$$

One representation of the F_k is obtained by putting

$$F_k = \tfrac{1}{2}\lambda_k \tag{4.7}$$

4.2.1 SU(2)

Setting $n = 2$, we have the three λ matrices:

$$\lambda_1 = \begin{pmatrix} 0 & 1 \\ 1 & 0 \end{pmatrix}, \quad \lambda_2 = \begin{pmatrix} 0 & -i \\ i & 0 \end{pmatrix}, \quad \lambda_3 = \begin{pmatrix} 1 & 0 \\ 0 & -1 \end{pmatrix}$$

Of course these are the familiar Pauli matrices, and the $F_k = \tfrac{1}{2}\lambda_k$ are the three spin-$\tfrac{1}{2}$ operators that satisfy the commutation relations:

$$[F_i, F_j] = i\epsilon_{ijk} F_k \tag{4.8}$$

where ϵ_{ijk} is the completely antisymmetric unit three-tensor. Relations (4.8) are satisfied not only by the 2×2 matrices F (the "fundamental" representation, spin $\tfrac{1}{2}$) but also by matrices of dimension $(2I + 1) \times (2I + 1)$, corresponding to spin $I = 1, \tfrac{3}{2}, 2$, and so on.

In the fundamental representation we can distinguish two kinds of vectors:

(i) Covariant, or column, spinors $\psi = \begin{pmatrix} \psi_1 \\ \psi_2 \end{pmatrix}$, which transform under a unitary transformation as

$$\psi'_i = U_i^j \psi_j$$

(ii) Contravariant, or row, spinors $\bar{\phi} = (\bar{\phi}^1, \bar{\phi}^2)$, which transform as

$$\bar{\phi}^{j\prime} = \bar{\phi}^i U_i^{j\dagger}$$

A covariant two-spinor is transformed to a contravariant two-spinor and vice versa with the aid of the completely antisymmetric unit tensors ϵ_{ij} and ϵ^{ij}. Thus, for example, $\phi_i = \epsilon_{ij} \bar{\phi}^j$ transforms as a covariant spinor and has the form

$$\phi_i = \begin{pmatrix} \bar{\phi}^2 \\ -\bar{\phi}^1 \end{pmatrix}$$

134 4 General properties of hadronic weak currents

Higher representations of $SU(2)$ are systematically built up from the fundamental 2×2 representation by forming tensor products of the basic spinors ψ_i, $\bar{\phi}^j$, or ϕ_i and then by symmetrizing and antisymmetrizing. Here we utilize a basic theorem that states that once a multispinor has been broken into its different symmetry parts, it has been decomposed uniquely into irreducible representations of the group $SU(n)$.

For example, consider the tensor product of two covariant spinors ψ_i and η_j:

$$\psi_i \eta_j = \tfrac{1}{2}(\psi_i \eta_j + \psi_j \eta_i) + \tfrac{1}{2}(\psi_i \eta_j - \psi_j \eta_i) \tag{4.9}$$

On the left-hand side of (4.9) we have four distinct elements, since i and j independently can take the values 1 and 2. On the right-hand side, the first (symmetric) term in parentheses has three independent elements and transforms as a spin-1-object. The second (antisymmetric) term in parentheses is nonzero only if $i \neq j$. It thus consists of only one component and transforms as a spin-0 object. We thus obtain the familiar result:

$$J = 1: \begin{cases} J_3 = +1 & \text{for } \psi_1 \eta_1 \\ J_3 = 0 & \text{for } 1\sqrt{2}(\psi_1 \eta_2 + \psi_2 \eta_1) \\ J_3 = -1 & \text{for } \psi_2 \eta_2 \end{cases}$$

$$J = 0: \quad J_3 = 0 \quad \text{for } 1/\sqrt{2}(\psi_1 \eta_2 - \psi_2 \eta_1)$$

as in the theory of angular momentum. Alternatively, we may combine

$$\psi_i \phi_j = \tfrac{1}{2}(\psi_i \phi_j + \psi_j \phi_i) + \tfrac{1}{2}(\psi_i \phi_j - \psi_j \phi_i)$$
$$= \tfrac{1}{2}(\psi_i \epsilon_{jk} \bar{\phi}^k + \psi_j \epsilon_{ik} \bar{\phi}^k) + \tfrac{1}{2}(\psi_i \epsilon_{jk} \bar{\phi}^k - \psi_j \epsilon_{ik} \bar{\phi}^k)$$

as in the theory of isotopic spin. For example, one might consider the Fermi–Yang model in which a nucleon and an antinucleon were imagined to form a pion-bound state. Then, ψ would represent the nucleon doublet and ϕ the antinucleon isodoublet; that is, $\bar{\phi} = (\bar{p}, \bar{n})$, $\phi = \begin{pmatrix} \bar{n} \\ -\bar{p} \end{pmatrix}$, and $\psi = \begin{pmatrix} p \\ n \end{pmatrix}$. We thus obtain an isospin-1 triplet, which was interpreted as the three pions, and also an isospin-0 object:

$$I = 1: \begin{cases} I_3 = +1 & \text{for } \pi^+ = p\bar{n} \\ I_3 = 0 & \text{for } \pi^0 = 1/\sqrt{2}(p\bar{p} - n\bar{n}) \\ I_3 = -1 & \text{for } \pi^- = n\bar{p} \end{cases}$$

$$I = 0: \quad I_3 = 0 \quad \text{for } 1/\sqrt{2}(p\bar{p} + n\bar{n})$$

4.2.2 SU(3)

Let us now turn to $SU(3)$. For $n = 3$, we have $3^2 - 1 = 8$ λ matrices:

4.2 Unitary symmetry and the quark model

$$\lambda_1 = \begin{pmatrix} 0 & 1 & 0 \\ 1 & 0 & 0 \\ 0 & 0 & 0 \end{pmatrix}, \quad \lambda_2 = \begin{pmatrix} 0 & -i & 0 \\ i & 0 & 0 \\ 0 & 0 & 0 \end{pmatrix}, \quad \lambda_3 = \begin{pmatrix} 1 & 0 & 0 \\ 0 & -1 & 0 \\ 0 & 0 & 0 \end{pmatrix}$$

$$\lambda_4 = \begin{pmatrix} 0 & 0 & 1 \\ 0 & 0 & 0 \\ 1 & 0 & 0 \end{pmatrix}, \quad \lambda_5 = \begin{pmatrix} 0 & 0 & -i \\ 0 & 0 & 0 \\ i & 0 & 0 \end{pmatrix}, \quad \lambda_6 = \begin{pmatrix} 0 & 0 & 0 \\ 0 & 0 & 1 \\ 0 & 1 & 0 \end{pmatrix} \quad (4.10)$$

$$\lambda_7 = \begin{pmatrix} 0 & 0 & 0 \\ 0 & 0 & -i \\ 0 & i & 0 \end{pmatrix}, \quad \lambda_8 = \frac{1}{\sqrt{3}} \begin{pmatrix} 1 & 0 & 0 \\ 0 & 1 & 0 \\ 0 & 0 & -2 \end{pmatrix}$$

We sometimes also write

$$\lambda_0 = \sqrt{\tfrac{2}{3}} \begin{pmatrix} 1 & 0 & 0 \\ 0 & 1 & 0 \\ 0 & 0 & 1 \end{pmatrix}$$

Taking $F_k = \tfrac{1}{2}\lambda_k$ as before, simple manipulations lead us to the commutation relations:

$$[F_i, F_j] = if_{ij}{}^k F_k \quad (4.11)$$

where the nonzero $SU(3)$ structure constants $f_{ij}{}^k$ are given in Table 4.1. It is easy to show that these quantities satisfy the following identities:

$$f_{ij}{}^k = -f_{ji}{}^k \quad (4.12)$$

Table 4.1. *Antisymmetric structure constants* $f_{ij}{}^k$ *and symmetric constants* d_{ijk} *in* SU(3)

i	j	k	$f_{ij}{}^k$	i	j	k	d_{ijk}
1	2	3	1	1	1	8	$1/\sqrt{3}$
1	4	7	$\tfrac{1}{2}$	1	4	6	$\tfrac{1}{2}$
1	5	6	$-\tfrac{1}{2}$	1	5	7	$\tfrac{1}{2}$
2	4	6	$\tfrac{1}{2}$	2	2	8	$1/\sqrt{3}$
2	5	7	$\tfrac{1}{2}$	2	4	7	$-\tfrac{1}{2}$
3	4	5	$\tfrac{1}{2}$	2	5	6	$\tfrac{1}{2}$
3	6	7	$-\tfrac{1}{2}$	3	3	8	$1/\sqrt{3}$
4	5	8	$\tfrac{1}{2}\sqrt{3}$	3	4	4	$\tfrac{1}{2}$
6	7	8	$\tfrac{1}{2}\sqrt{3}$	3	5	5	$\tfrac{1}{2}$
				3	6	6	$-\tfrac{1}{2}$
				3	7	7	$-\tfrac{1}{2}$
				4	4	8	$-1/\sqrt{12}$
				5	5	8	$-1/\sqrt{12}$
				6	6	8	$-1/\sqrt{12}$
				7	7	8	$-1/\sqrt{12}$
				8	8	8	$-1/\sqrt{3}$

and
$$f_{ij}^l f_{lk}^m + f_{ki}^l f_{lj}^m + f_{jk}^l f_{li}^m = 0 \tag{4.13}$$
We may also verify that the following equation holds:
$$\{F_i, F_j\} = \tfrac{1}{3}\delta_{ij} + d_{ijk}F^k \tag{4.14}$$
where $\{F_i, F_j\} = F_i F_j + F_j F_i$ and the coefficients d_{ijk} are totally symmetric and are given in Table 4.1.

We also find it very useful to define the following operators:
$$I_\pm = F_1 \pm iF_2 \tag{4.15}$$
$$I_3 = F_3 \tag{4.16}$$
$$U_\pm = F_6 \pm iF_7 \tag{4.17}$$
$$U_3 = -\tfrac{1}{2}F_3 + \sqrt{\tfrac{3}{4}}F_8 \tag{4.18}$$
$$V_\pm = F_4 \mp iF_5 \tag{4.19}$$
$$V_3 = -\tfrac{1}{2}F_3 - \sqrt{\tfrac{3}{4}}F_8 \tag{4.20}$$

Only eight of these nine operators are independent, since $I_3 + U_3 + V_3 = 0$. Commutation relations (4.11) can be rewritten in terms of the I, U, and V operators given in Table 4.2. We see from this table that these operators generate three noncommuting $SU(2)$ subgroups in $SU(3)$.

The irreducible representations of $SU(n)$ are conveniently displayed on a *weight diagram*, the dimension of which is equal to the number of generators that can be simultaneously diagonalized. For $SU(2)$, there is only one such generator and the diagram is one-dimensional. For $SU(3)$, there are two diagonal generators, F_3 and F_8, and the weight

Table 4.2. SU(3) *commutation relations for* I, U, *and* V *spin operators*

$[I_+, I_-] = 2I_3$	$[U_+, U_-] = 2U_3$	$[V_+, V_-] = 2V_3$
$[I_+, I_3] = -I_+$	$[U_+, U_3] = -U_+$	$[V_+, V_3] = -V_+$
$[I_-, I_3] = +I_-$	$[U_-, U_3] = +U_-$	$[V_-, V_3] = -\tfrac{1}{2}I_-$
$[I_3, U_3] = 0$	$[I_+, U_3] = \tfrac{1}{2}I_+$	$[I_-, U_3] = -\tfrac{1}{2}I_-$
$[I_3, U_+] = -\tfrac{1}{2}U_+$	$[I_+, U_+] = V_-$	$[I_-, U_+] = 0$
$[I_3, U_-] = \tfrac{1}{2}U_-$	$[I_+, U_-] = 0$	$[I_-, U_-] = -V_+$
$[I_3, V_3] = 0$	$[I_+, V_3] = \tfrac{1}{2}I_+$	$[I_-, V_3] = -\tfrac{1}{2}I_-$
$[I_3, V_+] = -\tfrac{1}{2}U_+$	$[I_+, V_+] = -U_-$	$[I_-, V_+] = 0$
$[I_3, V_-] = +\tfrac{1}{2}V_-$	$[I_+, V_-] = 0$	$[I_-, V_-] = U_+$
$[U_3, V_3] = 0$	$[U_+, V_3] = \tfrac{1}{2}U_+$	$[U_-, V_3] = -\tfrac{1}{2}U_-$
$[U_3, V_+] = -\tfrac{1}{2}V_+$	$[U_+, V_+] = I_-$	$[U_-, V_+] = 0$
$[U_3, V_-] = \tfrac{1}{2}V_-$	$[U_+, V_-] = 0$	$[U_-, V_-] = -I_+$

4.2 Unitary symmetry and the quark model

diagram is planar. A given irreducible representation consists of a set of eigenstates, each one of which has definite F_3 and F_8 eigenvalues and thus corresponds to a point on the diagram.

The simplest representation in $SU(3)$ is the singlet, or "1," which is represented by a single point at the origin (see Figure 4.1). The next representations, the 3 and $\bar{3}$, each consist of three independent states, and are displayed in Figures 4.2a and b, respectively. The 3 and $\bar{3}$ are described by covariant spinor ψ_i and contravariant spinor $\bar{\phi}^j$, respectively, with $i, j = 1, 2, 3$. In flavor $SU(3)$, or $SU(3)_f$, ψ consists of the quark states u, d, and s and $\bar{\phi}$ consists of the antiquark states $\bar{u}$, $\bar{d}$,

Figure 4.1. Weight diagram for $SU(3)$, showing the singlet representation and the I_3, U_3 and V_3 axes.

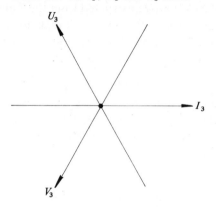

Figure 4.2. Weight diagrams for (a) the 3 and (b) the $\bar{3}$ representations.

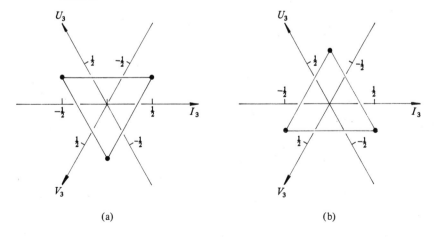

(a)   (b)

and $\bar{s}$:

$$\psi = \begin{pmatrix} u \\ d \\ s \end{pmatrix}, \quad \bar{\phi} = (\bar{u}, \bar{d}, \bar{s})$$

For $SU(3)_f$, it is convenient to define the hypercharge Y:

$$Y = S + B$$

where S is the strangeness and B is the baryon number. Then Y is related to F_8 by the formula:

$$Y = \sqrt{\tfrac{4}{3}} F_8 \tag{4.21}$$

and in $SU(3)_f$, I is the isotopic spin. In Figures 4.3a and b, we give the quantum numbers of u, d, s and $\bar{u}, \bar{d}, \bar{s}$ directly on the weight diagrams.

Higher representations of $SU(3)$ are constructed from the 3 and $\bar{3}$ by utilization of the invariant tensors δ_i^j, ϵ^{ijk}, and ϵ_{ijk} to break down an arbitrary tensor into its irreducible parts. For example, consider the mixed reducible tensor $\psi_i \bar{\phi}^j$ with nine components:

$$\psi_i \bar{\phi}^j = (\psi_i \bar{\phi}^j - \tfrac{1}{3} \delta_i^j \psi_k \bar{\phi}^k) + \tfrac{1}{3} \delta_i^j \psi_k \bar{\phi}^k \tag{4.22}$$

In this expression, we separated out the trace of $\psi_i \bar{\phi}^j$; this last quantity is a tensor of zero rank with one component (singlet). The remaining irreducible traceless second-rank tensor has eight independent components (octet). Symbolically we have:

$$3 \otimes \bar{3} = 8 + 1 \tag{4.23}$$

The "8" is represented by the weight diagram of Figure 4.4.

Figure 4.3. Quantum numbers of (a) u, d, s and (b) $\bar{u}, \bar{d}, \bar{s}$.

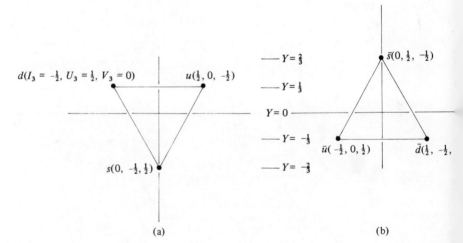

4.2 Unitary symmetry and the quark model

As another example, consider the tensor $\psi_i \eta_j$ composed of two covariant three-spinors. This is broken down as follows:

$$\psi_i \eta_j = \tfrac{1}{2}(\psi_i \eta_j + \psi_j \eta_i) + \tfrac{1}{2}(\psi_i \eta_j - \psi_j \eta_i) \tag{4.24}$$

or, symbolically, as

$$3 \otimes 3 = 6 + \bar{3} \tag{4.25}$$

The symmetric tensor on the right-hand side of (4.24) has six independent elements and the antisymmetric tensor $\bar{3}$ has three. One can build more complicated representations in a similar manner. The task is facilitated by the use of Young diagrams (see, e.g., Messiah 62, Matthews 67). Below, we summarize some tensor products of irreducible representations of $SU(3)$ that frequently appear in applications:

$$3 \otimes 3 = 6 + \bar{3}$$
$$3 \otimes \bar{3} = 8 + 1$$
$$3 \otimes 3 \otimes 3 = 10 + 8 + 8 + 1$$
$$8 \otimes 8 = 27 + 10 + \overline{10} + 8 + 8 + 1$$

In the context of $SU(3)_f$, let us now consider the valence quark composition of mesons ($q\bar{q}$) and baryons (qqq). As noted, the quarks u, d, and s form a 3 in $SU(3)_f$ and the antiquarks $\bar{u}$, $\bar{d}$, and $\bar{s}$ form a $\bar{3}$. It follows from the preceding tabulation that mesons should form singlet and octet representations of $SU(3)_f$ and baryons should appear in singlets, octets, and decuplets. The lightest mesons are those with

Figure 4.4. Weight diagram for the octet representation in $SU(3)$. The symbol at the origin indicates double degeneracy.

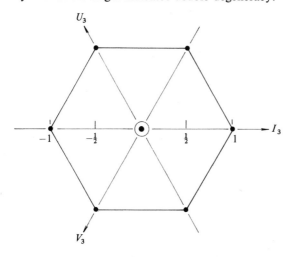

140   4 General properties of hadronic weak currents

$J^P = 0^-$, namely, the pseudoscalar mesons $K^\pm$, K^0, $\bar{K}^0$, $\pi^\pm$, π^0, and η^0. These do indeed form an octet, which is shown in Figure 4.5. The masses of the various mesons and their quark compositions are also given in the figure.

As a second example, we consider the $J^P = 1^-$ (vector) mesons (which are unstable for decay by strong or electromagnetic interactions). These form an octet and a singlet (although there is mixing between the $SU(3)$ singlet and the isosinglet component of the octet, so

Figure 4.5. The $SU(3)_f$ octet of pseudoscalar mesons (masses in MeV/c^2).

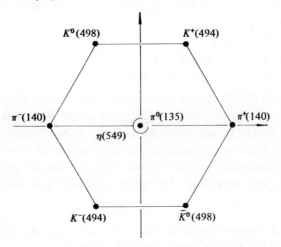

Figure 4.6. The $SU(3)_f$ nonet of vector mesons.

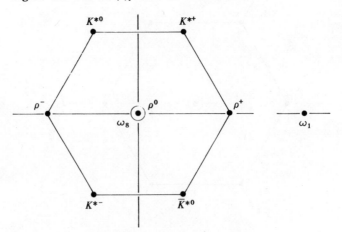

4.2 Unitary symmetry and the quark model

it is more appropriate to speak of the vector meson nonet). These particles are displayed in Figure 4.6. The essential difference between the pseudoscalar mesons and vector mesons is that the former consist of a $q\bar{q}$ pair in a $^1S_0$ state and the latter are each composed of a $q\bar{q}$ pair in a $^3S_1$ state.

Higher meson resonances consisting of up, down, and/or strange quarks and antiquarks are grouped in similar octets and singlets distinguished from one another by spin, parity, and G parity.

Turning next to the baryons with u, d, and/or s quarks, we consider

Figure 4.7. The $J^P = \frac{1}{2}^+$ metastable baryon octet.

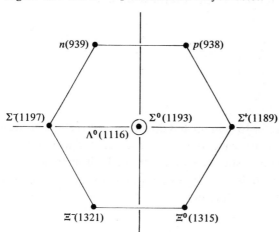

Figure 4.8. The $\frac{3}{2}^+$ baryon decuplet.

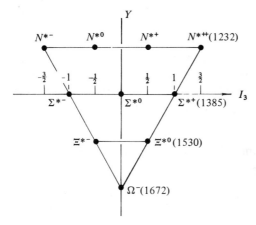

first the $J^P = \frac{1}{2}^+$ octet, which includes the nucleons and the metastable hyperons (Figure 4.7). The next baryon states are the $J^P = \frac{3}{2}^+$, which form a decuplet (Figure 4.8). Except for Ω^-, which decays by weak interaction, all of the $\frac{3}{2}^+$ resonances are unstable for decay by strong interaction.

4.2.3 $SU(6)_f$

Each quark possesses spin $\frac{1}{2}$ and thus has two possible spin orientations with respect to any given axis. Taking spin into account, we are led to consider, in place of $SU(3)_f$, the approximate symmetry group $SU(6)_f$ with fundamental representation "6":

$$6 = \begin{pmatrix} u_1 \\ u_2 \\ d_1 \\ d_2 \\ s_1 \\ s_2 \end{pmatrix} \tag{4.26}$$

where 1 and 2 refer to spin up and down, respectively. This description would appear to be valid only to the extent to which we can treat quarks as nonrelativistic objects within the hadron because, for a relativistic spin-$\frac{1}{2}$ particle, the spin operator does not commute with the Hamiltonian. Despite this limitation, the $SU(6)_f$ quark model we are about to describe turns out to be remarkably effective.

Since a meson contains a valence $q\bar{q}$ pair, the possible meson multiplets in $SU(6)_f$ are given by the formula:

$$SU(6)_f: \quad 6 \otimes \bar{6} = 35 + 1 \tag{4.27}$$

The "35" contains the following $SU(3)_f \times$ spin $SU(2)$ multiplets:

$$35 = (8 \otimes 1) + (8 \otimes 3 + 1 \otimes 3)$$

The first term in parentheses on the right-hand side is identified as the pseudoscalar octet and the second as the vector meson nonet.

Baryon multiplets in $SU(6)_f$ are obtained from the tensor product:

$$SU(6)_f: \quad 6 \otimes 6 \otimes 6 = 56 + 70 + 70 + 20 \tag{4.28}$$

The "56" is completely symmetric in spin–unitary spin indexes and contains the following $SU(3)_f \times$ spin $SU(2)$ multiplets:

$$56 = (8 \otimes 2) + (10 \otimes 4) \tag{4.29}$$

The "20" is completely antisymmetric and has the composition:

$$20 = (8 \otimes 2) + (1 \otimes 4) \tag{4.30}$$

The "70" has mixed spin–unitary spin exchange symmetry.

4.2.4 Color

Let us now consider the lowest-energy baryon states. We expect that the spatial wavefunction for relative quark motion should be s-wave and, therefore, exchange-symmetric, in order to minimize the quark kinetic energy. The Pauli principle would then appear to demand that the spin-unitary spin portion of the wavefunction be antisymmetric and, thus, that the lowest-energy baryons belong to the 20. However, from (4.30), this implies that there should exist an $SU(3)_f$ singlet with spin $\frac{3}{2}$ and fairly low energy, but no such multiplet appears. Moreover, the theoretical magnetic moments of the spin-$\frac{1}{2}$ octet in the 20 have the wrong signs and magnitudes compared with the experimental magnetic moments of the metastable $\frac{1}{2}^+$ octet.

On the other hand, the 56 contains just those $SU(3)_f$ multiplets that actually appear, and the magnetic moments of the spin-$\frac{1}{2}$ octet in the 56 are in reasonable agreement with experiment. Evidently, then it is the symmetric 56 that describes the lowest-energy baryon states. To satisfy the Pauli principle, we must have another degree of freedom in which the quarks within a baryon are exchange-antisymmetric.

That degree of freedom is color (Greenberg 64,78). A quark of any given flavor possesses color R, B, or G, which are states of a "3" in color $SU(3)$, or $SU(3)_c$. The color states of an antiquark – $\bar{R}$, $\bar{B}$ or $\bar{G}$ – form a $\bar{3}$ in $SU(3)_c$. It must be emphasized that $SU(3)_c$ is an entirely distinct symmetry from $SU(3)_f$; in contrast to the latter, $SU(3)_c$ is presumed to be an *exact* and *fundamental* symmetry of nature.

The color state of a baryon is the completely antisymmetric singlet that occurs in the tensor product:

$$SU(3)_c: \quad 3 \otimes 3 \otimes 3 = 10 + 8 + 8 + 1 \tag{4.31}$$

Mesons are also assumed to be in color-singlet states:

$$\text{``1''} = (R\bar{R} + B\bar{B} + G\bar{G})/\sqrt{3} \tag{4.32}$$

Although hadrons possess no color, this does not imply that there are no testable experimental consequences of the color hypothesis. For example, we already saw that the decay rates for the τ lepton to $e\nu\bar{\nu}$, to $\mu\nu\bar{\nu}$, and to hadrons are roughly in the ratio $1:1:3$. This follows from the fact that quarks appear in three colors. Similarly, it can be shown (Adler 70) that the color hypothesis is an essential ingredient for obtaining agreement between theory and experiment in the decay $\pi^0 \to \gamma\gamma$. Also, the numerical value of the ratio

$$R = \frac{\sigma(e^+e^- \to \text{hadrons})}{\sigma(e^+e^- \to \mu^+\mu^-)}$$

observed in e^+e^- collisions depends on color; and agreement between theory and experiment is achieved only if one assumes that quarks have three colors. We shall encounter further consequences of the color hypothesis in our study of hadronic weak interactions.

4.2.5 Weak and electromagnetic currents in $SU(3)_f$

Let us consider the pseudoscalar meson octet, written explicitly in matrix form:

$$M_i^j = \begin{pmatrix} \frac{2}{3}\bar{u}u - \frac{1}{3}\bar{d}d - \frac{1}{3}\bar{s}s & \bar{d}u & \bar{s}u \\ \bar{u}d & -\frac{1}{3}\bar{u}u + \frac{2}{3}\bar{d}d - \frac{1}{3}\bar{s}s & \bar{s}d \\ \bar{u}s & \bar{d}s & -\frac{1}{3}\bar{u}u - \frac{1}{3}\bar{d}d + \frac{2}{3}\bar{s}s \end{pmatrix}$$
(4.33)

We now set about to modify this expression without changing its $SU(3)$ transformation properties. First, we replace each quark by its corresponding field operator, for example, $u \to U$ and $\bar{u} \to \bar{U}$, where U and $\bar{U}$ are the up-quark Dirac field and its Dirac conjugate, respectively. Then we insert γ^μ matrices. The result is an octet of vector currents, j_μ:

$$\begin{pmatrix} \frac{2}{3}\bar{U}\gamma_\mu U - \frac{1}{3}\bar{D}\gamma_\mu D - \frac{1}{3}\bar{S}\gamma_\mu S & \bar{D}\gamma_\mu U & \bar{S}\gamma_\mu U \\ \bar{U}\gamma_\mu D & -\frac{1}{3}\bar{U}\gamma U + \frac{2}{3}\bar{D}\gamma_\mu D - \frac{1}{3}\bar{S}\gamma_\mu S & \bar{S}\gamma_\mu D \\ \bar{U}\gamma_\mu S & \bar{D}\gamma_\mu S & -\frac{1}{3}\bar{U}\gamma_\mu U - \frac{1}{3}\bar{D}\gamma_\mu D + \frac{2}{3}\bar{S}\gamma_\mu S \end{pmatrix}$$
(4.34)

This octet may be written in the convenient notation:

$$j_k^\mu(x) = \bar{\psi}(x)\gamma^\mu \tfrac{1}{2}\lambda_k \psi(x)$$
$$= \begin{pmatrix} j_3 + j_8/\sqrt{3} & j_1 + ij_2 & j_4 + ij_5 \\ j_1 - ij_2 & -j_3 + j_8/\sqrt{3} & j_6 + ij_7 \\ j_4 - ij_5 & j_6 - ij_7 & -\sqrt{\tfrac{2}{3}}j_8 \end{pmatrix}$$
(4.35)

where ψ is the quark field with three flavors and four space-time components. The first entry of the principal diagonal of (4.34) is obviously the electromagnetic current apart from a factor $|e|$; thus we may write:

$$j_{EM}^\mu = |e|[j_3^\mu(x) + j_8^\mu(x)/\sqrt{3}] \tag{4.36}$$

which is a linear combination of the third component of the isospin current j_3 and an isoscalar current j_8. It is readily verified from comparison with (4.18) that j_{EM} is a U-spin scalar.

The other isospin current components are $j_1 \pm ij_2$. These clearly correspond to the $\Delta S = 0$ charged hadronic weak vector current and its Hermitian conjugate. Also, $j_4 \pm ij_5$ correspond to the $|\Delta S| = 1$ charged hadronic weak vector current and its Hermitian conjugate.

4.2 Unitary symmetry and the quark model

Similarly we may define an octet of axial currents:

$$g_k^\mu(x) = \bar{\psi}(x)\gamma^\mu\gamma_5 \tfrac{1}{2}\lambda_k \psi(x) \tag{4.37}$$

The components $g_1 \pm ig_2$ and $g_4 \pm ig_5$ are clearly associated with the axial portions of the charged hadronic weak current for $\Delta S = 0$ and $|\Delta S| = 1$, respectively. However, the component $g_3 + g_8/\sqrt{3}$ has no immediate physical significance.

4.2.6 Baryon magnetic moments and quark masses

From the foregoing properties of the electromagnetic current and the fact that the metastable spin-$\tfrac{1}{2}$ baryons form an octet (Figure 4.7), one may derive certain relationships among the baryon magnetic moments that should be valid in the limit of perfect $SU(3)_f$ symmetry. For example, p and Σ^+ belong to the same U-spin doublet, as do Σ^- and Ξ^-, whereas the electromagnetic current is a U-spin scalar. Therefore we expect the magnetic moments of p and Σ^+ to be equal, and also the magnetic moments of Σ^- and Ξ^- should be equal. By the same argument, $\mu(n) = \mu(\Xi^0)$. The complete set of these relations, originally derived by Coleman and Glashow (61), is as follows:

$$\mu(\Sigma^+) = \mu(p) \tag{4.38}$$

$$\mu(\Lambda^0) = \tfrac{1}{2}\mu(n) \tag{4.39}$$

$$\mu(\Xi^0) = \mu(n) \tag{4.40}$$

$$\mu(\Xi^-) = \mu(\Sigma^-) = -[\mu(p) + \mu(n)] \tag{4.41}$$

$$\mu(\Sigma^0) = -\tfrac{1}{2}\mu(n) \tag{4.42}$$

and for the transition moment in $\Sigma^0 \to \Lambda^0\gamma$ decay:

$$\mu_{\Sigma\Lambda} = \sqrt{\tfrac{3}{2}}\mu(n) \tag{4.43}$$

If we assume that the baryons are described by the color-symmetric quark model, then the quark spin configurations of the various baryon states in the 56 are specified. For example, ignoring color indexes and permutation symmetry and employing the subscript 1(2) for spin up (down), we have:

$$N^{*++}(J_3 = \tfrac{3}{2}) = u_1 u_1 u_1 \tag{4.44}$$

$$\Sigma^{*+}(J_3 = \tfrac{3}{2}) = u_1 u_1 s_1 \tag{4.45}$$

$$\Xi^{*0}(J_3 = \tfrac{3}{2}) = u_1 s_1 s_1 \tag{4.46}$$

$$\Omega^-(J_3 = \tfrac{3}{2}) = s_1 s_1 s_1 \tag{4.47}$$

By application of suitable spin and/or isospin lowering operators, we find:

$$p(J_3 = \tfrac{1}{2}) = (18)^{-1/2}(2u_1u_1d_2 - u_1u_2d_1 - u_2u_1d_1 + \text{permutations}) \tag{4.48}$$

$$n(J_3 = \tfrac{1}{2}) = (18)^{-1/2}(2d_1d_1u_2 - d_1d_2u_1 - d_2d_1u_1 + \cdots) \tag{4.49}$$

$$\Sigma^+(J_3 = \tfrac{1}{2}) = (18)^{-1/2}(2u_1u_1s_2 - u_1u_2s_1 - u_2u_1s_1 + \cdots) \tag{4.50}$$

$$\Sigma^-(J_3 = \tfrac{1}{2}) = (18)^{-1/2}(2d_1d_1s_2 - d_1d_2s_1 - d_2d_1s_1 + \cdots) \tag{4.51}$$

$$\Lambda^0(J_3 = \tfrac{1}{2}) = (12)^{-1/2}(u_1d_2s_1 - u_2d_1s_1 - d_1u_2s_1 + d_2u_1s_1 + \cdots) \tag{4.52}$$

$$\Xi^0(J_3 = \tfrac{1}{2}) = (18)^{-1/2}(2s_1s_1u_2 - s_1s_2u_1 - s_2s_1u_1 + \cdots) \tag{4.53}$$

$$\Xi^-(J_3 = \tfrac{1}{2}) = (18)^{-1/2}(2s_1s_1d_2 - s_1s_2d_1 - s_2s_1d_1 + \cdots) \tag{4.54}$$

These formulas yield simple expressions for the baryon magnetic moments in terms of the magnetic moments of the quarks themselves. For example,

$$\begin{aligned} \mu(p) &= \langle p(J_3 = \tfrac{1}{2}) | \mu_u \sigma_z^{\,1} + \mu_u \sigma_z^{\,2} + \mu_d \sigma_z^{\,3} | p(J_3 = \tfrac{1}{2}) \rangle \\ &= \tfrac{1}{3}[4\mu(u) - \mu(d)] \end{aligned} \tag{4.55}$$

Similarly,

$$\mu(n) = \tfrac{1}{3}[4\mu(d) - \mu(u)] \tag{4.56}$$

Let us now assume that each quark possesses a normal Dirac magnetic moment $\mu(q) = Q(q)\hbar/2m_q c$, where $Q(q)$ is the quark electric charge and m_q an effective, or constituent, quark mass. If we assume that u and d quarks have the same mass, then taking into account the u and d quark charges and (4.55) and (4.56), we arrive at the prediction:

$$\mu(n)/\mu(p) = -\tfrac{2}{3}$$

which is in striking agreement with experiment, the actual value being -0.685. Alternatively, we may take the experimental values of proton and neutron moments and (4.55) and (4.56) to determine m_u and m_d:

$$m_u \simeq m_d \simeq 336 \quad \text{MeV}/c^2 \tag{4.57}$$

Furthermore, let us assume that the mass difference between Λ^0 and the proton is accounted for by the mass difference between the s quark and the u quark. From this we find the constituent mass of the s quark to be

$$m_s \simeq 510 \quad \text{MeV}/c^2 \tag{4.58}$$

It is important to emphasize that the constituent quark masses are not at all the same as those masses that appear in the $SU(3)_c$ Yang–Mills Langangian of quantum chromodynamics. The latter, called "current-quark" masses, have been estimated by Weinberg (77b) to be

$$\begin{aligned} m_u^{\,0} &= 4.3 \quad \text{MeV}/c^2 \\ m_d^{\,0} &= 7.5 \quad \text{MeV}/c^2 \\ m_s^{\,0} &= 150 \quad \text{MeV}/c^2 \end{aligned} \tag{4.59}$$

4.2 Unitary symmetry and the quark model

A clear understanding of the discrepancy between the more fundamental current-quark mass and the constituent-quark mass must await clarification of the relationship between QCD and the semiempirical quark model. In connection with the current-quark masses, we limit ourselves to the remark that $SU(3)_f$ appears to be a reasonably useful symmetry because the current-quark masses of u, d, and s are much less than those of the heavier quarks and also much less than the masses of hadrons composed of u, d, and/or s.

We conclude our discussion of the baryon magnetic moments by noting that, since the u and d quarks form a singlet spin state in the Λ^0 [see (4.52)], the magnetic moment of this baryon should be equal to that of the s quark. From (4.58), we arrive at a prediction for the magnetic moment that agrees with experiment to about 1 percent. The magnetic moments of other spin-$\frac{1}{2}$ baryons are in more-or-less reasonable agreement with experiment, but there are discrepancies that are difficult to explain (Lipkin 81a). We shall return to the subject of baryon magnetic moments in Section 6.5.

4.2.7 Charm and SU(4): Beyond charm

Of course there are more than three quarks. Charm was proposed as early as 1964, and as we noted in Chapter 1, Glashow, Iliopoulos, and Maiani (70) made use of the notion of charm to generalize

Figure 4.9. Observed charmonium ($c\bar{c}$) levels (masses in MeV/c^2).

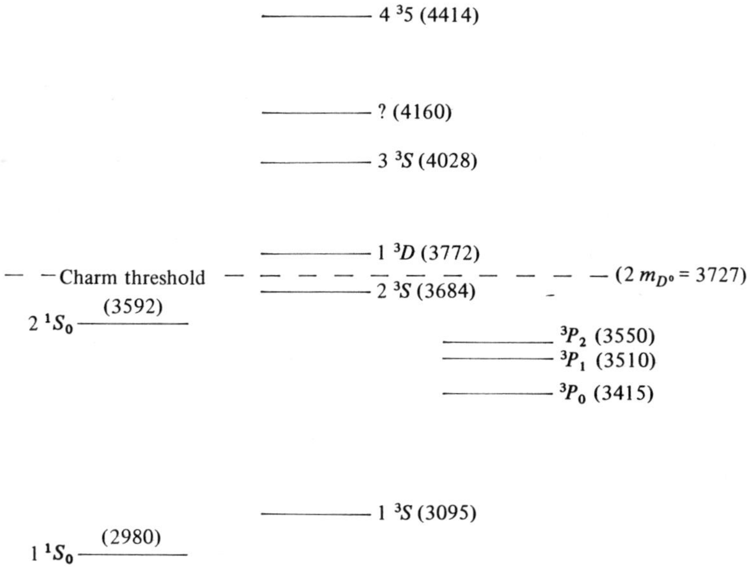

$4\ ^3S\ (4414)$

$?\ (4160)$
$3\ ^3S\ (4028)$

$1\ ^3D\ (3772)$
Charm threshold $\qquad\qquad\qquad 2\ ^3S\ (3684) \qquad\qquad (2\ m_{D^0} = 3727)$
(3592)
$2\ ^1S_0$
$\qquad\qquad\qquad\qquad\qquad\qquad\qquad ^3P_2\ (3550)$
$\qquad\qquad\qquad\qquad\qquad\qquad\qquad ^3P_1\ (3510)$
$\qquad\qquad\qquad\qquad\qquad\qquad\qquad ^3P_0\ (3415)$

$1\ ^3S\ (3095)$

(2980)
$1\ ^1S_0$

Cabibbo's hypothesis. In 1974, the narrow resonance ψ/J was observed simultaneously at SLAC and at Brookhaven, and subsequent experimental work revealed that ψ is the lowest $^3S_1$ (vector meson) bound state of a $c\bar{c}$ pair and is but the first of a complex sequence of resonant states (Figure 4.9) that are to be interpreted as $c\bar{c}$ bound states with quantum numbers as indicated. Moreover, above the charm threshold, e^+e^- collisions result in the production of meson pairs D^+D^-, $D^0\bar{D}^0$, and so on, in which each meson contains a single charmed quark, c or $\bar{c}$. The quantum numbers of c, u, d, and s are summarized in Table 4.3.

When charm is included, we must employ $SU(4)_f$ in place of $SU(3)_f$. This symmetry is badly broken because the mass of the charmed quark is large:

$$m_c^0 = 1.8 \quad \text{GeV}/c^2 \tag{4.60}$$

Nevertheless, $SU(4)_f$ is useful in some applications.

The fundamental representations in $SU(4)$ are the 4 and the $\bar{4}$, corresponding in $SU(4)_f$ to the covariant spinor and contravariant spinor, respectively:

$$\begin{pmatrix} u \\ d \\ s \\ c \end{pmatrix} \quad \text{and} \quad (\bar{u}, \bar{d}, \bar{s}, \bar{c})$$

Irreducible representations are built in the same general way here as they were for $SU(3)$. We summarize here those that appear most frequently in applications.[1]

Table 4.3. *Quantum numbers of quarks and antiquarks*

Flavor	J^P	I	I_3	Y	S	Z^a	C	$SU(3)_f$	$SU(4)_f$	Q	B	$SU(3)_c$
u	$\tfrac{1}{2}^+$	$\tfrac{1}{2}$	$\tfrac{1}{2}$	$\tfrac{1}{3}$	0	$\tfrac{1}{4}$	0	3	4	$\tfrac{2}{3}$	$\tfrac{1}{3}$	3
d	$\tfrac{1}{2}^+$	$\tfrac{1}{2}$	$-\tfrac{1}{2}$	$\tfrac{1}{3}$	0	$\tfrac{1}{4}$	0	3	4	$-\tfrac{1}{3}$	$\tfrac{1}{3}$	3
s	$\tfrac{1}{2}^+$	0	0	$-\tfrac{2}{3}$	-1	$\tfrac{1}{4}$	0	3	4	$-\tfrac{1}{3}$	$\tfrac{1}{3}$	3
c	$\tfrac{1}{2}^+$	0	0	0	0	$-\tfrac{3}{4}$	1	1	4	$\tfrac{2}{3}$	$\tfrac{1}{3}$	3
$\bar{u}$	$\tfrac{1}{2}^-$	$\tfrac{1}{2}$	$-\tfrac{1}{2}$	$-\tfrac{1}{3}$	0	$-\tfrac{1}{4}$	0	$\bar{3}$	$\bar{4}$	$-\tfrac{2}{3}$	$-\tfrac{1}{3}$	$\bar{3}$
$\bar{d}$	$\tfrac{1}{2}^-$	$\tfrac{1}{2}$	$\tfrac{1}{2}$	$-\tfrac{1}{3}$	0	$-\tfrac{1}{4}$	0	$\bar{3}$	$\bar{4}$	$\tfrac{1}{3}$	$-\tfrac{1}{3}$	$\bar{3}$
$\bar{s}$	$\tfrac{1}{2}^-$	0	0	$\tfrac{2}{3}$	1	$-\tfrac{1}{4}$	0	$\bar{3}$	$\bar{4}$	$\tfrac{1}{3}$	$-\tfrac{1}{3}$	$\bar{3}$
$\bar{c}$	$\tfrac{1}{2}^-$	0	0	0	0	$\tfrac{3}{4}$	-1	1	$\bar{4}$	$-\tfrac{2}{3}$	$-\tfrac{1}{3}$	$\bar{3}$

^{a} Supercharge Z in $SU(4)_f$ is analogous to hypercharge Y in $SU(3)_f$.

---

1 S, symmetric; A, antisymmetric; and M, mixed symmetry.

4.2 Unitary symmetry and the quark model

$$4 \otimes \bar{4} = 15 + 1$$
$$4 \otimes 4 = 10 + \bar{6}$$
$$4 \otimes 4 \otimes 4 = 20_S + 20_M + 20_M + \bar{4}_A$$
$$15 \otimes 15 = 1 + 15_S + 15_A + 20_S + 45 + \overline{45} + 84$$

Let us now consider meson and baryon states in which a charmed quark appears. The lowest-energy $^1S_0$ "15" includes, in addition to the pseudoscalar noncharmed mesons of Figure 4.5, also the following mesons:

$$D^0 = c\bar{u}, \quad \bar{D}^0 = \bar{c}u, \quad m = 1863 \text{ MeV}/c^2$$
$$D^+ = c\bar{d}, \quad D^- = \bar{c}d, \quad m = 1868 \text{ MeV}/c^2$$
$$F^+ = c\bar{s}, \quad F^- = \bar{c}s, \quad m = 2.04 \text{ GeV}/c^2$$

A systematic treatment of the baryon multiplets requires that quark spin be taken into account and thus that we employ the approximate symmetry group $SU(8)_f$ in place of $SU(4)_f$. Rather than presenting any of these details, we merely provide Table 4.4, which summarizes the properties of the charmed baryons. Of all the baryons listed in this table, only $\Lambda_1^+ = \Lambda_c^+$ has been observed in any detail.

Resonances similar to those in the $c\bar{c}$ system were first observed in 1977:

$$Y(9.40 \text{ GeV}), \quad Y'(10.01 \text{ GeV}), \quad Y''(10.40 \text{ GeV})$$

and are interpreted as $b\bar{b}$ bound states. Above the "bottom" threshold, one can form pairs of mesons, each one of which includes a single b ($\bar{b}$). The bottom mesons:

$$B_{\bar{u}}^- = b\bar{u}, \quad B_d^0 = b\bar{d}, \quad B_s^0 = b\bar{s}, \quad B_{\bar{c}} = b\bar{c}$$

Table 4.4. *Properties of charmed baryons*

Quark content	I	I_3	S	Y	C	Z	Q	Symbol[a]
cuu	1	1	0	$\frac{2}{3}$	1	$-\frac{1}{4}$	2	Σ_1^{++}
cud	1	0	0	$\frac{2}{3}$	1	$-\frac{1}{4}$	1	Σ_1^+
	0	0	0	$\frac{2}{3}$	1	$-\frac{1}{4}$	1	Λ_1^+
cdd	1	-1	0	$\frac{2}{3}$	1	$-\frac{1}{4}$	0	Σ_1^0
cus	$\frac{1}{2}$	$\frac{1}{2}$	-1	$-\frac{1}{3}$	1	$-\frac{1}{4}$	1	Ξ_1^+
cds	$\frac{1}{2}$	$-\frac{1}{2}$	-1	$-\frac{1}{3}$	1	$-\frac{1}{4}$	0	Ξ_1^0
css	0	0	-2	$-\frac{4}{3}$	1	$-\frac{1}{4}$	0	Ω_1^0
ccu	$\frac{1}{2}$	$\frac{1}{2}$	0	$\frac{1}{3}$	2	$-\frac{5}{4}$	2	Ξ_2^{++}
ccd	$\frac{1}{2}$	$-\frac{1}{2}$	0	$\frac{1}{3}$	2	$-\frac{5}{4}$	1	Ξ_2^+
ccs	0	0	-1	$-\frac{2}{3}$	2	$-\frac{5}{4}$	1	Ω_2^+
ccc	0	0	0	0	3	-2	2	Ω_3^{++}

[a] See, for example, Lichtenberg (75).

should have a rich and complex pattern of weak decays, study of which is just beginning. We shall consider the decays of hadrons containing c and b quarks in Chapter 6.

4.3 The six-quark model

Let us now recall the scheme of Glashow, Iliopoulos, and Maiani (70), or GIM, in which only the four quarks u, c, d, and s are considered. Each is assumed to possess definite mass, but the quark states entering into the charged weak current are linear combinations of d and s:

$$d_C = d \cos \theta_C + s \sin \theta_C$$
$$s_C = -d \sin \theta_C + s \cos \theta_C \quad (4.61)$$

where θ_C is the Cabibbo angle. This simple idea shall now be put in a form that lends itself easily to generalization. That u, c, d and s have definite masses is expressed by writing the mass terms in the Lagrangian in matrix form:

$$\mathscr{L}_{\text{mass}} = (\bar{P}_R M_P P_L + \bar{P}_L M_P^\dagger P_R) + (\bar{N}_R M_N N_L + \bar{N}_L M_N^\dagger N_R) \quad (4.62)$$

where

$$P_{R,L} = \begin{pmatrix} u \\ c \end{pmatrix}_{R,L}, \qquad N_{R,L} = \begin{pmatrix} d \\ s \end{pmatrix}_{R,L}$$

and

$$M_P = \begin{pmatrix} m_u & 0 \\ 0 & m_c \end{pmatrix}, \qquad M_N = \begin{pmatrix} m_d & 0 \\ 0 & m_s \end{pmatrix}$$

are real diagonal matrices containing the various quark masses.[2] Expression (4.62) is gauge-invariant, as we know from a similar discussion in Section 2.6.

The (Hermitian conjugate) charged weak current of quarks may be written as

$$J_\mu^\dagger = \bar{P}_L \gamma_\mu U N_L$$

where, in the GIM model,

$$U = \begin{pmatrix} \cos \theta_C & \sin \theta_C \\ -\sin \theta_C & \cos \theta_C \end{pmatrix}$$

For the sake of discussion, however, let us only require U to be unitary and drop all other restrictions on U. A 2×2 unitary matrix contains

---

[2] The mass matrix for leptons is not assumed to be diagonal; see Sections 2.6 and 10.1.

4.3 The six-quark model

four complex numbers, or eight real parameters, but the condition $U^\dagger U = I$ imposes four constraints. In addition, an overall phase is arbitrary, so the number of independent real parameters is only three. Let us perform unitary transformations on the vectors $P_{L,R}$ and $N_{L,R}$ in an attempt to simplify U further. We write:

$$P_{L,R} \to P'_{L,R} = e^{i\lambda_P \theta_P} P_{L,R} \tag{4.63}$$

$$N_{L,R} \to N'_{L,R} = e^{i\lambda_N \theta_N} N_{L,R} \tag{4.64}$$

$$M_{P,N} \to M'_{P,N} = e^{i\lambda_{P,N} \theta_{P,N}} M_{P,N} e^{-i\lambda_{P,N} \theta_{P,N}} \tag{4.65}$$

$$U \to U' = e^{i\lambda_P \theta_P} U e^{-i\lambda_N \theta_N} \tag{4.66}$$

The λ's are Hermitian matrices of the same dimension as U and M (2×2 at present) and the θ_P and θ_N are real numbers. We require $M'_{P,N}$ to be diagonal, which means that the λ's must also be diagonal. In the 2×2 case, the only such nontrivial transformation is $e^{i\sigma_3 \theta_P} U e^{-i\sigma_3 \theta_N}$, where $\sigma_3 = \begin{pmatrix} 1 & 0 \\ 0 & -1 \end{pmatrix}$ is the Pauli spin matrix. In (4.66), there are thus two parameters, θ_P and θ_N, that remove 2 degrees of freedom. Thus, finally, U contains only one free real parameter, so that the GIM model with one Cabibbo angle yields the most general form for U in the four-quark case.

The argument may now be extended to six quarks. The vectors $P_{L,R}$ and $N_{L,R}$ are now

$$P_{L,R} = \begin{pmatrix} u \\ c \\ t \end{pmatrix}_{L,R}, \quad N_{L,R} = \begin{pmatrix} d \\ s \\ b \end{pmatrix}_{L,R} \tag{4.67}$$

and U has $3 \times 3 = 9$ complex numbers or 18 real parameters. Unitarity removes 9, and one more goes into an arbitrary overall phase; thus $N^2 - 1 = 8$ are left. There are 8 independent 3×3 Hermitian matrices of which 2:

$$T_3 = \begin{pmatrix} 1 & 0 & 0 \\ 0 & 0 & 0 \\ 0 & 0 & -1 \end{pmatrix} \quad \text{and} \quad Y = \begin{pmatrix} 1 & 0 & 0 \\ 0 & -2 & 0 \\ 0 & 0 & 1 \end{pmatrix} \tag{4.68}$$

are diagonal [see (4.10)]. The most general transformation of form (4.66) is then

$$U' = e^{i(aT_3 + bY)} U e^{-i(cT_3 + dY)} \tag{4.69}$$

where a, b, c, and d are real parameters. Therefore, $8 - 4 = 4$ degrees of freedom remain in U. However, a real orthogonal 3×3 matrix is characterized by three real parameters, for example, the Euler angles. Thus such a matrix is inadequate to describe the most general charged current of six quarks. Instead we need three Cabibbo-like angles and an additional parameter (which may be interpreted as a *CP*-violating phase; see Chapter 7). The generalization of this argument to more quark generations is given in Table 4.5.

The standard form[3] of the charged current in the six-quark model was first given by Kobayashi and Maskawa (73):

$$J = (\bar{u}, \bar{c}, \bar{t})_L \gamma_\mu \begin{pmatrix} c_1 & +s_1 c_3 & +s_1 s_3 \\ -s_1 c_2 & c_1 c_2 c_3 - s_2 s_3 e^{i\delta} & c_1 c_2 s_3 + s_2 c_3 e^{i\delta} \\ -s_1 s_2 & c_1 s_2 c_3 + c_2 s_3 e^{i\delta} & c_1 s_2 s_3 - c_2 c_3 e^{i\delta} \end{pmatrix} \begin{pmatrix} d \\ s \\ b \end{pmatrix}_L$$

(4.70)

In this formula, c_i (or s_i) = $\cos \theta_i$ (or $\sin \theta_i$) with $i = 1, 2, 3$ and δ the *CP*-violating phase. The original Cabibbo angle is $\theta_1 = \theta_C$. If $s_2 = s_3 = 0$, then (4.70) reduces to the GIM model, and there is no charged weak coupling of b and t quarks to the other quarks.

As we shall see, analyses of nuclear β decay, hyperon semileptonic decays, and K_{e3} decay may be used to determine the $\bar{u}d$ and $\bar{u}s$ couplings. One finds:

$$|c_1| = 0.9737 \pm 0.0025, \qquad |s_1 c_3| = 0.219 \pm 0.011 \qquad (4.71)$$

which may be combined to yield

$$|s_3| = 0.28 \begin{cases} +0.21 \\ -0.28 \end{cases} \qquad (4.72)$$

Table 4.5. *Generalization to* N × N *model*

Argument	Real parameters
General $N \times N$ matrix	$2N^2$
Unitarity	$-N^2$
Overall phase	-1
remaining	$N^2 - 1$
2 × number of diagonal matrices	$-2(N - 1)$
remaining	$(N - 1)^2$
Real mixing angles	$\frac{1}{2}N(N - 1)$
remaining *CP*-violating phases	$\frac{1}{2}(N - 1)(N - 2)$

---

[3] Actually, in (4.70) we adopted a somewhat different choice of phases than that given by Kobayashi and Maskawa.

4.4 Decay of charged pions

In principle, measurements of branching ratios in the decays of D mesons might be used to determine the remaining angle θ_2, but the experimental results, as of mid-1981, were inadequate for this purpose (see Chapter 6). An estimate of the various remaining matrix elements in (4.70) has been given by Sakurai (81) but the entries in the third row or third column are very uncertain.

$$\begin{pmatrix} 0.974 & 0.219 & 0.059 \\ -0.213 & 0.845 & 0.488 \\ 0.057 & -0.489 & 0.870 \end{pmatrix} \quad (4.73)$$

4.4 Decay of charged pions

Perhaps the simplest hadronic weak transition is pion decay $\pi_{l2}: \pi^- \to l^- \bar{\nu}_l$, illustrated in Figure 4.10. The amplitude is

$$\mathcal{M} = \frac{G_F}{\sqrt{2}} \bar{\mu}_l \lambda^\alpha (1 - \gamma_5) v_{\nu l} [-(2E_\pi)^{1/2} \langle 0 | J_\alpha^\dagger | \pi^- \rangle] \quad (4.74)$$

where $J_\alpha^\dagger$ is given by (4.70).

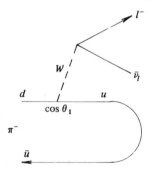

Figure 4.10. Pion decay π_{l2}.

Since the leptonic factor in (4.74) is a vector plus an axial vector, $\langle 0 | J_\alpha^\dagger | \pi^- \rangle$ must be a vector, an axial vector, or a combination of the two and it must be constructed from available kinematic quantities. However, since the pion is a pseudoscalar particle, the only available vector is the momentum transfer $q_\alpha = k_\alpha + p_\alpha$, where k_α and p_α are the neutrino and charged lepton 4-momenta, respectively. Thus

$$-(2E_\pi)^{1/2} \langle 0 | J_\alpha^\dagger | \pi^- \rangle = if_\pi q_\alpha \cos \theta_1 \quad (4.75)$$

where f_π is some real numerical constant, called the pion decay constant, which characterizes the effects of strong interactions. Note that since π^- is pseudoscalar and q_α a polar vector, only the *axial* portion of $J_\alpha^\dagger$, namely $A_\alpha^\dagger$, is operative in (4.75).

Substituting (4.75) in (4.74), we obtain

$$\mathcal{M} = if_\pi \cos\theta_1 \frac{G_F}{\sqrt{2}} \bar{u}_l(\not{p} + \not{k})(1 - \gamma_5)v_{\bar{\nu}} \tag{4.76}$$

However, $\bar{u}_l \not{p} = m_l \bar{u}_l$ and $\not{k}(1 - \gamma_5)v_{\bar{\nu}} = (1 + \gamma_5)\not{k} v_{\bar{\nu}} = 0$. Therefore,

$$\mathcal{M} = i \frac{G_F}{\sqrt{2}} f_\pi \cos\theta_1 \, m_l \bar{u}_l (1 - \gamma_5)v_{\bar{\nu}} \tag{4.77}$$

Thus

$$|\mathcal{M}|^2 = \frac{G_F^2}{2} f_\pi^2 \cos^2\theta_1 \, m_l^2 \frac{1}{4m_l m_\nu} \mathrm{tr}[(\not{p} + m_l)(1 - \gamma_5)\not{k}(1 + \gamma_5)]$$

$$= \frac{G_F^2}{2} f_\pi^2 \cos^2\theta_1 \frac{m_l}{m_\nu} \mathrm{tr}[\not{p}\not{k}(1 + \gamma_5)]$$

$$= G_F^2 f_\pi^2 \cos^2\theta_1 \frac{m_l}{m_\nu} pk$$

We calculate the transition probability per unit time in the rest frame as

$$dW = \frac{G_F^2 f_\pi^2 \cos^2\theta_1 m_l^2 p \cdot k}{2m_\pi} \frac{d^3\mathbf{k}\, d^3\mathbf{p}}{(2\pi)^6 E_l E_\nu} (2\pi)^4 \, \delta(m_\pi - E_l - E_\nu)\, \delta^3(\mathbf{k} + \mathbf{p})$$

Integrating over **p**, we obtain

$$W = \frac{2\pi G_F^2 f_\pi^2 m_l^2 \cos^2\theta_1}{(2\pi)^2 m_\pi} \int \frac{(E_l E_\nu + E_\nu^2) E_\nu^2 \, dE_\nu}{E_l E_\nu}$$

$$\times \delta[m_\pi - E_\nu - (E_\nu^2 + m_l^2)^{1/2}]$$

or, finally,

$$W(\pi_{l2}) = \frac{G_F^2 f_\pi^2}{8\pi} \cos^2\theta_1 m_l^2 m_\pi \left(1 - \frac{m_l^2}{m_\pi^2}\right)^2 \tag{4.78}$$

In the case of $\pi_{\mu 2}$, we may solve this equation for f_π by inserting the known values of m_μ, m_π, G_F [see (3.64)], and θ_1 and also the experimental value of $W^{-1} = \tau = (2.603 \pm 0.002) \times 10^{-8}$ sec, since the branching ratios for all other decay modes are so small. We thus obtain $f_\pi = 0.94\, m_\pi$. We may also obtain the transition rate for $\pi \to e\nu_e$ from the ratio

$$R_0 = \frac{W(\pi \to e\nu_e)}{W(\pi \to \mu\nu_\mu)} = \left(\frac{m_e}{m_\mu}\right)^2 \left(\frac{m_\pi^2 - m_e^2}{m_\pi^2 - m_\mu^2}\right)^2 \tag{4.79}$$

This must be corrected for radiative effects before it can be compared with experiment. One finds (Kinoshita 59):

4.6 Other decays related to $\pi_{\mu 2}$

$$R_{corr} = R_0(1 - 16.9\alpha/\pi) = 1.24 \times 10^{-4} \qquad (4.80)$$

and the observed ratio is (DeCapua et al. 64, Bryman and Picciotto 75)

$$R_{expt} = (1.274 \pm 0.024) \times 10^{-4} \qquad (4.81)$$

The excellent agreement between theory and experiment gives support to the idea of electron–muon universality and to the V–A coupling of the leptonic current.

4.5 $K_{\mu 2}$ and K_{e2} decays

The transitions $K^- \to \mu^- \bar{\nu}_\mu$ and $K^- \to e^- \bar{\nu}_e$ are similar to the decays $\pi_{\mu 2}$ and π_{e2}, respectively, except that in the present case we deal with a quark transformation $s \to u$. Thus, from (4.70), the factor c_1 must be replaced by $+s_1 c_3$. In the following, we shall assume $c_3 = 1$, which is a good approximation. Also, we write:

$$-(2E_K)^{1/2}\langle 0|J_\alpha^\dagger|K^-\rangle = if_K q_\alpha \sin\theta_1 \qquad (4.82)$$

Then we obtain

$$W(K^\pm \to l^\pm \nu_l) = \frac{G_F^2 f_K^2 \sin^2\theta_1}{8\pi} m_l^2 m_K \left(1 - \frac{m_l^2}{m_K^2}\right)^2 \qquad (4.83)$$

Comparing with pion decay, we have

$$\frac{W(K \to \mu\nu)}{W(\pi \to \mu\nu)} = \tan^2\theta_1 \frac{f_K^2 m_K}{f_\pi^2 m_\pi} \frac{[1 - (m_\mu/m_K)^2]^2}{[1 - (m_\mu/m_\pi)^2]^2} \qquad (4.84)$$

From the known masses, experimental transition rates, and θ_1 determined from other experiments, one finds

$$f_K/f_\pi = 1.28 \qquad (4.85)$$

Unfortunately, the experimental value of the ratio of K_{l2} and $K_{\mu 2}$ rates is so uncertain that it does not provide a good test of the prediction

$$R = \frac{W(K_{e2})}{W(K_{\mu 2})}\bigg|_{theor, corr} = 2.09 \times 10^{-5} \qquad (4.86)$$

4.6 Other decays related to $\pi_{\mu 2}$

The pion mass is too small to permit the decay $\pi \to \tau\nu$. However, the inverse transition proceeds with the theoretical rate:

$$W(\tau^- \to \pi^- \nu_\tau) = \frac{G_F^2 f_\pi^2 \cos^2\theta_1 \, m_\tau^3}{16\pi}\left(1 - \frac{m_\pi^2}{m_\tau^2}\right)^2 \qquad (4.87)$$

Observations of the branching ratio (~9 percent) for this decay are in accord with this formula. The transition $\tau^- \to K^- \nu_\tau$ is strongly inhibited by the s_1^2 factor and has not yet been observed. However, the decay

$\tau \to \rho \nu_\tau$ has a large branching ratio (~20 percent), and observations are in good agreement with theory, providing a striking confirmation of the CVC hypothesis to be discussed later (Tsai 71, Gilman and Miller 78, Abrams et al. 79).

The rate for $D^\pm \to l\nu$ should be very small because of the low lepton mass and an unfavorable Cabibbo factor; however, in the case $F^\pm \to \tau\nu$, the large τ mass and the Cabibbo factor are favorable, and it is possible that this transition may be observed in the coming decade.

The simple quark model affords some insight into the numerical values of f_π, f_K, and so forth. Here one views the u and $\bar{d}$ quarks in π^+, for example, as annihilating to produce a W^+, somewhat as e^+ and e^- annihilate in positronium. The amplitude for annihilation is proportional to the wavefunction for relative quark motion at the origin. Early calculations by van Royen and Weisskopf (67a,b) and more recent calculations by Poggio and Schnitzer (79) and Krasemann (80) show that such an approach yields a very good account of the magnetic dipole radiative decay rates of vector mesons, and also reasonable order of magnitude estimates for the decay constants such as f_π.

4.7 Restrictions on weak amplitudes from proper Lorentz invariance

We now discuss amplitudes for more complicated weak transitions, such as meson → meson + leptons, baryon → baryon + leptons, nonleptonic hyperon decays, and so forth.

4.7.1 Meson → meson + leptons ($M_a \to M_b l\nu$)

$\pi^+ \to \pi^0 l\nu$, $K^+ \to \pi^0 l\nu$, $K_L^0 \to \pi^+ l\nu$, and so forth are examples of meson → meson + leptons transitions. The hadronic amplitude must be formed from the available vectors and must itself be a vector, an axial vector, or a linear combination of the two. In the present case, the only available 4-vectors are the initial and final meson momenta k_a and k_b. We cannot construct an axial vector from these alone. Thus the hadronic matrix element has the form

$$\langle M_b | J^\alpha | M_a \rangle = N(f_a k_a^\alpha + f_b k_b^\alpha) \tag{4.88}$$

where N is a Cabibbo factor obtained from (4.70) and f_a and f_b are form factors, that is, scalar functions of the various kinematic invariants. In the present case, the latter are $k_a^2 = m_a^2$, $k_b^2 = m_b^2$, and $q^2 = (k_a - k_b)^2$, which is the only variable at our disposal. In all cases of interest, the initial and final mesons are pseudoscalar; therefore only

4.7 Restrictions on weak amplitudes

the *vector* portion V^α of J^α is operative. In practice it is most convenient to write

$$\langle M_b|V^\alpha|M_a\rangle = N[f_+(q^2)(k_a^\alpha + k_b^\alpha) + f_-(q^2)(k_a^\alpha - k_b^\alpha)] \quad (4.89)$$

where $f_+(q^2)$ and $f_-(q^2)$ are linear combinations of f_a and f_b.

4.7.2 Meson → 2 mesons + leptons ($M_a \to M_b M_c l\nu$)

Here we can form a polar vector

$$\langle M_b M_c|A^\alpha|M_a\rangle = N[a_1(k_b^\alpha + k_c^\alpha) + a_2(k_b^\alpha - k_c^\alpha) + a_3(k_a^\alpha - k_b^\alpha - k_c^\alpha)] \quad (4.90)$$

an an axial vector

$$\langle M_b M_c|V^\alpha|M_a\rangle = Na_4\epsilon^{\alpha\beta\gamma\phi}k_{a\beta}k_{b\gamma}k_{c\phi} \quad (4.91)$$

where a_1, a_2, a_3, and a_4 are form factors depending on the kinematic invariants constructed from k_a, k_b, and k_c. The most general amplitude is a linear combination of (4.90) and (4.91).

4.7.3 Baryon semileptonic transitions (B → B'lν, etc.)

In this case polar vectors can be formed from the momentum transfer $q = p_B - p'_B$, the baryon spinors, and the γ matrices. The quantities

$$\bar{u}(B')\gamma^\alpha u(B), \quad \bar{u}(B')\sigma^{\alpha\nu}q_\nu u(B), \quad \bar{u}(B')u(B)q^\alpha \quad (4.92)$$

are in fact the only three independent polar vectors, since all others reduce to a linear combination of these by suitable application of the Dirac equation. Similarly, the only three independent axial vectors are

$$\bar{u}(B')\gamma^\alpha\gamma_5 u(B), \quad \bar{u}(B')\sigma^{\alpha\nu}q_\nu\gamma_5 u(B), \quad \bar{u}(B')\gamma_5 u(B)q^\alpha \quad (4.93)$$

The most general hadronic amplitude for charged weak interactions is then

$$\langle B'|J^\alpha|B\rangle = N\bar{u}(B')[f_1(q^2)\gamma^\alpha + if_2(q^2)\sigma^{\alpha\nu}q_\nu + f_3(q^2)q^\alpha \\ - g_1(q^2)\gamma^\alpha\gamma_5 - ig_2(q^2)\sigma^{\alpha\nu}q_\nu\gamma_5 - g_3(q^2)\gamma_5 q^\alpha]u(B) \quad (4.94)$$

where $f_{1,2,3}$ and $g_{1,2,3}$ are form factors.[4] A similar expression holds for the neutral weak hadronic current.

4.7.4 Baryon → baryon + meson

In this case we deal with the nonleptonic hyperon ($J^P = \frac{1}{2}^+$) decays:

---

[4] Many different notations are employed in the literature for these form factors, and this can easily lead to confusion, especially since some authors employ the "Pauli" metric and others use the metric and relativistic notation of this text.

$$\Lambda^0 \to p\pi^-, \quad \Sigma^+ \to p\pi^0, \quad \Xi^0 \to \Lambda^0 \pi^0$$
$$\Lambda^0 \to n\pi^0, \quad \Sigma^+ \to n\pi^+, \quad \Xi^- \to \Lambda^0 \pi^- \qquad (4.95)$$
$$\Sigma^- \to n\pi^-$$

and the decays of $\Omega^-(J^P = \tfrac{3}{2}^+)$

$$\Omega^- \to \Lambda K^-$$
$$\to \Xi^0 \pi^-$$
$$\to \Xi^- \pi^0$$

There also exist the decays of the charmed baryon Λ_c^+ ($J_P = \tfrac{1}{2}^+$):

$$\Lambda_c^+ \to \Lambda \pi^+ \pi^+ \pi^-$$
$$\to pK^- \pi^+$$
$$\to pK^*$$
$$\to \Delta(1232)^{++} K^-$$

For the nonleptonic hyperon decays (4.95), the initial and final baryons have spin $\tfrac{1}{2}$ and the meson has spin 0. The most general proper Lorentz-invariant amplitude is written, according to standard convention, as

$$\mathcal{M} = G_F m_{\pi^+}^2 \bar{u}_f (A - B\gamma_5) u_i \qquad (4.96)$$

where u_i and u_f are Dirac spinors for the initial and final baryons, respectively, and A and B are dimensionless constants. Other scalar and pseudoscalar terms constructed from the available kinematic quantities might at first appear possible, but in each case one can show that such terms are already included in (4.96) by suitable application of the Dirac equation. Since the pion is pseudoscalar, the A term in (4.96) is also pseudoscalar and corresponds to zero orbital angular momentum (s-wave) in the final state and the B term is scalar and corresponds to a p-wave. According to convention, the Cabibbo factor is included in A and B.

To describe the nonleptonic decays of Ω^-, we use the fact that a particle of spin $\tfrac{3}{2}$ is described by a wavefunction $\psi_\mu(x)$, which obeys

$$(i\gamma^\lambda \partial_\lambda - m)\psi_\mu(x) = 0 \qquad (4.97)$$

together with the subsidiary conditions

$$\gamma^\mu \psi_\mu(x) = 0, \quad \partial^\mu \psi_\mu(x) = 0 \qquad (4.98)$$

The most general proper Lorentz-invariant amplitude is

$$\mathcal{M} = G_F \frac{k^\lambda}{m_\Omega} \bar{u}(B' - A'\gamma_5) w_\lambda \qquad (4.99)$$

where w_λ and u are spin-$\tfrac{3}{2}$ and -$\tfrac{1}{2}$ spinors for Ω and the final baryon,

4.7 Restrictions on weak amplitudes

respectively, k^λ the 4-momentum of the final meson, and A' and B' the coefficients corresponding to d- and p-wave final states, respectively.

4.7.5 Meson → mesons

Consider the transition $K_S^0 \to \pi^+\pi^-$ and assume that the effective Hamiltonian is formed from the product of the $\Delta S = 0$ and $|\Delta S| = 1$ hadronic weak currents. Since the parities of the initial and final states are opposite, the operative portions of the current × current products are $A \cdot V$ and $V \cdot A$.

In the case of three final mesons, for example, $K \to \pi\pi\pi$, and indeed for analysis of three-body decays in general, it is very convenient to describe the kinematics in terms of a Dalitz diagram. For purposes of illustration, we consider $K_{\pi 3}$ decays and denote the energies and 3-momenta of the three pions by E_1, E_2, and E_3 and $\mathbf{p}_1, \mathbf{p}_2, \mathbf{p}_3$ respectively. Then

$$m_K = E_1 + E_2 + E_3$$
$$= T_1 + T_2 + T_3 + m_1 + m_2 + m_3$$

or

$$Q = m_K - m_1 - m_2 - m_3 = T_1 + T_2 + T_3 \qquad (4.100)$$

where the T's are the pion kinetic energies and $m_1 = m_2 = m$. Figure 4.11 shows an equilaterial triangle of height Q and an arbitrary point P within the triangle that represents a 3π final state with kinetic energies T_1, T_2, and T_3. The sum of the distances T_1, T_2, T_3 from point P to the three sides of the triangle is equal to Q. Relative to the center of the triangle O, we assign to P the polar coordinates $(Q/3)r$, θ. Then,

Figure 4.11. Equilateral triangle representing $K_{\pi 3}$ decay.

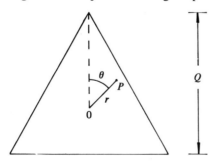

$$T_3 = \frac{Q}{3}(1 + r \cos \theta)$$

$$T_1 = \frac{Q}{3}\left[1 + r \cos\left(\theta + \frac{2\pi}{3}\right)\right]$$

$$T_2 = \frac{Q}{3}\left[1 + r \cos\left(\theta - \frac{2\pi}{3}\right)\right]$$

Conservation of linear momentum implies that

$$2\mathbf{p}_1 \cdot \mathbf{p}_2 = (E_3^2 - m_3^2) - (E_1^2 - m^2) - (E_2^2 - m^2)$$

and thus that

$$(E_1^2 - m^2)(E_2^2 - m^2) \geq [m_K^2 - 2m_K(E_1 + E_2) + 2E_1E_2 + 2m^2 - m_3^2]^2 \tag{4.101}$$

which defines a region interior to a closed curve (see Figure 4.12). This result is expressed in polar coordinates as

$$r(\theta) \leq r_0(\theta) \tag{4.102}$$

where

$$r_0^2 = (1 + \epsilon_0)^{-1}(1 - \epsilon_0 r_0^3 \cos 3\theta)$$

and

$$\epsilon_0 = 2Qm_K/(2m_K - Q)^2$$

The amplitude A_i for a particular three-meson decay can always be expressed as a function of the kinetic energy of two of the particles from conservation of energy:

$$A_i = A_i(T_1, T_2) = A_i(r, \theta) \tag{4.103}$$

Dalitz diagrams are very useful for analysis of data bcause of the following result, which is easily shown. If an experiment is performed to observe a particular decay and each observed event is plotted as a

Figure 4.12. Dalitz plot boundary; see (4.101).

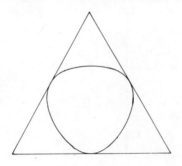

point in the Dalitz diagram, then the density of events in the neighborhood of a given point with coordinates r, θ is proportional to $|A_i(r, \theta)|^2$.

4.8 Time-reversal invariance; *CPT* invariance

Let us consider a general transition $A(\mathbf{p}_i, \mathbf{s}_i) \rightarrow B(\mathbf{p}_f, \mathbf{s}_f)$, where A stands for one or more initial particles and B for final particles, with definite momenta and spins $\mathbf{p}_i$, $\mathbf{s}_i$ and $\mathbf{p}_f$, $\mathbf{s}_f$. This transition is described by the S-matrix element (see Appendix B):

$$S_{fi} = \delta_{fi} - i\mathcal{T}_{fi}$$

where

$$\begin{aligned}\mathcal{T}_{fi} &= \langle B|\mathcal{T}|A\rangle \\ &= (2\pi)^4\,\delta^4(p_f - p_i)\Pi(m_i/E_i V)^{1/2}\Pi(m_f/E_f V)^{1/2}\mathcal{M}\end{aligned} \quad (4.104)$$

if the particles are fermions (or with the replacement $(m/EV)^{1/2} \rightarrow (1/2EV)^{1/2}$ for each particle that is a boson). We now reverse all spins and momenta and interchange initial and final states. The amplitude for this "time-reversed" process will be called $\mathcal{M}'$ and time-reversal invariance holds by definition if $\mathcal{M}' = \mathcal{M}^*$.

For certain transitions involving strong or electromagnetic interactions, it is feasible to test for T invariance directly by comparing the forward and inverse reaction rates, for example,

$$n + p \rightleftarrows d + \gamma \quad (4.105)$$

Such detailed balance comparisons are obviously a practical impossibility for weak interactions, but here we can rely on a more indirect argument, which depends on the fact that in first-order perturbation theory the operator $\mathcal{T}$ appearing in (4.104) is Hermitian: $\mathcal{T}^\dagger = \mathcal{T}$.

To illustrate, we consider neutron β decay with the initial neutron at rest and spin in a definite direction (Figure 4.13a) and the final particle momenta also specified. In our hypothetical experiment, we do not observe the final spins and these may be disregarded. The amplitude for this decay is $\mathcal{M}$, as in (4.104). Now we take the complex conjugate of both sides of (4.104), and since $\mathcal{T} = \mathcal{T}^\dagger$,

$$\langle f|\mathcal{T}|i\rangle^* = \langle i|\mathcal{T}|f\rangle \quad (4.106)$$

Physically we have interchanged initial and final plane-wave states without altering spins or momenta, and the new process is described by amplitude $\mathcal{M}^*$ (see Figure 4.13b). Next we perform a time-reversal transformation on the system of Figure 4.13b. This has the effect of reversing spins and momenta and interchanging initial and final states once again. The result (Figure 4.13c) is described by some new amplitude $\mathcal{M}'^*$; and after a rotation of 180° (Figure 4.13d), it is seen to be equivalent to the configuration of Figure 4.13a, except that the neutron spin is reversed. If $|\mathcal{M}'| \neq |\mathcal{M}^*|$, then T invariance is violated and the probabilities for decays with $\boldsymbol{\sigma}_n \cdot \mathbf{p}_e \times \mathbf{p}_\nu > 0$ and $\boldsymbol{\sigma}_n \cdot \mathbf{p}_e \times \mathbf{p}_\nu < 0$ are unequal. The test for T invariance is thus equivalent to a test for the existence of a *triple correlation* such as $\boldsymbol{\sigma}_n \cdot \mathbf{p}_e \times \mathbf{p}_\nu$ in neutron decay or, by an analogous argument, $\boldsymbol{\sigma}_\mu \cdot \mathbf{p}_\pi \times \mathbf{p}_\mu$ in K_{l3} decay, and so forth. It is important to point out that this simple argument is valid only in the absence of final-state interactions such as the Coulomb interaction of e^- and p after neutron decay. Clearly these are not accountable for in the simple first-order description and their inclusion introduces phase shifts that can masquerade as violations of T invariance.

The time-reversal transformation operator $\hat{T}$ on the Hilbert space of state vectors is antiunitary. This follows, for example, from the requirement that

$$\hat{T}\hat{\mathbf{j}}\hat{T}^{-1} = -\hat{\mathbf{j}} \quad \text{and} \quad \hat{T}\hat{\mathbf{p}}\hat{T}^{-1} = -\hat{\mathbf{p}}$$

Figure 4.13. (a) Original state, amplitude $\mathcal{M}$. Vectors p, e^-, $\bar{\nu}_e$ in plane of page. Neutron spin out of page. (b) Hermitian conjugation; amplitude $\mathcal{M}^*$. (c) Time reversal of (b); amplitude $\mathcal{M}'^*$. (d) Rotation of state (c) by 180° about neutron spin axis; amplitude $\mathcal{M}'^*$. This state is identical to (a) except for neutron spin-flip. If time reversal invariance holds, $|\mathcal{M}'| = |\mathcal{M}^*|$.

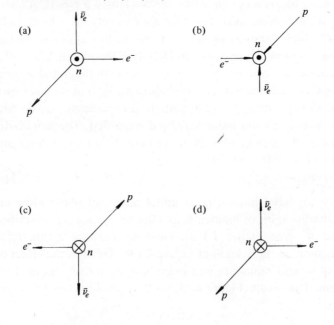

4.8 Time-reversal invariance; CPT invariance

where $\hat{\mathbf{J}}$ and $\hat{\mathbf{p}}$ are angular momentum and linear momentum operators, respectively, together with the requirement that $\hat{T}$ leave invariant the fundamental quantum-mechanical commutation relations $[\hat{\mathbf{x}}, \hat{\mathbf{p}}] = i$ and $\hat{\mathbf{J}} \times \hat{\mathbf{J}} = i\hat{\mathbf{J}}$. It can be shown (see Appendix A) that in the Pauli–Dirac representation $\hat{T}$ has the following effect on a single-particle Dirac wavefunction $\psi(\mathbf{x}, t)$:

$$\hat{T}\psi(\mathbf{x}, t) = i\gamma^1\gamma^3\psi^*(\mathbf{x}, t) \tag{4.107}$$

As a result, under time reversal, the bilinear form $\bar{u}_b(p', s')F u_a(p, s)$, where F is a 4×4 matrix, undergoes the transformation

$$\bar{u}_b(p', s')Fu_a(p, s) \rightarrow \bar{u}_a(p, s)\gamma_0\gamma_3\gamma_1 \tilde{F} \gamma_1\gamma_3\gamma_0 u_b(p', s') \tag{4.108}$$

Thus, under $\hat{T}$,

$$\bar{u}_b u_a \rightarrow \bar{u}_a u_b \tag{4.109a}$$

$$\bar{u}_b \gamma^\mu u_a \rightarrow \bar{u}_a \gamma_\mu u_b \tag{4.109b}$$

$$\bar{u}_b \sigma^{\mu\nu} u_a \rightarrow -\bar{u}_a \sigma_{\mu\nu} u_b \tag{4.109c}$$

$$\bar{u}_b \gamma_5 u_a \rightarrow \bar{u}_a \gamma_5 u_b \tag{4.109d}$$

$$\bar{u}_b \gamma^\mu \gamma_5 u_a \rightarrow \bar{u}_a \gamma_\mu \gamma_5 u_b \tag{4.109e}$$

$$\bar{u}_b \sigma^{\mu\nu} \gamma_5 u_a \rightarrow \bar{u}_a \sigma_{\mu\nu} \gamma_5 u_b \tag{4.109f}$$

We note that under T a shift from contravariant to covariant indexes occurs in (4.109b, c, e, and f).

Both P and C invariance are separately violated in weak interactions, and CP invariance also fails in decay of neutral kaons, but it is widely believed that CPT invariance holds strictly for all interactions. Thus CP violation must be accompanied by T violation in neutral kaon decay, and indeed this is known to be true from experiment. On the other hand, no incontrovertible evidence for T violation has been found in any other process, and we shall therefore find it useful to assume the validity of T invariance except where explicitly noted.

Under C, particles are transformed to antiparticles and spins and momenta are left unchanged; under P, 3-momenta are reversed; and under the antiunitary T transformation, initial and final states are interchanged and spins and momenta are reversed. Thus the net effect of a CPT transformation is

$$\langle B|\mathcal{T}|A\rangle \xrightarrow{CPT} \langle \bar{A}'|\mathcal{T}|\bar{B}'\rangle = \langle \bar{B}'|\mathcal{T}\dagger|\bar{A}'\rangle^*$$

where the overbar means antiparticle and the prime signifies that spins are reversed. Since the weak interaction is assumed to be CPT invariant, we conclude that

$$\langle B|\mathcal{T}|A\rangle = \langle \bar{B}'|\mathcal{T}\dagger|\bar{A}'\rangle^* \tag{4.110}$$

We now consider the implications of T and CPT invariance for the processes discussed in the previous section. According to (4.89), the amplitude for a transition meson → meson + leptons is

$$\mathcal{M} = N\frac{G_F}{\sqrt{2}}[f_+(q^2)(k_a{}^\alpha + k_b{}^\alpha) + f_-(q^2)(k_a{}^\alpha - k_b{}^\alpha)]\bar{u}_l\gamma_\alpha(1 - \gamma_5)v_\nu \tag{4.111}$$

The time-reversed amplitude is

$$\mathcal{M}' = N\frac{G_F}{\sqrt{2}}[f_+(q^2)(k_{a\alpha} + k_{b\alpha}) + f_-(q^2)(k_{a\alpha} - k_{b\alpha})]\bar{v}_\nu\gamma^\alpha(1 - \gamma_5)u_l \tag{4.112}$$

where we have employed formulas (4.109). Meanwhile, the Hermitian conjugate amplitude is

$$\mathcal{M}^* = N\frac{G_F}{\sqrt{2}}[f_+^*(q^2)(k_a{}^\alpha + k_b{}^\alpha) + f_-^*(q^2)(k_a{}^\alpha - k_b{}^\alpha)]\bar{v}_\nu\gamma_\alpha(1 - \gamma_5)u_l \tag{4.113}$$

Comparison of (4.112) and (4.113) shows that for T invariance f_+ and f_- must be relatively real.

For the hyperon nonleptonic decays, the amplitude is given by (4.96). According to (4.109a and d), the time-reversed amplitude is

$$\mathcal{M}' = G_F m_{\pi^+}^2 \bar{u}_i(A + B\gamma_5)u_f \tag{4.114}$$

and the Hermitian conjugate amplitude is

$$\mathcal{M}^* = G_F m_{\pi^+}^2 \bar{u}_i(A^* + B^*\gamma_5)u_f \tag{4.115}$$

Again, T invariance requires A and B to be relatively real.

Similarly, formulas (4.109) lead us to the conclusion that in a baryon semileptonic transition the six form factors $f_{1,2,3}$ and $g_{1,2,3}$ must be relatively real for T invariance.

The CPT-invariance condition (4.110) is useful in cases where analogous transitions can be observed for particles and antiparticles. For example, consider the transition $K^+ \to \pi^0\mu^+\nu$ or $K^+ \to \pi^0 e^+\nu$ and denote the hadronic amplitude by $A_+{}^\alpha$:

$$A_+{}^\alpha = \langle \pi^0|J^\alpha|K^+\rangle$$

Then, from (4.110), we have

$$A_-{}^\alpha \equiv \langle \pi^0|J^{\alpha\dagger}|K^-\rangle = A_+^* \tag{4.116}$$

In similar fashion, we define f and g for $K^0 \to \pi^- l^+\nu$ or $K^0 \to \pi^+ l^-\bar{\nu}$ by the following equations:

$$f^\alpha \equiv \langle \pi^-|J^\alpha|K^0\rangle, \qquad \Delta S = \Delta Q \tag{4.117a}$$

$$g^\alpha \equiv \langle \pi^+|J^{\alpha\dagger}|K^0\rangle, \qquad \Delta S = -\Delta Q \tag{4.117b}$$

Then from (4.110), we obtain

$$\langle \pi^+|J^{\alpha\dagger}|\bar{K}^0\rangle = f^{\alpha*}, \quad \Delta S = \Delta Q \quad (4.118a)$$

$$\langle \pi^-|J^{\alpha}|\bar{K}^0\rangle = g^{\alpha*}, \quad \Delta S = -\Delta Q \quad (4.118b)$$

Similar remarks apply to $K^+ \to \pi^+\pi^+\pi^-$ and $K^- \to \pi^-\pi^-\pi^+$.

4.9 The conserved vector current hypothesis

We recall that, according to the standard model, the vector portion of the charged weak current, its Hermitian conjugate, and the isovector electromagnetic current form a single isospin triplet of conserved currents (see Section 2.5). Here we explore the consequences of this "CVC hypothesis." The matrix element of $J_{EM}{}^\alpha(x)$ between two proton states $|p, s\rangle$ and $|p', s'\rangle$ is given by the general expression

$$\langle p's'|J_{EM}{}^\alpha(x)|ps\rangle = e^{i(p'-p)\cdot x}\bar{u}(p', s')[C_p(q^2)\gamma^\alpha + iM_p(q^2)\sigma^{\alpha\nu}q_\nu + F_{3p}(q^2)q^\alpha]u(p, s) \quad (4.119)$$

where C_p, M_p, and F_{3p} are form factors. These are real if T invariance holds, but otherwise, they must be determined. One very useful restriction follows from $\partial_\alpha J_{EM}{}^\alpha(x) = 0$. Taking the divergence of both sides of (4.119), we find that

$$0 = \bar{u}(p', s')[C_p(q^2)(\not{p}' - \not{p}) - iM_p(q^2)\sigma^{\alpha\nu}q_\nu q_\alpha - F_{3p}(q^2)q^2]u(p, s) \quad (4.120)$$

However, $\bar{u}'(\not{p}' - \not{p})u = (m_p - m_p)\bar{u}'u = 0$ and also $\sigma^{\alpha\nu}q_\nu q_\alpha = 0$ since $\sigma^{\alpha\nu}$ is an antisymmetric tensor. Therefore, $q^2 F_{3p}(q^2) = 0$, which implies $F_{3p}(q^2) = 0$. Thus we have

$$\langle p's'|J_{EM}{}^\alpha(x)|ps\rangle = e^{i(p'-p)\cdot x}\bar{u}(p', s')[C_p(q^2)\gamma^\alpha + iM_p(q^2)\sigma^{\gamma\nu}q_\nu]u(p, s) \quad (4.121)$$

Furthermore, in the limit as $q^2 \to 0$, the proton interacts with an external static electromagnetic field. Since the proton charge is $+|e|$, we have the further restriction $C_p(0) = 1$. As is well known, the C_p term accounts not only for the proton charge but also for its normal, or Dirac, magnetic moment of value $\mu_0 = |e|\hbar/2m_p c$. However, the proton magnetic moment is actually $2.79\mu_0$, and the anomalous part is accounted for by the "Pauli moment term": $M_p(0) = 1.79/2m_p$.

In the same fashion we may construct a matrix element for the electromagnetic current between two neutron states:

$$\mathcal{M}_n = e^{-iq \cdot x}\bar{u}_n(p', s')[C_n(q^2)\gamma^\alpha + iM_n\sigma^{\alpha\nu}q_\nu]u_n(p, s) \quad (4.122)$$

Since the neutron has no net charge and its magnetic moment is entirely anomalous ($\mu_n = -1.91\mu_0$), we have $C_n(0) = 0$ and $M_n(0) = -1.91/2m_p$, where we assume $m_p = m_n$.

The two matrix elements $\mathcal{M}_p$ and $\mathcal{M}_n$ may now be united by regarding n and p as $-\frac{1}{2}$ and $+\frac{1}{2}$ components of the nucleon isodoublet, described by a spinor u without suffix. The spinors u_p and u_n are recovered with the aid of the projection operators:

$$\tfrac{1}{2}(1 + \tau_3)u = u_p, \qquad \tfrac{1}{2}(1 - \tau_3)u = u_n \tag{4.123}$$

Then defining the isoscalar and isovector form factors by

$$C^{(0)}(q^2) = C_p(q^2) + C_n(q^2) \quad \text{(isoscalar)} \tag{4.124}$$
$$C^{(1)}(q^2) = C_p(q^2) - C_n(q^2) \quad \text{(isovector)} \tag{4.125}$$
$$M^{(0)}(q^2) = M_p(q^2) + M_n(q^2) \quad \text{(isoscalar)} \tag{4.126}$$
$$M^{(1)}(q^2) = M_p(q^2) - M_n(q^2) \quad \text{(isovector)} \tag{4.127}$$

we easily obtain, in place of (4.121) and (4.122),

$$\mathcal{M} = \bar{u}'\{\tfrac{1}{2}[C^{(0)}(q^2)\gamma^\alpha + iM^{(0)}(q^2)\sigma^{\alpha\nu}q_\nu] \\ + [C^{(1)}(q^2)\gamma^\alpha + iM^{(1)}(q^2)\sigma^{\alpha\nu}q_\nu]t_3\}u \tag{4.128}$$

We now return to the weak interaction and write the matrix element of the vector hadronic weak current between neutron and proton states:

$$\langle p|V^{\alpha\dagger}|n\rangle = \cos\theta_1 \, e^{iq\cdot x}\bar{u}'[f_1(q^2)\gamma^\alpha + if_2(q^2)\sigma^{\alpha\nu}q_\nu + f_3(q^2)q^\alpha]t_+u \tag{4.129}$$

The essential point of CVC is that the form factors in (4.129) must be identical to the corresponding isovector form factors in (4.128) because the V^α, $V^{\alpha\dagger}$, and $j_{\text{EM}}^{\text{isov}}$ form an isotriplet. We thus have

$$f_1(q^2) = C^{(1)}(q^2) \tag{4.130}$$
$$f_2(q^2) = M^{(1)}(q^2) \tag{4.131}$$
$$f_3(q^2) = 0 \tag{4.132}$$

In the limit as $q^2 \to 0$, f_1 is just the vector coupling constant C_V of nuclear β decay. Since $C^{(1)}(0) = C_p(0) = 1$, we have the result $C_V = 1$, which is borne out by experiment (see Chapter 5). In fact, it was the remarkable equality of C_V and $C_p(0)$ that first led to the CVC hypothesis, long before the invention of $SU(2) \times U(1)$ electroweak theories, which naturally incorporate CVC. The vector coupling constant C_V is thus the same in weak decays of hadrons as it is in purely leptonic weak decays (muon and τ decay). Relation (4.130) implies that $f_1(q^2)$ and $C^{(1)}(q^2)$ are also the same at large q^2, a result found by high-energy neutrino–nucleon scattering experiments.

Equation (4.131) requires that

$$f_2(0) = [1.79 - (-1.91)]/2m_p = 3.70/2m_p \tag{4.133}$$

This formula relates the vector coupling in β decay and other charged weak interactions to the magnetic moments of proton and neutron, a

4.9 The CVC hypothesis

most remarkable result called weak magnetism (see Chapter 5 for further details). Finally, (4.132) states that the "induced scalar term" proportional to $\bar{u}'uq^\alpha$ vanishes in β decay as well as in electromagnetism.

Pion β decay $\pi^+ \to \pi^0 e^+ \nu_e$ furnishes another interesting example of the CVC hypothesis. The matrix element between two π^+ states of j_{EM}^α is

$$\mathcal{M}_{EM}^\alpha = \langle \pi^+(k') | j_{EM}^\alpha(x) | \pi^+(k) \rangle$$
$$= [F_1(q^2)(k'^\alpha + k^\alpha) + F_2(q^2)(k'^\alpha - k^\alpha)] e^{-iq \cdot x} \quad (4.134)$$

We take the 4-divergence of both sides of (4.134) to obtain

$$0 = F_1(q^2)(k'^2 - k^2) + F_2(q^2)(k' - k)^2 \quad (4.135)$$

However $k'^2 = k^2 = m_\pi^2$, so $F_2(q^2) = 0$. We thus find $\mathcal{M}_{EM} = F_1(q^2)(k'^\alpha + k^\alpha)e^{-iq \cdot x}$, which can be rewritten in terms of the three-component isospinors ϕ and ϕ' for the pions and the matrix

$$I_3 = \begin{pmatrix} 1 & 0 & 0 \\ 0 & 0 & 0 \\ 0 & 0 & -1 \end{pmatrix} \quad (4.136)$$

as follows:

$$\mathcal{M}_{EM}^\alpha = F_1(q^2)(k'^\alpha + k^\alpha)[\phi'^\dagger I_3 \phi] e^{-iq \cdot x} \quad (4.137)$$

According to CVC, the hadronic portion of the amplitude for pion β decay is then

$$\mathcal{M}^\alpha = \cos \theta_1 \, F_1(q^2)(k'^\alpha + k^\alpha)[\phi'^\dagger I_- \phi] e^{-iq \cdot x} \quad (4.138)$$

where

$$I_- = \begin{pmatrix} 0 & 0 & 0 \\ \sqrt{2} & 0 & 0 \\ 0 & \sqrt{2} & 0 \end{pmatrix}$$

In $\pi^+ \to \pi^0 e^+ \nu_e$, the momentum transfer is small ($\Delta \equiv m_{\pi^+} - m_{\pi^0} = 4.60$ MeV/c^2). Thus we must consider $\lim_{q^2 \to 0} F_1(q^2)$, which is unity since π^+ has positive unit charge. Thus

$$\mathcal{M}(\pi^+ \to \pi^0 e^+ \nu_e) = \frac{G_F}{\sqrt{2}} \cos \theta_1 \sqrt{2}(k' + k)^\alpha \bar{u}_\nu \gamma_\alpha (1 - \gamma_5) v_e \quad (4.139)$$

In this formula the factor $\sqrt{2}$ in the numerator is particularly significant; it arises from CVC, as we have noted.

Since $\Delta \ll m_{\pi^0}$, we may ignore the π^0 kinetic energy in the π^+ rest frame to a good approximation. Then, (4.139) becomes

$$\mathcal{M} = 2G_F \cos \theta_1 \, m_\pi \bar{u}_\nu \gamma^0 (1 - \gamma_5) v_e \quad (4.140)$$

168   4 General properties of hadronic weak currents

It is then easy to show that the total decay rate is given by

$$W = \frac{G_F^2 \cos^2\theta_1 \Delta^5}{30\pi^3}\left(1 - \frac{3}{2}\frac{\Delta}{m_{\pi^+}}\frac{5m_e^2}{\Delta^2} + \cdots\right) \qquad (4.141)$$

or

$$W_{\text{theor}} = 0.393 \quad \text{sec}^{-1} \qquad (4.142)$$

Experimentally (De Pommier et al. 68) one finds

$$W_{\text{expt}} = 0.38 \pm 0.03 \quad \text{sec}^{-1} \qquad (4.143)$$

which gives excellent agreement.

4.10 Second-class currents and G parity

A G-parity transformation is compounded of a charge conjugation C and an isospin rotation of π about the I_2 axis: $G = Ce^{i\pi I_2}$. The strong interaction is separately invariant under C and arbitrary isospin rotations, so it is G-invariant. Let us consider the G-parity transformation properties of the hadronic portion of the amplitude for a baryon semileptonic transition (4.94). We shall merely be interested in how the various terms transform relative to one another. Now, under rotation of π about the I_2 axis, all six terms obviously transform in the same way: as isovectors. Thus we need only consider the question of how they transform under C. Now under C, a Dirac spinor u transforms to $i\gamma^2\gamma^0\tilde{\bar{u}}$, where we employ the usual phase convention. Thus, if F is a 4×4 matrix, the bilinear form $\bar{u}_b F u_a$ undergoes the transformation

$$\bar{u}_b F u_a \xrightarrow{C} \bar{u}_a(\gamma^0\gamma^2\tilde{F}\gamma^2\gamma^0)u_b \qquad (4.144)$$

Thus, under C,

$$\bar{u}_p u_n \to -\bar{u}_n u_p \qquad (f_3) \qquad (4.145)$$
$$\bar{u}_p \gamma_5 u_n \to -\bar{u}_n \gamma_5 u_p \qquad (g_3) \qquad (4.146)$$
$$\bar{u}_p \gamma^\mu \gamma_5 u_n \to \bar{u}_n \gamma^\mu \gamma_5 u_p \qquad (g_1) \qquad (4.147)$$

while

$$\bar{u}_p \gamma^\mu u_n \to \bar{u}_n \gamma^\mu u_p \qquad (f_1) \qquad (4.148)$$
$$\bar{u}_p \sigma^{\mu\nu} u_n \to \bar{u}_n \sigma^{\mu\nu} u_p \qquad (f_2) \qquad (4.149)$$
$$\bar{u}_p \sigma^{\mu\nu} \gamma_5 u_n \to \bar{u}_n \sigma^{\mu\nu} \gamma_5 u_p \qquad (g_2) \qquad (4.150)$$

Thus of the three vector terms, those in f_1 and f_2 transform in the same way under G, whereas the f_3 term transforms oppositely. Also, the terms in g_1 and g_3 transform in the same way, whereas the g_2 term transforms oppositely. We call the f_1, f_2, g_1, and g_3 terms *first class* and the f_3 and g_2 terms *second class*. Now, in the absence of strong interactions, one would have only f_1 and g_1 terms; the other being "induced" by strong interaction. Therefore, since the latter is G-invariant, one might expect f_3 and g_2 to be zero. Indeed, f_3 is already zero, assuming CVC. However, we cannot hastily conclude from the superficial argument just given that $g_2 = 0$, since the axial current is not conserved.

On the other hand, Langacker (76,77) has shown that, within the context of renormalizable field theory, second-class currents constructed from fermion (i.e., quark) fields require the existence of two distinct isomultiplets of such fields with identical quantum numbers except for isospin. (See also Halprin et al, 76). Such currents can emerge from a gauge theory of the weak interaction, but if they are to contribute observably to β decay, severe restrictions must be placed on the field theory of strong interactions. That is, there must exist strongly coupled scalar or pseudoscalar fields or the gauge groups for weak and strong interactions cannot commute. In the latter case, if the strong interaction gauge symmetry is not broken, the hadron states cannot be singlets. Either of these conditions conflicts with the conventional scheme according to which strong interactions are described by quark–quark–gluon couplings according to color $SU(3)$, in which hadron states are singlets and electroweak interactions are described by $SU(2)_L \times U(1)$. Therefore, any observable second-class current effects in β decay, as manifested by $g_2 \neq 0$, would have ominous implications for the standard theory. In fact, however, experiments in recent years have succeeded in placing very small upper limits on g_2 and are consistent with $g_2 = 0$ (see Section 5.5).

4.11 The partially conserved axial current hypothesis

If we assume $g_2 = 0$, the axial vector contribution to the neutron β decay matrix element is

$$\langle p|A_0^{\alpha\dagger}(x)|n\rangle = -\cos\theta_1\, e^{i(p_P-p_N)\cdot x}\bar{u}(p)[g_1(q^2)\gamma^\alpha\gamma_5 + g_3(q^2)q^\alpha\gamma_5)]u(n) \quad (4.151)$$

We may analyze the various contributions to (4.151) if we ignore quarks for the moment, think first of a "bare" neutron, and then add various corrections for strong interactions.

Bare neutron decay corresponds to Figure 4.14a and contributes a term $-\bar{u}(p)\gamma^\alpha\gamma^5 u(n)$ to (4.151). To this we have to add pion vertex corrections such as shown in Figure 4.14b. Of course no one knows how to calculate the total contribution of Figure 4.14b and similar higher-order corrections, but we do know that all such terms contribute to

$g_1(q^2)$. On the other hand, there is also the one-pion exchange diagram (Figure 4.15) and higher-order corrections associated with it. According to the usual rules, Figure 4.15 corresponds to the amplitude:

$$A = (n \to p\pi^- \text{ vertex}) \times (\pi^- \text{ propagator}) \times (\pi^- \to e^-\bar{\nu}_e \text{ vertex})$$
(4.152)

Now

$$(n \to p\pi^- \text{ vertex}) = ig_0\sqrt{2}\bar{u}(p)\gamma_5 u(n)$$
(4.153)

where g_0 is the physical pion–nucleon coupling constant and

$$(\pi^- \to e^-\bar{\nu}_e \text{ vertex}) = \cos\theta_1 \frac{G_F}{\sqrt{2}} if_\pi q_\mu L^\mu$$
(4.154)

where $L^\mu = \bar{u}_e \gamma^\mu (1 - \gamma_5) v_\nu$. Inserting all this in (4.152), we obtain

$$A = -\cos\theta_1 \frac{G_F}{\sqrt{2}} f_\pi g_0 \sqrt{2} \frac{1}{q^2 - m_\pi^2} \bar{u}(p) q^\alpha \gamma_5 u(n) L_\alpha$$
(4.155)

Clearly this expression contributes to the g_3 term in (4.151), and if there were no other corrections, we would have

$$g_3(q^2) = -f_\pi g_0 \sqrt{2}/(q^2 - m_\pi^2)$$
(4.156)

In fact, including vertex corrections, we obtain

$$g_3(q^2) = -f_\pi g_0 \sqrt{2} F(q^2)/(q^2 - m_\pi^2)$$
(4.157)

Figure 4.14. (a) Bare neutron decay. (b) Pion vertex corrections.

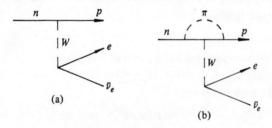

Figure 4.15. One-pion exchange.

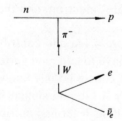

4.11 The partially conserved axial current hypothesis

where $F(q^2)$ is a smooth function of q^2 with $F(q^2 = m_\pi^2) = 1$, since g_0 is the physical pion–nucleon coupling constant. Multipion exchange diagrams would give an additive correction to (4.157), which we ignore. Thus (4.151) becomes

$$\langle p|A_0^{\dagger\alpha}(x)|n\rangle = -\cos\theta_1 \, e^{i(p_p - p_N)x}$$

$$\times \bar{u}(p)\left[g_1(q^2)\gamma^\alpha - \frac{f_\pi g_0 \sqrt{2}}{q^2 - m_\pi^2} F(q^2) q^\alpha\right]\gamma_5 u(u) \quad (4.158)$$

Now we know from Section 2.5 that the axial current is not ~~observed~~ conserved. If it were, the equation

$$\langle 0|A_0^{\alpha\dagger}(x)|\pi^-\rangle = i\cos\theta_1 \, f_\pi q^\alpha e^{-iq\cdot x} \quad (4.159)$$

would imply

$$0 = \langle 0|\partial A_0^{\alpha\dagger}(x)/\partial x^\alpha|\pi^-\rangle = \cos\theta_1 \, f_\pi m_\pi^2 e^{-iq\cdot x} \quad (4.160)$$

which is impossible, since the pion has finite mass and also $f_\pi \neq 0$ (the pion decays). Nevertheless, the axial current may be thought of as conserved in an approximate sense, namely, conserved in the limit when $m_\pi \to 0$. This is called the *partially conserved axial current* (PCAC) hypothesis. Assuming that

$$\lim_{m_\pi \to 0} \partial A_0^\alpha(x)/\partial x^\alpha = 0$$

we calculate the 4-divergence of both sides of (4.158) and obtain

$$0 = \lim_{m_\pi \to 0} \bar{u}(p)\left[g_1(q^2)(m_p + m_n) - \frac{f_\pi g_0 \sqrt{2}}{q^2 - m_\pi^2} F(q^2) q^2\right]\gamma_5 u(n) \quad (4.161)$$

If, in addition, we assume that $F(q^2)$ is a slowly varying function, then we may extrapolate from $q^2 = m_\pi^2$, where we know $F(q^2) = 1$, to $q^2 = 0$:

$$F(0) = 1$$

In this case we find, from (4.161), that

$$C_A \equiv g_1(0) \simeq f_\pi g_0 \sqrt{2}/(m_p + m_n) \simeq f_\pi g_0/\sqrt{2} m_p \quad (4.162)$$

This result is the Goldberger–Treiman relation (58a,b). It and PCAC may be expressed in terms of the renormalized π^- field operator $\phi_1 + i\phi_2$, which satisfies

$$\langle 0|\phi_{\pi^-}(x)|\pi^-\rangle = \frac{1}{\sqrt{2}}\langle 0|\phi_1(x) + i\phi_2(x)|\pi^-\rangle = e^{-iq\cdot x} \quad (4.163)$$

Comparing (4.160) with (4.163) and employing (4.162), we obtain

$$\partial_\alpha A_0^{\alpha\dagger}(x) \simeq \cos\theta_1 \frac{m_p m_\pi^2 g_1(0)}{g_0}(\phi_1 + i\phi_2) \quad (4.164)$$

172   4 General properties of hadronic weak currents

Finally, (4.158) yields

$$g_3(q^2) \simeq -2m_p C_A/(q^2 - m_\pi^2) \tag{4.165}$$

which is a useful approximate formula for the "induced pseudoscalar" form factor g_3. As we shall see, this quantity has been measured in muon capture experiments and the results are in agreement with PCAC.

4.12 Form factors for neutral currents

Let us now recall the expression for the neutral weak current (2.104):

$$J_Z^\mu = \tfrac{1}{2} J_3^\mu - 2 \sin^2 \theta_W J_{EM} \tag{4.166}$$

We wish to obtain matrix elements between initial and final proton states or initial and final neutron states. These will be useful in discussions of neutrino–nucleon neutral current elastic scattering at relatively low energies and also for parity violation in atomic physics. As we know, the most general forms for such matrix elements are

$$\langle p'|J_Z^\mu|p\rangle = e^{-iq\cdot x}\bar{u}_p'[f_{1p}(q^2)\gamma^\mu + if_{2p}(q^2)\sigma^{\mu\nu}q_\nu \\ - g_{1p}(q^2)\gamma^\mu\gamma_5 - g_{3p}(q^2)q^\mu\gamma_5]u_p \tag{4.167}$$

and

$$\langle n'|J_Z^\mu|n\rangle = e^{iq\cdot x}\bar{u}_n'[f_{1n}(q^2)\gamma^\mu + if_{2n}(q^2)\sigma^{\mu\nu}q_\nu \\ - g_{1n}(q^2)\gamma^\mu\gamma_5 - g_{3n}(q^2)q^\mu\gamma_5]u_n \tag{4.168}$$

where the form factors are real for T invariance and we have discarded second-class terms. Now, since

$$\langle p'|J_Z^\mu|p\rangle = \tfrac{1}{2}\langle p'|J_3^\mu|p\rangle - 2\sin^2\theta \langle p'|J_{EM}|p\rangle \tag{4.169}$$

and J_3^μ is just the third component of the isospin current, whereas $\langle p'|J_{EM}|p\rangle$ is given by (4.121), we may easily write the neutral-current form factors by inspection (Weinberg 72). They are:

$$f_{1p}(q^2) = \tfrac{1}{2}f_1(q^2) - 2\sin^2\theta_W C_p(q^2) \tag{4.170}$$

$$-f_{1n}(q^2) = \tfrac{1}{2}f_1(q^2) - 2\sin^2\theta_W C_n(q^2) \tag{4.171}$$

$$f_{2p}(q^2) = \tfrac{1}{2}f_2(q^2) - 2\sin^2\theta_W M_p(q^2) \tag{4.172}$$

$$-f_{2n}(q^2) = \tfrac{1}{2}f_2(q^2) - 2\sin^2\theta_W M_n(q^2) \tag{4.173}$$

$$g_{1p}(q^2) = \tfrac{1}{2}g_1(q^2) = -g_{1n}(q^2) \tag{4.174}$$

$$g_{3p}(q^2) = \tfrac{1}{2}g_3(q^2) = -g_{3n}(q^2) \tag{4.175}$$

These form factors are important in high-energy elastic scattering, to be discussed in Chapter 8. However, at low q^2, we may ignore the

4.12 Form factors for neutral currents

terms in f_{2p}, f_{2n}, g_{3p}, and g_{3n} and take the limit as $q^2 \to 0$ in $f_{1p,n}$ and $g_{1p,n}$. Thus we obtain

$$f_{1p}(0) = \tfrac{1}{2}f_1(0) - 2\sin^2\theta_W\, C_p(0) = \tfrac{1}{2}(1 - 4\sin^2\theta_W)$$
$$-f_{1n}(0) = \tfrac{1}{2}f_1(0) - 2\sin^2\theta_W\, C_n(0) = \tfrac{1}{2}$$
$$g_{1p}(0) = -g_{1n}(0) = \tfrac{1}{2}(1.25) = 0.63$$

Thus, for example, consider the low-energy neutral weak coupling between an electron and a proton (Figure 4.16a) or a neutron (Figure 4.16b). We easily find the amplitudes in these two cases to be

$$\mathcal{M}(e,p) = -\frac{G_F}{2\sqrt{2}}\bar{u}'_e\gamma^\mu(1 - 4\sin^2\theta_W - \gamma_5)u_e$$
$$\times \bar{u}'_p\gamma_\mu(1 - 4\sin^2\theta_W - 1.25\gamma^5)u_p \qquad (4.176)$$

$$\mathcal{M}(e,n) = \frac{G_F}{2\sqrt{2}}\bar{u}'_e\gamma^\mu(1 - 4\sin^2\theta_W - \gamma_5)u_e\bar{u}'_n\gamma_\mu(1 - 1.25\gamma^5)u_n$$
$$\qquad (4.177)$$

These formulas will be used later when we discuss parity violation in atoms.

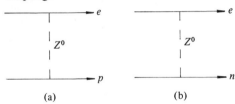

Figure 4.16. (a) Electron–proton coupling. (b) Electron–neutron coupling.

4.13 The empirical selection rules $\Delta S = \Delta Q$ and $|\Delta I| = \tfrac{1}{2}$

4.13.1 $\Delta S = \Delta Q$

This rule applies to leptonic and semileptonic $|\Delta S| = 1$ transitions and refers to the change in S and Q of the transforming hadron. It predicts, for example, that decays such as $\Sigma^+ \to nl^+\nu_l$ and $K^+ \to \pi^+\pi^+e^-\bar{\nu}_e$ are forbidden; these are in fact not observed. From the point of view of the quark model, the $\Delta S = \Delta Q$ rule is seen to be a simple consequence of the quantum numbers of u, d, and s quarks. Similarly, there exists the selection rule $\Delta C = -\Delta Q$ for decay of the charmed quark c to a d or s quark.

Let us consider the K_{l3} decays in terms of the $\Delta S = \Delta Q$ rule. We see that $K^+ \to \pi^0 l^+ \nu$ and $K^- \to \pi^0 l^- \bar{\nu}$ satisfy it automatically. However, K_L^0 is a superposition of K^0 and $\bar{K}^0$ states; therefore, in transitions $K_L^0 \to \pi^{\pm} l^{\mp} \nu$, one can have contributions from amplitudes f or f^* [$\Delta S = \Delta Q$, (4.117a) and (4.118a)], and g or g^* [(4.117b) and (4.118b)] a priori. If we assume CP invariance, we can write

$$|K_S^0\rangle = \frac{1}{\sqrt{2}} [|K^0\rangle - |\bar{K}^0\rangle]$$

and

$$|K_L^0\rangle = \frac{1}{\sqrt{2}} [|K^0\rangle + |\bar{K}^0\rangle]$$

Moreover, since CP invariance and CPT invariance imply T invariance, f and g must be real. Consider a neutral K^0 beam formed in the state $|K^0\rangle$ at time $t = 0$. Then, at a later time t,

$$|K^0(t)\rangle = \frac{1}{\sqrt{2}} [|K_S^0(t)\rangle + |K_L^0(t)\rangle]$$

$$= \frac{1}{\sqrt{2}} [|K_S^0(0)\rangle e^{-\lambda_S t} + |K_L^0(0)\rangle e^{-\lambda_L t}]$$

$$= \tfrac{1}{2} [|K^0(0) - \bar{K}^0(0)\rangle e^{-\lambda_S t} + |K^0(0) + \bar{K}^0(0)\rangle e^{-\lambda_L t}]$$

We take the matrix elements corresponding to the transitions $K^0 \to \pi^- e^+ \nu_e$ and $K^0 \to \pi^+ e^- \bar{\nu}_e$, and from (4.117) and (4.118), we have

$$\langle \pi^- | V^\lambda | K^0(t) \rangle = \tfrac{1}{2} [(f - g) e^{-\lambda_S t} + (f + g) e^{-\lambda_L t}]$$
$$\langle \pi^+ | V^{\lambda\dagger} | K^0(t) \rangle = \tfrac{1}{2} [(g - f) e^{-\lambda_S t} + (g + f) e^{-\lambda_L t}]$$

Therefore the rate of production of $e^{\pm}$ at time t is proportional to

$$R^{\pm}(t) = (1 + x)^2 e^{-\lambda_S t} + (1 - x)^2 e^{-\lambda_L t}$$
$$\pm 2(1 - x^2) e^{-[(\lambda_S + \lambda_L)/2]t} \cos(m_L - m_S)t \quad (4.178)$$

where $x = -g/f$. More generally, if we do not assume CP invariance, then x can be complex. One finds

$$R^{\pm}(t) = |1 + x|^2 e^{-\lambda_S t} + |1 - x|^2 e^{-\lambda_L t} \pm 2(1 - |x|^2) e^{-(\lambda_S + \lambda_L)t/2} \cos \Delta mt$$
$$- 4\text{Im}(x) e^{-(\lambda_S + \lambda_L)t/2} \sin \Delta mt \quad (4.179)$$

with $x = -g^*/f$. In either case, the experimental data are consistent with $x = 0$.

4.13.2 The $|\Delta I| = \tfrac{1}{2}$ rule

Whereas the strangeness-conserving portion of the hadronic current J_0 has the structure $\bar{d}u$ and transforms like an isovector

($|\Delta I| = 1$), the strangeness-violating ($|\Delta S| = 1$) hadronic current J_1 has the structure $\bar{s}u$ and transforms like an isospinor ($|\Delta I| = \frac{1}{2}$). Therefore, in leptonic and semileptonic decays of strange hadrons, we expect the rule $|\Delta I| = \frac{1}{2}$ for the transforming hadron.

In nonleptonic decays of strange hadrons, the effective current–current Lagrangian has the structure

$$\mathcal{L} \simeq J_1^\dagger J_0 + J_0^\dagger J_1 \simeq \bar{u}s \cdot \bar{d}u + \bar{s}u \cdot \bar{u}d$$

In these cases, since both $|\Delta I| = \frac{1}{2}$ and $|\Delta I| = 1$ currents appear, we expect $|\Delta I| = \frac{1}{2}$ and $|\Delta I| = \frac{3}{2}$ amplitudes for the decays themselves.

Nevertheless, experiment shows that the $|\Delta I| = \frac{3}{2}$ portion is suppressed and the $|\Delta I| = \frac{1}{2}$ portion is enhanced. Thus, the rule $|\Delta I| = \frac{1}{2}$ is approximately valid even in these nonleptonic transitions. We defer a detailed discussion of the rule until Chapter 6, where we will include a plausible dynamical explanation for it.

4.14 Current algebra

We conclude this chapter with a brief discussion of certain useful algebraic relations that are satisfied by the hadronic currents and that follow from $SU(3)_f$ symmetry and the anticommutation relations satisfied by the quark Dirac fields. Let us recall the vector and axial vector octets of current operators, defined in (4.35) and (4.37):

$$j_k^\mu(x) = \bar{\psi}(x)\gamma^\mu \tfrac{1}{2}\lambda_k \psi(x) \tag{4.35}$$

$$g_k^\mu(x) = \bar{\psi}(x)\gamma^\mu \gamma_5 \tfrac{1}{2}\lambda_k \psi(x) \tag{4.37}$$

The anticommutation relations satisfied by the Dirac field ψ are as follows:

$$\{\psi_s^\beta(x), \psi_r^{\alpha\dagger}(y)\}_{x^0 = y^0} = \delta_{\beta\alpha}\, \delta_{rs}\, \delta^3(\mathbf{x} - \mathbf{y})$$

$$\{\psi_s^\beta(x), \psi_r^\alpha(y)\} = \{\psi_s^{\beta\dagger}(x), \psi_r^{\alpha\dagger}(y)\} = 0 \tag{4.180}$$

where α and β are space-time spinor indexes and s and r are flavor indexes.

From (4.180) and (4.181), it is possible to show, by straightforward manipulations, that the following "equal-time current commutation relations" are satisfied:

$$[j_k^0(x), j_l^0(y)]_{x^0 = y^0} = i\,\delta^3(\mathbf{x} - \mathbf{y}) f_{kl}{}^m j_m^0(x) \tag{4.182}$$

$$[j_k^0(x), g_l^0(y)]_{x^0 = y^0} = i\,\delta^3(\mathbf{x} - \mathbf{y}) f_{kl}{}^m g_m^0(x) \tag{4.183}$$

$$[g_k^0(x), g_l^0(y)]_{x^0 = y^0} = i\,\delta^3(\mathbf{x} - \mathbf{y}) f_{kl}{}^m j_m^0(x) \tag{4.184}$$

where the $f_{kl}{}^m$ are the $SU(3)$ structure constants. It is also convenient to define, as in Section 2.5, the chiral currents:

$$j_k^{L\mu} = \tfrac{1}{2}(j_k^\mu - g_k^\mu) \tag{4.185}$$

$$j_k^{R\mu} = \tfrac{1}{2}(j_k^\mu + g_k^\mu) \tag{4.186}$$

where L is left and R is right. It is easy to show from (4.184) that these chiral currents satisfy the "$SU(3)_f$" algebra:

$$[j_k^{L_0}(x), j_l^{L_0}(y)]_{x^0=y^0} = i\,\delta^3(\mathbf{x} - \mathbf{y}) f_k^{\ m} j_m^{L_0}(x) \tag{4.187}$$

$$[j_k^{R_0}(x), j_l^{R_0}(y)]_{x^0=y^0} = i\,\delta^3(\mathbf{x} - \mathbf{y}) f_k^{\ m} j_m^{R_0}(x) \tag{4.188}$$

$$[j_k^{L_0}(x), j_l^{R_0}(y)]_{x^0=y^0} = 0 \tag{4.189}$$

We may also define the vector and axial vector *charges* by

$$F_i(x^0) = \int d^3\mathbf{x}\, j_i^0(x) \tag{4.190}$$

$$F_i^5(x^0) = \int d^3\mathbf{x}\, g_i^0(x) \tag{4.191}$$

From formulas (4.187) to (4.189), it follows that

$$[F_k(x^0), F_m(x^0)] = if_{km}^{\ \ n} F_n(x_0) \tag{4.192}$$

$$[F_k(x^0), F_m^5(x^0)] = if_{km}^{\ \ n} F_n^5(x_0) \tag{4.193}$$

$$[F_k^5(x^0), F_m^5(x^0)] = if_{km}^{\ \ n} F_n(x_0) \tag{4.194}$$

More generally, one can obtain analogous relations involving the commutators of the vector and axial charges with the space components of the vector and axial currents.

The current and charge commutation relations just given may be used to derive a number of very useful approximate results in the theory of weak interactions. For example, Adler (65) and Weisberger (66) independently showed how to calculate the numerical value of the axial vector coupling constant in neutron β decay in terms of known low-energy pion–nucleon scattering cross sections, with results in very good agreement with experiment. The method employs PCAC and also makes use of a sum rule derived from the relation

$$[I_+^5, I_-^5] = 2I_3 \tag{4.195}$$

which is a particular case of (4.194). Actually (4.195) does not require the full apparatus of $SU(3)_f \times SU(3)_f$ but only $SU(2) \times SU(2)$ for the first quark generation. The derivation of the Adler–Weisberger result is outlined in Appendix D.

Problems

4.1 Starting with the $SU(3)$ commutation relations (4.11), show that identity (4.13) is satisfied.

4.2 By explicit use of the invariant tensors $\delta_i^{\ j}$, ϵ^{ijk}, and ϵ_{ijk}, obtain expressions for the 27, the 10, the $\overline{10}$, the symmetric 8, the antisymmetric 8, and the singlet, which are the irreducible representations appearing in the tensor product of two $SU(3)$ octets $B_i^j - \frac{1}{3}\delta_i^j B_k^k$ and $C_m^n - \frac{1}{3}\delta_m^n C_k^k$.

4.3 Equations (4.55) and (4.56) give the proton and neutron spin magnetic moments, respectively, in terms of the up and down quark magnetic moments, assuming that p and n belong to the spin-$\frac{1}{2}$ octet, which is included in the symmetric 56 of $SU(6)_f$. Obtain analogous formulas, assuming that the baryon spin-$\frac{1}{2}$ octet is included instead in the antisymmetric 20 of $SU(6)_f$. This would be the natural assumption we would be forced to if the color degree of freedom did not exist.

4.4 Derive the transition probability for the transition $\tau^- \to \pi^- \nu_\tau$ (4.87).

4.5 Consider the problem of forming polar vectors from the initial and final baryon spinors, the momentum transfer, and the γ matrices, in connection with a transition of the form $B \to B' l \nu$. Prove that (4.92) are the three independent polar vectors that one can form. Prove the same thing for the axial vectors of (4.93).

4.6 Consider the density of events in the neighborhood of a point with polar coordinates r, θ in a Dalitz diagram. Prove that this density is proportional to the absolute square of the transition amplitude $A(r, \theta)$.

4.7 Starting with (4.40), derive the total decay rate of pion β decay (4.141).

4.8 Starting from (4.180) and (4.181), derive the current commutation relations (4.182)–(4.184). Then obtain (4.192)–(4.194).

5

Nuclear β decay and muon capture

5.1 The impulse approximation

In this chapter we concentrate on those essential features of nuclear β decay and muon capture that are directly relevant to the development of general understanding in weak interactions. Because of space limitations, it is necessary to give only passing reference to many important topics and omit others entirely. Readers interested in more details are urged to consult the monographs by Konopinski (66), Schopper (66), Wu and Moszkowski (66), and Morita (73), as well as the review of electron capture by Bambynek et al. (77). Double β decay is discussed in Section 10.7.

The most direct approach to nuclear β decay is the *impulse approximation,* in which the nucleus is treated as a collection of free independent nucleons. The β-decay amplitude is obtained simply by summing the amplitude for neutron decay (or its conjugate for e^+ emission, e^- capture) over all nucleons. Meson exchange and higher-order many-body effects are neglected completely. The impulse approximation is adequate to describe the main effects in allowed decays, in particular, those that were crucial for establishing the V–A law. For dealing properly with certain small but important effects of "recoil order" in which many-body corrections are significant, it is convenient to employ a more sophisticated approach in which the nucleus is treated as an elementary particle (Kim and Primakoff 1965, Holstein, 1974). However, we shall adhere to the impulse approximation.

In β decay the nucleus is well localized and it is necessary to use spatial wavefunctions $\psi(\mathbf{x}, t) = \psi(\mathbf{x})e^{-iEt}$ for the various particles instead of plane waves. We begin our analysis of β decay by writing the transition matrix element for neutron decay in terms of the wavefunctions

5.1 The impulse approximation

for n, e^-, and $\bar{\nu}_e$ and the vector and axial form factors introduced in (4.35). Taking into account that $q = p_n - p_p = p_e + p_\nu$ and that $iq_\lambda e^{iq\cdot x} = \partial_\lambda e^{iq\cdot x}$, we have

$$\mathcal{T}_{fi} = \frac{G_F}{\sqrt{2}} \cos\theta_1 \int d^4x \bar{\psi}_p(\mathbf{x}, t)[f_1\gamma^\lambda j_\lambda(\mathbf{x}, t) + f_2\sigma^{\lambda\nu}\partial_\nu j_\lambda(\mathbf{x}, t)$$
$$- if_3 \partial^\lambda j_\lambda(\mathbf{x}, t) - g_1\gamma^\lambda\gamma_5 j_\lambda(\mathbf{x}, t) - g_2\sigma^{\lambda\nu}\gamma_5 \partial_\nu j_\lambda(\mathbf{x}, t)$$
$$+ ig_3\gamma_5 \partial^\lambda j_\lambda(\mathbf{x}, t)]\psi_n(\mathbf{x}, t) \tag{5.1}$$

where $j_\lambda(x, t) = \bar{\psi}_e(x, t)\gamma_\lambda(1 - \gamma_5)\psi_{\bar{\nu}}(x, t)$.

Now the kinetic energy of a nucleon within a typical nucleus is of order 8 MeV and the kinetic energy of recoil of the nucleus itself in β decay is of order 1 keV or less. Therefore, in the rest frame of the initial nucleus, all the initial and final nucleons are nonrelativistic. It is thus appropriate to carry out a two-component reduction of the nucleon wavefunctions in (5.1) and to discard terms of order $(v_{\text{nucleon}}/c)^2$ or higher. To this end, we write for neutron and proton:

$$\psi_n(\mathbf{x}) = \begin{pmatrix} \chi_n(\mathbf{x}) \\ -\dfrac{i\boldsymbol{\sigma}\cdot\boldsymbol{\nabla}}{2m_N}\chi_n(\mathbf{x}) \end{pmatrix}, \quad \psi_p(\mathbf{x}) = \begin{pmatrix} \chi_p(\mathbf{x}) \\ -\dfrac{i\boldsymbol{\sigma}\cdot\boldsymbol{\nabla}}{2m_N}\chi_p(\mathbf{x}) \end{pmatrix}$$

where χ_n and χ_p are two-component spinors and $m_N = m_p \approx m_n$.

Carrying out the two-component reduction, integrating over t, replacing the form factors by their values at $q^2 = 0$, and writing $\Delta = m_n - E_p$, we find that (5.1) becomes

$$\mathcal{T}_{fi} = 2\pi\,\delta(E_e + E_{\bar{\nu}} - \Delta)A(n \to p) \tag{5.2}$$

where

$$A(n \to p) = \frac{G_F}{\sqrt{2}} \cos\theta_1 \int d^3x\,\chi_p^\dagger(\mathbf{x})O\chi_n(\mathbf{x}) \tag{5.3}$$

and:

$$\chi_p^\dagger O\chi_n = \chi_p^\dagger \left\{ f_1(0)\left[j_0\chi_n + \frac{1}{2m_N}(\boldsymbol{\sigma}\cdot\boldsymbol{\nabla}\boldsymbol{\sigma}\cdot\mathbf{j}\chi_n + \boldsymbol{\sigma}\cdot\mathbf{j}\boldsymbol{\sigma}\cdot\boldsymbol{\nabla}\chi_n)\right] \right.$$
$$- f_2(0)\boldsymbol{\sigma}\cdot(\boldsymbol{\nabla}\times\mathbf{j})\chi_n + f_3(0)[(E_e + E_{\bar{\nu}})j_0\chi_n + i(\boldsymbol{\nabla}\cdot\mathbf{j})\chi_n]$$
$$- g_1(0)\left[\boldsymbol{\sigma}\cdot\mathbf{j}\chi_n - \frac{i}{2m_N}(\boldsymbol{\sigma}\cdot\boldsymbol{\nabla}j_0\chi_n + j_0\boldsymbol{\sigma}\cdot\boldsymbol{\nabla}\chi_n)\right]$$
$$- g_2(0)[(E_e + E_\nu)\boldsymbol{\sigma}\cdot\mathbf{j}\chi_n - i(\boldsymbol{\sigma}\cdot\boldsymbol{\nabla}j_0)\chi_n]$$
$$+ \frac{i}{2m_N}g_3(0)[(E_e + E_\nu)(\boldsymbol{\sigma}\cdot\boldsymbol{\nabla}j_0\chi_n - j_0\boldsymbol{\sigma}\cdot\boldsymbol{\nabla}\chi_n)$$
$$\left. + (\boldsymbol{\sigma}\cdot\boldsymbol{\nabla}\boldsymbol{\nabla}\cdot\mathbf{j}\chi_n - \boldsymbol{\nabla}\cdot\mathbf{j}\boldsymbol{\sigma}\cdot\boldsymbol{\nabla}\chi_n)]\right\} \tag{5.4}$$

In (5.4) only the lowest-order momentum-transfer terms have been retained for f_2 and g_2. According to the impulse approximation, (5.3) may be generalized to *nuclear* β^- decay if we allow Δ to denote the energy difference between initial and final *nuclei* and write

$$A_{\mathrm{fi}} = \frac{G_F}{\sqrt{2}} \cos\theta_1 \sum_{i=1}^{A} \int d^3\mathbf{x}\, \chi'^\dagger O_i \tau_i^+ \chi \tag{5.5}$$

where χ and χ' are the initial and final nuclear wavefunctions, respectively (assumed to be expressible as sums of products of nucleon wavefunctions), τ_i^+ the isospin raising operator for the ith nucleon, and O_i the operator of (5.4) referred to the ith nucleon. Neglecting recoil of the final nucleus, we obtain the transition probability per unit time for β decay from (5.2), (5.3), and (5.5) by Fermi's golden rule:

$$dW = (2\pi)^{-5}\, d^3\mathbf{p}_e\, d^3\mathbf{p}_{\bar{\nu}}\, \delta(E_e + E_\nu - \Delta)|A_{\mathrm{fi}}|^2 \tag{5.6}$$

5.2 Allowed β decay

Since the momentum of either lepton in β decay is of order $m_e c$, each lepton de Broglie wavelength $\sim \hbar/m_e c \simeq 4 \times 10^{-11}$ cm is much larger than the nuclear radius R. Consequently, the lepton wavefunctions vary only very slightly over the nuclear volume. In the *allowed approximation*, we ignore this variation and also discard all terms in (5.4) that are proportional to $E_e + E_\nu$ (these are of relative order Δ/m_N) or which contain ∇ applied to the nucleon spinor ($\sim$ nucleon momentum). Then (5.5) reduces to

$$A_{\mathrm{fi}} = \frac{G_F}{\sqrt{2}} \cos\theta_1 [C_V \langle 1 \rangle j_0(0) - C_A \langle \boldsymbol{\sigma} \rangle \cdot \mathbf{j}(0)] \tag{5.7}$$

where $C_V = f_1(0)$, $C_A = g_1(0)$, and

$$\langle 1 \rangle = \sum_{i=1}^{A} \int \chi'^\dagger(\mathbf{x}) \tau_i^+ \chi(\mathbf{x})\, d^3\mathbf{x} \tag{5.8}$$

$$\langle \boldsymbol{\sigma} \rangle = \sum_{i=1}^{A} \int \chi'^\dagger(\mathbf{x}) \boldsymbol{\sigma}_i \tau_i^+ \chi(\mathbf{x})\, d^3\mathbf{x} \tag{5.9}$$

The "Fermi" matrix element $\langle 1 \rangle$ is, from the point of view of the theory of angular momentum, the matrix element of a tensor of zeroth rank (scalar) between initial and final nuclear angular momentum states. Therefore it satisfies the selection rule: $J_i = J_f$. Also, since $\sum_{i=1}^{A} \tau_i^+ = T^+$ is the total isospin raising operator, which has no non-zero matrix elements between states of different total isospin, the Fermi matrix element satisfies the rule $\Delta T = 0$. The "Gamow–Teller"

5.2 Allowed β decay

matrix element $\langle\boldsymbol{\sigma}\rangle$ contains a first-rank tensor $\Sigma\,\sigma_i\tau_i^+$ with respect to angular momentum and satisfies the rule

$$J_i = J_f, J_f \pm 1$$
$$J_i = 0 \not\to J_f = 0$$

Also in this case we have $|\Delta T| = 0$ or 1. For either $\langle 1 \rangle$ or $\langle\boldsymbol{\sigma}\rangle$, there is no change in orbital angular momentum of the nucleons, so both matrix elements satisfy the rule for nuclear parity:

$$\pi_i \pi_f = +1$$

Let us give some examples of the various types of allowed transitions. The $0^+ \to 0^+$ "superallowed" decays: ^{10}C $\to$ ^{10}B, ^{14}O $\to$ ^{14}N*, and so forth, are pure Fermi. Transitions such as ^{6}He(1^+) $\to$ ^{6}Li(0^+), in which $|\Delta J| = 1$, are pure Gamow–Teller, and $n(\tfrac{1}{2}^+) \to p(\tfrac{1}{2}^+)$, ^{3}H($\tfrac{1}{2}^+$) $\to$ ^{3}He($\tfrac{1}{2}^+$) are "mixed," that is, they contain Fermi and Gamow–Teller contributions.

It is frequently useful to express $\langle\boldsymbol{\sigma}\rangle$ in terms of a reduced matrix element $\langle\sigma\rangle$ and Clebsch–Gordan coefficients. Writing the various components of $\boldsymbol{\sigma}$ in a spherical unit vector basis, we have

$$\langle \sigma_M \rangle = \langle \sigma \rangle \langle J_i, m_i, 1, M | J_f, m_i + M \rangle$$

We consider, in passing, the various corrections to the allowed approximation and return to this question in Section 5.5. There are first of all the spatial variations of the currents $j(\mathbf{x})$ and $j_0(\mathbf{x})$ in the terms already considered in (5.4), as well as the additional terms in $f_1(0) = C_V$ and $g_1(0) = C_A$. These contribute to "forbidden" transitions, which violate the selection rules just discussed, or yield "second-forbidden" corrections to allowed transitions. The f_2 ("weak magnetism") term also yields a measurable contribution; as we know, it should be determined by CVC. The g_2 term is second class; as we shall see, unambiguous tests for its existence require clear-cut determination of the f_2 term. The term in f_3 is expected to be zero because it is second class and also because of CVC. Finally, the term in g_3 ("induced pseudoscalar") is altogether negligible in β decay but does manifest itself in muon capture (see Section 5.6).

Returning now to our discussion of allowed β decay, we consider the Coulomb correction factor $F(Z, E)$, which accounts for distortion of the electron wavefunction $\psi_e(\mathbf{r})$ by the Coulomb field of the nucleus and atomic electrons and enters in

$$|A|^2 = F(Z, E)|A_0|^2 m_e m_\nu / E_e E_\nu \qquad (5.10)$$

where

$$A_0 = \frac{G_F \cos\theta_1}{\sqrt{2}} [C_V \langle 1\rangle \bar{u}_e \gamma^0 (1 - \gamma_5) v_{\bar{\nu}} - C_A \langle \boldsymbol{\sigma}\rangle \cdot \bar{u}_e \boldsymbol{\gamma}(1 - \gamma_5) v_{\bar{\nu}}]$$

(5.11)

u_e and $v_{\bar{\nu}}$ are conventional lepton plane-wave spinors and $F(Z, E)$ is the average over the nucleus of $|\psi_e(\mathbf{r})/\psi_{e0}(\mathbf{r})|^2$, where $\psi_{e0}(\mathbf{r})$ is the electron wavefunction without Coulomb distortion. For a point nucleus, one finds

$$\left|\frac{\psi_e(\mathbf{r})}{\psi_{e0}(\mathbf{r})}\right|^2 = 2(1 + \kappa_0)(2p_e r)^{-2(1-\kappa_0)} e^{\pi n} \frac{|\Gamma(\kappa_0 + in)|^2}{|\Gamma(1 + 2\kappa_0)|^2} \quad (5.12)$$

where $n = \pm Ze^2/v_e$ for $e^\pm$ emission and $\kappa_0 = (1 - Z^2\alpha^2)^{1/2}$. The right-hand side of (5.12) diverges at $r = 0$, but if the nucleus is taken to be of finite extent, (5.12) must be modified and the divergence disappears. For a number of important applications, it is insufficient to replace $F(Z, E)$ by (5.12) evaluated at the nuclear radius. Instead, one must integrate the Dirac equation numerically, taking into account a realistic nuclear charge distribution and atomic screening (Behrens and Jänecke 69).

5.3 The transition probability for allowed decays

We now consider in some detail the differential transition probability per unit time for allowed β decay:

$$dW = (2\pi)^{-5} \delta(E_e + E_\nu - \Delta) \frac{m_e}{E_e} \frac{m_\nu}{E_\nu} F(Z, E) |A_0|^2 d^3\mathbf{p}_e \, d^3\mathbf{p}_\nu \quad (5.13)$$

with A_0 as in (5.11). It is useful here to distinguish between two broad categories of experiments:

(i) Those in which the longitudinal polarization of electrons is observed in decay from unpolarized samples of initial nuclei.

(ii) Those in which the initial nucleus may be polarized and we sum over final spins but otherwise examine the decay in detail.

In the case of longitudinal electron polarization, all the information we seek is contained in the factors $\bar{u}_e \gamma_\lambda (1 - \gamma_5) v_{\bar{\nu}}$. The electron spinor thus appears in the form

$$u_{eL}^\pm = \frac{1}{2}(1 - \gamma_5) \begin{pmatrix} \chi_e^\pm \\ \dfrac{\boldsymbol{\sigma}\cdot\mathbf{p}}{E_e + m_e} \chi_e^\pm \end{pmatrix}$$

where $\chi_e^\pm$ is a two-component spinor with spin up (down) with respect to $\mathbf{p}$. Taking the latter vector along $\hat{z}$, the probability of generating an electron in β decay with spinor $u_{eL}^\pm$ is

5.3 The transition probability for allowed decays

$$P_\pm = (u_{eL}^\pm)^\dagger (u_{eL}^\pm)$$

$$= \left(\chi_e^\dagger, \pm \frac{p}{E_e + m_e} \chi_e^\dagger\right) \frac{1}{2}(1 - \gamma_5) \begin{pmatrix} \chi_e \\ \pm \frac{p}{E_e + m_e} \chi_e \end{pmatrix}$$

$$= \frac{1}{2}\left[1 \mp \frac{2p}{E_e + m_e} + \frac{p^2}{(E_e + m_e)^2}\right]$$

$$= \left[\frac{E \mp p}{E + m_e}\right]$$

where we have set $(\chi_e^\pm)^\dagger \chi_e^\pm = 1$. The polarization is, in natural units:

$$\mathcal{P} = \frac{P_+ - P_-}{P_+ + P_-} = \frac{(E - p) - (E + p)}{(E - p) + (E + p)} = -\frac{p}{E} = -v \quad (5.14)$$

Thus we obtain the important result for V–A coupling that the longitudinal polarization of e^- emitted by unpolarized nuclei must be $\mathcal{P}(e^-) = -v/c$, regardless of the type of decay [whereas for e^+ emission, $\mathcal{P}(e^+) = +v/c$].

The experimental e^- polarizations are summarized in Figure 5.1

Figure 5.1. Measured degree of longitudinal polarization for allowed e^- decay. Λ is a small correction for Coulomb and screening effects. The various data points represent measurements by a number of workers. (From Koks and Van Klinken 76. Reprinted with permission.)

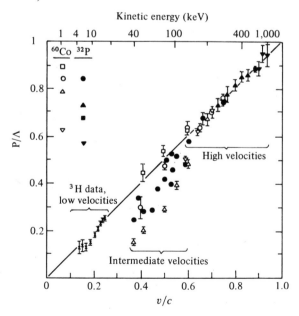

(Koks and Van Klinken 76). Almost all measurements shown were carried out by Mott scattering, which yields a more precise result than Møller scattering but requires that the longitudinal polarization of the electron be converted to transverse polarization by an external electrostatic field (see, for example, Kofoed-Hansen and Christensen 62). For $v/c > 0.6$, only a representative sample of data is shown in Figure 5.1. For that region, the agreement with $\mathcal{P}_{\text{theor}} = -v/c$ is excellent. The agreement is also very good at low velocities (^{3}H decay). However, at intermediate velocities, difficulties arise from various causes: poor Mott scattering analyzing power, thin scattering foils, depolarization in the source material, and so forth. It is to be noted that for high-Z nuclei an appreciable correction Λ is required because of Coulomb and screening effects. This correction is negligible for ^{3}H (<0.1 percent), but it is 3.3 percent for ^{60}Co at $v = 0.37c$ and 9 percent for ^{147}Pm at the same velocity.

From the data of Figure 5.1, one finds for e^- decay:

$$\mathcal{P}/\mathcal{P}_{\text{V-A}} = \begin{cases} 1.005 \pm 0.026 & \text{for low } v/c \quad (5.15a) \\ 1.001 \pm 0.008 & \text{for high } v/c \quad (5.15b) \end{cases}$$

For e^+, the results are fewer and less accurate. A measurement of ^{68}Ga decay (Gerber et al. 77) employs a novel positron polarimeter and obtains a result in agreement with the V-A prediction to about 10 percent, which is about the best precision that has been achieved for e^+.

We consider next the second category of experiments, where most of the remaining essential features of allowed decays are revealed. These are described by a formula originally derived by Jackson, Treiman, and Wyld (57) for dW of (5.13) with $|A_0|^2$ of (5.11) summed over final spins but with given initial nuclear polarization:

$$dW = \frac{G_F^2 \cos^2 \theta_1}{(2\pi)^5} \delta(E_e + E_\nu - \Delta) F(Z, E) \, d^3\mathbf{p}_e \, d^3\mathbf{p}_\nu$$

$$\times \xi \left\{ 1 + a \frac{\mathbf{p}_e \cdot \mathbf{p}_\nu}{E_e E_\nu} + \frac{\langle \mathbf{J}_i \rangle}{J_i} \cdot \left[A \frac{\mathbf{p}_e}{E_e} + B \frac{\mathbf{p}_\nu}{E_\nu} + D \frac{\mathbf{p}_e \times \mathbf{p}_\nu}{E_e E_\nu} \right] \right.$$

$$\left. + c \left[\frac{\mathbf{p}_e \cdot \mathbf{p}_\nu}{3 E_e E_\nu} - \frac{(\mathbf{p}_e \cdot \mathbf{j})(\mathbf{p}_\nu \cdot \mathbf{j})}{E_e E_\nu} \frac{J_i(J_i + 1) - 3\langle \mathbf{J}_i \cdot \mathbf{j}\rangle^2}{J_i(2J_i - 1)} \right] \right\} \quad (5.16)$$

We now discuss each of the terms in (5.16), defining the various new symbols as we go along.

5.3.1 Total decay rate: ft values

The first term inside the curly brackets of (5.16) is the only one that survives if we integrate over electron and neutrino solid angles. If we also integrate over neutrino energy, we find

5.3 The transition probability for allowed decays

$$dW = \frac{G_F^2 \cos^2 \theta_1}{2\pi^3} \xi F(Z, E_e) E_e p_e \, dE_e (\Delta - E_e)^2 \tag{5.17}$$

where

$$\xi = |C_V|^2 |\langle 1 \rangle|^2 + |C_A|^2 |\langle \sigma \rangle|^2 \tag{5.18}$$

In (5.17), which gives the electron energy spectrum in allowed decay, we assumed that the neutrino mass is zero. Modifications for finite m_ν mass and experiments to determine m_ν are described in Chapter 10. When (5.17) is integrated over electron energy, we obtain

$$W = \frac{G_F^2 \cos^2 \theta_1}{2\pi^3} \xi f_0(Z, \Delta) \tag{5.19}$$

where

$$f_0(Z, \Delta) = \int_{E_e = m_e}^{\Delta} F(Z, E) p_e E (\Delta - E)^2 \, dE \tag{5.20}$$

is the so-called Fermi integral. It must be corrected for radiative and "second-forbidden" effects and has been tabulated (Behrens and Jänecke 69) for a wide range of values of Δ and Z. The theoretical mean-life is then

$$\tau = \frac{2\pi^3}{G_F^2 \cos^2 \theta_1} \frac{1}{\xi f_{0,\text{corr}}} \tag{5.21}$$

One defines the comparative half-life $f_{0,\text{corr}} t_{1/2}$, or simply ft, by

$$ft = \frac{2\pi^3 \ln 2}{G_F^2 \cos^2 \theta_1 \, \xi} \tag{5.22a}$$

in natural units or more explicitly by

$$ft = \frac{(2\pi)^3 \ln 2 \cdot \hbar^7}{m_e^5 c^4} \frac{1}{G_F^2 \cos^2 \theta_1} \frac{1}{\xi} \tag{5.22b}$$

For the superallowed $0^+ \to 0^+$ transitions $^{10}C \to {}^{10}B$, $^{14}O \to {}^{14}N^*$, and so forth, the initial and final nuclei are, in each case, members of the same $T = 1$ isomultiplet. One then has $\langle \sigma \rangle = 0$ and $\langle 1 \rangle = \sqrt{2}(1 - \delta)$. Here $\delta \simeq 1\%$ is due to radiative corrections and isospin impurities in both initial and final nuclear states and can be calculated reliably (Towner et al. 77, Wilkinson 77, Towner and Hardy 78), up to a radiative correction $\Delta_R^{(v)}$ involving Z^0 exchange. Figure 5.2 shows the results of ft determinations for the $0^+ \to 0^+$ transitions. The values are remarkably consistent and give a weighted average (Raman et al. 75):

$$ft = 3088.6 \pm 2.1 \quad \text{sec} \tag{5.23}$$

Since G_F is known from muon decay (3.64) and $\cos \theta_1$ is determined from hyperon semileptonic decays, result (5.23) and formula (5.22) may be employed to determine $\Delta_R^{(v)}$. If the unresolved discrepancy is indeed due to Z^0 radiative corrections, the experimental results imply the existence of a Z^0 boson with mass between 70 and 150 GeV/c^2 (Sirlin 74).

In the case of neutron β decay, $\langle 1 \rangle = 1$ and $\langle \sigma \rangle = +\sqrt{3}$. Here careful measurements of the half-life lead to a determination of ξ and thus to the result (Kelly et al. 80)

$$C_A/C_V = 1.250 \pm 0.009 \tag{5.24}$$

5.3.2 Electron–neutrino angular correlation

If we average (5.16) over nuclear spin but do not integrate over the angle between e and ν_e, we obtain

$$dW = \frac{G_F^2 \cos^2 \theta_1}{2\pi^3} F(Z, E) p_e E (\Delta - E)^2 \, dE \, \xi \left(1 + a \frac{\mathbf{p}_e \cdot \mathbf{p}_\nu}{E_e E_\nu} \right) \tag{5.25}$$

where

$$\xi a = |C_V|^2 |\langle 1 \rangle|^2 - \tfrac{1}{3} |C_A|^2 |\langle \sigma \rangle|^2 \tag{5.26}$$

The quantity a is the electron–neutrino angular correlation coefficient

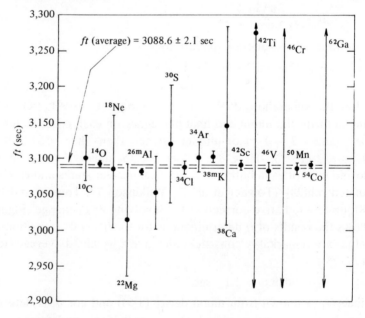

Figure 5.2. ft values for the superallowed $0^+ \to 0^+$ transitions. All corrections except $\Delta_R^{(v)}$ included. (From Raman et al. 75. Reprinted with permission.)

5.3 The transition probability for allowed decays

and represents a parity-conserving term in dW (since $\mathbf{p}_e$ and $\mathbf{p}_\nu$ are both polar vectors). This correlation has been measured for the decays of Table 5.1, in all cases by observation of the spectrum of recoil ions.

In Figure 5.3 we plot a versus the "Fermi fraction" $x = (1 + |\rho|^2)^{-1}$, where $\rho = C_A\langle\sigma\rangle/C_V\langle 1\rangle$. The data are clearly consistent with V and A couplings only. It is easy to see why a takes the values shown for pure S, V, T, or A couplings. For example, let us consider ^{35}Ar decay, where $\langle\sigma\rangle$ happens to be very small and the transition is essentially pure vector. Since e^+ has helicity $\sim +1$ and the neutrino is left-handed, the latter must be radiated in the same direction as e^+ ($a \simeq +1$) since the total angular momentum carried off by the leptons in a Fermi transition is zero.

Another very elegant method for determining neutrino helicity is the

Table 5.1. *Correlation coefficient* a_{expt} *for various decays*

Decay	J_i	J_f	a_{expt}	Reference
$n \rightarrow pe^-\bar{\nu}_e$	$\frac{1}{2}$	$\frac{1}{2}$	-0.1017 ± 0.0051	Stratowa et al. (78)
^{19}Ne $\rightarrow$ ^{19}Fe$^+\nu_e$	$\frac{1}{2}$	$\frac{1}{2}$	0.00 ± 0.08	Allen et al. (59)
^{35}Ar $\rightarrow$ ^{35}Cl$e^+\nu_e$	$\frac{3}{2}$	$\frac{3}{2}$	0.97 ± 0.14	Allen et al. (59)
^{6}He $\rightarrow$ ^{6}Li$e^-\bar{\nu}_e$	0	1	-0.3343 ± 0.0030	Johnson et al. (63)
^{23}Ne $\rightarrow$ ^{23}Na$e^-\bar{\nu}_e$	$\frac{3}{2}$	$\frac{3}{2}$	-0.33 ± 0.03	Johnson et al. (63)

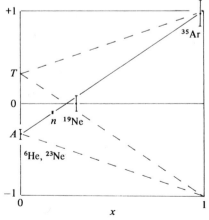

Figure 5.3. The e–ν correlation coefficient a.

experiment of Goldhaber, Grodzins, and Sunyar (58). Here we are concerned with the decay

$$e^- + {}^{152}\text{Eu}(J = 0) \rightarrow {}^{152}\text{Sm}^*(J = 1) + \nu_e$$
$$\phantom{e^- + {}^{152}\text{Eu}(J = 0) \rightarrow {}^{152}}\hookrightarrow {}^{152}\text{Sm}(J = 0) + \gamma$$

Since the initial angular momentum is zero, ν_e and ^{152}Sm* have opposite projections of angular momentum on the recoil axis and that of ^{152}Sm* must be transferred to the γ ray emitted in the de-excitation. A measurement of the latter's circular polarization results in the determination $h(\nu_e) = -1$ [see Section 5.6 for a discussion of $h(\nu_\mu)$].

In the case of neutron decay, the measurement (Stratowa et al. 78)

$$a = -0.1017 \pm 0.0051 \tag{5.27}$$

yields the result

$$|C_A/C_V| = 1.259 \pm 0.017 \tag{5.28}$$

which agrees well with the value obtained from the neutron half-life (5.24).

5.3.3 β and neutrino asymmetry coefficients A and B and time-reversal coefficient D

We now consider those terms in (5.16) that depend on nuclear polarization. First, we dispense with the term in c, which is a tensor polarization correlation that appears only for $J_i, J_f \geq 1$ and has no particular interest. It will be ignored from now on.

The β-decay asymmetry parameter A is given by

$$\xi A = \pm \kappa |C_A|^2 |\langle\sigma\rangle|^2 - (C_V C_A^* + C_A C_V^*)\langle 1\rangle\langle\sigma\rangle \left(\frac{J_i}{J_i + 1}\right)^{1/2} \tag{5.29}$$

where

$$\kappa = \begin{cases} 1 & \text{for } J_f = J_i - 1 \\ (J_i + 1)^{-1} & \text{for } J_f = J_i \\ -J_i(J_i + 1)^{-1} & \text{for } J_f = J_i + 1 \end{cases}$$

and where we employ $\pm$ for $e^\pm$.[1] Measurements of A have been carried

---

[1] Since (5.16) was derived in the allowed approximation, we ignore small but significant corrections to A, B, and D in (5.29)–(5.31). Actually A varies slightly with electron energy; see, e.g., Holstein 74, Calaprice et al. 75.

out for many transitions and have yielded a great deal of interesting information. Parity violation in weak interactions was first detected experimentally by observation of A in the decay of polarized ^{60}Co (Wu et al. 57). Observations of A in $n \to p$, ^{19}Ne $\to$ ^{19}F, ^{35}Ar $\to$ ^{35}Cl, together with measurement of the ft values, yield determinations of $G_F \cos \theta_1$. The results for n and ^{19}Ne are consistent with the $0^+ \to 0^+$ decays, but there is a significant discrepancy for ^{35}Ar. It is interesting to note that the discrepancy would disappear if the Cabibbo angle were set equal to zero in that case (Hagberg et al. 78). Salam and Strathdee (74,75) have speculated that "symmetry restoration" might occur for very large electromagnetic fields, and the critical field they propose is the same order of magnitude as that found in any nucleus. But why should such an effect occur only in ^{35}Ar? This little puzzle remains unsolved.

In addition, $A(^{19}\text{Ne})$ may be used to set limits on right-handed currents that might be expected to occur in "left–right symmetric" gauge theories (see Section 2.12). Finally, the β-decay asymmetries of ^{12}B and ^{12}N have been instrumental in various experiments to test CVC and PCAC and to search for second-class currents (see Sections 5.5 and 5.6).

The neutrino asymmetry parameter is given by

$$\xi B = \mp \kappa |C_A|^2 |\langle \sigma \rangle|^2 - (C_V C_A^* + C_A C_V^*) \langle 1 \rangle \langle \sigma \rangle \left(\frac{J_i}{J_i + 1} \right)^{1/2} \tag{5.30}$$

and the coefficient D by the formula

$$\xi D = i(C_V C_A^* - C_A C_V^*) \langle 1 \rangle \langle \sigma \rangle \left(\frac{J_i}{J_i + 1} \right)^{1/2} \tag{5.31}$$

Without loss of generality, we may take $\langle 1 \rangle$ and $\langle \sigma \rangle$ to be real, but D can only be nonzero if C_A/C_V is complex, that is, if T invariance fails. This last remark must be qualified, however: final-state interactions can introduce phase shifts and thus masquerade as T violation.

Both B and D have been measured for $n \to p$ and ^{19}Ne $\to$ ^{19}F by observation of electron–recoil ion coincidences in the decay of polarized nuclei. The latest measurements of $D(^{19}\text{Ne})$ are at a level of experimental precision that is almost sufficient to detect an expected final-state contribution arising mainly from the weak magnetism form factor (Callan and Treiman 67). Table 5.2 summarizes the measured values of A, B, and D for $n \to p$, ^{19}Ne $\to$ ^{19}F, and ^{35}Ar $\to$ ^{35}Cl.

Table 5.2. *Measured values of* A, B, *and* D *for* $n \to p$, $^{19}Ne \to {}^{19}F$, *and* $^{35}Ar \to {}^{35}Cl$

Transition	A	B	D	$(C_A/C_V)^a$	Reference
$n \to p$	-0.115 ± 0.008	1.01 ± 0.04	—	$+1.26 \pm 0.02$	Christensen et al. (69, 70)
$n \to p$	-0.118 ± 0.010	0.995 ± 0.034	—	$+1.27 \pm 0.025$	Erozolimskii et al. (70)
$n \to p$	—	—	-0.01 ± 0.01	—	Erozolimskii et al. (74)
$n \to p$	—	—	-0.0027 ± 0.0033	—	Steinberg et al. (76)
$n \to p$	-0.113 ± 0.006	—	-0.0011 ± 0.0017	—	Steinberg et al. (76)
$n \to p$	-0.113 ± 0.006	—	—	1.258 ± 0.015	Krohn and Ringo (75)
$n \to p$	-0.115 ± 0.006	—	—	1.263 ± 0.015	Erozolimskii et al. (76)
$^{19}Ne \to {}^{19}F$	-0.039 ± 0.002	—	0.001 ± 0.003	—	Calaprice et al. (74)
$^{19}Ne \to {}^{19}F$	—	-0.90 ± 0.13	-0.0005 ± 0.0010	—	Baltrusaitis and Calaprice (77)
$^{19}Ne \to {}^{19}F$	-0.04^b	—	—	—	Calaprice et al. (75)
$^{35}Ar \to {}^{35}Cl$	0.16 ± 0.04	—	—	—	Calaprice et al. (65)

^{a} From A and B.
^{b} In this experiment, the variation of A with e^+ energy was measured.

5.4 β-decay coupling constants: Validity of the V–A law

As we have seen, the V–A law accounts extremely well for allowed β decay, but it is of considerable interest to set quantitative limits on this validity. Such an exercise is not only generally useful but it is further motivated by the possibility that left–right symmetric gauge theories with right-handed (V + A) currents may be tested in β decay.

Writing the most general derivative-free four fermion β-decay amplitude consistent with proper Lorentz invariance and lepton conservation, we have

$$A' = \frac{G_F \cos \theta_1}{\sqrt{2}} \sum_i \int d^3x \, [\bar{\psi}_p O_i \psi_n][\bar{\psi}_e O_i (C_i - C'_i \gamma_5) \psi_\nu] \quad (5.32)$$

where the C_i and C'_i are to be determined by experiment. We find that the coefficients are restricted as follows:

(i) C_p and C'_p are eliminated in the nonrelativistic nucleon limit.

(ii) Electron polarizations $\mathcal{P}(e^\pm) = \pm v/c$ imply

$$C_V = C'_V, \quad C_A = C'_A, \quad C_S = -C'_S, \quad C_T = -C'_T \quad (5.33)$$

for we may write

$$(C_i - C'_i \gamma_5) = (C_i - C'_i) \frac{1 + \gamma_5}{2} + (C + C'_i) \left(\frac{1 - \gamma_5}{2}\right) \quad (5.34)$$

Furthermore,

$$O_i \left(\frac{1 \pm \gamma_5}{2}\right) = \left(\frac{1 \pm \gamma_5}{2}\right) O_i, \quad i = S, T, P \quad (5.35)$$

and

$$O_i \left(\frac{1 \pm \gamma_5}{2}\right) = \left(\frac{1 \mp \gamma_5}{2}\right) O_i, \quad i = V, A \quad (5.36)$$

Thus,

$$\bar{\psi}_e O_i (C_i - C'_i \gamma_5) \psi_\nu$$
$$= \begin{cases} \bar{\psi}_{eR}(C_i + C'_i) O_i \psi_\nu + \bar{\psi}_{eL}(C_i - C'_i) O_i \psi_\nu & \text{for S, T, P} \quad (5.37) \\ \bar{\psi}_{eL}(C_i + C'_i) O_i \psi_\nu + \bar{\psi}_{eR}(C_i - C'_i) O_i \psi_\nu & \text{for V, A} \quad (5.38) \end{cases}$$

(iii) Retaining the S and T couplings a priori, we are led to the expectation of an allowed spectrum that departs from the usual statistical shape by an extra factor

$$1 \pm b[1 - Z^2 \alpha^2]^{1/2} E^{-1} \quad \text{for} \quad e^\pm \quad (5.39)$$

where b, the *Fierz interference coefficient*, is given by

$$b = (\xi_F b_F + \xi_{GT} b_{GT})\xi^{-1} \tag{5.40}$$

with

$$\xi = \xi_F + \xi_{GT} \tag{5.41a}$$
$$\xi_F = 2(|C_S|^2 + |C_V|^2)|\langle 1 \rangle|^2 \tag{5.41b}$$
$$\xi_{GT} = 2(|C_A|^2 + |C_T|^2)|\langle \sigma \rangle|^2 \tag{5.41c}$$
$$\xi_F b_F = (C_S^* C_V + cc)|\langle 1 \rangle|^2 \tag{5.41d}$$
$$\xi_{GT} b_{GT} = (C_T^* C_A + cc)|\langle \sigma \rangle|^2 \tag{5.41e}$$

If no assumption is made about neutrino helicity, measured allowed spectrum shapes lead to the conclusion (Paul 70) that

$$C_S/C_V = 0.08 \pm 1.2, \quad C_T/C_A = 0.006 \pm 0.2 \tag{5.42}$$

However, if we take the measurements summarized in Figure 5.3 and the ^{152}Eu result (Goldhaber et al. 58) to imply $h(\nu_e) = -1$, then

$$C_S/C_V = -0.001 \pm 0.006, \quad C_T/C_A = -0.004 \pm 0.003 \tag{5.43}$$

5.5 Tests of CVC and searches for second-class currents

We now consider the effects of "recoil order" in allowed β decay that are used to verify CVC and search for second-class currents (SCC). One particular case, that of the $A = 12$ "triad" shown in Figure 5.4, proves to be of crucial importance in the whole discussion. Here we have the β decays $^{12}\text{B} \rightarrow {}^{12}\text{C}\ e^-\bar{\nu}_e$ and $^{12}\text{N} \rightarrow {}^{12}\text{C}\ e^+\nu_e$, as well as

Figure 5.4. The decay scheme for $A = 12$. (Energies in MeV.)

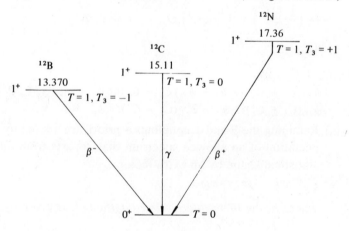

the magnetic dipole transition $^{12*}\text{C} \to {}^{12}\text{C} + \gamma$. In the allowed approximation, each β decay is pure axial vector, but the energy release is so large that appreciable second-forbidden corrections must be taken into account, as well as contributions from f_2 and (if it exists) g_2. In certain key experiments that we shall describe, comparison of the ^{12}B and ^{12}N decays enables one to isolate the weak magnetism and SCC effects and to determine them unambiguously. These experiments involve measurements of the $e^\pm$ energy distributions (spectral shapes) and $e^\pm$ anisotropies in decay of oriented ^{12}B and ^{12}N. Relevant information is also provided by measurement of the polarization of the ^{12}B recoil nucleus following μ^- capture on ^{12}C (see Section 5.6).

We begin with the effect of weak magnetism on the spectral shape. Gell-Mann (58) first demonstrated that the electron energy distribution for the $A = 12$ β decays is given by

$$dW = \text{const } F(Z, E)pE(\Delta - E)^2(1 \mp \tfrac{8}{3}aE) \tag{5.44}$$

for $e^\pm$. Apart from the last factor, (5.44) is the usual allowed spectral shape formula. We shall now indicate how the correction factor $(1 \mp \tfrac{8}{3}aE)$ is derived and show its connection to weak magnetism.

Our starting point is the transition amplitude (5.2) with A as in (5.5) and O_i as in (5.4) for the ith nucleon. In what follows we suppress the index i and the isospin operators $\tau_i^\pm$ for simplicity of notation. In addition to the allowed axial vector contribution, there is also the weak magnetism term in $f_2(0)$ and the forbidden correction in $f_1(0) = C_V = 1$:

$$A = \int d^3x \chi'^\dagger \left[-C_A \boldsymbol{\sigma} \cdot \mathbf{j}\chi - f_2(0)\boldsymbol{\sigma} \cdot (\boldsymbol{\nabla} \times \mathbf{j})\chi \right.$$
$$\left. + i\frac{C_V}{2m_N}(\boldsymbol{\sigma} \cdot \boldsymbol{\nabla}\boldsymbol{\sigma} \cdot \mathbf{j}\chi + \boldsymbol{\sigma} \cdot \mathbf{j}\boldsymbol{\sigma} \cdot \boldsymbol{\nabla}\chi) \right] \tag{5.45}$$

Now,

$$\frac{i}{2m_N}(\boldsymbol{\sigma} \cdot \boldsymbol{\nabla}\boldsymbol{\sigma} \cdot \mathbf{j}\chi + \boldsymbol{\sigma} \cdot \mathbf{j}\boldsymbol{\sigma} \cdot \boldsymbol{\nabla}\chi)$$

$$= \frac{i}{2m_N}[\boldsymbol{\nabla} \cdot (\mathbf{j}\chi) + \mathbf{j} \cdot \boldsymbol{\nabla}\chi + i\boldsymbol{\sigma} \cdot (\boldsymbol{\nabla} \times \mathbf{j}\chi) + i\boldsymbol{\sigma} \cdot \mathbf{j} \times \boldsymbol{\nabla}\chi]$$

$$= -\frac{1}{2m_N}\boldsymbol{\sigma} \cdot (\boldsymbol{\nabla} \times \mathbf{j})\chi + \frac{i}{m_N}\mathbf{j} \cdot \boldsymbol{\nabla}\chi + \frac{i}{2m_N}\boldsymbol{\nabla} \cdot \mathbf{j}\chi \tag{5.46}$$

Also $\mathbf{j}(\mathbf{x}) = \mathbf{j}(0)\exp(-i\mathbf{q} \cdot \mathbf{r}) \simeq (1 - i\mathbf{q} \cdot \mathbf{r})\mathbf{j}(0)$ and $\boldsymbol{\nabla} \times \mathbf{j} \simeq -i\mathbf{q} \times \mathbf{j}(0)$. Therefore, integrating (5.46) over $\mathbf{x}$ and noting that in a $1^+ \to 0^+$ transition the contributions of $\int \chi'^\dagger \mathbf{j}(0) \cdot \boldsymbol{\nabla}\chi$ and $\int \chi'^\dagger \boldsymbol{\nabla} \cdot \mathbf{j}\chi$ vanish, we obtain, in place of (5.45),

$$A = -C_A\langle\sigma\rangle \cdot \mathbf{j} + if_2(0)\langle\sigma\rangle \times \mathbf{q} \cdot \mathbf{j} + \frac{i}{2m_N}\langle\sigma\rangle \times \mathbf{q} \cdot \mathbf{j}$$

$$+ \frac{1}{m_N} \int \chi'^\dagger(\mathbf{q}\cdot\mathbf{r}\,\mathbf{j}(0) \cdot \nabla)\chi \, d^3x \qquad (5.47)$$

In (5.47), the second and third terms may be combined. Since

$$f_2(0) = +(\mu_p - \mu_n - 1)/2m_N$$

they yield

$$\frac{i}{2m_N}(\mu_p - \mu_n)\langle\sigma\rangle \times \mathbf{q} \cdot \mathbf{j}$$

Next we analyze the fourth term of (5.47). For this purpose, we employ the identity

$$\mathbf{q}\cdot\mathbf{r}\,\mathbf{j}\cdot\nabla = \tfrac{1}{3}\mathbf{q}\cdot\mathbf{j}\,\mathbf{r}\cdot\nabla + \tfrac{1}{2}\mathbf{q}\times\mathbf{j}\cdot\mathbf{r}\times\nabla$$
$$+ \tfrac{1}{4}[q_i j_k + q_k j_i - \tfrac{2}{3}\delta_{ij}\mathbf{q}\cdot\mathbf{j}]$$
$$\times [r_i\nabla_k + r_k\nabla_i - \tfrac{2}{3}\delta_{ij}\mathbf{r}\cdot\nabla] \qquad (5.48)$$

Only the middle term $\tfrac{1}{2}\mathbf{q}\times\mathbf{j}\cdot\mathbf{r}\times\nabla = \tfrac{1}{2}i\mathbf{L}\times\mathbf{q}\cdot\mathbf{j}$, where $\mathbf{L} = -i\mathbf{r}\times\nabla$, can contribute to the matrix element for a $1^+ \to 0^+$ transition. Therefore (5.47) becomes

$$A_\pm = \left\{-C_A\langle\sigma\rangle + \frac{i}{2m_N}[(\mu_p - \mu_N)\langle\sigma\rangle + \langle\mathbf{L}\rangle] \times \mathbf{q}\right\} \cdot \mathbf{j}(0)_\pm \qquad (5.49)$$

where $j_+ = \bar{u}_\nu\gamma_\lambda(1 - \gamma_5)v_e$ and $j_- = \bar{u}_e\gamma_\lambda(1 - \gamma_5)v_\nu$. Actually, it can be shown (Morita et al. 76) from explicit calculations of nuclear wavefunctions that $\langle\mathbf{L}\rangle \ll (\mu_p - \mu_n)\langle\sigma\rangle$, namely, that

$$x \equiv 1 + \frac{\langle\mathbf{L}\rangle}{(\mu_p - \mu_n)\langle\sigma\rangle} = 0.981 \qquad (5.50)$$

The allowed shape factor arises of course from the square of $-C_A\langle\sigma\rangle \cdot \mathbf{j}(0)_\pm$, whereas the correction $\mp\tfrac{8}{3}\underline{a}E$ in (5.44), where

$$\underline{a} = \frac{x(\mu_p - \mu_n)}{2m_p}\left|\frac{C_V}{C_A}\right| \qquad (5.51)$$

originates from interference between the first and second terms in (5.49). It is easy to complete the derivation of (5.44) by standard methods and, in particular, to show that for $e^\pm$ the corrections have opposite signs.

In 1962–3, several research groups measured the β spectra of ^{12}B and ^{12}N. In particular, Lee, Mo, and Wu (63) obtained results in very good agreement with formulas (5.44) and (5.51), and CVC thus seemed firmly established. Subsequently some systematic errors were found in

5.5 Tests of CVC and searches for second-class currents

tabulated positron wavefunctions, and a redetermination of the ^{12}N end-point energy was made. This seemed to indicate that the results claimed by Lee, Mo, and Wu were not as satisfactory as first thought (Calaprice and Holstein 76). However, a reexamination of the data, in which revised branching ratios for inner β groups were also adopted, achieved consistency with CVC (Wu et al. 77). Furthermore, a new measurement of the $A = 12$ spectral shapes also confirmed CVC (Kaina et al. 77), and observations of the β spectrum of ^{20}F corroborated it further.

In the late 1960s, it was suggested that a finite SCC might manifest itself in an asymmetry in the ft values of mirror pairs of decays (e.g., for $A = 8, 9, 12, 13, 17, 18, 20, 24, 28$, and 30) (Wilkinson and Alburger 70). However, it was eventually established that even in the most favorable case ($A = 8$) the complications of nuclear structure make it impossible to interpret the ft asymmetry unambiguously (Kubodera and Arima 77).

On the other hand, the angular distribution of β radiation from oriented nuclei depends on weak magnetism and SCC effects in a clear-cut manner. It can be shown (Holstein 74, Morita et al. 76) that in $A = 12$ the angular distribution of decay electrons is given by the formula, for $e^{\pm}$,

$$W(\theta) = \text{const}[1 \pm P(1 + \alpha_{\pm}E)P_1(\cos\theta) + A\alpha_{\pm}EP_2(\cos\theta)] \quad (5.52)$$

where θ is the angle between the electron momentum and nuclear orientation axis, $P_l(\cos\theta)$ the Legendre polynomials, P the nuclear polarization, and A the alignment. These quantities are defined in terms of the Zeeman sublevel weights a_{-1}, a_0, a_1 for $m_J = -1, 0, +1$, respectively, with $a_{-1} + a_0 + a_1 = 1$:

$$P = a_1 - a_{-1}, \quad A = 1 - 3a_0$$

The parameter $\alpha_{\pm}$ is given by

$$\alpha_{\pm} = \mp \tfrac{2}{3}(\underline{a} - b) \quad (5.53)$$

where $\underline{a}$ represents the effects of weak magnetism and is given by (5.51) and b contains contributions from SCC and forbidden corrections.

It has been shown (Morita et al. 76) that

$$b = g_2/C_A \pm y/2m_N \quad (5.54)$$

where

$$y = 1 + 2i\frac{\int r(\boldsymbol{\sigma}\cdot\mathbf{p})}{\langle\boldsymbol{\sigma}\rangle} \simeq 3.55 \quad (5.55)$$

In (5.54), the first term is the one of interest and the second is the "forbidden" correction. It is important to note that when one takes the difference between α_+ and α_-, this latter contribution cancels out.

Theoretical calculations by Hwang and Primakoff (77) lead to the prediction (for $g_2 = 0$):

$$\alpha_- = (0.08 \pm 0.03)/\text{GeV} \tag{5.56}$$

$$\alpha_+ = -(2.75 \pm 0.03)/\text{GeV} \tag{5.57}$$

Originally, measurements of α_- were carried out (Sugimoto et al. 75) with polarized nuclei, and there seemed to be a large SCC effect. However, subsequent measurements were done with *aligned* nuclei in which the quantity $\alpha_\pm$ can be measured more accurately, as is evident from (5.52) (Brändl et al. 78a,b, Lebrun et al. 78, Masuda et al. 79).

Although these experiments vary in detail, they all have common features. For example, in the work of Masuda et al. (79), polarized ^{12}B were produced in a very thin target in the ^{11}B$(d, p)^{12}$B reaction at 1.5 MeV with a polarization $P \simeq 10\%$, whereas ^{12}N were produced in the ^{10}B$(^3\text{He}, n)^{12}$N at 3.0 MeV with a polarization $P \simeq 20\%$ by selecting recoil nuclei at the appropriate angle. Conversion of the nuclear polarization to alignment was achieved by means of an NMR technique in which unequal Zeeman splittings caused by an electric quadrupole interaction superimposed on the magnetic interaction was utilized to allow selective transitions $-1 \leftrightarrow 0$ and $+1 \leftrightarrow 0$ on magnetic sublevels. For this purpose, the polarized recoil nuclei emerging from the back of the target were implanted in a thin Mg single crystal (≤ 150 μm in thickness), the "c" axis of which was set parallel to an external magnetic field $B \simeq 1$ kG. The alignment A could be reversed at will by means of an ingenious sequence of applied radiofrequency pulses.

The experimental results are as follows:

(i) $\alpha_- = -0.07 \pm 0.20/\text{GeV}$ (Lebrun et al. 78),

(ii) $\alpha_- = 0.24 \pm 0.44/\text{GeV}$ (Brändl et al. 78a),

(iii) $\alpha_+ = -2.73 \pm 0.39/\text{GeV}$ (Brändl et al. 78b),

(iv) $\alpha_- = 0.25 \pm 0.34/\text{GeV}$ (Masuda et al. 79),

(v) $\alpha_+ = -2.77 \pm 0.52/\text{GeV}$ (Masuda et al. 79).

These are in very good agreement with theory [(5.56) and (5.57)] and yield the result $g_2 = -(0.4 \pm 0.9)g_1/2m_p$. The conclusion of this section is thus that CVC is valid and there are no measurable second-class currents, a conclusion not altered when many-body nuclear effects are taken into account (Hwang 79b, Oka and Kubodera 80).

5.6 Muon capture

Muon capture by a proton

$$\mu^- + p \to n + \nu_\mu \tag{5.58}$$

or by a nucleus is characterized by two essential features that distinguish it from the analogous process of orbital electron capture in the theory of nuclear β decay: competition with muon decay, and the relatively large momentum transfer imparted to the nucleus resulting from capture of the muon ($m_\mu \simeq 0.1 m_p$). The capture almost invariably occurs from a $1s$ hydrogenic orbital. For light nuclei, where the Bohr radius is much larger than the nuclear radius, this orbital is well described by the spatial wavefunction

$$\phi_\mu = (Z^3/\pi a_0^3)^{1/2} e^{-Zr/a_0} \tag{5.59}$$

where $a_0 = \hbar^2/m'e^2$ and m' is the reduced mass of the muon–nucleus system. The probability of capture is proportional to the number of nuclear protons Z and to the probability density of the muon at the nucleus $\phi_\mu^2(0) = Z^3/\pi a_0^3$. Thus the capture rate varies as Z^4, and although it is much less probable in hydrogen than muon decay, the capture and decay rates are about equal for $Z = 6$ (carbon). From a theoretical point of view muon capture by a proton is attractive because of its simplicity, but muon capture in hydrogen is difficult to observe experimentally because of the small rate and the complications that ensue from the fact that hydrogen exists in molecular form (Quaranta et al. 67).

Muon capture on a complex nucleus has its own complications, however, because of the large momentum transfer that, except for very small Z, results in a profusion of excited nuclear states. Sometimes these can be separated experimentally from capture to the ground state. For example, this is the case for

$$\mu^- + {}^{12}\mathrm{C} \to {}^{12}\mathrm{B} + \nu \tag{5.60}$$

Otherwise analysis is difficult and one must resort to the "closure" approximation first introduced by Primakoff (59). The large momentum transfer in muon capture also implies that there are substantial contributions to the amplitude from weak magnetism $[f_2(q^2)]$ and induced pseudoscalar $[g_3(q^2)]$ terms.

In this section our limited goal is formulation of the simplest theory of muon capture in the framework of the impulse approximation and a discussion of several recent experiments in which the standard theory, in particular, the induced pseudoscalar term, is verified. For more detailed treatments, the reader is again referred to the monographs by Konopinski (66) and Morita (73) and to reviews by Primakoff (59) and others.

The differential transition probability per unit time for muon capture is

$$dW = \frac{2\pi\,\delta(E_i - E_f + E_\nu - m_\mu)\,d^3\mathbf{p}_\nu}{(2\pi)^3}|A|^2 \tag{5.61}$$

where E_i and E_f are the initial and final nuclear energies, respectively. In the delta function, which expresses energy conservation, we neglect the atomic binding energy of the muon. Also, the neutrino, the muon, and the nuclei are normalized to one per large arbitrary volume V. The quantity A is analogous to the corresponding expression in nuclear β decay. For example, in muon capture by a proton, we have

$$A = \frac{G_F \cos\theta_1}{\sqrt{2}} \int d^3x\,\bar{\psi}_n(x)\hat{O}\psi_p(x) \tag{5.62}$$

where

$$\hat{O} = f_1(q^2)\gamma^\alpha j_\alpha + if_2(q^2)\sigma^{\alpha\beta}q_\beta j_\alpha - g_1(q^2)\gamma^\alpha\gamma_5 j_\alpha - g_3(q^2)q^\alpha j_\alpha \tag{5.63}$$

and

$$j_\alpha = \bar{\psi}_\nu(x)\gamma_\alpha(1-\gamma_5)\psi_\mu(x) \tag{5.64}$$

(Here and henceforth we omit second-class terms.) We must be careful to evaluate the form factors at $q^2 \simeq -0.9m^2$ for capture by a proton and suitable values of q^2 for other captures. One finds that

$$f_1(-0.9m_\mu^2) = 0.97$$
$$f_2(q^2) = (0.97)3.70/2m_N$$

but that $g_1(q^2)$ varies slowly enough that we can safely take $g_1(q^2) = C_A = 1.25$ (Fujii and Primakoff 59). As for $g_3(q^2)$, it should be given by the Goldberger–Treiman relation:

$$g_3(q^2) = -2m_N C_A/(q^2 - m_\pi^2) = 2m_N C_A/(0.9m_\mu^2 + m_\pi^2) \tag{5.65}$$

It is convenient to define

$$g_p = -m_\mu g_3 \simeq -7C_A \tag{5.66}$$

The muon is described by

$$\psi_\mu(\mathbf{r}) = \phi_\mu u_\mu = \phi_\mu \begin{pmatrix} \chi_\mu \\ 0 \end{pmatrix} \tag{5.67}$$

and the neutrino by

5.6 Muon capture

$$\psi_\nu(\mathbf{r}) = \frac{1}{\sqrt{2}} \exp(i\mathbf{p}_\nu \cdot \mathbf{r}) \begin{pmatrix} \chi_\nu \\ \boldsymbol{\sigma} \cdot \hat{\mathbf{p}}_\nu \chi_\nu \end{pmatrix} \tag{5.68}$$

where χ_μ and χ_ν are two-component spinors normalized to unity.

Formula (5.62) for A may be reduced to a convenient form by taking advantage of the fact that the nucleon motions are nonrelativistic even after muon absorption. We write

$$j_0 = \bar{\psi}_\nu \gamma^0 (1 - \gamma_5) \psi_\mu$$
$$= \frac{1}{\sqrt{2}} [\chi_\nu^\dagger (1 - \boldsymbol{\sigma} \cdot \hat{\mathbf{p}}_\nu) \chi_\mu] e^{-i\mathbf{p} \cdot \mathbf{x}} \phi_\mu \tag{5.69}$$

and

$$\mathbf{j} = -\frac{1}{\sqrt{2}} [\chi_\nu^\dagger (1 - \boldsymbol{\sigma} \cdot \hat{\mathbf{p}}_\nu) \boldsymbol{\sigma} \chi_\mu] \exp(-i\mathbf{p}_\nu \cdot \mathbf{x}) \phi_\mu \tag{5.70}$$

Then, carrying out a two-component reduction of the nuclear wavefunctions as in β decay and retaining only lowest-order terms in f_2 and g_3, we find, after some algebra, that

$$A = \tfrac{1}{2} \int \chi_n^\dagger(\mathbf{x}) \chi_\nu^\dagger \mathcal{H} \chi_\mu \chi_p(\mathbf{x}) \exp(-i\mathbf{p}_\nu \cdot \mathbf{x}) \phi_\mu(\mathbf{x}) \, d^3\mathbf{x} \tag{5.71}$$

where the effective Hamiltonian $\mathcal{H}$ is (Primakoff 59)

$$\mathcal{H} = (1 - \boldsymbol{\sigma} \cdot \hat{\mathbf{p}}_\nu)[G_V + G_A \boldsymbol{\sigma}_A \cdot \boldsymbol{\sigma} + G_P \boldsymbol{\sigma}_A \cdot \hat{\mathbf{p}}_\nu] \tag{5.72}$$

and where

$$G_V = G_F \cos\theta_1 f_1(q^2) \left(1 + \frac{E_\nu}{2m_N}\right) \tag{5.73}$$

$$G_A = G_F \cos\theta_1 \left\{-C_A - \frac{E_\nu}{2m_N}[f_1(q^2) + 2m_N f_2(q^2)]\right\} \tag{5.74}$$

$$G_P = G_F \cos\theta_1 \frac{E_\nu}{2m_N} \{C_A + g_P - [f_1(q^2) + 2m_N f_2(q^2)]\} \tag{5.75}$$

In (5.72), $\boldsymbol{\sigma}_A$ and $\boldsymbol{\sigma}$ are Pauli spin matrices for the nucleons and leptons, respectively. In the more general case of muon capture by a nucleus, the right-hand side of (5.71) must be summed over all nuclear protons.

From (5.61) and (5.71)–(5.74), we obtain the total transition probability per unit time:

$$W = \frac{1}{(2\pi)^2} \int d\Omega_\nu \left\{\int E_\nu^2 \, dE_\nu \, \delta[E_\nu + (M_f^2 + E_\nu^2)^{1/2} - M_i - m_\mu]\right\} |A|^2$$

$$= \frac{1}{4} \frac{E_\nu^2}{1 + E_\nu/(E_\nu^2 + M_f^2)^{1/2}} \int \frac{d\Omega_\nu}{4\pi^2} \left(\frac{Z^3}{\pi a_0^3}\right) \tag{5.76}$$

$$\times |\chi_\nu^\dagger (1 - \boldsymbol{\sigma} \cdot \hat{\mathbf{p}}_\nu)[G_V \langle 1 \rangle + (G_A \boldsymbol{\sigma} + G_P \hat{\mathbf{p}}_\nu) \cdot \langle \boldsymbol{\sigma}_A \rangle] \chi_\mu|^2$$

where

200   5 Nuclear β decay and muon capture

$$\langle 1 \rangle = (\pi a_0^3/Z^3)^{1/2} \left\langle f \left| \sum_A \tau_A^- \exp(-i\mathbf{p}_\nu \cdot \mathbf{x}_A) \phi(\mathbf{x}_A) \right| i \right\rangle \quad (5.77)$$

and

$$\langle \boldsymbol{\sigma}_A \rangle = (\pi a_0^3/Z^3)^{1/2} \left\langle f \left| \sum_A \tau_A^- \exp(-i\mathbf{p}_\nu \cdot \mathbf{x}_A) \phi(\mathbf{x}_A) \boldsymbol{\sigma}_A \right| i \right\rangle \quad (5.78)$$

These last two quantities would reduce to matrix elements $\langle 1 \rangle$ and $\langle \boldsymbol{\sigma}_A \rangle$ appropriate for the description of positron emission or orbital electron capture in the theory of β decay if we could ignore the variation of the lepton wavefunctions within the nucleus.

We now consider the transition

$$\mu^- + {}^{12}C(0^+) \to {}^{12}B_{\text{gr.st.}}(1^+) + \nu_\mu \quad (5.79)$$

This is of special interest because of its relative theoretical simplicity (Foldy and Walecka 65, Devanathan et al. 72, Hwang 79a) and also because several decisive experiments employing the reaction have been carried out. These are:

(i) The total capture rate, which has been observed and corrected for captures leading to excited ^{12}B states by observation of deexcitation γ rays (Maier et al. 64). Thus one determines the total ground-state capture rate to be

$$\Gamma_{\text{expt}}(\text{gr.st.}) = 6750 \begin{Bmatrix} +300 \\ -750 \end{Bmatrix} \text{sec}^{-1} \quad (5.80)$$

(ii) The polarization P_z of ^{12}B recoils in the direction z of muon polarization (Possoz et al. 77). This measurement, achieved by observation of the ^{12}B β-decay asymmetry, may be used in two ways: It yields a determination of the muon helicity $h(\mu^-)$ in $\pi^- \to \mu^- \nu_\mu$ decay, and the result

$$h(\mu^-) \geq +0.90 \quad (5.81)$$

is in accord with the V–A law. In addition, it may be combined with (5.68) and compared with theory (Hwang 79b) to test the weak magnetism and induced pseudoscalar terms and to place a limit on SCC. One finds agreement with CVC, obtains the value $g_P = (-7.1 \pm 2.7)C_A$ in conformity with the Goldberger–Treiman relation (5.66), and finds no evidence for SCC.

(iii) The polarization $P_{z'}$ of ^{12}B in the direction z' of its recoil momentum. This quantity depends only on the helicity $h(\nu_\mu)$. The measurements yield

5.6 Muon capture

$$h(\nu_\mu) = -1.06 \pm 0.11 \tag{5.82}$$

in agreement with the V–A law (Roesch et al. 82).

Let us carry out a simple analysis of the polarization experiments in order to see the essential features. Figure 5.5 illustrates the various vectors of interest: the muon polarization $\mathbf{p}_\mu$ directed along z, the ^{12}B recoil momentum along z', and the neutrino momentum and spin along $-z'$ and z', respectively, assuming $h(\nu_\mu) = -1$.

Suppose at first that $p_{\mu z} = +1$. Then the muon spin state is

$$\alpha_z^\mu = \cos{\tfrac{1}{2}\theta}\, \alpha_{z'}^\mu + \sin{\tfrac{1}{2}\theta}\, \beta_{z'}^\mu \tag{5.83}$$

where α and β denote spin up and spin down, respectively. Then, assuming $h(\nu_\mu) = -1$, the neutrino has spin state $\alpha_{z'}^\nu$. Let us suppose that the reduced matrix elements for coupling $\alpha_{z'}^\mu$ and $\beta_{z'}^\mu$ to ^{12}C to form ^{12}B $+ \alpha_{z'}^\nu$ are M_0 and M_{-1}, respectively. We define $X = M_0/M_{-1}$, and it is easy to show from (5.59) and (5.60) that $X = (G_A - G_P)/G_A$. Then, by a simple application of the Wigner–Eckhart theorem, one finds the ^{12}B(1^+) state to be

$$|B, \theta\rangle'_+ = -(1/3\Gamma)^{1/2}[\cos{\tfrac{1}{2}\theta}\, M_0 |10\rangle_{z'} + \sqrt{2} \sin{\tfrac{1}{2}\theta}\, M_{-1}|1-1\rangle_{z'}] \tag{5.84}$$

In (5.84), the superscript on the left-hand side indicates that z' is the quantization axis and the subscript indicates that $p_{\mu z} = +1$. Also Γ is a normalization factor: $\Gamma = \tfrac{1}{3}M_-^2(2 + X^2)$. If the muon polarization were along $-z$ instead, $|B, \theta\rangle$ would become:

$$|B, \theta\rangle'_- = (1/3\Gamma)^{1/2} M_-[\sin{\tfrac{1}{2}\theta}\, X|10\rangle' - \sqrt{2} \cos{\tfrac{1}{2}\theta}|1-1\rangle'] \tag{5.85}$$

Figure 5.5. Vectors for the $\mu^- + {}^{12}\text{C} \to {}^{12}\text{B} + \nu_\mu$ polarization experiments.

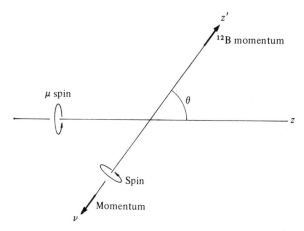

Thus we find for the expectation value of ^{12}B polarization along z':
$$\langle B, \theta | J_z' | B, \theta \rangle_\pm' = -(M_-^2/3\Gamma)(1 \mp \cos\theta) \tag{5.86}$$
and for an arbitrary muon polarization:
$$\langle B, \theta | J_z' | B, \theta \rangle' = -(M_-^2/3\Gamma)(1 - p_\mu \cos\theta) \tag{5.87}$$
Also, we find:
$$\langle B, \theta | J_x' | B, \theta \rangle' = (1/3\Gamma) M_- M_0 p_\mu \sin\theta \tag{5.88}$$
Therefore, making explicit the neutrino helicity, we obtain
$$\langle J_z \rangle = \cos\theta \langle J_{z'} \rangle + \sin\theta \langle J_{x'} \rangle$$
$$= \frac{1}{X^2 + 2} [p_\mu (X \sin^2\theta + \cos^2\theta) + h_\nu \cos\theta] \tag{5.89}$$

In the experiment to measure $\langle J_z \rangle$, one stops μ^- in a thick graphite target ($10 \times 10 \times 1$ cm^3). The ^{12}B recoils are emitted over a 4π solid angle and come to rest within the graphite, and one detects the β-decay asymmetry with counters in front and in back of the target. To compare (5.89) with the results obtained, it is necessary to integrate $\langle J_z \rangle$ over the entire sphere, in which case the term in h contributes nothing and we obtain

$$\overline{\langle J_z \rangle} = \frac{\frac{4}{3}X + \frac{3}{2}}{2 + X^2} p_\mu \tag{5.90}$$

The kinematic and atomic depolarization of muons in graphite is corrected for by an auxiliary measurement of the μ^--decay asymmetry expressed as a fraction of the μ^+-decay asymmetry. The depolarization of ^{12}B recoils is minimized by imposition of a longitudinal magnetic field of about 1 kG and is corrected for by calibration measurements involving production of ^{12}B of known polarization in the ^{11}B(d, p)^{12}B reaction. The final result is

$$\overline{\langle J_z \rangle}_{\text{expt}} = 0.537 \pm 0.049 \tag{5.91}$$

which, as mentioned earlier, yields good agreement with theory for standard values of the μ^- capture couplings and also gives a value of $h(\mu^-)$ in accord with the V–A law.

In order to measure $h(\nu_\mu)$, it is evident from (5.89) that one must separate the forward (F) and backward (B) recoil nuclei (that is, those emitted at angles $\theta < \pi/2$ or $\theta > \pi/2$, respectively). One has:

$$F \equiv \int_0^{\pi/2} \sin\theta \, d\theta \, \langle J_z(\theta) \rangle = \left[\frac{1}{3} \frac{1 + 2X}{2 + X^2} + \frac{h(\nu)}{2 + X^2} \right] \tag{5.92}$$

$$B \equiv \int_{\pi/2}^\pi \sin\theta \, d\theta \, \langle J_z(\theta) \rangle = \left[\frac{1}{3} \frac{1 + 2X}{2 + X^2} - \frac{h(\nu)}{2 + X^2} \right] \tag{5.93}$$

The experiment employs 3,000 identical modules, each consisting of a thin C layer (60 μm) sandwiched between a 1.5-μm-thick Al foil and a thin layer of silver (1,200 μg/cm^2). The ^{12}B recoils are produced in the carbon but come to rest in the aluminum (F or B) or the silver (B or F). It has been shown by separate calibration experiments that these are completely depolarizing and polarization-maintaining, respectively. The entire assembly can be flipped mechanically so that the F and B roles of the Al and Ag layers are reversed. Measurement of the ^{12}B β-decay asymmetry in this elegant experiment yields (5.82).

Problems

5.1 Starting with (5.1), carry out the two-component reduction indicated in the text to obtain (5.2) with (5.3) and (5.4).

5.2 Consider neutron β decay.
(a) Derive (5.16) from (5.7) for this special case with ξ, a, A, B, and D given by (5.18), (5.26), (5.29), (5.30), and (5.31), respectively.
(b) Calculate the recoil momentum spectrum of protons in the neutron rest frame.

5.3 Design an experiment to observe the β-decay asymmetry A (5.29) in the transition ^{37}K $\rightarrow$ ^{37}Ar (ground state) + e^+ + ν_e. How would you polarize the ^{37}K nuclei and measure the polarization accurately?

5.4 Verify the identity (5.48). Complete the derivation of (5.44), which is left incomplete in the text.

5.5 Carry out the two-component reduction for muon capture to obtain (5.71) with (5.72) to (5.75).

6
Weak decays of mesons and hyperons

6.1 Introduction

In this chapter we shall study the weak decays of hadrons containing s, c, and/or b quarks. These include the semileptonic and nonleptonic decays of K mesons, charmed mesons, B mesons, strange hyperons, and charmed baryons. A wide variety of complex phenomena are encountered here, and many features will be seen to depend critically on aspects of quark–gluon dynamics that are not well understood. To study these problems, we shall require all of the tools developed in Chapter 4 and others besides. Two very important subjects associated with K decay – strangeness oscillations and CP violation – are deferred to Chapter 7; here we shall generally assume that CP is conserved.

6.2 K_{l3} decays

K mesons decay in many modes – leptonic, semileptonic, and nonleptonic (see Table 6.1). We have already considered the leptonic decays $K_{\mu 2}$ and K_{e2} in Section 4.5. Let us therefore begin by discussing the K_{l3} decays:

(a) $K^+ \to \pi^0 \mu^+ \nu_\mu$ (c) $K_L^0 \to \pi^- \mu^+ \nu_\mu$
(b) $K^- \to \pi^0 \mu^- \bar{\nu}_\mu$ (d) $K_L^0 \to \pi^+ \mu^- \bar{\nu}_\mu$ (6.1)
(e) $K^+ \to \pi^0 e^+ \nu_e$ (g) $K_L^0 \to \pi^- e^+ \nu_e$
(f) $K^- \to \pi^0 e^- \bar{\nu}_e$ (h) $K_L^0 \to \pi^+ e^- \bar{\nu}_e$

It was established [(4.89)] that for a V–A interaction the K_{l3} amplitude has the general form

$$\mathcal{M} = \frac{G_F}{\sqrt{2}} s_1 c_3 [f_+(q^2)(k_K^\alpha + k_\pi^\alpha) + f_-(q^2)(k_K^\alpha - k_\pi^\alpha)] \\ \times \bar{u}_l \gamma_\alpha (1 - \gamma_5) v_\nu \qquad (6.2)$$

6.2 K_{l3} decays

where $f_+(q^2)$ and $f_-(q^2)$ are form factors and k_K and k_π are the meson 4-momenta. The form factors are restricted to be relatively real by T invariance and are further restricted by CPT invariance [(4.116)–(4.118b)] and the $\Delta S = \Delta Q$ rule. The $|\Delta I| = \frac{1}{2}$ rule applied to semileptonic transitions is simply a manifestation of the quark isospin change $s(I = 0) \leftrightarrow u(I = \frac{1}{2})$. It is more general than the $\Delta S = \Delta Q$ rule and implies the latter. Assuming the $|\Delta I| = \frac{1}{2}$ rule, one may easily show

Table 6.1. *Decay of K mesons*

Particle	Mass (MeV/c^2)	Mean lifetime (sec)	Mode	Branching fraction
$K^\pm$	493.669(0.015)	1.2371×10^{-8} (0.0026)	$\mu^+\nu$	63.50(0.16)%
			$\pi^+\pi^0$	21.16(0.15)%
			$\pi^+\pi^+\pi^-$	5.59(0.03)%
			$\pi^+\pi^0\pi^0$	1.73(0.05)%
			$\mu^+\nu\pi^0$	3.20(0.09)%
			$e^+\nu\pi^0$	4.82(0.05)%
			+ miscellaneous modes with very small branching ratios	
$K^0, \bar{K}^0$	497.76(0.13)			
K_S^0		0.8923×10^{-10} (0.0022)	$\pi^+\pi^-$	68.61(.24)%
			$\pi^0\pi^0$	31.39(.24)%
			$\mu^+\mu^-$	$<3.2 \times 10^{-7}$
			e^+e^-	$<3.4 \times 10^{-4}$
			$\pi^+\pi^-\gamma$	$1.85(0.10) \times 10^{-3}$
			$\gamma\gamma$	$<0.9 \times 10^{-3}$
K_L^0		5.183×10^{-8} (0.040)	$\pi^0\pi^0\pi^0$	21.5(0.7)%
			$\pi^+\pi^-\pi^0$	12.39(0.18)%
			$\pi^\pm\mu\nu$	27.0(0.5)%
			$\pi^\pm e^\pm\nu$	38.8(0.5)%
			$\pi e\nu\gamma$	1.3(0.8)%
			$\pi^+\pi^{-a}$	0.203(0.005)%
			$\pi^0\pi^{0a}$	0.094(0.018)%
			$\pi^+\pi^-\gamma$	$6.0(2.0) \times 10^{-5}$
			$\pi^0\gamma\gamma$	$<2.4 \times 10^{-4}$
			$\gamma\gamma$	$4.9(0.5) \times 10^{-4}$
			$e\mu$	$<2.0 \times 10^{-9}$
			$\mu^+\mu^-$	$(9.1 \times 1.9) \times 10^{-9}$
			$\mu^+\mu^-\nu$	$<7.8 \times 10^{-6}$
			$\mu^+\mu^-\pi^0$	$<5.7 \times 10^{-5}$
			e^+e^-	$<2.0 \times 10^{-5}$
			$e^+e^-\gamma$	$<2.8 \times 10^{-5}$
			$\pi^+\pi^-e^+e^-$	$<8.8 \times 10^{-6}$
			$\pi^0\pi^\pm e^\pm\nu$	$<2.2 \times 10^{-3}$

[a] *CP* violation.

that $A_+ = A_- = f/\sqrt{2}$ and $g = 0$. Thus, if we assume CP invariance and correct for differing phase-space factors and radiative effects, we find that the rates for transitions (6.1a–d) should be the same as should the rates for (6.1e–h). (In fact, CP violation causes a very small difference in the rates for $K_L^0 \to \pi^- l^+ \nu$ and $K_L^0 \to \pi^+ l^- \nu$; see Chapter 7.) We have thus succeeded in relating the various K_{l3} transitions to one another, but we must still discuss the form factors $f_\pm$ themselves.

It will be recalled from Section 4.9 that the amplitude for $\pi^+ \to \pi^0 e^+ \nu_e$ is

$$\mathcal{M}(\pi^+ \to \pi^0 e^+ \nu_e) = G_F \cos\theta_1 [F_1(q^2)(k'^\alpha + k^\alpha) + F_2(q^2)(k'^\alpha - k^\alpha)]\bar{u}_p \gamma_\alpha (1 - \gamma_5) v_e \quad (6.3)$$

where $F_1(0) = 1$ and $F_2(0) = 0$ because of CVC. In this transition and also for K_{l3}, it is the vector portion V^α of the charged hadronic weak current that is operative. Furthermore, according to Cabibbo's hypothesis, V^α may be expressed as

$$V^\alpha = (j_1 - ij_2)^\alpha \cos\theta_1 + (j_4 - ij_5)^\alpha \sin\theta_1 \quad (6.4)$$

where the j's all belong to the same octet of vector current operators, and, for the present, we ignore the other Cabibbo angles θ_2 and θ_3. It follows that, in the limit of perfect $SU(3)_f$ symmetry, $f_+(0) = 1$ and $f_-(0) = 0$. These relations are not exactly true because of departures from perfect $SU(3)_f$ symmetry, but according to the Ademollo–Gatto theorem (Ademollo and Gatto 64; see also Appendix D), the change induced in $f_+(q^2)$ is only of second order in the symmetry-breaking parameter. Indeed, Langacker and Pagels (73) have shown that one expects $f_+(0) = 0.98$, which is very close to unity.

In a more general analysis of K_{l3} decay, we might drop the assumption of V–A coupling. We would then be obliged to include in (6.2) the following additional terms:

$$\frac{G_F}{\sqrt{2}} s_1 c_3 \left[2m_K f_S \bar{u}_l (1 - \gamma_5) v_\nu + \frac{2f_T}{m_K} p_K^\lambda p_\pi^\mu \bar{u}_l \sigma_{\lambda\mu} (1 - \gamma_5) v_\nu \right] \quad (6.5)$$

with scalar (f_S) and tensor (f_T) form factors. However, as we shall see momentarily, in K_{e3} decay, the term in f_- makes a negligible contribution, and analysis of the $\pi^0 \nu_e$ angular correlation reveals that the scalar and tensor contributions are very small (see Figure 6.1):

$$f_S/f_+(0) = 0.00 + 0.10, \quad f_T/f_+(0) = 0.02 \pm 0.12 \quad (6.6)$$

Therefore, in what follows, we shall ignore f_S and f_T.

Now, returning to the V–A amplitude (6.2), we substitute $p_K - p_\pi = p_l + p_{\bar{\nu}}$ and obtain

6.2 K_{l3} decays

$$\begin{aligned}\mathcal{M} &= \frac{G_F}{\sqrt{2}} s_1 c_3 [f_+(q^2)(p_K + p_\pi)^\alpha \bar{u}_l \gamma_\alpha (1 - \gamma_5) v_{\bar{\nu}} \\ &\quad + f_-(q^2) \bar{u}_l (\not{p}_l + \not{p}_{\bar{\nu}})(1 - \gamma_5) v_{\bar{\nu}}] \\ &= \frac{G_F}{\sqrt{2}} s_1 c_3 [f_+(q^2)(p_K + p_\pi)^\alpha \bar{u}_l \gamma_\alpha (1 - \gamma_5) v_{\bar{\nu}} \\ &\quad + f_-(q^2) m_l \bar{u}_l (1 - \gamma_5) \bar{v}_\nu]\end{aligned} \quad (6.7)$$

Defining $\xi = f_-(q^2)/f_+(q^2)$ and substituting in (6.7), we have

$$\mathcal{M} = \frac{G_F}{\sqrt{2}} s_1 c_3 f_+(q^2)[2p_K^\alpha \bar{u}_l \gamma_\alpha (1 - \gamma_5) v_{\bar{\nu}} + m_l(\xi - 1)\bar{u}_l(1 - \gamma_5)v_{\bar{\nu}}] \quad (6.8)$$

In K_{e3} decay, the term with $(\xi - 1)$ is negligible because of the small value of the electron mass. However, this term must be taken into account for $K_{\mu 3}$ decay.

It is customarily assumed that one may express each form factor f_+ and f_- as a constant plus a term linear in q^2, the q^2 dependence arising from strong interaction. The coefficients $\lambda_\pm$, λ_0, and $f_0(q^2)$ are defined by the following equations:

Figure 6.1. Cosine of the angle between π^0 and ν in the dilepton CM system for K_{e3} decay. Shown are predictions for S, V, and T couplings and the experimental results of Braun et al. (75). (Reprinted with permission.)

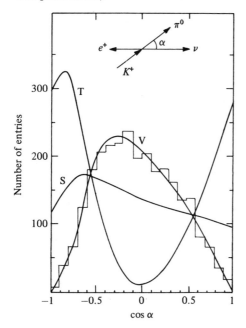

$$f_\pm(q^2) = f_\pm(0)(1 + \lambda_\pm q^2/m_\pi^2) \tag{6.9}$$

$$f_0(q^2) = f_+(q^2) + \frac{q^2}{m_K^2 - m_\pi^2} f_-(q^2) \tag{6.10}$$

$$\lambda_0 = \frac{m_\pi^2}{q^2}\left[\frac{f_0(q^2)}{f_0(0)} - 1\right] \tag{6.11}$$

There exist several methods for determining the quantities λ.

(a) For K_{e3} decay, a measurement of the pion energy spectrum yields the q^2 dependence of f_+. If we neglect the term with $(\xi - 1)$, it may easily be shown that (6.8) leads to the following differential transition probability, integrated over electron and neutrino momenta and summed over electron spins:

$$dW = G_F^2 s_1^2 c_3^2 \frac{m_K f_+^2(q^2)|\mathbf{p}_\pi|^3 \, dE_\pi}{12\pi^3} \tag{6.12}$$

(b) The Dalitz plot density for $K_{\mu 3}$ (see Section 4.7) is (Chounet et al. 72):

$$\rho(E_\pi, E_\mu) \propto f_+^2(q^2)[A + B\xi + C\xi^2] \tag{6.13}$$

where

$$\begin{aligned} A &= m_K[2(E_\mu E_\nu - m_K E_\pi') + m_\mu^2(\tfrac{1}{4}E_\pi' - E_\nu)] \\ B &= m_\mu^2(E_\nu - \tfrac{1}{2}E_\pi') \\ C &= \tfrac{1}{4}m_\mu^2 E_\pi' \end{aligned} \tag{6.14}$$

$$E_\pi' = E_\pi^{\max} - E_\pi = \frac{m_K^2 + m_\pi^2 - m_\mu^2}{2m_K} - E_\pi$$

The density ρ is fit to the data to determine λ_+ and ξ (or equivalently λ_+ and λ_0).

(c) The branching ratio $\Gamma(K_{\mu 3})/\Gamma(K_{e3})$ can be measured and compared with theory (Fearing et al. 70).

(d) For muon polarization $\mathscr{P}$ in $K_{\mu 3}$ decay it can be shown (Cabibbo and Maksymovicz 64a,b,65) that in the K rest frame this is given by

$$\mathscr{P} = \mathbf{A}/|\mathbf{A}|$$

where

$$\mathbf{A} = a_1(\xi)\mathbf{p}_\mu - a_2(\xi)\left\{\frac{\mathbf{p}_\mu}{m_\mu}\left[m_K - E_\pi + \frac{\mathbf{p}_\pi \cdot \mathbf{p}_\mu}{|\mathbf{p}_\mu|^2}(E_\mu - m_\mu)\right] + \mathbf{p}_\pi\right\} + m_K \, \mathrm{Im}\, \xi (\mathbf{p}_\pi \times \mathbf{p}_\mu) \tag{6.15}$$

and where

6.3 Kaon nonleptonic decays

$$a_1(\xi) = (2m_K^2/m_\mu)[E_\nu + E'_\pi \operatorname{Re} b(q^2)]$$
$$a_2(\xi) = m_K^2 + 2 \operatorname{Re} b(q^2) m_K E_\mu + m_\mu^2 |b(q^2)|^2 \qquad (6.16)$$
$$b(q^2) = \tfrac{1}{2}[\xi(q^2) - 1]$$

If time-reversal invariance is valid, ξ is real and the polarization vector lies entirely in the $\mathbf{p}_\pi - \mathbf{p}_\mu$ plane and is defined by the angle it makes with the muon momentum.

Unfortunately, the experimental results obtained by methods (b), (c), and (d) for $K_{\mu 3}$ decays are not entirely consistent. The accompanying tabulation summarizes the average results.

	$K_{\mu 3}^+$	$K_{\mu 3}^0$	K_{e3}^+	K_{e3}^0
λ_+	+0.026(8)	+0.034(5)	0.0285(43)	0.0301(16)
λ_0	−0.003(7)	+0.022(6)		
$\xi(0)$	−0.35(14)	−0.14(11)		

From the experimental branching ratios for K_{e3}^+ and K_{e3}^0 and the renormalization value of $f_+(0) = 0.98$, and using G_F from muon decay, one can extract a value of $|V_{12}| = |s_1 c_3|$ in the six-quark matrix (4.70). It is found (Shrock and Wang 78) that

$$K_{e3}: \quad |V_{12}| = 0.219 \pm 0.003 \qquad (6.17)$$

This result is in good agreement with that obtained from hyperon decays, as we shall see.

6.3 Kaon nonleptonic decays

6.3.1 $K_{2\pi}$ decays

From K_{l3} we pass next to the following decays:

$$\begin{aligned} K^+ &\to \pi^+ \pi^0 \\ K_S^0 &\to \pi^+ \pi^- \\ K_S^0 &\to \pi^0 \pi^0 \\ K_L^0 &\to \pi^+ \pi^- \\ K_L^0 &\to \pi^0 \pi^0 \end{aligned} \qquad (6.18)$$

The transitions $K_L^0 \to \pi^+ \pi^-$ and $K_L^0 \to \pi^0 \pi^0$ are CP-violating and discussion of them is postponed until Chapter 7. In $K_S^0 \to \pi^+ \pi^-$, $K_S^0 \to \pi^0 \pi^0$, and $K^+ \to \pi^+ \pi^0$, the two-pion final state always has zero angular momentum and is spatially symmetric. Since pions are bosons, this implies that the isospin part of the two-pion wavefunction must also be symmetric. Now $I(\pi) = 1$ and so, a priori, the possible isospin of two pions is $I = 2$, 1, or 0. Of these, the $I = 2$ and 0 states are symmetric with respect to pion exchange, whereas $I = 1$ is antisymmetric, and therefore excluded. Hence only an $I = 0$ final state is possible if $|\Delta I| = \frac{1}{2}$. Now

$$|I = 0, I_3 = 0\rangle = \frac{1}{\sqrt{3}} |\pi_1^+ \pi_2^- + \pi_1^- \pi_2^+ - \pi_1^0 \pi_2^0\rangle \qquad (6.19)$$

Taking into account that the probability for $K_S^0 \to \pi^+ \pi^-$ is proportional to

$$\tfrac{1}{3}|\pi_1^+ \pi_2^-|^2 + \tfrac{1}{3}|\pi_1^- \pi_2^+|^2 = \tfrac{2}{3}$$

and the probability for $K_S^0 \to \pi^0 \pi^0$ is proportional to $\tfrac{1}{3}|\pi_1^0 \pi_2^0|^2 = \tfrac{1}{3}$, the $|\Delta I| = \frac{1}{2}$ rule implies

$$\gamma(K_S^0 \to \pi^+\pi^-)/\gamma(K_S^0 \to \pi^0\pi^0)|_{\Delta I = 1/2} = 2 \qquad (6.20)$$

where γ is the transition rate divided by the phase-space factor. In the case $K^+ \to \pi^+ \pi^0$, the final two-pion state has $I_3 = +1$, so $I_f = 0$ is excluded. Therefore $I_f = 2$, and we must have $|\Delta I| = \frac{3}{2}$ and/or $\frac{5}{2}$. The $K^+ \to \pi^+ \pi^0$ rate is very slow, however, compared with $K_S^0 \to \pi^+ \pi^-$. Let us denote the $|\Delta I| = \frac{1}{2}, \frac{3}{2}$, and $\frac{5}{2}$ transition amplitudes by a_1, a_3, and a_5, respectively. Also we define the amplitudes

$$A_0 = \langle (2\pi)^S_{I=0} | H_{wk} | K^0 \rangle \qquad (6.21)$$

and

$$A_2 = \langle (2\pi)^S_{I=2} | H_{wk} | K^0 \rangle \qquad (6.22)$$

where the superscript S means stationary, that is, no final-state interaction between the pions. We then have

$$a_1 = A_0 e^{i\delta_0}$$
$$a_3 + a_5 = A_2 e^{i\delta_2}$$

where δ_0 and δ_2 are the 2π final-state phase shifts for $I = 0$ and $I = 2$, respectively. It can then be shown that

$$\frac{\gamma(K^+ \to \pi^+ \pi^0)}{\gamma(K_S^0 \to \pi^+ \pi^-) + \gamma(K_S^0 \to 2\pi^0)} = \frac{3}{4} \frac{|a_3 - \tfrac{2}{3} a_5|^2}{A_0^2 + (\operatorname{Re} A_2)^2} \qquad (6.23)$$

The experimental values of this ratio and that of (6.20) allow us to conclude that

$$\begin{aligned}|a_3/a_1| &= 0.045 \pm 0.005 \\ |a_5/a_1| &= 0.001 \pm 0.003\end{aligned} \qquad (6.24)$$

One finds that a_3 is comparable to the amplitude for semileptonic K decay and that a_1 is enhanced by a factor of about 20 above that.

Let us now consider this $|\Delta I| = \frac{1}{2}$ enhancement from the point of

6.3 Kaon nonleptonic decays

view of $SU(3)_f$. From Section 4.1, we know that the hadronic weak current is composed of vector and axial vector components that belong to two distinct octets of current operators. If the effective weak Hamiltonian maintains its current × current structure in the presence of strong interactions, then, since K_S^0, π^+, and π^- are pseudoscalars and $J_i = J_f = 0$, the effective Hamiltonian must be of the form:

$$H \simeq V_8 \times A_8 + A_8 \times V_8 \qquad (6.25)$$

which is a symmetric combination of two octets. Now when one takes the tensor product of two octets in $SU(3)$, one obtains the irreducible representations given by the following equation:

$$8 \otimes 8 = 27 + 10 + \overline{10} + 8_S + 8_A + 1 \qquad (6.26)$$

Here the 27, 8_S, and 1 are symmetric, whereas the 10, $\overline{10}$, and 8_A are antisymmetric. Thus, in the symmetric combination (6.25), only the 27, 8_S, and 1 appear. However, we are interested in the portion of H_{wk} that can induce a weak transition with $|\Delta S| = 1$; therefore the singlet, which has no such component, cannot contribute, and we are left with 8_S and 27. The 8_S contains only a $|\Delta I| = \frac{1}{2}$ component. Therefore results (6.24) imply that the 27 is suppressed and/or the 8_S is enhanced. This is *octet dominance*, and in general, the $|\Delta I| = \frac{1}{2}$ rule for all nonleptonic strangeness-changing decays appears to be caused by octet dominance.

It will now be demonstrated that if the effective weak Hamiltonian is assumed to be of form (6.25) and *CP* invariance is valid, then the amplitude for $K_S^0 \to \pi^+\pi^-$ or $\pi^0\pi^0$ *vanishes* in the limit of perfect $SU(3)_f$ symmetry. This follows because the effective Hamiltonian (6.25) contains a combination of octet components that transform under $SU(3)$ as follows:

$$\begin{aligned} H_{\text{eff}} &\simeq (\lambda_1 - i\lambda_2)(\lambda_4 + i\lambda_5) + (\lambda_1 + i\lambda_2)(\lambda_4 - i\lambda_5) \\ &+ (\lambda_4 + i\lambda_5)(\lambda_1 - i\lambda_2) + (\lambda_4 - i\lambda_5)(\lambda_1 + i\lambda_2) \\ &\propto \lambda_6 \end{aligned} \qquad (6.27)$$

where the λs are $SU(3)$ matrices and the last step can be verified from Table 4.1.

Thus H_{eff} transforms under $SU(3)_f$ like λ_6, and since it is a product of vector and axial vector currents, it is odd under parity. Furthermore, the vector and axial vector currents transform oppositely under charge conjugation, so H_{eff} itself is odd under C and, therefore, even under *CP*. On the other hand, K_S^0 transforms like λ_7 if *CP* is conserved.

We may now pretend that the weak transition occurs by absorption of a fictitious "spurion" h, which is odd under C and P, even under CP, and transforms like λ_6 under $SU(3)_f$:

$$h + K_S^0 \to \pi^+\pi^- \tag{6.28}$$

The "interaction" of (6.28) may now be assumed $SU(3)_f$- and C-invariant. Now the most general $SU(3)_f$-invariant amplitude we can write to account for (6.28) must be a linear combination of traces of products of $SU(3)$ matrices containing creation and destruction operators for the particles of (6.28). For K_S^0, we have the meson matrix M and for $\pi^+\pi^-$, we have $\overline{M}M$. Taking into account that each matrix has zero trace, the amplitude must be

$$\mathcal{M} = a\,\mathrm{tr}(\overline{M}\overline{M}hM) + b\,\mathrm{tr}(\overline{M}\overline{M}Mh) + c\,\mathrm{tr}(\overline{M}h\overline{M}M) \\ + d\,\mathrm{tr}(\overline{M}h)\,\mathrm{tr}(\overline{M}M) + e\,\mathrm{tr}(\overline{M}M)\,\mathrm{tr}(Mh) \tag{6.29}$$

where $a, \ldots, e$ are constants.

However, under C, $M \to \widetilde{M}$, $\overline{M} \to \widetilde{\overline{M}}$, and $h \to -h$. Thus we find that the most general amplitude takes the form $\mathcal{M} \simeq \mathrm{tr}[(\lambda_6\lambda_7 - \lambda_7\lambda_6)\overline{M}M]$, which has no terms with $\pi^+\pi^-$ or $\pi^0\pi^0$ (Cabibbo 64, Gell-Mann 64). Then, why is the amplitude for K_S^0 decay so large? A possible explanation is given in Section 6.6.

6.3.2 $K_{3\pi}$ decays

Next we consider the transitions

$$K^+ \to \pi^+\pi^+\pi^- \tag{6.30}$$
$$K^- \to \pi^-\pi^-\pi^+ \tag{6.31}$$
$$K^\pm \to \pi^0\pi^0\pi^\pm \tag{6.32}$$
$$K_L^0 \to \pi^0\pi^0\pi^0 \tag{6.33}$$
$$K_L^0 \to \pi^+\pi^-\pi^0 \tag{6.34}$$

In (6.30) to (6.33), the first two pions are identical. The final three-pion state must therefore be symmetric with respect to exchange of these pions, and we have $A(r, \theta) = A(r, -\theta)$, where r and θ are the Dalitz plot coordinates [see (4.102)]. For $K_L^0 \to \pi^+\pi^-\pi^0$, CP invariance, which is valid to a good approximation, implies $A(r, \theta) = A(r, -\theta)$ also.

Assuming then that each amplitude A is an even function of θ, we expand A as a function of T_1, T_2, and T_3 about the origin. Taking into account that $A_i(T_1, T_2, T_3) = A_i(T_2, T_1, T_3)$ and keeping only first-order terms, since $Q/3 \ll m_\pi$, we have

6.3 Kaon nonleptonic decays

$$A_i(T_1, T_2, T_3) = A_i(0) + \left.\frac{\partial A_i}{\partial T_1}\right|_0 \left(T_1 - \frac{Q}{3}\right) + \left.\frac{\partial A_i}{\partial T_2}\right|_0 \left(T_2 - \frac{Q}{3}\right)$$

$$+ \left.\frac{\partial A_i}{\partial T_3}\right|_0 \left(T_3 - \frac{Q}{3}\right)$$

$$= A_i(0) + a_i \left[r \cos\left(\theta + \frac{2\pi}{3}\right) + r \cos\left(\theta - \frac{2\pi}{3}\right)\right]$$

$$+ b_i r \cos \theta$$

where a_i and b_i are two constants. Thus introducing the slope parameter σ_i and the quantity $y = r \cos \theta$, we have

$$A_i = A_i(0)[1 + (2m_K Q/3m_\pi^2)\sigma_i y] \tag{6.35}$$

The isospin of a two-pion state may be $I_{ab} = 2, 1$, or 0. When we include the third pion, we obtain one $I = 3$ state, two $I = 2$ states, three $I = 1$ states, and, if the total charge of the three pions is zero, one $I = 0$ state. Of these, the $I = 3$ state is completely symmetric, the $I = 0$ state is completely antisymmetric, and an appropriate linear combination of the $I = 1$ states yields a completely symmetric $I = 1$ state.

Since pions and kaons have zero spin, the zero orbital angular momentum of the three-pion final state is the sum of the orbital angular momentum l of two pions (dipion) about their CM and the orbital angular momentum L of the third pion about the three-pion CM:

$$L + l = 0$$

It follows that $l = L$. Since pions are bosons, the total three-pion wavefunction must be symmetric in spatial coordinates and charge coordinates with respect to interchange of any two pions. Since $I = 3$ and $I = 1$ isospin states are symmetric, these must correspond to a symmetric spatial wavefunction for which $l = L = 0, 2, 4, \ldots$. Centrifugal barrier effects should inhibit $l = L = 2$ and 4 and also the odd $l = L$ states corresponding to the antisymmetric state $I = 0$ and the mixed symmetry $I = 2$ case. If the $|\Delta I| = \frac{1}{2}$ rule is ignored, the preceding argument implies that only $I = 1$ and 3 final states are allowed.

Although CP violation occurs in $K_L^0 \to 2\pi$ decay, it is a very small effect and can be ignored in discussing $K_{3\pi}$ decays in the first approximation. Thus with CP invariance assumed, K_S^0 and K_L^0 can decay to three-pion final states only with $CP = +1$ and -1, respectively. Now the neutral three-pion $J = 0$ final states have $P = -1$ and $C = (-1)^l$; thus $CP = (-1)^{l+1}$. It follows that for $K_S^0 \to 3\pi$ only $I = 0$ and $I = 2$ are permitted, on CP-invariance grounds alone, whereas for $K_L^0 \to 3\pi$ only $I = 1$ and $I = 3$ can occur.

214   6 *Weak decays of mesons and hyperons*

Finally, the $|\Delta I| = \frac{1}{2}$ rule implies that $I = 0$ or $I = 1$ final states are allowed but not $I = 2$ or $I = 3$. Thus when all these restrictions are combined we find that the allowed $K_{3\pi}$ transitions are $K^\pm \to (I = 1)$ and $K_L^0 \to (I = 1)$.

Now, as can easily be shown, if we assume that only $I = 1$ and $I = 3$ states contribute, we obtain

$$|\pi^+\pi^+\pi^-\rangle = \frac{2}{\sqrt{5}}|I = 1, I_3 = 1\rangle + \frac{1}{\sqrt{5}}|3, 1\rangle$$

$$|\pi^+\pi^0\pi^0\rangle = \frac{1}{\sqrt{5}}|1, 1\rangle + \frac{2}{\sqrt{5}}|3, 1\rangle$$

$$|\pi^+\pi^-\pi^0\rangle = \sqrt{\tfrac{2}{5}}|1, 0\rangle + \sqrt{\tfrac{3}{5}}|3, 0\rangle$$

$$|\pi^0\pi^0\pi^0\rangle = -\sqrt{\tfrac{3}{5}}|1, 0\rangle + \sqrt{\tfrac{2}{5}}|3, 0\rangle$$

Thus, assuming an $I = 1$ final state and applying the Wigner–Eckhart theorem, we find

$$\langle \pi^+\pi^+\pi^-|H_{\text{wk}}|K^+\rangle = \frac{2}{\sqrt{5}}\left\langle \frac{1}{2}, \frac{1}{2}; \frac{1}{2}, \frac{1}{2}\bigg|11\right\rangle \mathcal{M}_0 = \frac{2}{\sqrt{5}}\mathcal{M}_0$$

$$\langle \pi^+\pi^0\pi^0|H_{\text{wk}}|K^+\rangle = -\frac{1}{\sqrt{5}}\mathcal{M}_0$$

$$\langle \pi^+\pi^-\pi^0|H_{\text{wk}}|K^0\rangle = -\frac{1}{\sqrt{5}}\mathcal{M}_0$$

$$\langle \pi^0\pi^0\pi^0|H_{\text{wk}}|K^0\rangle = \sqrt{\tfrac{3}{10}}\mathcal{M}_0$$

where $\mathcal{M}_0$ is a reduced matrix element. Then, taking into account that, for CP invariance,

$$K^0 = \frac{1}{\sqrt{2}}|K_L^0 + K_S^0\rangle$$

we predict

$$\frac{\gamma(K^+ \to \pi^+\pi^+\pi^-)}{\gamma(K^+ \to \pi^+\pi^0\pi^0)} = 4 \tag{6.36}$$

and

$$\frac{\gamma(K_L^0 \to \pi^+\pi^-\pi^0)}{\gamma(K_L^0 \to \pi^0\pi^0\pi^0)} = \frac{2}{3} \tag{6.37}$$

(which require only that $|\Delta I| = \frac{5}{2}$ and $\frac{7}{2}$ amplitudes are zero) and

$$\frac{\gamma(K_L \to \pi^+\pi^-\pi^0)}{\gamma(K^+ \to \pi^+\pi^0\pi^0)} = 2 \tag{6.38}$$

$$\frac{\gamma(K_L \to \pi^0\pi^0\pi^0)}{\gamma(K^+ \to \pi^+\pi^+\pi^-) - \gamma(K^+ \to \pi^+\pi^0\pi^0)} = 1 \tag{6.39}$$

$$\frac{\sigma(K^+ \to \pi^+\pi^0\pi^0)}{\sigma K^+ \to \pi^+\pi^+\pi^-)} = -2 \tag{6.40}$$

$$\frac{\sigma(K^+ \to \pi^+\pi^0\pi^0)}{\sigma(K_L^0 \to \pi^+\pi^-\pi^0)} = +1 \tag{6.41}$$

if $|\Delta I| = \tfrac{1}{2}$.

The experimental data show that, for $K_{3\pi}$,

$$|A_{3/2}/A_{1/2}| = 0.06 \pm 0.01 \tag{6.42}$$

which again demonstrates that the $|\Delta I| = \tfrac{1}{2}$ component is much larger than the $|\Delta I| = \tfrac{3}{2}$ component, although the latter does not vanish.

6.4 Hyperon semileptonic decays

6.4.1 Transition probabilities

We now study the hyperons, the general properties of which are summarized in Table 6.2. Our first task is to consider the various semileptonic decays (Table 6.3). These transitions, analogous in many respects to neutron β decay, have small branching ratios and are difficult to observe. We describe the typical transition by means of the amplitude

$$\mathcal{M} = \frac{G_F}{\sqrt{2}} \begin{pmatrix} \cos\theta_1 \\ \sin\theta_1 \end{pmatrix} \bar{u}(B')[f_1(q^2)\gamma^\lambda + if_2(q^2)\sigma^{\lambda\nu}q_\nu - g_1(q^2)\gamma^\lambda\gamma_5$$
$$- g_3(q^2)\gamma_5 q^\lambda]u(B)\bar{u}_l\gamma_\lambda(1-\gamma_5)v_\nu \tag{6.43}$$

Table 6.2. *General properties of hyperons*

Hyperon	Mass (MeV/c^2)	Mean lifetime (sec)	Decay length (cm per GeV/c)	Magnetic moment ($|e|\hbar/2m_p c$)
$\Lambda^0(\tfrac{1}{2}^+)$	1115.60(05)	2.632(20) × 10^{-10}	6.93	−0.6129(45)
$\Sigma^+(\tfrac{1}{2}^+)$	1189.37(06)	0.802(5) × 10^{-10}	2.02	2.33(13)
$\Sigma^0(\tfrac{1}{2}^+)^a$	1192.47(08)	0.58(13) × 10^{-19}	—	—
$\Sigma^-(\tfrac{1}{2}^+)$	1197.35(06)	1.483(15) × 10^{-10}	3.71	−1.41(25)
$\Xi^0(\tfrac{1}{2}^+)$	1314.9(6)	2.90(10) × 10^{-10}	6.59	−1.253(014)
$\Xi^-(\tfrac{1}{2}^+)$	1321.32(13)	1.654(21) × 10^{-10}	3.75	−2.1(8)
$\Omega^-(\tfrac{3}{2}^+)$	1672.2(4)	0.82(6) × 10^{-10}	1.47	—

[a] Decays by electromagnetic interaction.

Table 6.3. Baryon semileptonic decays

ΔS	Transition	Branching ratio		C_V	C_A	Experiment
				Cabibbo's hypothesis		
0	$n \to p e^- \bar{\nu}$	100%	$\cos\theta_1$	1	$D + F$	$C_V = 1, C_A = 1.25, C_A/C_V = 1.25$,
	$\Sigma^- \to \Sigma^0 l^- \bar{\nu}$	—	$\cos\theta_1$	$\sqrt{2}$	$\sqrt{2}F$	—
	$\Sigma^- \to \Lambda^0 l^- \bar{\nu}$	$e^- 0.60(06) \times 10^{-4}$	$\cos\theta_1$	0	$\sqrt{\frac{2}{3}} D$	$C_V/C_A = 0.24(23)$
	$\Sigma^+ \to \Sigma^0 l^+ \nu$	$e^+ 2.02(47) \times 10^{-5}$	$\cos\theta_1$	0	$\sqrt{\frac{2}{3}} D$	—
	$\Xi^- \to \Sigma^0 l^- \bar{\nu}$	—	—	-1	$D - F$	—
1	$\Sigma^- \to n l^- \bar{\nu}$	$e^- 1.08(04) \times 10^{-3}$ $\mu^- 0.45(04) \times 10^{-3}$	$\sin\theta_1$	-1	$D - F$	$e^-: C_A/C_V = 0.385(070)$
	$\Lambda^0 \to p l^- \bar{\nu}$	$e^- 8.13(29) \times 10^{-4}$ $\mu^- 1.57(35) \times 10^{-4}$	$\sin\theta_1$	$\sqrt{\frac{3}{2}}$	$-\sqrt{\frac{1}{6}}(D + 3F)$	$C_V = 1.229(035), C_A = -0.903(046),$ $C_A/C_V = -0.734(031)$
	$\Xi^- \to \Lambda^0 l^- \bar{\nu}$	$(0.69 \pm 0.18) \times 10^{-3}$	$\sin\theta_1$	$\sqrt{\frac{3}{2}}$	$-\sqrt{\frac{1}{6}}(D - 3F)$	—
	$\Xi^0 \to \Sigma^+ l^- \bar{\nu}$	1.1×10^{-3}	$\sin\theta_1$	1	$D + F$	—
	$\Xi^- \to \Sigma^0 l^- \bar{\nu}$	—	$\sin\theta_1$	$\sqrt{\frac{1}{2}}$	$\sqrt{\frac{1}{2}}(D + F)$	—
	$\Omega^- \to \Lambda e^- \bar{\nu}_e$	$\sim 10^{-2}$	$\sin\theta_1$	—	—	

6.4 Hyperon semileptonic decays

or its analog for l^+ emission.[1] Here and henceforth we ignore second-class terms. The g_3 term is altogether negligible for transitions in which electrons are emitted. It gives a very small correction for muon modes, which can be estimated by PCAC; however, experiments so far are inadequate to detect it. Recoil effects include contributions from f_2 and are taken into account sufficiently by considering terms of first order in

$$\delta = \frac{m(B) - m(B')}{m(B) + m(B')} \tag{6.44}$$

As in neutron – or allowed nuclear β – decay, we find it useful to consider the total decay rate, the lepton–neutrino angular correlation, and the asymmetry coefficients for the decay of polarized hyperons. In addition, terms arise in the transition probability involving the polarization of the *final* baryon in its own rest frame for unpolarized initial baryons. These are of interest because, in principle, such quantities can be measured by observation of asymmetry in subsequent nonleptonic decay of the final baryon. The following formulas are readily derived from (6.43) by standard means (Frampton and Tung, 71) and are analogous to the β-decay formulas of Jackson, Treiman, and Wyld (see Section 5.3).

(a) Total decay rate:

$$R = G_F^2 \begin{pmatrix} \cos^2 \theta_1 \\ \sin^2 \theta_1 \end{pmatrix} \frac{(\Delta m)^5}{60\pi^3} [|f_1|^2 + 3|g_1|^2](1 - 3\delta) \tag{6.45}$$

(b) Electron–neutrino angular correlation:

$$W(\cos \theta_{ev}) = \tfrac{1}{2}(1 + a_{ev} \cos \theta_{ev}) \tag{6.46}$$

where

$$a_{ev} = \frac{|f_1|^2 - |g_1|^2}{|f_1|^2 + 3|g_1|^2} - 2\delta$$

In practice, this quantity is very useful for determination of the ratio $|g_1/f_1|$.

(c) Polarization of the decay baryon in its own rest frame for unpolarized baryons:

---

[1] In (6.43), we make the replacement $s_1 c_3 \to s_1$, which is a very good approximation.

$$P_{B'} = \frac{1}{|f_1|^2 + 3|g_1|^2} \times \{\tfrac{8}{3}\mathrm{Re}(f_1^* g_1)\hat{\boldsymbol{\alpha}} + \tfrac{8}{3}|g_1|^2\hat{\boldsymbol{\beta}} + \tfrac{1}{2}\pi\,\mathrm{Im}(f_1^* g_1)\hat{\boldsymbol{\gamma}}\} \qquad (6.47)$$

where

$$\hat{\boldsymbol{\alpha}} = \frac{\mathbf{k}_e + \mathbf{k}_\nu}{|\mathbf{k}_e + \mathbf{k}_\nu|} \qquad (6.48)$$

$$\hat{\boldsymbol{\beta}} = \frac{\mathbf{k}_e + \mathbf{k}_\nu}{|\mathbf{k}_e - \mathbf{k}_\nu|} \qquad (6.49)$$

$$\hat{\boldsymbol{\gamma}} = \frac{\mathbf{k}_e \times \mathbf{k}_\nu}{|\mathbf{k}_e \times \mathbf{k}_\nu|} \qquad (6.50)$$

(d) Asymmetry parameters for decay of polarized hyperons for $l^-\bar{\nu}$ (otherwise replace α_l by α_{lepton} and α_ν by $\alpha_{\text{antilepton}}$):

$$\alpha_l \simeq \frac{2(\mathrm{Re}\, f_1 g_1^* - |g|^2) - \tfrac{1}{3}\delta[2(f_1 + g_1)^2 + 2f_2(f_1 + g_1)]}{|f_1|^2 + 3|g_1|^2} \qquad (6.51)$$

$$\alpha_\nu \simeq \frac{2(\mathrm{Re}\, f_1 g_1^* + |g|^2) - \tfrac{1}{3}\delta[2(f_1 - g_1)^2 + 2f_2(f_1 - g_1)]}{|f_1|^2 + 3|g_1|^2} \qquad (6.52)$$

$$\alpha_{B'} \simeq \frac{-\tfrac{5}{2}\mathrm{Re}\, f_1 g_1^* + \tfrac{5}{6}\delta[2(f_1 + f_2)g_1]}{|f_1|^2 + 3|g_1|^2} \qquad (6.53)$$

For comparison with high-precision experiments, it is necessary in these formulas to modify the form factors at $q^2 = 0$ by the usual "dipole" q^2 dependence and also to apply appropriate radiative corrections (Yokoo et al. 73). Experimental results will be considered in Section 6.4.3.

6.4.2 Cabibbo's hypothesis

For the moment, let us ignore f_2 and g_3 and assume that $f_1(q^2)$ and $g_1(q^2)$ can be replaced by C_V and C_A, respectively. Our goal in this section is to calculate C_V and C_A for each of the baryon semileptonic transitions of Table 6.3, with the aid of Cabibbo's hypothesis. As a starting point, we assume perfect $SU(3)_f$ symmetry.

We know that the hadronic portion of the matrix element for the $\Delta S = 0$ decays takes the form

$$\mathcal{M}_\lambda^0 = \cos\theta_1 \langle B'|(j_1 \pm ij_2)_\lambda + (g_1 \pm ig_2)_\lambda|B\rangle \qquad (6.54)$$

and the $|\Delta S| = 1$ decays are associated with the form

$$\mathcal{M}_\lambda^1 = \sin\theta_1 \langle B'|(j_4 \pm ij_5)_\lambda + (g_4 \pm ig_5)_\lambda|B\rangle \qquad (6.55)$$

where we ignore θ_3. Here the symbols j and g refer to the vector and axial vector octets of current operators, respectively, as defined in

6.4 Hyperon semileptonic decays

(4.35) and (4.37). Each operator j or g contains sums of products of annihilation operators B for the initial baryon and creation operators $\bar{B}$ for the final baryon. The annihilation operators B transform under $SU(3)_f$ as members of the baryon octet:

$$B_i^j = \begin{pmatrix} \frac{1}{\sqrt{2}}\left(\Sigma^0 + \frac{\Lambda^0}{\sqrt{3}}\right) & \Sigma^+ & p \\ \Sigma^- & \frac{1}{\sqrt{2}}\left(-\Sigma^0 + \frac{1}{\sqrt{3}}\Lambda^0\right) & n \\ \Xi^- & \Xi^0 & -\frac{2}{\sqrt{6}}\Lambda^0 \end{pmatrix} \quad (6.56)$$

and the creation operators $\bar{B}$ transform as members of the antibaryon octet:

$$\bar{B}_i^j = \begin{pmatrix} \frac{1}{\sqrt{2}}\left(\overline{\Sigma^0} + \frac{1}{\sqrt{3}}\overline{\Lambda^0}\right) & \overline{\Sigma^-} & \overline{\Xi^-} \\ \overline{\Sigma^+} & \frac{1}{\sqrt{2}}\left(-\overline{\Sigma^0} + \frac{1}{\sqrt{3}}\overline{\Lambda^0}\right) & \overline{\Xi^0} \\ \bar{p} & \bar{n} & -\frac{2}{\sqrt{6}}\overline{\Lambda^0} \end{pmatrix} \quad (6.57)$$

In (6.56) and (6.57), indexes i and j refer to the row and column, respectively.

Now as we know, the tensor product of two octets is

$$8 \otimes 8 = 27 + 10 + \overline{10} + 8_S + 8_A + 1 \quad (6.58)$$

where one of the resulting octets is symmetric and the other antisymmetric:

$$8_S: \quad S_i^n = \bar{B}_i^k B_k^n + \bar{B}_k^n B_i^k - \tfrac{2}{3}\delta_i^n \bar{B}_m^k B_k^m \quad (6.59)$$

$$8_A: \quad A_i^n = \bar{B}_k^n B_i^k - \bar{B}_i^k B_k^n \quad (6.60)$$

Since, according to Cabibbo's hypothesis, the j and g operators transform as members of V and A octets and since they are made up of products of $\bar{B}$ and B octets, it follows that operators j_i^n and g_i^n are independently expressible as linear combinations of the S_i^n and A_i^n.

In fact, we can easily show that the vector operators j are made up exclusively of the antisymmetric A_i^n, for the $\Delta S = 0$ charged hadronic vector current transforms like the isovector lowering operator I_- and such an operator does not connect an $I = 1$ state with an $I = 0$ state. Therefore the vector contribution to the $\Delta S = 0$ transition $\Sigma^+ \to \Lambda e^+ \nu$ must be zero. (This is similar to the $\Delta T = 0$ selection rule for pure

Fermi-allowed β decays obtained in Chapter 5.) However, one may easily verify from (6.56), (6.57), (6.59), and (6.60) that

$$S_1^2 = \left(\frac{2\bar{\Lambda}\Sigma^+}{\sqrt{6}} + \frac{2\bar{\Sigma}^-\Lambda}{\sqrt{6}} + \bar{n}p + \bar{\Xi}^-\Xi^0\right) \tag{6.61}$$

$$S_1^3 = \left(\frac{\bar{\Xi}^-\Sigma^0}{\sqrt{2}} + \frac{\bar{\Sigma}^0 p}{\sqrt{2}} - \frac{\bar{\Lambda}p}{\sqrt{6}} - \frac{\bar{\Xi}^-\Lambda^0}{\sqrt{6}} + \bar{\Sigma}^-n - \bar{\Xi}^0\Sigma^+\right) \tag{6.62}$$

$$A_1^2 = (\sqrt{2}\,\bar{\Sigma}^-\Sigma^0 + \sqrt{2}\,\bar{\Sigma}^0\Sigma^+ - \bar{\Xi}^-\Xi^0 + \bar{n}p) \tag{6.63}$$

$$A_1^3 = \left(\bar{\Xi}^0\Sigma^+ - \bar{\Sigma}^-n - \frac{\bar{\Xi}^-\Sigma^0}{\sqrt{2}} - \frac{\bar{\Sigma}^0 p}{\sqrt{2}} + 3\frac{\bar{\Xi}^-\Lambda}{\sqrt{6}} - 3\frac{\bar{\Lambda}^0 p}{\sqrt{6}}\right) \tag{6.64}$$

As (6.61) shows, S_1^2 contains a contribution $\bar{\Lambda}\Sigma^+$ and therefore, it cannot appear in j_1^2.

On the other hand, the axial octet operators g_i^j can contain contributions from S_i^j and A_i^j. Thus, employing (6.61)–(6.64), we find that

$$\mathcal{M}_\lambda^0(B', B) = \cos\theta_1\, \bar{u}(B')\{a_1^2(B', B)\gamma_\lambda$$
$$- [Ds_1^2(B', B) + Fa_1^2(B', B)]\gamma_\lambda\gamma_5\}u(B) \tag{6.65}$$

$$\mathcal{M}_\lambda^1(B'B) = \sin\theta_1\, \bar{u}(B')\{a_1^3(B', B)\gamma_\lambda$$
$$- [Ds_1^3(B', B) + Fa_1^3(B', B)]\gamma_\lambda\gamma_5\}u(B) \tag{6.66}$$

where the $a_i^j(B', B)$ and $s_i^j(B', B)$ are the numerical coefficients multiplying the corresponding combinations $\bar{B}$ and B in the A_i^j and S_i^j, respectively, of (6.61)–(6.64), and D and F are constants to be determined. Thus we arrive at the predictions of Table 6.3, columns 5 and 6 ("Cabibbo's hypothesis"), assuming perfect $SU(3)_f$ symmetry.

Of course this symmetry is very imperfect. However, another application of the Ademollo–Gatto theorem demonstrates that, to first order in the $SU(3)_f$ symmetry breaking, the coefficients of Table 6.3, columns 5 and 6, remain unaffected.

The values of C_A and C_V predicted by Cabibbo can be understood more intuitively from the point of view of the color-symmetric quark model. Here we start by assuming the hadronic charged weak current to be given by

$$J_\lambda = \cos\theta_1\, \bar{D}\gamma_\lambda(1 - \gamma_5)U + \sin\theta_1\, \bar{S}\gamma_\lambda(1 - \gamma_5)U \tag{6.67}$$

We are interested in matrix elements of this operator or its Hermitian conjugate between baryon states at zero momentum transfer. As in neutron β decay, the vector coupling constant will be unrenormalized, but we expect a modification of the axial coupling because of strong interactions. Thus, the vector portion of $J_\lambda^\dagger$ makes a contribution to the hadronic matrix element

6.4 Hyperon semileptonic decays

$$\begin{pmatrix} \cos\theta_1 \\ \sin\theta_1 \end{pmatrix} \langle B'|\Sigma\tau_i^+|B\rangle \tag{6.68}$$

where the sum is over three quarks, and the axial portion of $J_\lambda^\dagger$ yields

$$\begin{pmatrix} \cos\theta_1 \\ \sin\theta_1 \end{pmatrix} a\langle B'(J_3 = +\tfrac{1}{2})|\Sigma\sigma_{3i}\tau_i^+|B(J_3 = +\tfrac{1}{2})\rangle \tag{6.69}$$

where a is the renormalization factor. It is now a simple, if tedious, matter to insert the various quark spin configurations [(4.48)–(4.54)] and compute matrix elements (6.68) and (6.69). For neutron β decay we find:

$$\langle p(J_3 = +\tfrac{1}{2})|\Sigma\tau_i^+|n(J_3 = +\tfrac{1}{2})\rangle = 1 \tag{6.70}$$

and

$$a\langle p(J_3 = +\tfrac{1}{2})|\Sigma\tau_i^+\sigma_{3i}|n(J_3 = +\tfrac{1}{2})\rangle = \tfrac{5}{3}a = D + F \tag{6.71}$$

For $\Sigma^+ \to \Lambda l^+ \nu$, we obtain

$$\langle \Lambda(J_3 = +\tfrac{1}{2})|\Sigma\tau_i^-|\Sigma^+(J_3 = +\tfrac{1}{2})\rangle = 0 \tag{6.72}$$

and

$$a\langle \Lambda(J_3 = +\tfrac{1}{2})|\Sigma\tau_i^-\sigma_{3i}|\Sigma^+(J_3 = +\tfrac{1}{2})\rangle = \sqrt{\tfrac{2}{3}}\,a = \sqrt{\tfrac{2}{3}}\,D \tag{6.73}$$

Thus we have $a = D$ and $F = \tfrac{2}{3}D$. Also, since $C_A = 1.25$ in neutron β decay, we obtain the prediction $D = 0.75$ and $F = 0.50$.

6.4.3 Experimental results for semileptonic decays

Let us consider one example of experimental observations on hyperon semileptonic decays and then summarize the various results. At the Brookhaven AGS, the absolute rate (Wise et al. 1980) and e–$\bar{\nu}_e$

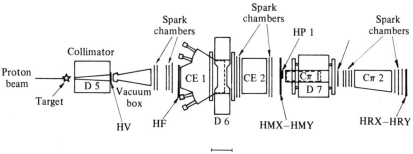

Figure 6.2. Plan view of Brookhaven experimental apparatus for measurement of $\Lambda\,\beta$ decay. (From Wise et al. 80, 81.) (Reprinted with permission.)

angular correlation (Wise et al. 1981) have been measured in $\Lambda^0 \to pe^-\bar{\nu}_e$. The 2.85-GeV/c external proton beam produced Λ^0's by colliding with an iridium target (Figure 6.2). A sweeping magnet (D5) removed charged particles and a 10-radiation-length Pb filter removed photons, leaving a neutral beam consisting mainly of Λ's, kaons, and neutrons. The Λ-momentum spectrum extended from about 4 to 24 GeV/c with a maximum at 12 GeV/c. The spectrometer consisted of two analyzing magnets (D6 and D7) to measure the laboratory momenta of the two charged particles in Λ β decay with comparable precision. The trigger utilized information from several banks of scintillator hodoscopes: HV, HF, HMX–HMY, HP1, HP2, and HRX–HRY. Four gas-filled atmospheric pressure threshold Čerenkov counters (CE1, CE2, Cπ1, and Cπ2) were used for particle identification. The $\Lambda^0 \to p\pi^-$ decay was employed for normalization in the measurement of the $e\nu$ correlation, the results of which are shown in Figure 6.3.

On the basis of 10^4 $\Lambda \to pe^-\bar{\nu}_e$ events, one finds, using (6.45), (6.46), and appropriate corrections:

$$|f_1(0)| = 1.229 \pm 0.035 \tag{6.74}$$

$$|g_1(0)| = 0.903 \pm 0.046 \tag{6.75}$$

From Table 6.3 we note that, according to Cabibbo's hypothesis, $|f_1| =$

Figure 6.3. e–ν angular correlation for $\Lambda \to pe\nu$; Brookhaven experiment. The solid line represents the data and the circles show the result of the best fit. The peaking of the distribution near $\cos\theta_{e\nu} = -1$ is due to the method of selecting the Λ^0 momentum. The acceptance of the spectrometer is flat over the entire angular region. (From Wise et al. 81. Reprinted with permission.)

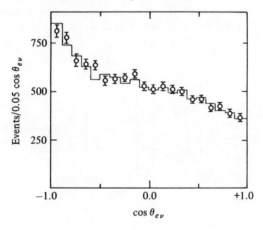

6.5 Hyperon nonleptonic decays

$\sqrt{\tfrac{3}{2}} = 1.22$. The agreement of result (6.74) with this prediction is impressive.

Results for the other hyperon semileptonic decays are summarized in Table 6.3, column 7. Shrock and Wang (78) have made fits of the branching ratio and C_A/C_V data in order to determine $\sin \theta_1$ and $\alpha_D = D/(D + F)$, assuming G_F given by muon decay [(3.64)].

On the basis of the $|\Delta S| = 1$ hyperon transitions alone, Shrock and Wang found

$$|V_{12}| = 0.220 \pm 0.003 \qquad (6.76)$$

$$\alpha_D = 0.654 \pm 0.008 \qquad (6.77)$$

where $|V_{12}| = s_1 c_3$ is defined by (4.70). A fit to $\Delta S = 0$ and $\Delta S = 1$ hyperon decays yields:

$$|V_{12}| = 0.222 \pm 0.003 \qquad (6.78)$$

$$\alpha_D = 0.645 \pm 0.008 \qquad (6.79)$$

For comparison we note that the results of the Brookhaven $\Lambda^0 \to pe^-\bar{\nu}_e$ experiment alone imply

$$\alpha_D = 0.62 \pm 0.04 \qquad (6.80)$$

whereas the color-symmetric quark model requires $\alpha_D = 0.60$.

6.5 Hyperon nonleptonic decays

The hyperon nonleptonic decays are listed in Table 6.4. We noted in Chapter 4 that, for the $J^P = \tfrac{1}{2}^+$ hyperons, the most general

Table 6.4. *Hyperon nonleptonic decays*

Particle	Mode	Branching ratio (%)	α	ϕ^0	γ	Δ^0
Λ	$p\pi^-$	64.2 ± 0.5	$+0.647(13)$	$-6.5(3.5)$	0.76	7.7(4.0)
	$n\pi^0$	35.8 ± 0.5	$+0.646(44)$	—	—	—
Σ^+	$p\pi^0$	51.6 ± 0.7	$-0.979(16)$	36(34)	0.17	187(6)
	$n\pi^+$	48.4 ± 0.7	$+0.072(15)$	167(20)	-0.97	$-72\{^{+132}_{-11}$
Σ^0	$\Lambda\gamma$	100	Electromagnetic decay			
Σ^-	$n\pi^-$	100	$-0.069(8)$	10(15)	0.98	$249\{^{+12}_{-115}$
Ξ^0	$\Lambda\pi^0$	100	$-0.478(35)$	21(12)	0.84	$216\{^{+13}_{-19}$
Ξ^-	$\Lambda\pi^-$	100	$-0.392(21)$	2(6)	0.92	185(13)
Ω^-	ΛK^-	67.0 ± 2.2	0.06(0.14)	—	—	—
	$\Xi^0\pi^-$	2.46 ± 1.9	—	—	—	—
	$\Xi^-\pi^0$	8.4 ± 1.1	—	—	—	—

decay amplitude is
$$\mathcal{M} = G_F m_{\pi^+}^2 \bar{u}_f (A - B\gamma_5) u_i \tag{6.81}$$
where A and B are constants. Starting from this expression, we may easily calculate the transition probability in the rest frame of the initial baryon. We have
$$u_i = \begin{pmatrix} \chi_i \\ 0 \end{pmatrix}$$
$$u_f = \left(\frac{E_f + m_f}{2m_f}\right)^{1/2} \begin{pmatrix} \chi_f \\ \dfrac{\boldsymbol{\sigma} \cdot \mathbf{p}_f}{E_f + m_f} \chi_f \end{pmatrix}$$
where χ_i and χ_f are two-component spinors. Thus,
$$\mathcal{M} = G_F m_{\pi^+}^2 \left(\frac{E_f + m_f}{2m_f}\right)^{1/2} \left(A\chi_f^\dagger \chi_i + B\chi_f^\dagger \frac{\boldsymbol{\sigma} \cdot \mathbf{p}_f}{E_f + m_f} \chi_i\right) \tag{6.82}$$
We now define $\hat{\mathbf{n}} = \mathbf{p}_f/|\mathbf{p}_f|$, $s = A$, and $p = |\mathbf{p}_f| B/(E_f + m_f)$. Thus (6.82) becomes:
$$\mathcal{M} = G_f m_{\pi^+}^2 \left(\frac{E_f + m_f}{2m_f}\right)^{1/2} [\chi_f^\dagger (s + p\boldsymbol{\sigma} \cdot \hat{\mathbf{n}})\chi_i] \tag{6.83}$$
and the transition probability is proportional to
$$|\chi_f^\dagger (s + p\boldsymbol{\sigma} \cdot \hat{\mathbf{n}})\chi_i|^2$$
$$= \mathrm{tr}\left[\left(\frac{1 + \boldsymbol{\sigma} \cdot \hat{\boldsymbol{\omega}}_f}{2}\right)(s + p\boldsymbol{\sigma} \cdot \hat{\mathbf{n}})\left(\frac{1 + \boldsymbol{\sigma} \cdot \hat{\boldsymbol{\omega}}_i}{2}\right)(s^* + p^*\boldsymbol{\sigma} \cdot \hat{\mathbf{n}})\right] \tag{6.84}$$
where $\hat{\boldsymbol{\omega}}_i$ and $\hat{\boldsymbol{\omega}}_f$ are unit vectors in the directions of initial and final baryon spins, respectively. Evaluating the trace by standard means, we find that the transition rate is proportional to
$$R = 1 + \gamma \hat{\boldsymbol{\omega}}_f \cdot \hat{\boldsymbol{\omega}}_i + (1 - \gamma)\hat{\boldsymbol{\omega}}_f \cdot \hat{\mathbf{n}} \hat{\boldsymbol{\omega}}_i \cdot \hat{\mathbf{n}}$$
$$+ \alpha(\hat{\boldsymbol{\omega}}_f \cdot \hat{\mathbf{n}} + \hat{\boldsymbol{\omega}}_i \cdot \hat{\mathbf{n}}) + \beta \hat{\mathbf{n}} \cdot (\hat{\boldsymbol{\omega}}_f \times \hat{\boldsymbol{\omega}}_i) \tag{6.85}$$
where
$$\alpha = 2 \operatorname{Re} \frac{sp^*}{|s|^2 + |p|^2} \tag{6.86}$$
$$\beta = 2 \operatorname{Im} \frac{sp^*}{|s|^2 + |p|^2} \tag{6.87}$$
and
$$\gamma = \frac{|s|^2 - |p|^2}{|s|^2 + |p|^2} \tag{6.88}$$

6.5 Hyperon nonleptonic decays

One also refers to the parameter ϕ, defined by

$$\beta = (1 - \alpha^2)^{1/2} \sin \phi \tag{6.89}$$

Note that α, β, and γ are not independent, since

$$\alpha^2 + \beta^2 + \gamma^2 = 1 \tag{6.90}$$

Let us consider the transition rate averaged over initial hyperon polarizations $\hat{\omega}_i$ (decay of an unpolarized hyperon). Then (6.85) becomes

$$R = 1 + \alpha \hat{\omega}_f \cdot \hat{n} \tag{6.91}$$

which implies that the final baryon possesses longitudinal polarization α. On the other hand, we may consider decay of a polarized hyperon and sum over final baryon polarizations $\hat{\omega}_f$. Then (6.85) becomes

$$R = 1 + \alpha \hat{\omega}_i \cdot \hat{n} \tag{6.92}$$

which implies that the final baryon (and meson) is emitted anisotropically with asymmetry coefficient α.

In recent years techniques have been developed to generate hyperon beams of very high intensity in

$$p + \text{nucleus} \rightarrow \text{hyperon} + X$$

with the aid of high-energy protons. In some circumstances, the hyperon emerges with a polarization perpendicular to the reaction plane (see Figure 6.4 and Heller et al. 78, Wilkinson et al. 81; also see the review by Lach and Pondrom 79).[2] The hyperon magnetic moment may be determined by allowing polarized hyperons to undergo Larmor spin precession in a known magnetic field and detecting the change in decay asymmetry [(6.92)]. With the aid of these methods, the measurements of $\mu(\Lambda^0)$ and $\mu(\Xi^0)$ have become remarkably precise (see Table 6.2 column 5; also Cox et al. 81 and references therein.)

As we saw in Chapter 4, time-reversal invariance implies, in the absence of final-state interaction, that coefficients s and p be relatively real and therefore that $\beta = 0$. However, in the present case, the final-state interaction is strong, thus,

$$s = |s|e^{i\delta_s}, \quad p = |p|e^{i\delta_p}$$

where δ_s and δ_p are the pion–baryon s- and p-wave strong-interaction phase shifts. We therefore have

$$\beta = \frac{2|s||p|}{|s|^2 + |p|^2} \sin(\delta_s - \delta_p) \tag{6.93}$$

One defines $\Delta = -\arctan \beta/\alpha$. For the decay $\Lambda \rightarrow p\pi^-$, the value of Δ

---

[2] It is easy to see that this does not violate parity.

Figure 6.4. Polarization of Λ and $\bar\Lambda$ produced by 400-GeV protons on Be. (From Lach and Pondrom 79. Reprinted with permission.)

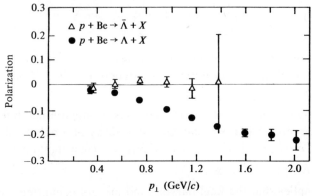

may be compared with the s- and p-wave phase shifts in low-energy π^-p scattering and the results are consistent with T invariance.

The $|\Delta I| = \frac{1}{2}$ rule implies that the effective Hamiltonian for $J^P = \frac{1}{2}^+$ hyperon nonleptonic decays transforms as an isospinor under isotopic rotations. We may then employ the Wigner–Eckhart theorem to derive certain useful relations between the various decay amplitudes, which are defined as follows:

Decay	Amplitude name
$\Lambda^0 \to p\pi^-$	Λ_-^0
$\Lambda^0 \to n\pi^0$	Λ_0^0
$\Sigma^+ \to n\pi^+$	Σ_+^+
$\Sigma^+ \to p\pi^0$	Σ_0^+
$\Sigma^- \to n\pi^-$	Σ_-^-
$\Xi^0 \to \Lambda^0\pi^0$	Ξ^0
$\Xi^- \to \Lambda^0\pi^-$	Ξ^-

Using a table of Clebsch–Gordan coefficients $\langle I_1 m_1; I_2 m_2 | I, m_1 + m_2 \rangle$, we easily find:

$$\Lambda_-^0/\Lambda_0^0 = \frac{\langle 1, -1; \frac{1}{2}, \frac{1}{2} | \frac{1}{2}, -\frac{1}{2} \rangle}{\langle 1, 0; \frac{1}{2}, -\frac{1}{2} | \frac{1}{2}, -\frac{1}{2} \rangle} = -\sqrt{2} \tag{6.94}$$

$$\Sigma_+^+ - \Sigma_-^- = -\sqrt{2}\,\Sigma_0^+ \tag{6.95}$$

$$\Xi^0 = \frac{1}{\sqrt{2}}\,\Xi^- \tag{6.96}$$

6.5 Hyperon nonleptonic decays

These results are well satisfied experimentally, which may be seen if we characterize the departure from the $|\Delta I| = \frac{1}{2}$ rule by comparing amplitudes for $|\Delta I| = \frac{1}{2}$ and $|\Delta I| = \frac{3}{2}$ or greater:

$$\Lambda \to N\pi: \quad \left|\frac{A_{3/2}}{A_{1/2}}\right| = 0.03 \begin{cases} \pm 0.01 & \text{for } s\text{-wave} \\ \pm 0.03 & \text{for } p\text{-wave} \end{cases} \quad (6.97)$$

$$\Sigma \to N\pi: \quad \left|\frac{A(I \geq \frac{3}{2})}{A_{1/2}}\right| = 0.07 \pm 0.03 \quad (6.98)$$

$$\Xi \to \Lambda\pi: \quad \left|\frac{A_{3/2}}{A_{1/2}}\right| = \begin{cases} 0.035 \pm 0.02 & \text{for } s\text{-wave} \\ 0.12 \pm 0.15 & \text{for } p\text{-wave} \end{cases} \quad (6.99)$$

It turns out that $A_{3/2}$ is roughly comparable to semileptonic hyperon decay amplitudes and $A_{1/2}$ is much larger (by a factor 15 to 30).

In the case of nonleptonic Ω^- decays, the transitions are of the form:

$$\text{spin } \tfrac{3}{2} \to \text{spin } \tfrac{1}{2} + \text{spin } 0 \quad (6.100)$$

and here we have p- and d-waves. An analysis similar to that carried out for the $J^P = \frac{1}{2}^+$ hyperons reveals that α, β, and γ may be defined as in (6.86) to (6.88) except that s and p are replaced by p and d, respectively. It may be shown that the asymmetry parameter in the decay of polarized Ω^- is $\alpha_f \alpha_\Omega$, where α_f refers to the final hyperon.

According to the $|\Delta I| = \frac{1}{2}$ rule, the amplitudes for $\Omega^- \to \Xi^0 \pi^-$ and $\Omega^- \to \Xi^- \pi^0$ ought to be in the ratio

$$\frac{\Omega^- \to \Xi^0 \pi^-}{\Omega^- \to \Xi^- \pi^0} = \frac{\langle 1, -1; \tfrac{1}{2}, \tfrac{1}{2} | \tfrac{1}{2}, -\tfrac{1}{2}\rangle}{\langle 1, 0; \tfrac{1}{2}, -\tfrac{1}{2} | \tfrac{1}{2}, -\tfrac{1}{2}\rangle} = -\sqrt{2} \quad (6.101)$$

Thus taking into account the slight difference in m_{π^-} and m_{π^0}, we are led to the prediction

$$R = \Gamma(\Xi^0 \pi^-)/\Gamma(\Xi^- \pi^0)|_{|\Delta I|=1/2} = 2.03 \quad (6.102)$$

However, experiment shows that $R = 2.94$, which suggests that there may be some violation of the rule in this case.

In addition to (6.94)–(6.96), which follow from the $|\Delta I| = \frac{1}{2}$ rule, we may obtain one further relation among the hyperon nonleptonic decay amplitudes by assuming octet dominance [i.e., that H_{wk} transforms as an octet under $SU(3)_f$]. The argument is similar to that employed in Section 6.3.1 [(6.27)–(6.29)]. The result is the Lee–Sugawara relation (Lee 64, Sugawara 64):

$$2\Xi^- + \Lambda^0 = \sqrt{3}\Sigma_0^+ \quad (6.103)$$

which turns out to be satisfied quite well by both s and p amplitudes.

6.6 Possible dynamical origins of the $|\Delta I| = \tfrac{1}{2}$ rule

We now consider how strong-interaction effects may modify the effective weak Hamiltonian to yield octet dominance. The essential ideas depend on renormalization group methods in quantum chromodynamics that are beyond the scope of this book. Therefore we can provide no more than a brief and superficial summary, and refer the interested reader to the original literature and numerous reviews for more details.

The starting point of quantum chromodynamics is the color–$SU(3)$ Yang–Mills theory embodied in the Lagrangian

$$\mathscr{L} = \bar{\psi}(x)[i\gamma^\mu D_\mu - m_0]\psi(x) - \tfrac{1}{4} F_{\mu\nu}{}^a(x) F^{\mu\nu a}(x) \tag{6.104}$$

where the quark field $\psi(x)$ contains $4 \times 3 \times N$ components corresponding to four space-time indexes, three colors, and $N = 6(?)$ flavors (Altarelli 74, Politzer 74, Marciano and Pagels 78, Peterman 79, Buras 80, Reya 81). In (6.104), D_μ is the covariant derivative

$$D_\mu = \partial_\mu + igF^a A_\mu{}^a \tag{6.105}$$

where $F^a = \tfrac{1}{2}\lambda^a$ are the eight $SU(3)_c$ generators [see (4.10) and (4.11)] and the $A_\mu{}^a$ are eight color–gluon gauge fields. The $F_{\mu\nu}{}^a$ satisfy

$$F_{\mu\nu}{}^a = \partial_\mu A_\nu{}^a - \partial_\nu A_\mu{}^a - g f_{ab}{}^c A_\mu{}^b A_\nu{}^c \tag{6.106}$$

where $f_{ab}{}^c$ are the $SU(3)$ structure constants (see Table 4.1). Thus the second term of (6.104) plays a role somewhat analogous to $\mathscr{L}_{YM}$ [(2.40)] in the electroweak standard model. As in that case, the non-Abelian term proportional to g gives rise to nonlinear gluon self-interactions that are of crucial importance for the short-distance behavior of the quark–quark coupling ("asymptotic freedom").

Let us attempt to explain this by considering first the electron–electron interaction in electrodynamics. We know that the interaction between two electrons separated by a distance r much greater than their Compton wavelength $\hbar/m_e c = m_e^{-1}$ is

$$V(r) = \alpha r^{-1} \tag{6.107}$$

where $\alpha = e^2/4\pi\hbar c = (137.036)^{-1}$. However, if we are to consider the behavior of $V(r)$ for very short distances, we must take into account the phenomenon of vacuum polarization, in which a bare electron charge e_0 polarizes the vacuum and creates a spherically symmetric positive charge density around it. The result is that the bare charge is shielded, and the effective charge of the electron observed in laboratory experiments is much smaller in magnitude than e_0.

When our two electrons are separated by a distance r much less than

6.6 Possible dynamical origins of the $|\Delta I| = \frac{1}{2}$ rule

m_e^{-1}, the shielding effect of vacuum polarization is reduced and the effective coupling strength in $V(r)$ is therefore increased. As is well known, this leads to the "one-loop" corrected potential:

$$V(r) = \frac{\alpha}{r}\left[1 + \frac{\alpha}{3\pi}\ln(rm_e)^{-1}\right] \tag{6.108}$$

Since small distances are equivalent to large momentum transfers, we may say that the effective coupling constant $\alpha(-q^2)$ is an increasing function of $-q^2 > 0$ and only takes the "fine-structure-constant" value $(137.036)^{-1}$ at $q^2 = 0$.

In QCD we consider the interaction between two quarks via gluon exchange and attempt to describe it in terms of a potential $V(r)$. As in electrodynamics, there is a "vacuum polarization" term analogous to the logarithmic term of (6.108), which is also positive. However, in addition, there is a term corresponding to the "A^3" self-interaction of the gluon field, which is larger in magnitude and has opposite sign. The result is that as $r \to 0$ the coupling between the quarks becomes very weak. An effective strong coupling constant $\alpha_S(-q^2)$ may be defined, somewhat analogous to $\alpha(-q^2)$ in electrodynamics. However, $\alpha_S(-q^2)$ is a *decreasing* function of $-q^2$ (provided that $N \leq 16$) and as $-q^2 \to +\infty$, $\alpha_S \to 0$ ("asymptotic freedom"). By means of the Callan-Symanzik renormalization-group equation (Callan 70, Symanzik 70, Reya 81), it may be shown that

$$\alpha_s(-q^2) = \frac{12\pi}{(33 - 2N)\ln(-q^2/\Lambda^2)} \tag{6.109}$$

where Λ is the only free parameter in QCD, a number that must be fixed by experiment. From the results of deep inelastic lepton–nucleon scattering experiments, it can be determined that

$$0.2 \text{ GeV} \lesssim \Lambda \lesssim 0.7 \text{ GeV} \tag{6.110}$$

with $\Lambda \simeq 0.3$ GeV a reasonable value for the sake of our discussion.

We now return to the nonleptonic weak interactions and consider the effective weak Hamiltonian in the absence of strong interactions and in the low-q^2 (four-fermion contact-interaction) limit. We also neglect b and t quarks. Then

$$H = \frac{G_F}{\sqrt{2}} J_\lambda^\dagger J^\lambda \tag{6.111}$$

where, according to the GIM scheme,

$$J^\lambda = \bar{D}_C \gamma^\lambda (1 - \gamma_5)U + \bar{S}_C \gamma^\lambda (1 - \gamma_5)C \tag{6.112}$$

Now restricting ourselves to the C-conserving $\Delta S = -1$ portion of H and writing $\gamma_{\lambda L} = \frac{1}{2}\gamma_\lambda(1 - \gamma_5)$, we have

$$H_{\text{eff}} = 2\sqrt{2}G_F \sin\theta_C \cos\theta_C [\bar{S}\gamma_L{}^\lambda U \cdot \bar{U}\gamma_{L\lambda}D - \bar{S}\gamma_L{}^\lambda C \cdot \bar{C}\gamma_{L\lambda}D] \quad (6.113)$$

In this expression and those that follow, each factor such as $\bar{U}\gamma_{L\lambda}D$ really means $\Sigma_{i=1}^3 \bar{U}_i\gamma_{L\lambda}D_i$, that is, a sum over colors.

It is convenient to rewrite (6.113) by defining the operators:

$$O^+ = [(\bar{S}U)(\bar{U}D) + (\bar{S}D)(\bar{U}U)] - [(\bar{S}C)(\bar{C}D) + (\bar{S}D)(\bar{C}C)] \quad (6.114)$$

$$O^- = [(\bar{S}U)(\bar{U}D) - (\bar{S}D)(\bar{U}U)] - [(\bar{S}C)(\bar{C}D) - (\bar{S}D)(\bar{C}C)] \quad (6.115)$$

where space-time indexes are now suppressed. Then, (6.113) becomes

$$H_{\text{eff}} = \sqrt{2}G_F \sin\theta_C \cos\theta_C [O^+ + O^-] \quad (6.116)$$

The reason for decomposing H_{eff} into $O^\pm$ is that these operators are eigenstates of the "renormalization group equations" of QCD (Wilson 69, Callen 70, Altarelli and Maiani 74, Gaillard and Lee 74a, de Rujula et al. 75, Wilczek et al. 75, Kingsley et al. 76, Shifman et al. 77, Vainshtein et al. 78). When we allow for gluon-radiative corrections, as in Figure 6.5 (which enters into $K_S^0 \to \pi^+\pi^-$ decay), it can be shown that (6.116) is changed to

$$H_{\text{eff}} = \sqrt{2}G_F \sin\theta_C \cos\theta_C [c^+O^+ + c^-O^-] \quad (6.117)$$

where the coefficients $c^\pm$ are given by

$$c^\pm = [\alpha_S(m_W^2)/\alpha_S(\mu^2)]^{a_\pm} \quad (6.118)$$

and

$$a_+ = 6/(33 - 2N) \quad (6.119)$$

$$a_- = -12/(33 - 2N) \quad (6.120)$$

are the "anomalous dimensions" of $O^\pm$, whereas μ is the QCD renormalization parameter, taken to be 1 GeV. Ignoring t and b quarks, we have $N = 4$. Then, employing $m_W = 70$ GeV, one obtains the result: $c^+ \simeq 0.68$, $c^- \simeq 2.15$ (Altarelli and Maiani 74, Gaillard and Lee 74a).

Figure 6.5. Gluon-radiative corrections.

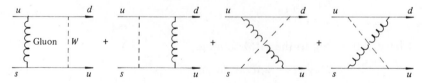

6.6 Possible dynamical origins of the $|\Delta I| = \frac{1}{2}$ rule

Thus (6.117) becomes

$$H_{\text{eff}} \simeq \sqrt{2} G_F \sin \theta_C \cos \theta_C [0.68 \, O^+ + 2.15 \, O^-] \tag{6.121}$$

We may now define the operators

$$O_1^\pm = (\bar{S}U)(\bar{U}D) \pm (\bar{S}D)(\bar{U}U) \tag{6.122}$$

$$O_2^\pm = (\bar{S}C)(\bar{C}D) \pm (\bar{S}D)(\bar{C}C) \tag{6.123}$$

and we have $O^\pm = O_1^\pm - O_2^\pm$. Let us see how $O_1^\pm$ and $O_2^\pm$ transform under isospin rotations. First, consider O_1^-, which plainly has $I_3 = -\frac{1}{2}$. Applying the isospin-lowering operator to O_1^-, we obtain zero, which means that $I(O_1^-) = \frac{1}{2}$. Next, operator $\bar{S}D \cdot \bar{U}D$ has $I_3 = -\frac{3}{2}$, so it must have $I = \frac{3}{2}$. Applying the isospin-raising operator to $\bar{S}D \cdot \bar{U}D$, we then obtain a component with $I = \frac{3}{2}$ and $I_3 = -\frac{1}{2}$:

$$I^+[\bar{S}D \cdot \bar{U}D] = \bar{S}U \cdot \bar{U}D + \bar{S}D \cdot \bar{U}U - \bar{S}D \cdot \bar{D}D = O_1^+ - \bar{S}D \cdot \bar{D}D$$

Thus, $O_1^+ = \bar{S}D \cdot \bar{D}D + O(\frac{3}{2})$, that is, O_1^+ contains a portion with $I = \frac{3}{2}$. By similar arguments, one can show that the $O_2^\pm$ correspond to $I = \frac{1}{2}$. Therefore (6.121) shows that $|\Delta I| = \frac{3}{2}$ is suppressed and $|\Delta I| = \frac{1}{2}$ is enhanced by the mechanism of Figure 6.5, although this effect by itself is much too small to account quantitatively for the experimental results. Moreover, we still have no explanation for the large $K_S^- \to \pi^+\pi^-, \pi^0\pi^0$ amplitudes. Recall they should be zero (Section 6.3.1).

However, as pointed out by Shifman et al. (Shifman et al. 77, Vainshtein et al. 78), QCD effects lead to additional terms in the weak Hamiltonian, which are represented by the so-called "penguin" diagrams of the form of Figure 6.6. These additional terms are

$$H_{\text{eff}} = 2\sqrt{2} G_F \sin \theta_C \cos \theta_C (c_1 P_1 + c_2 P_2) \tag{6.124}$$

where

$$P_1 = \left(\bar{D}\gamma_{\mu L} \frac{\lambda^a}{2} S\right)\left(\bar{U}\gamma_R^\mu \frac{\lambda^a}{2} U + U \to D + U \to S\right) \tag{6.125}$$

$$P_2 = (\bar{D}\gamma_{\mu L} S)(\bar{U}\gamma_R^\mu U + U \to D + U \to S) \tag{6.126}$$

Figure 6.6. An example of a "penguin" diagram.

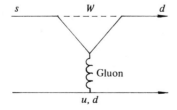

and where the color–$SU(3)$ matrices are denoted by λ^a. Note that since a gluon can couple to left- *or* right-handed quarks, such quark fields appear in P_1 and P_2. Coefficients c_1 and c_2 are evaluated by methods similar to those employed for $c^\pm$, and they turn out to be very small:

$$c_1 \simeq -0.28, \quad c_2 \simeq -0.10$$

However, it is still necessary to calculate the matrix elements of $H_{\text{eff}} + H'_{\text{eff}}$, and in spite of the smallness of c_1 and c_2, these turn out to be very sizable. Since diagrams such as Figure 6.6 yield only $|\Delta I| = \frac{1}{2}$, the $|\Delta I| = \frac{1}{2}$ amplitudes are thus further enhanced.

Indeed, it appears quite possible that the major portion of the $K_S^0 \to \pi^+\pi^-$ or $K_S^0 \to \pi^0\pi^0$ amplitude arises in just this way, the numerical value obtained by Shifman et al. being within about 20 percent of the experimental result. Also, analysis of the $J^P = \frac{1}{2}^+$ hyperon decays carried out on this basis yields results in good agreement with experiment (Tadić and Trampetić 81).

Finally, we noted earlier that the $|\Delta I| = \frac{1}{2}$ rule may be violated to some extent in Ω^- decay [see (6.102) and the accompanying experimental result]. Application of similar arguments to this case (Finjord 78) suggests that suppression of $|\Delta I| = \frac{3}{2}$ terms in the Ω^- decay amplitude is not nearly so pronounced as for $K_S^0 \to \pi\pi$ or $J^P = \frac{1}{2}^+$ hyperon decay.

While the methods just outlined do seem plausible, it must be emphasized that they take into account only "hard-gluon" exchanges (gluons with energies $\gtrsim 1$ GeV), since only in such cases is $\alpha_S(-q^2)$ small enough to justify perturbation theory. It seems much more difficult to treat "soft-gluon" effects as well, but these might be of central importance for nonleptonic weak decays.

6.7 Weak decays of charmed hadrons

The "charmonium atom" consists of a bound $c\bar{c}$ system (see Figure 4.9), which has been studied in great detail in e^+e^- colliding-beam experiments.[3] Somewhat above the "charm threshold" of charmonium, one has sufficient energy in e^+e^- collisions to produce pairs of mesonic states $D^+ = c\bar{d}$ and $D^- = \bar{c}d$ or $D^0 = c\bar{u}$ and $\bar{D}^0 = \bar{c}u$, which are "long-lived" and can decay only by weak interaction. D-meson

---

[3] Recent reviews relevant to this section are by Chinowsky 77, Applequist et al. 78, Quigg and Rosner 79, Goldhaber and Wiss 80, Hitlin 80, and Trilling 81. Many of the ideas about charmed particles stem from the pioneering article by Gaillard et al. 75.

6.7 Weak decays of charmed hadrons

production is especially copious at the $\psi''(3770)$ resonance (see Figure 6.7). At somewhat higher energies, it becomes possible to produce F^+F^- pairs ($F^+ = c\bar{s}$). In addition, charmed mesons are produced in high-energy neutrino–nucleon collisions (see Chapter 8), and evidence for D-meson production has even been found in pp collisions.

Let us discuss the weak decays of $D^\pm$, $F^\pm$, D^0, and $\bar{D}^0$. We start by making the simplest possible assumptions, namely, that the c quark decays to an s quark as if it were free, while the remaining $\bar{q}$ (e.g., $\bar{d}$ in D^+) plays the role of a mere passive "spectator" (Figure 6.8).

In this approximation we can easily calculate the semileptonic decay rate for $D^+ \to l\nu X$. It is given by the following formula, borrowed from muon decay:

$$\Gamma(D^+ \to l\nu X) = \frac{G_F^2 m_c^5}{192\pi^3} g(\epsilon) \left[1 - \frac{2\alpha_s}{3\pi} f(\epsilon)\right] \qquad (6.127)$$

where $\epsilon = m_s/m_c$ is the ratio of masses of s and c quarks, $g(\epsilon)$ a phase-space correction factor accounting for finite ϵ, and the factor in square brackets a gluon-radiative correction (Cabibbo and Maiani 78c,

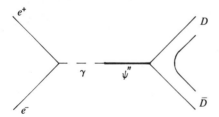

Figure 6.7. $D\bar{D}$ production in e^+e^- collisions at the $\psi''(3770)$ resonance where ψ'' is thought to be a 1 $^3D_1$ state mixed with 2 $^3S_1$. It decays strongly to $D\bar{D}$.

Figure 6.8. Semileptonic D^+ decay according to the spectator model.

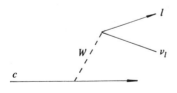

Suzuki 79). Employing the effective c-quark mass: $m_c = 1.75$ GeV/c^2, one finds

$$\Gamma(D^+ \to l\nu X) \simeq 2 \times 10^{11} \text{ sec}^{-1} \qquad (6.128)$$

a result that applies equally to $e\nu X$ and $\mu\nu X$. The purely leptonic decay rates (analogous to $\pi_{\mu 2}$) are quite negligible by comparison. As for the hadronic decay modes, it is evident within the framework of the four-quark model that these can be separated into three categories: (a) Cabibbo-favored, (b) Cabibbo-suppressed, and (c) very suppressed, according to the Cabibbo factor appearing as in Figure 6.9. Since only one combination of final quark flavors is Cabibbo-allowed and since

Figure 6.9. Allowed and suppressed hadronic decays of D^+, according to the GIM scheme.

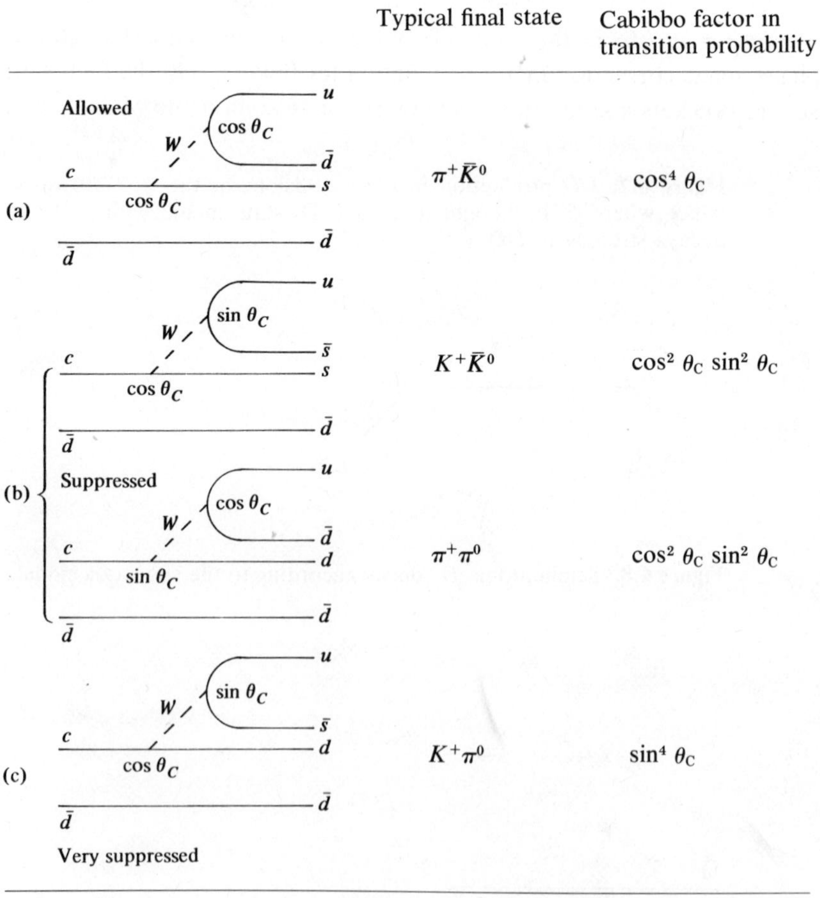

	Typical final state	Cabibbo factor in transition probability
(a) Allowed	$\pi^+ \bar{K}^0$	$\cos^4 \theta_C$
(b) Suppressed	$K^+ \bar{K}^0$	$\cos^2 \theta_C \sin^2 \theta_C$
	$\pi^+ \pi^0$	$\cos^2 \theta_C \sin^2 \theta_C$
(c) Very suppressed	$K^+ \pi^0$	$\sin^4 \theta_C$

quarks have three colors, we conclude that in the "spectator approximation" the decay rates to $e\nu X$, to $\mu\nu X$, and to hadrons should be in the ratio 1:1:3. Thus the total lifetime of D^+ should be $\tau(D^+) \simeq 10^{-12}$ sec. Moreover, the spectator approximation obviously implies equal lifetimes for D^+, D^0, and F^+.

Now let us compare these simple predictions with experimental data. Since the lifetime of a charmed meson is of order 10^{-12} sec, the experimental determination of this quantity is extremely difficult. The most promising current method for lifetime measurement employs high-resolution photographic emulsions in neutrino–nucleon collisions (see Chapter 8). By late 1982, the following results had been obtained (Bingham 81; see also Ushida et al. 82):

Meson	Mean lifetime (10^{-13} sec)	Number of events
D^+	$7.6\begin{cases}+3.6\\-1.5\end{cases}$	17
D^0	$2.7\begin{cases}+0.8\\-0.4\end{cases}$	29
F^+	$1.8\begin{cases}+1.2\\-0.6\end{cases}$	4

Evidently the spectator approximation gives a reasonable result for $\tau(D^+)$, and in addition, the measured values of the semileptonic branching ratio for D^+ yield

$$B_e(D^+) = \frac{\Gamma(D^+ \to e\nu X)}{\Gamma(D^+ \to \text{all})} = 19\begin{cases}+4\\-3\end{cases}\%$$

in accord with this approximation. However, the lifetime of D^0 seems considerably shorter than that of D^+ and what evidence we do have suggests that $B_e(D^0) \simeq 5\%$, although this quantity is very difficult to measure directly. Hence it seems possible that there is considerable enhancement of hadronic decay modes for D^0 and F^+.

In order to account for this, we note first that even within the spectator approximation there exist hard-gluon-radiative corrections, as in the analysis given in connection with Figure 6.5 (Ellis et al. 75, Cabibbo and Maini 78a,b, Fakirov and Stech 78).[4] These would not af-

---

[4] Note that penguin diagrams (analogous to Figure 6.6) are not significant for Cabibbo-allowed charmed-meson decays.

236   6 Weak decays of mesons and hyperons

fect D^+ differently from D^0 and F^+, but additional diagrams may contribute significantly to D^0 or F^+ decay and not to D^+ decay (Figure 6.10a and b).

Figure 6.10. (a) D^0 decay by W exchange; (b) F^+ decay by W exchange.

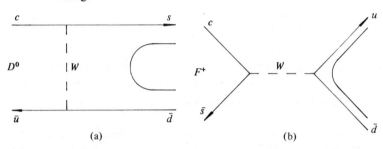

A great deal of theoretical analysis incorporating QCD effects has been performed on hadronic decay modes of charmed mesons, but since the experimental data are still so preliminary, it seems best that we forego a presentation of these details. Instead we confine ourselves to a discussion of the $SU(3)_f$ and $SU(4)_f$ transformation properties of the effective weak Hamiltonian and also take into account the possibility that the amplitudes for these decays are influenced by the existence of heavy quark states (Quigg 80).

Let us then consider N generations of left-handed color-triplet quark doublets:

$$\begin{pmatrix} q_{1+} \\ q_{1-} \end{pmatrix}_L, \begin{pmatrix} q_{2+} \\ q_{2-} \end{pmatrix}_L, \ldots, \begin{pmatrix} q_{N+} \\ q_{N-} \end{pmatrix}_L$$

which we arrange in the column matrix

$$\phi^\alpha = \begin{pmatrix} q_{1+} \\ q_{2+} \\ \cdot \\ \cdot \\ \cdot \\ q_{N+} \\ q_{1-} \\ q_{2-} \\ \cdot \\ \cdot \\ \cdot \\ q_{N-} \end{pmatrix}$$

6.7 Weak decays of charmed hadrons

The charged current can then be written as $J = \bar{\phi}\hat{O}\phi$, where

$$\hat{O} = \begin{pmatrix} 0 & V \\ 0 & 0 \end{pmatrix}$$

and V is the $N \times N$ unitary quark-mixing matrix [a generalization of the six-quark matrix, (4.70)]. Denoting $\phi^{\alpha\dagger}$ by $\bar{\phi}_\alpha$, we have

$$J = \hat{O}_{\alpha\beta}\bar{\phi}_\alpha\phi^\beta \tag{6.129}$$

which is a linear combination of states $\bar{\phi}_\alpha\phi^\beta$ that transform under $SU(2N)_f$ as the direct product

$$\overline{(2N)} \otimes (2N) = 1 + (4N^2 - 1)$$

However, $\hat{O}_{\alpha\beta}$ is traceless, so the singlet does not appear, and J transforms as the $(4N^2 - 1)$ representation of $SU(2N)_f$. For $N = 2$ (the first two generations), we have $SU(4)_f$, J transforms as a "15", and it is represented by the mixed tensor $T_\alpha{}^\beta = \bar{\phi}_\alpha\phi^\beta$.

Apart from an irrelevant constant factor, the effective weak Hamiltonian is given by

$$H_{wk} = \tfrac{1}{2}\{J, J^+\} = \tfrac{1}{2}\hat{O}_{\alpha\beta}\hat{O}_{\sigma\delta}\dagger\{T_\alpha{}^\beta, T_\sigma{}^\delta\} \tag{6.130}$$

Also we know that in $SU(4)$

$$15 \otimes 15 = 1 + 15_S + 15_A + 20 + 45 + \overline{45} + 84$$

However, since H_{wk} is symmetric, we need only consider terms $1 + 15_S + 20 + 84$, and in fact, it can easily be shown that, since $\hat{O}$ is traceless, the 15_S does not enter. Also, in the part of H_{wk} that interests us ($|\Delta C| = 1$ transitions), the singlet makes no contribution. The result is that the relevant portion of H_{wk} transforms as a linear combination of a 20 and an 84 in $SU(4)_f$.

We now employ the following convenient notation. Let

$$\psi^\alpha = \begin{pmatrix} \psi^0 \\ \psi^1 \\ \psi^2 \\ \psi^3 \end{pmatrix} = \begin{pmatrix} c \\ u \\ d \\ s \end{pmatrix}$$

and

$$T_{ij}{}^k = \{\psi^k\psi_i, \psi^0\psi_j\} - \tfrac{1}{3}\delta_i^k\{\psi^m\psi_m, \psi^0\psi_j\}$$

Then, ignoring space-time indexes, (6.130) becomes

$$\begin{aligned} H_{wk}(\Delta C = 1) &= \{\bar{C}S, \bar{D}U\}V_{11}V_{22} + \{\bar{C}S, \bar{S}U\}V_{12}V_{22} \\ &\quad + \{\bar{C}D, \bar{D}U\}V_{11}V_{21} + \{\bar{C}D, \bar{S}U\}V_{12}V_{21} \\ &= T_{31}{}^2 V_{11}V_{22} + (T_{31}{}^3 - T_{21}{}^2)\Sigma \\ &\quad + (T_{13}{}^3 + T_{12}{}^2)\Delta + T_{21}{}^3 V_{12}V_{21} \end{aligned} \tag{6.131}$$

where
$$\Sigma = \tfrac{1}{2}[V_{12}V_{22} - V_{11}V_{21}] \tag{6.132}$$
and
$$\Delta = \tfrac{1}{2}[V_{12}V_{22} + V_{11}V_{21}] \tag{6.133}$$

Under $SU(3)_f$, the product of charm-changing and charm-conserving currents transforms as

$$\overline{3} \otimes 8 = \overline{15} + 6 + \overline{3}$$

On the right-hand side, the $\overline{3}$ and $\overline{15}$ are contained in the $SU(4)_f$ 84 and the 6 is in the $SU(4)_f$ 20. Thus, finally, one obtains, from straightforward manipulations, that

$$\begin{aligned}
H_{\text{wk}}(\Delta C = 1) = &\tfrac{1}{2}\{[\overline{15}]_{31}{}^2 + \tfrac{1}{4}[6]^{22}\}V_{11}V_{22} \\
&+ \tfrac{1}{2}\{[\overline{15}]_{31}{}^3 - [\overline{15}]_{21}{}^2 + [6]^{23}\}\Sigma \\
&+ \{-\tfrac{1}{2}[\overline{15}]_{11}{}^1 + \tfrac{3}{4}[\overline{3}]_1\}\Delta \\
&+ \{\tfrac{1}{2}[\overline{15}]_{21}{}^3 - \tfrac{1}{4}[6]^{33}\}V_{12}V_{21}
\end{aligned} \tag{6.134}$$

which is the essential formula summarizing the $SU(3)_f$ content of H_{wk}. It can easily be shown that the terms in Σ, in $V_{11}V_{12}$, and in $V_{12}V_{21}$ transform as U-spin vectors in $SU(3)_f$, whereas the term in Δ is a U-spin scalar.

We are now in a position to consider the decays of charmed mesons into two pseudoscalar mesons, for example, $D^0 \to \pi^+ K^-$. We write the amplitude in the general form

$$\langle PP|H_{\text{wk}}|P_c\rangle \tag{6.135}$$

where P stands for pseudoscalar. Bose symmetry requires the final state to be symmetric in $SU(3)_f$ indexes; therefore, 1, 8_S, and 27 are the only possibilities. There are then five possible amplitudes with distinct reduced matrix elements, called G, F, S, E, and T:

$$G: \quad \langle[1]\|[\overline{3}]\|[3]\rangle \tag{6.136}$$
$$F, S, E: \quad \langle[8]\|[\overline{3}], [6], [\overline{15}]\|[3]\rangle \tag{6.137}$$
$$T: \quad \langle[27]\|[15]\|[3]\rangle \tag{6.138}$$

One may now work out, in a straightforward way, the amplitudes for charmed meson decays in terms of the five reduced matrix elements and suitable Clebsch–Gordan coefficients. The results are summarized in Table 6.5 for D^0 decay as an illustration. A complete table of results for D^0, D^+, and F^+ is given in Quigg (80).

If we assume "20" dominance in $SU(4)_f$, analogous to "8" dominance in $SU(3)_f$, then the [6] in H_{wk}, which belongs to the $SU(4)_f$ 20, should be most important, and "S"-type amplitudes should be largest.

6.7 Weak decays of charmed hadrons

Quite apart from knowledge of $G, F, S, E,$ or T, however, the results of Table 6.5 permit several interesting predictions about ratios of amplitudes. For example, we consider the Cabibbo-suppressed transitions $D^0 \to K^-K^+$ and $D^0 \to \pi^-\pi^+$ compared with the allowed transition $D^0 \to K^-\pi^+$. Defining $\alpha_1 = 2T + E - S$ and $\beta_1 = \frac{1}{2}(3T + 2G + F - E)$, we see that

$$R_1 \equiv \frac{A(D^0 \to K^-K^+)}{A(D^0 \to K^-\pi^+)} = \frac{\alpha_1 \Sigma + \beta_1 \Delta}{\alpha_1 V_{11} V_{22}} = \frac{1}{V_{11} V_{22}}\left(\Sigma + \frac{\beta_1}{\alpha_1}\Delta\right) \quad (6.139)$$

and

$$R_2 \equiv \frac{A(D^0 \to \pi^-\pi^+)}{A(D^0 \to K^-\pi^+)} = \frac{1}{V_{11} V_{22}}\left(-\Sigma + \frac{\beta_1}{\alpha_1}\Delta\right) \quad (6.140)$$

Experiment shows that

$$\frac{\Gamma(D^0 \to K^-K^+)}{\Gamma(D^0 \to K^-\pi^+)} = 0.113 \pm 0.030 \quad (6.141)$$

Table 6.5. *Amplitudes for D^0 hadronic weak decays to two pseudoscalar mesons*[a,b]

Decay	Amplitude	
$K^-\pi^+$	$2T + E - S$	
$\bar{K}^0\pi^0$	$1/\sqrt{2}(3T - E + S)$	$V_{11}V_{22}$
$\bar{K}^0\eta$	$1/\sqrt{6}(3T - E + S)$	
$\bar{K}^0X$	$2(E - S)/\sqrt{3}$	
K^+K^-	$(2T + E - S)\Sigma + \frac{1}{2}(3T + 2G + F - E)\Delta$	
$\pi^+\pi^-$	$-(2T + E - S)\Sigma + \frac{1}{2}(3T + 2G + F - E)\Delta$	
$\pi^0\pi^0$	$\frac{1}{2}(3T - E + S)\Sigma + \frac{1}{4}(-7T + 2G + F - E)\Delta$	
$K^0\bar{K}^0$	$\frac{1}{2}(-T + 2G - 2F + 2E)\Delta$	
$\eta\eta$	$-\frac{1}{6}(3T - E + S)\Sigma + \frac{1}{4}(-3T + 2G - F + E)\Delta$	
XX	$\frac{1}{2}G\Delta$	
$\pi^0\eta$	$-1/\sqrt{3}\,(3T - E + S)\Sigma + (1/2\sqrt{3})(-6T + 3F - 3E)\Delta$	
ηX	$\sqrt{2}(S - E)\Sigma + 1/\sqrt{2}(F - E)\Delta$	
$\pi^0 X$	$\sqrt{\frac{2}{3}}(E - S)\Sigma - \sqrt{\frac{1}{6}}(3F - 3E)\Delta$	
$K^+\pi^-$	$(2T + E - S)$	
$K^0\pi^0$	$1/\sqrt{2}(3T - E + S)$	$V_{12}V_{21}$
$K^0\eta$	$1/\sqrt{6}(3T - E + S)$	
K^0X	$2(E - S)/\sqrt{3}$	

[a] $X = \eta'(958)$ is treated as an $SU(3)_f$ singlet, which completes the pseudoscalar nonet.
[b] Bose statistics convention: For decays into pairs of identical particles, reduced rates are given by $2 \times |\text{amplitude}|^2$.

and

$$\frac{\Gamma(D^0 \to \pi^-\pi^+)}{\Gamma(D^0 \to K^-\pi^+)} = 0.033 \pm 0.015 \qquad (6.142)$$

Taking into account (6.139), (6.140), and (6.133), the inequality of these experimental ratios implies that Δ is nonzero and, therefore, that

$$\begin{pmatrix} V_{11} & V_{12} \\ V_{21} & V_{22} \end{pmatrix}$$

is not a unitary matrix. Hence matrix elements V_{i3} and V_{3i} must be taken into account and heavy quarks play a significant role in charmed meson decays. To state this alternatively, there is an appreciable $SU(3)_f$ U-spin scalar contribution to the effective weak Hamiltonian.

As the foregoing discussion shows, the hadronic decays of charmed mesons are quite involved, and improved data as well as further theoretical analysis are obviously required. For more detailed theoretical discussions, the reader is referred to the original literature (see, for example, Donoghue and Holstein 80, Lipkin 81a, Matsuda 81, Shizuya 81, Suzuki 81). The current experimental situation is summarized in the excellent review by Trilling (81).

Of the many possible baryon states containing a c quark (see Table 4.4), the Λ_c^+ is the only one observed in some detail so far, although several events are suggestive of Σ_c^+ and Σ_c^{++} as well.

Production of Λ_c^+ is much more difficult than that of charmed mesons because the rate for $e^+e^- \to$ baryon–antibaryon is low. A small number of completely reconstructed Λ_c^+ events have been observed in neutrino–nucleon collisions, for example:

$$\nu_\mu p \to \mu^- \Lambda \pi^+ \pi^+ \pi^+ \pi^-$$

which is interpreted as

$$\nu_\mu p \to \Sigma_c^{++} \mu^-$$
$$\; \hookrightarrow \Lambda_c^+ \pi^+$$
$$\phantom{\nu_\mu p \to \Sigma_c^{++}}\; \hookrightarrow \Lambda \pi^+ \pi^+ \pi^-$$

and

$$\nu_\mu p \to \mu^- p K^- \pi^+ \pi^+$$

which is interpreted as

$$\nu_\mu p \to \mu^- \Lambda_c^+ \pi^+$$
$$\; \hookrightarrow p K^- \pi^+$$

In addition, Λ_c^+ has been produced in various inclusive reactions, generated by neutrino, photon, and proton collisions with nucleons and in e^+e^- annihilation. A combination of these various experiments yields the mass determinations:

$$m(\Lambda_c^+) = 2285 \pm 5 \quad \text{MeV}/c^2$$
$$m(\Sigma_c) - m(\Lambda_c) = 168 \pm 3 \quad \text{MeV}/c^2$$

The Λ_c^+ lifetime is measured by its decay path in nuclear emulsions. One finds

$$\tau(\Lambda_c^+) \simeq 2 \times 10^{-13} \quad \text{sec}$$

which is comparable to $\tau(D^0)$ or $\tau(F^+)$ rather than $\tau(D^+)$. A relatively large hadronic decay amplitude is in fact expected because the W-exchange diagram (Figure 6.11) is Cabibbo-favored. In $\Lambda_c^+ \to pK^-\pi^+$, a considerable fraction of the observed events occur in the $\bar{K}^{*0}p$ and $\Delta^{++}K^-$ resonances. Finally, observations of semileptonic decays of Λ_c^+ have been carried out at SPEAR (Vella et al. 82). One finds $B(\Lambda_c \to e^+X) = (4.5 \pm 1.7)$ percent.

Figure 6.11. Decay of Λ_c^+ to $\Lambda^0\pi^+$.

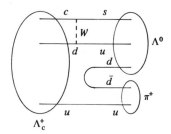

6.8 Bottom mesons

Experiment shows that quarks b and $\bar{b}$ form long-lived bound states of "quarkonium," which are analogous to the $c\bar{c}$ state of charmonium. There exist the narrow resonances $\Upsilon(9.46 \text{ GeV})$, $\Upsilon'(10.02 \text{ GeV})$, and $\Upsilon''(10.38 \text{ GeV})$, which are interpreted as $1\,^3S_1$, $2\,^3S_1$, and $3\,^3S_1$ states of $b\bar{b}$, respectively. A fourth resonance, $\Upsilon'''(10.55 \text{ GeV})$, presumed to be a $4\,^3S_1$ state, is broader because it has sufficient energy to decay strongly to a $B\bar{B}$ pair. Here B stands for the pseudoscalar meson $B_{\bar{u}} = b\bar{u}$ or $B_d^0 = b\bar{d}$. At somewhat higher energies, one expects to be able to produce $B_s^0 = b\bar{s}$, and at still higher energies, $B_{\bar{c}} = b\bar{c}$.

We consider briefly the salient features of B-meson weak decays expected on the basis of various theoretical ideas developed so far. First of all, to account for weak transitions of a b quark, we must employ the six-quark matrix of (4.70):

$$J \simeq (\bar{u}, \bar{c}, \bar{t}) \begin{pmatrix} V_{ud} & V_{us} & V_{ub} \\ V_{cd} & V_{cs} & V_{cb} \\ V_{td} & V_{ts} & V_{tb} \end{pmatrix} \begin{pmatrix} d \\ s \\ b \end{pmatrix}$$

Since $V_{ub} \ll V_{cb}$, it is clear that $b \to u$ is quite negligible and $b \to c$ dominates, with coefficient $V_{cb} = c_1 c_2 s_3 + s_2 c_3 e^{i\delta} \simeq s_3 + s_2$. The most important hadronic decays should be of the following type:

Decay	Cabibbo amplitude factor N
$b \to cd\bar{u}$	$c_1(c_1 c_2 s_3 + s_2 c_3 e^{i\delta}) \simeq s_2 + s_3$
$b \to cs\bar{c}$	$(c_1 c_2 c_3 - s_2 s_3 e^{i\delta})(c_1 c_2 s_3 + s_2 c_3 e^{i\delta}) \simeq s_2 + s_3$

Hard-gluon exchange should bring about new terms in the effective Hamiltonian for hadronic decays, which are calculated in a similar fashion to the case of strange-particle decay (Figure 6.5). Thus one has

$$H_{\text{eff}} = \frac{G_F}{\sqrt{2}} N \left\{ \frac{f_+ + f_-}{2} [(\bar{C}B)(\bar{D}U) + (\bar{C}B)(\bar{S}C)] \right.$$
$$\left. + \frac{f_+ - f_-}{2} [(\bar{C}U)(\bar{D}B) + (\bar{C}C)(\bar{S}B)] \right\} \quad (6.143)$$

where N is the Cabibbo factor and $f_+ = f_- = 1.0$ in the absence of QCD effects. Taking into account such effects, one finds $f_+ \simeq 0.78$ and $f_- \simeq 1.65$. This leads to a modest enhancement of nonleptonic modes. Departures from the simple spectator model are also possible, but the analysis is complicated and also quite speculative at this time, in the absence of data.

Observation of the weak decays of B mesons are still in their infancy. So far it has been possible to observe enhancements in the rates for $e^+ e^- \to eX$ (Bebek et al. 81) and $e^+ e^- \to \mu X$ (Chadwick et al. 81) in the neighborhood of Υ''' resonance, using the Cornell $e^+ e^-$ colliding-beam facility CESR. Also, the energy spectrum of outgoing electrons in $e^+ e^- \to eX$ has been measured (Spencer et al. 81). These results reveal that, as expected, $b \to cl\nu$ is favored for semileptonic decays and that

$$\frac{B \to Xe}{B \to \text{all}} = 13 \pm 3(\text{stat.}) \pm 3(\text{syst.})\% \quad (6.144)$$

$$\frac{B \to X\mu\nu}{B \to \text{all}} = 9.4 \pm 3.6\% \quad (6.145)$$

Problems

6.1 Verify formulas (6.61)–(6.64).

6.2 Verify formulas (6.72) and (6.73).

6.3 Show that in the reaction

p + nucleus $\to$ hyperon + X

the production of hyperon polarization perpendicular to the reaction plane does not violate parity.

6.4 From (6.84), derive (6.85) with (6.86)–(6.88).

6.5 Verify (6.95) and (6.96).

6.6 Show that the asymmetry parameter in the decay of polarized Ω^- to $J^P = \frac{1}{2}^+$ hyperon plus pseudoscalar meson is $\alpha_f \alpha_\Omega$, where α_f refers to the final hyperon.

6.7 Derive the Lee–Sugawara relation (6.103).

7
Neutral K mesons and CP violation

7.1 Introduction

Let us consider the K^0 and $\bar{K}^0$ mesons, which are charge conjugates of one another and possess opposite strangeness ($+1$ and -1, respectively). As we noted in Chapter 1, it is appropriate to employ the states $|K^0\rangle$ and $|\bar{K}^0\rangle$, which describe these particles at rest, when discussing the strong and electromagnetic interactions, since these interactions conserve strangeness. However, K^0 and $\bar{K}^0$ do not have definite lifetimes for weak decay nor do they have definite masses, because the weak interactions do not conserve strangeness. Instead, there exist two independent linear combinations of the states $|K^0\rangle$ and $|\bar{K}^0\rangle$, namely, $|K_S^0\rangle$ and $|K_L^0\rangle$, which are associated with particles of definite and distinct mass m_S and m_L and mean lifetimes γ_S^{-1} and γ_L^{-1}. Of course, the states $|K_L^0\rangle$ and $|K_S^0\rangle$ do not have definite strangeness.

The short-lived K_S^0 decays in only two significant modes $\pi^+\pi^-$ and $\pi^0\pi^0$, each with CP eigenvalue $+1$, and K_L^0 decays in many modes including the fully allowed decay to $\pi^+\pi^-\pi^0$, which is predominantly an eigenstate of CP with eigenvalue -1 (see Table 7.1). If CP invariance were valid, it would follow that K_S^0 and K_L^0 are eigenstates of CP with eigenvalues $+1$ and -1, respectively. Now, with a conventional choice of phase, one has

$$CP|K^0\rangle = -|\bar{K}^0\rangle \qquad (7.1a)$$

$$CP|\bar{K}^0\rangle = -|K^0\rangle \qquad (7.1b)$$

and, therefore, the linear combinations

$$|K_1^0\rangle = \frac{1}{\sqrt{2}}[|K^0\rangle + |\bar{K}^0\rangle], \qquad CP = -1 \qquad (7.2a)$$

7.1 Introduction

$$|K_2^0\rangle = \frac{1}{\sqrt{2}}[|K^0\rangle - |\bar{K}^0\rangle], \quad CP = +1 \tag{7.2b}$$

are eigenstates of CP. Thus CP invariance would imply that $|K_S^0\rangle = |K_2^0\rangle$ and $|K_L^0\rangle = |K_1^0\rangle$.

However, in 1964, Cronin, Fitch and co-workers (Christenson 64) observed that there is a small but finite probability for the decay $K_L^0 \rightarrow \pi^+\pi^-$, in which the final state has $CP = +1$:

$$\Gamma(K_L^0 \rightarrow \pi^+\pi^-)/\Gamma(K_L^0 \rightarrow \text{all}) \simeq 2 \times 10^{-3} \tag{7.3}$$

(see Figures 7.1 and 7.2). This unexpected and important result means that we cannot identify $|K_S^0\rangle$ with $|K_2^0\rangle$ or $|K_L^0\rangle$ with $|K_1^0\rangle$. Instead we must write

Table 7.1. *Properties of neutral kaons*[a]

Kaon	Mean lifetime (sec)	Decays	Fraction
K_S^0	$(0.8923 \pm 0.0022) \times 10^{-10}$	$\pi^+\pi^-$	68.61% $\pm 0.24\%$
		$\pi^0\pi^0$	31.89%
		$\mu^+\mu^-$	$<3.2 \times 10^{-7}$
		e^+e^-	$<3.4 \times 10^{-4}$
		$\pi^+\pi^-\gamma$	$(1.85 \pm 0.10) \times 10^{-3}$
		$\gamma\gamma$	$<0.4 \times 10^{-3}$
K_L^0	$(5.183 \pm 0.040) \times 10^{-8}$	$\pi^0\pi^0\pi^0$	$21.5 \pm 0.7\%$
		$\pi^+\pi^-\pi^0$	$12.39 \pm 0.18\%$
		$\pi^\pm\mu^\mp\nu$	$27.0 \pm 0.5\%$
		$\pi^\pm e^\mp\nu$	$38.8 \pm 0.5\%$
		(incl. $\pi e\nu\gamma$)	
		$\pi e\nu\gamma$	$1.3 \pm 0.8\%$
		$\pi^+\pi^-$	$0.203 \pm 0.005\%$
		$\pi^0\pi^0$	$0.094 \pm 0.018\%$
		$\pi^+\pi^-\gamma$	$(6.0 \pm 2.0) \times 10^{-5}$
		$\pi^0\gamma\gamma$	$<2.4 \times 10^{-4}$
		$\gamma\gamma$	$(4.9 \pm 0.5) \times 10^{-4}$
		$e\mu$	$<2.0 \times 10^{-9}$
		$\mu^+\mu^-$	$(9.1 \pm 1.9) \times 10^{-9}$
		$\mu^+\mu^-\gamma$	$<7.8 \times 10^{-6}$
		$\mu^+\mu^-\pi^0$	$<5.7 \times 10^{-5}$
		e^+e^-	$<2.0 \times 10^{-9}$
		$e^+e^-\gamma$	$<2.8 \times 10^{-5}$
		$\pi^+\pi^-e^+e^-$	$<8.8 \times 10^{-6}$
		$\pi^0\pi^\pm e^\mp\nu$	$<2.2 \times 10^{-3}$

Source: Kelly et al., "Review of particle properties," *Rev. Mod. Phys.* 52, 518 (1980).
[a] $m = 497.67$ MeV/c^2; Δm $(0.5349 \pm 0.0022) \times 10^{10}$ $\hbar/c^2$ sec^{-1}.

Figure 7.1. Plan view of experimental apparatus used by Christenson et al. (64) to observe $K_L \to \pi\pi$. The K_L particles in the beam decay in a region filled with helium to reduce regeneration effects (see Sections 7.2 and 7.7). The charged decay products were analyzed in two symmetrically placed magnetic spectrometers, utilizing spark chambers that were triggered by scintillators and photographed for later reconstruction of the decay event. (Reprinted with permission.)

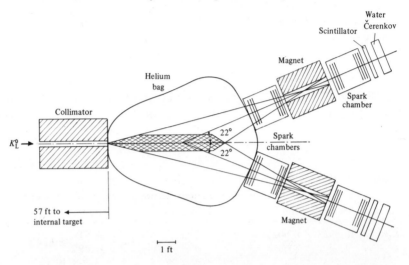

Figure 7.2. Experiment of Christenson et al. (64) in which two-pion final states could be distinguished from the common three-body decays ($\pi\pi\pi$) and ($\pi l \nu$) in two ways: the invariant 4-momentum of the pair of detected particles must be equal to the K_L mass for two-body decays and the vector sum of the 3-momenta must point along the incident beam ($\theta = 0$). (Reprinted with permission.)

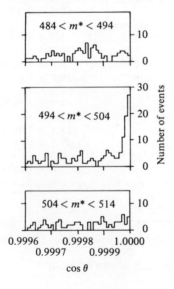

$$|K_L^0\rangle = (1 + |\epsilon_1|^2)^{-1/2}(|K_1\rangle + \epsilon_1|K_2\rangle) \quad (7.4a)$$

$$|K_S^0\rangle = (1 + |\epsilon_2|^2)^{-1/2}(|K_2\rangle + \epsilon_2|K_1\rangle) \quad (7.4b)$$

where ϵ_1 and ϵ_2 are small complex numbers.

In this chapter we shall study the peculiar and interesting "particle-mixture" phenomena associated with neutral K mesons, develop the phenomenology of CP violation, and discuss the experiments in which the various parameters are determined. We shall then examine the many theories proposed to explain CP violation, particularly those that incorporate ideas of the standard model. Finally, we shall consider prospects for observing CP violation in charm- and heavy-quark meson decays. We note at the onset that, although a very large amount of experimental effort has been expended in the study of CP violation and although the phenomenological parameters describing it have been measured quite precisely, the fundamental cause for CP violation is unknown and likely to remain a mystery for years to come.

7.2 Strangeness oscillations and regeneration

We introduce the phenomena associated with K^0–$\bar{K}^0$ mixing in the simplest possible way by ignoring CP violation at first. Suppose we produce K^0 at $t = 0$ by a typical strong production reaction:

$$\pi^- p \to K^0 \Lambda^0$$

The K^0 state at $t = 0$ is then

$$|\psi(0)\rangle = |K^0\rangle = \frac{1}{\sqrt{2}}(|K_S^0\rangle + |K_L^0\rangle) \quad (7.5)$$

Downstream, along the kaon beam, we then have

$$|\psi(t)\rangle = \frac{1}{\sqrt{2}}[|K_S^0(t)\rangle + |K_L^0(t)\rangle]$$

$$= \frac{1}{\sqrt{2}}[|K_S^0(0)\rangle e^{-\lambda_S t} + |K_L^0(0)\rangle e^{-\lambda_L t}] \quad (7.6)$$

where t is the time since production in the kaon rest frame:

$$t = d/\beta\gamma \quad (7.7)$$

d the distance, and *in the lab*

$$\lambda_S = \tfrac{1}{2}\gamma_S + im_S, \quad \lambda_L = \tfrac{1}{2}\gamma_L + im_L. \quad (7.8)$$

The K^0 fraction of the beam after a time t is just

$$I(K^0) = |\langle K^0|\psi(t)\rangle|^2 = \tfrac{1}{4}[e^{-\gamma_S t} + e^{-\gamma_L t} + 2e^{-(\gamma_S+\gamma_L)t/2}\cos\Delta mt] \quad (7.9a)$$

where $\Delta m = m_L - m_S$. Similarly,

$$I(\bar{K}^0) = \tfrac{1}{4}[e^{-\gamma_S t} + e^{-\gamma_L t} - 2e^{-(\gamma_S+\gamma_L)t/2}\cos\Delta mt] \tag{7.9b}$$

Clearly, the beam strangeness, defined as

$$\langle S\rangle = \frac{I(K^0) - I(\bar{K}^0)}{I(K^0) + I(\bar{K}^0)} \tag{7.10}$$

oscillates with a frequency $\Delta m/2\pi$. Since Δm is comparable to γ_1, these strangeness oscillations are clearly visible before they are damped away by the decay of the K_S^0 component, as shown in Figure 7.3.

For times t such that

$$1/\gamma_L = 5.2\times 10^{-8}\ \text{sec} \gg t \gg 1/\gamma_S = 8.9\times 10^{-11}\ \text{sec} \tag{7.11}$$

we have

$$\psi(t) \simeq \frac{1}{\sqrt{2}}K_L^0(0)e^{-\lambda_L t} \tag{7.12}$$

Thus, downstream from the source, the entire beam is converted into the long-lived component K_L^0. This is the regime where most experiments are carried out, since

$$c/\gamma_S = 2.7\ \text{cm} \tag{7.13a}$$

and

$$c/\gamma_L = 15.5\ \text{m} \tag{7.13b}$$

Figure 7.3. Strangeness oscillations and regeneration for K^0 and $\bar{K}^0$.

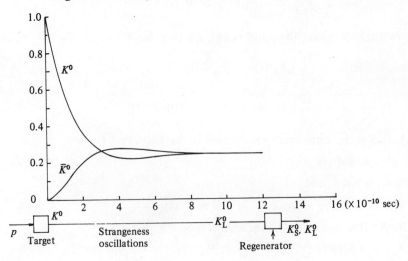

Next, let us consider the effect of a piece of matter of thickness l placed in the path of the beam. Neglecting time dependence, the wavefunction prior to collision is

$$\psi = \frac{1}{\sqrt{2}} K_L^0 = \frac{1}{2}(K^0 + \bar{K}^0) \tag{7.14}$$

However, the K^0 and $\bar{K}^0$ scatter from matter nuclei with very different amplitudes, because K^0 can only scatter elastically, whereas $\bar{K}^0$ may also be absorbed through reactions such as

$$\bar{K}^0 + p \rightarrow \Lambda^0 + \pi$$

This results in a change in the relative amplitude and phase of K^0 versus $\bar{K}^0$ and causes the K_S^0 component, which had died away before entrance into the matter, to be *regenerated*. As a result of the scattering, the wavefunction becomes:

$$\psi = \tfrac{1}{2}(aK^0 + b\bar{K}^0) \tag{7.15}$$

where $a \neq b$ are two complex numbers. Thus

$$\psi = \frac{1}{2\sqrt{2}}[(a-b)K_S^0 + (a+b)K_L^0] = \frac{a+b}{2\sqrt{2}}[K_L^0 + rK_S^0] \tag{7.16}$$

where

$$r = \frac{a-b}{a+b}$$

is a complex number called the *regeneration parameter*. It can be related to the physical properties of the regenerator (see Section 7.7).

7.3 The mass and decay matrices

We now begin a more detailed analysis of the K^0–$\bar{K}^0$ system, taking into account *CP* violation. Suppose that we have an arbitrary superposition of K^0 and $\bar{K}^0$ states with time-dependent coefficients A and B:

$$|\psi(t)\rangle = A(t)|K^0\rangle + B(t)|\bar{K}^0\rangle$$

or simply

$$\psi = \begin{pmatrix} A \\ B \end{pmatrix}$$

The effective Hamiltonian governing the time evolution of ψ is

$$H = H_{\text{Strong}} + H_{\text{Electromagnetic}} + H_{\text{Weak}} = H_S + H_{\text{EM}} + H_{\text{wk}}$$

and it satisfies the equation

$$H|\psi\rangle = i\frac{\partial}{\partial t}|\psi\rangle$$

In the K^0–$\bar{K}^0$ basis, H is represented by a 2×2 matrix:

$$H = M - i\Gamma \tag{7.17}$$

where M and Γ are Hermitian 2×2 matrices, called the *mass* and *decay matrices*, respectively (Wu and Yang 64). Because the neutral kaons decay, H itself is not Hermitian.

States $|K^0\rangle$ and $|\bar{K}^0\rangle$ are defined as eigenstates of $H_S + H_{EM}$ with strangeness eigenvalues ± 1, respectively. Therefore, since the strong and electromagnetic interactions conserve strangeness, we must have

$$\langle \bar{K}^0|H_S + H_{EM}|K^0\rangle = 0$$

If H_W were zero, then K^0 and $\bar{K}^0$ would not decay and Γ would also be zero. In this case, H would reduce to

$$H \to M = \begin{pmatrix} m_{K^0} & 0 \\ 0 & m_{\bar{K}^0} \end{pmatrix}$$

where

$$m_{K^0} = \langle K^0|H_S + H_{EM}|K^0\rangle \tag{7.18}$$

and

$$m_{\bar{K}^0} = \langle \bar{K}^0|H_S + H_{EM}|\bar{K}^0\rangle \tag{7.19}$$

CPT invariance of H_S and H_{EM} implies that $m_{K^0} = m_{\bar{K}^0}$.

Now, in fact, the weak interaction Hamiltonian is not zero, so $\Gamma \neq 0$. We postpone consideration of Γ for the moment, however, and consider in some detail the properties of M, which now contains nonzero off-diagonal elements. We have:

$$M_{11} = m_{K^0} + \langle K^0|H_{wk}|K^0\rangle + \sum_n{}' \frac{|\langle K^0|H_{wk}|n\rangle|^2}{m_{K^0} - E_n} + \cdots \tag{7.20}$$

$$M_{22} = m_{\bar{K}^0} + \langle \bar{K}^0|H_{wk}|\bar{K}^0\rangle + \sum_n{}' \frac{|\langle \bar{K}^0|H_{wk}|n\rangle|^2}{m_{\bar{K}^0} - E_n} + \cdots \tag{7.21}$$

$$M_{21} = M_{12}^* = \langle \bar{K}^0|H_{wk}|K^0\rangle$$
$$+ \sum_n{}' \frac{\langle \bar{K}^0|H_{wk}|n\rangle\langle n|H_{wk}|K^0\rangle}{m_{K^0} - E_n} + \cdots \tag{7.22}$$

If H_{wk} is *CPT*-invariant, then

$$\langle K^0|H_{wk}|n\rangle = \langle K^0|(CPT)^{-1}H_{wk}(CPT)|n\rangle$$
$$= -\langle \bar{n}'|H_{wk}|\bar{K}^0\rangle = -\langle \bar{K}^0|H_{wk}|\bar{n}'\rangle^*$$

7.3 The mass and decay matrices

where the prime denotes reversed spins. However, in (7.20) and (7.21), the sums are taken over all possible states n; thus *CPT* invariance of H_{wk} in addition to that of H_S and H_{EM} implies $M_{11} = M_{22}$.

Since first-order weak interactions obey the selection rule $|\Delta S| \leq 1$, the term $\langle \bar{K}^0|H_{wk}|K^0\rangle$ in (7.22) is zero. Also, if H_{wk} is time-reversal-invariant, $M_{12} = M_{21} = M_{12}^*$. In the discussion that follows, we shall find that the second-order terms Σ_n in (7.20)–(7.22) are important and must be retained, but we shall assume that it is legitimate to ignore terms of higher order.

Let us now consider the $K_L^0 - K_S^0$ mass difference $\Delta m = m_L - m_S$. For this purpose, we ignore *CP* violation at first and write

$$m_L = \langle K_L^0|H_S + H_{EM} + H_{wk}|K_L^0\rangle + \sum_n{}' \frac{\langle K_L^0|H_{wk}|n\rangle\langle n|H_{wk}|K_L^0\rangle}{m_{K^0} - E_n}$$

$$m_S = \langle K_S^0|H_S + H_{EM} + H_{wk}|K_S^0\rangle + \sum_n{}' \frac{\langle K_S^0|H_{wk}|n\rangle\langle n|H_{wk}|K_S^0\rangle}{m_{K^0} - E_n}$$

where $|K_L^0\rangle$ and $|K_S^0\rangle$ are given by (7.4a,b) with $\epsilon_1 = \epsilon_2 = 0$. Taking the difference $\Delta m = m_L - m_S$ and making use of *CPT* invariance and other results previously obtained, we find

$$\Delta m = m_L - m_S = 2\sum_n{}' \frac{\langle \bar{K}^0|H_{wk}|n\rangle\langle n|H_{wk}|K^0\rangle}{m_{K^0} - E_n} \quad (CP \text{ invariance})$$

(7.23)

Prior to the introduction of the GIM scheme, calculations of Δm based on (7.23) led to results that did not agree with Δm_{expt} unless an unreasonably small value of m_W (~ 2 GeV/c^2) was chosen. It is instructive to carry out this calculation in some detail, as we do now, and to see how it is modified by the GIM scheme to yield much more sensible results.

In the absence of charm, the diagrams that contribute to (7.23) in lowest order are those of Figures 7.4a and 7.4b. The amplitude corresponding to Figure 7.4a is

Figure 7.4. Contributions to Δm in the absence of charm.

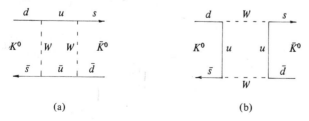

7 Neutral K mesons and CP violation

$$\mathcal{M}_a = i\left(\frac{-ig}{2\sqrt{2}}\sin\theta_C\right)^2 \left(\frac{-ig}{2\sqrt{2}}\cos\theta_C\right)^2 \int \frac{d^4k}{(2\pi)^4}\left[\bar{u}_s\gamma_\lambda(1-\gamma_5)\frac{\slashed{k}+m_u}{k^2-m_u^2}\right.$$

$$\times \gamma_\rho(1-\gamma_5)u_d\bar{v}_s\gamma_\alpha(1-\gamma_5)\frac{\slashed{k}+m_u}{k^2-m_u^2}\gamma_\sigma(1-\gamma_5)v_d$$

$$\left.\times\left(-i\frac{g^{\lambda\sigma}-k^\lambda k^\sigma/m_W^2}{k^2-m_W^2}\right)\left(-i\frac{g^{\alpha\rho}-k^\alpha k^\rho/m_W^2}{k^2-m_W^2}\right)\right] \quad (7.24)$$

We ignore the terms in $k^\lambda k^\sigma/m_W^2$ and $k^\alpha k^\rho/m_W^2$ and simplify (7.24) to obtain

$$\mathcal{M}_a = -ig^4 \frac{\sin^2\theta_C \cos^2\theta_C}{16}\int \frac{d^4k}{(2\pi)^4}\frac{k_\mu k_\nu}{(k^2-m_u^2)^2(k^2-m_W^2)^2}T^{\mu\nu} \quad (7.25)$$

where

$$T^{\mu\nu} = \bar{u}_s\gamma_\lambda\gamma^\mu\gamma_\rho(1-\gamma_5)u_d\bar{v}_s\gamma^\rho\gamma^\nu\gamma^\lambda(1-\gamma_5)v_d$$

It may be readily verified that the amplitude corresponding to Figure 7.4b is identical.

We now consider the integral

$$I_{\mu\nu} = \int \frac{d^4k}{(2\pi)^4}\frac{k_\mu k_\nu}{(k^2-m_W^2)^2(k^2-m_u^2)^2} \quad (7.26)$$

It may be shown (see Problem 7.2) that, when terms of order m_u^2/m_W^2 are ignored, (7.26) becomes

$$I_{\mu\nu} = g_{\mu\nu}/64\pi^2 i m_W^2 \quad (7.27)$$

Inserting (7.27) in (7.25), we obtain

$$\mathcal{M} = \mathcal{M}_a + \mathcal{M}_b = -\frac{g^4 \sin^2\theta_C \cos^2\theta_C}{8\cdot 64\pi^2 m_W^2}$$

$$\times [\bar{u}_s\gamma_\lambda\gamma_\mu\gamma_\rho(1-\gamma_5)u_d\bar{v}_s\gamma^\rho\gamma^\mu\gamma^\lambda(1-\gamma_5)v_d]$$

Now, simplifying the factor in square brackets, carrying out a Fierz transformation, and substituting $g = 2^{5/4}m_W G_F^{1/2}$, we find

$$\mathcal{M} = \sum_n{}' \frac{\langle \bar{K}^0|H_{wk}|n\rangle\langle n|H_{wk}|K^0\rangle}{m_{K^0}-E_n}$$

$$= \frac{G_F^2}{4\pi^2}m_W^2 \cos^2\theta_C \sin^2\theta_C \langle \bar{K}^0|\bar{S}\gamma_\lambda(1-\gamma_5)D\cdot\bar{S}\gamma^\lambda(1-\gamma_5)D|K^0\rangle$$

Between operators D and $\bar{S}$ we may insert a factor $1 = \Sigma'_n|n\rangle\langle n| \simeq |0\rangle\langle 0|$, where the last step is called the "vacuum insertion approximation." Thus we obtain

$$\mathcal{M} = \frac{G_F^2}{4\pi^2}m_W^2 \cos^2\theta_C \sin^2\theta_C[\langle \bar{K}^0|\bar{S}\gamma_\lambda(1-\gamma_5)D|0\rangle\langle 0|\bar{S}\gamma^\lambda(1-\gamma_5)D|K^0\rangle]$$

7.3 The mass and decay matrices

However, the factors in square brackets may be related to the kaon-decay constant f_K, which appears in the matrix element for K_{l2} decay [see (4.82)]. We thus find

$$\mathcal{M} = \frac{G_F^2}{4\pi^2} m_W^2 \cos^2\theta_C \sin^2\theta_C f_K^2 m_K^2 (2E_{K^0} 2E_{\bar{K}^0})^{-1/2}$$

which yields, for $E_{\bar{K}^0} = E_{\bar{K}^0} = m_K$,

$$\Delta m = m_L - m_S \simeq \frac{G_F^2}{4\pi^2} m_W^2 \cos^2\theta_C \sin^2\theta_C f_K^2 m_K \quad (7.28)$$

It is this formula that gives much too large a value of Δm if $m_W = 80$ GeV/c^2 is used or much too small a value of m_W if Δm_{expt} is employed.

We now modify the foregoing calculation by introducing the charmed quark according to the GIM scheme. Figure 7.4 must then be replaced by Figure 7.5. The amplitude corresponding to the diagram in Figure 7.5a is

$$\mathcal{M}_a = i \left(\frac{-ig}{2\sqrt{2}} \sin\theta_C\right)^2 \left(\frac{-ig}{2\sqrt{2}} \cos\theta_C\right)^2 \int \frac{d^4k}{(2\pi)^4}$$

$$\times \left[\bar{u}_s \gamma_\lambda (1-\gamma_5) \left(\frac{\slashed{k}+m_u}{k^2-m_u^2} - \frac{\slashed{k}+m_c}{k^2-m_c^2}\right) \gamma_\rho (1-\gamma_5) u_d \right.$$

$$\times \bar{v}_s \gamma_\alpha (1-\gamma_5) \left(\frac{\slashed{k}+m_u}{k^2-m_u^2} - \frac{\slashed{k}+m_c}{k^2-m_c^2}\right) \gamma_\sigma (1-\gamma_5) v_d$$

$$\left. \times \left(-i\frac{g^{\lambda\sigma} - k^\lambda k^\sigma/m_W^2}{k^2 - m_W^2}\right)\left(-i\frac{g^{\alpha\rho} - k^\alpha k^\rho/m_W^2}{k^2 - m_W^2}\right) \right]$$

where the terms in u and c have opposite signs because of the GIM

Figure 7.5. Contributions to Δm with charm.

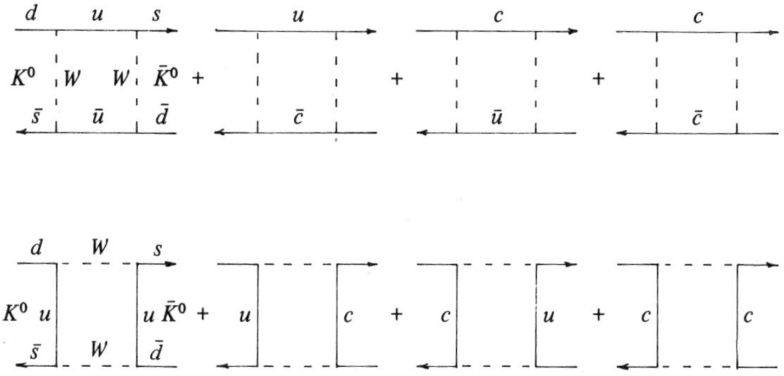

mechanism. Proceeding as in the previous discussion, we now encounter, in place of the integral $I_{\mu\nu}$ (7.26), the modified integral:

$$I'_{\mu\nu} \int \frac{d^4k}{(2\pi)^4} \frac{k_\mu k_\nu (m_c^2 - m_u^2)^2}{(k^2 - m_u^2)^2(k^2 - m_c^2)^2(k^2 - m_W^2)^2}$$

which converges very rapidly. For $m_W \gg m_c \gg m_u$ we find

$$I'_{\mu\nu} \simeq \frac{g_{\mu\nu}}{64\pi^2 i} \frac{1}{m_c^2} \frac{m_c^4}{m_W^4}$$

Now, completing the calculation as before, we arrive at the final result

$$\Delta m \simeq \frac{G_F^2}{4\pi^2} f_K^2 m_K m_c^2 \cos^2 \theta_C \sin^2 \theta_C \qquad (7.29)$$

This gives good agreement with the observed mass difference for m_c in the range $m_c \sim 1.5$ GeV/c^2. In this way, the mass of the charmed quark was actually predicted before charmonium was observed (see Gaillard and Lee 74b, Gaillard et al. 75).[1]

We now turn our attention to the matrix Γ. Conditions on its elements are derived from the equation that describes conservation of probability:

$$\frac{d}{dt}\langle\psi|\psi\rangle + 2\pi \sum_F \rho(F)|\langle F|H_W|\psi(t)\rangle|^2 = 0 \qquad (7.30)$$

Here $|F\rangle$ denotes a particular final state resulting from K^0 and $\bar{K}^0$ decay, and $\rho(F)$ is the density of final states. For an arbitrary state $|\psi\rangle = A|K^0\rangle + B|\bar{K}^0\rangle$, we have

$$\frac{d}{dt}\langle\psi|\psi\rangle = \langle\psi|\dot\psi\rangle + \langle\dot\psi|\psi\rangle$$

$$= -\langle\psi|\Gamma - iM|\psi\rangle - \langle\psi|\Gamma + iM|\psi\rangle = -2\langle\psi|\Gamma|\psi\rangle$$

Therefore, since A and B are arbitrary, we obtain

$$\Gamma_{11} = \langle K^0|\Gamma|K^0\rangle = \pi \sum_F \rho(F)|\langle F|H_{wk}|K^0\rangle|^2 \qquad (7.31a)$$

$$\Gamma_{22} = \langle \bar{K}^0|\Gamma|\bar{K}^0\rangle = \pi \sum_F \rho(F)|\langle F|H_{wk}|\bar{K}^0\rangle|^2 \qquad (7.31b)$$

$$\Gamma^*_{12} = \Gamma_{21} \qquad = \langle \bar{K}^0|\Gamma|K^0\rangle$$
$$= \pi \sum_F \rho(F)\langle \bar{K}^0|H_{wk}|F\rangle\langle F|H_{wk}|K^0\rangle \qquad (7.31c)$$

---

[1] In our calculation of Δm, we have ignored color. Its inclusion modifies the result slightly (see Problem 7.3).

7.4 Eigenvalues and eigenvectors of $\Gamma + iM$

Matrix elements Γ_{11} and Γ_{22} are thus expressed as sums over terms corresponding to distinct final states F. In principle it is possible to determine, by the measurement of partial decay rates, the contribution of each of these terms to Γ_{11} and Γ_{22} (and some restrictions can also be derived for Γ_{12}). However, no such separation is possible for the various contributions to M_{11} and M_{22} [(7.20) and (7.21)].

If CPT invariance holds for H_{wk}, then

$$\langle F|H_{wk}|K^0\rangle = \langle F|(CPT)^{-1}H_{wk}(CPT)|K^0\rangle$$
$$= -\langle \bar{K}^0|H_{wk}|\bar{F}'\rangle = -\langle \bar{F}'|H_{wk}|\bar{K}^0\rangle^* \quad (7.32)$$

where the prime denotes reversed spins. However, in (7.31a) and (7.31b), the sums are taken over all possible final states F; and so CPT invariance implies $\Gamma_{11} = \Gamma_{22}$. If T invariance holds for H_{wk}, then we can always adjust phases so that $\langle F|H_{wk}|K^0\rangle$ and $\langle F|H_{wk}|\bar{K}^0\rangle$ are relatively real. In this case, it follows that, for T invariance,

$$\Gamma_{12} = \Gamma_{21} = \Gamma_{12}^* \quad \text{and} \quad M_{12} = M_{21} = M_{12}^* \quad (7.33)$$

7.4 Eigenvalues and eigenvectors of $\Gamma + iM$

To examine the eigenvalues and eigenvectors of $\Gamma + iM$, we write

$$\Gamma + iM = DI + i\mathbf{E} \cdot \boldsymbol{\sigma} \quad (7.34)$$

where I and $\boldsymbol{\sigma}$ are the 2×2 identity and Pauli matrices, respectively, and D and $\mathbf{E}$ are a complex number and a complex vector, respectively:

$$\mathbf{E} = E_1\hat{\mathbf{i}} + E_2\hat{\mathbf{j}} + E_3\hat{\mathbf{k}} \quad (7.35)$$

It is convenient to write

$$E_1 = E \sin\theta \cos\phi$$
$$E_2 = E \sin\theta \sin\phi$$
$$E_3 = E \cos\theta$$

where θ and ϕ are, in general, complex. Thus we have

$$\Gamma + iM = \begin{pmatrix} D + iE_3 & i(E_1 - iE_2) \\ i(E_1 + iE_2) & D - iE_3 \end{pmatrix}$$
$$= \begin{pmatrix} D + iE\cos\theta & iE\sin\theta e^{-i\phi} \\ iE\sin\theta e^{i\phi} & D - iE\cos\theta \end{pmatrix} \quad (7.36)$$

which is readily diagonalized with the aid of some matrix C to yield

$$C(\Gamma + iM)C^{-1} = \begin{pmatrix} \lambda_S = D + iE & 0 \\ 0 & \lambda_L = D - iE \end{pmatrix} \quad (7.37)$$

The eigenvalues of $\Gamma + iM$ are thus

$$D + iE \equiv \lambda_S \equiv \tfrac{1}{2}\gamma_S + im_S \tag{7.38a}$$

$$D - iE \equiv \lambda_L \equiv \tfrac{1}{2}\gamma_L + im_L \tag{7.38b}$$

where γ_S, γ_L, m_S, and m_L are real quantities corresponding to the eigenvectors $|K_S^0\rangle$ and $|K_L^0\rangle$:

$$(\Gamma + iM)|K_S^0\rangle = \lambda_S|K_S^0\rangle = (D + iE)|K_S^0\rangle \tag{7.39a}$$

$$(\Gamma + iM)|K_L^0\rangle = \lambda_L|K_L^0\rangle = (D - iE)|K_L^0\rangle \tag{7.39b}$$

Thus, states $|K_S^0\rangle$ and $|K_L^0\rangle$ each have definite lifetimes $\tau_S = 1/\gamma_S$ and $\tau_L = 1/\gamma_L$ for weak decay:

$$|K_S^0(t)\rangle = |K_S^0(0)\rangle e^{-im_S t} e^{-(\gamma_S/2)t} \tag{7.40a}$$

$$|K_L^0(t)\rangle = |K_L^0(0)\rangle e^{-im_L t} e^{-(\gamma_L/2)t} \tag{7.40b}$$

Let us now express the eigenvectors $|K_S^0\rangle$ and $|K_L^0\rangle$ of $\Gamma + iM$ as linear combinations of the basis states $|K^0\rangle$ and $|\bar{K}^0\rangle$. To this end we write

$$|K_S^0\rangle = (|u|^2 + |v|^2)^{-1/2}[u|K^0\rangle + v|\bar{K}^0\rangle] \tag{7.41a}$$

$$|K_L^0\rangle = (|p|^2 + |q|^2)^{-1/2}[p|K^0\rangle + q|\bar{K}^0\rangle] \tag{7.41b}$$

Then we obtain from (7.36) and (7.39a,b):

$$iE(\cos\theta - 1)u + iE\sin\theta e^{-i\phi}v = 0$$

and

$$iE(\cos\theta + 1)p + iE\sin\theta e^{-i\phi}q = 0$$

or

$$\frac{v}{u} = \frac{1 - \cos\theta}{\sin\theta} e^{i\phi} = \tan\frac{\theta}{2} e^{i\phi} \tag{7.42}$$

and

$$\frac{q}{p} = -\cot\frac{\theta}{2} e^{i\phi} \tag{7.43}$$

It is customary to define the new complex quantities ϵ_1 and ϵ_2, where $\epsilon = (\epsilon_1 + \epsilon_2)/2$ and $\delta = (\epsilon_1 - \epsilon_2)/2$, by

$$\frac{1 - \epsilon_1}{1 + \epsilon_1} \equiv -\frac{v}{u} = -\tan\frac{\theta}{2} e^{i\phi} \tag{7.44}$$

and

$$\frac{1 - \epsilon_2}{1 + \epsilon_2} \equiv \frac{q}{p} = -\cot\frac{\theta}{2} e^{i\phi} \tag{7.45}$$

In terms of ϵ_1 and ϵ_2, (7.41a,b) becomes

$$|K_S^0\rangle = [2(1 + |\epsilon_1|^2)]^{-1/2}[(1 + \epsilon_1)|K^0\rangle - (1 - \epsilon_1)|\bar{K}^0\rangle] \tag{7.46a}$$

and

7.4 Eigenvalues and eigenvectors of $\Gamma + iM$

$$|K_L^0\rangle = [2(1 + |\epsilon_2|^2)]^{-1/2}[(1 + \epsilon_2)|K^0\rangle + (1 - \epsilon_2)|\bar{K}^0\rangle] \quad (7.46b)$$

If CPT invariance holds, then $(\Gamma + iM)_{11} = (\Gamma + iM)_{22}$. Thus, from (7.36), $E_3 = 0$ and $\theta = -\pi/2$. Therefore,

$$\frac{1 - \epsilon_1}{1 + \epsilon_1} = \frac{1 - \epsilon_2}{1 + \epsilon_2} = e^{i\phi} = \left[\frac{(\Gamma^* + iM^*)_{12}}{(\Gamma + iM)_{12}}\right]^{1/2}$$

and

$$\epsilon_1 = \epsilon_2 = \epsilon, \quad \delta = 0 \qquad CPT \text{ invariance} \quad (7.47)$$

Consequently, CPT invariance implies

$$|K_S^0\rangle = [2(1 + |\epsilon|^2)]^{-1/2}[(1 + \epsilon)|K^0\rangle - (1 - \epsilon)|\bar{K}^0\rangle] \quad (7.48a)$$
$$|K_L^0\rangle = [2(1 + |\epsilon|^2)]^{-1/2}[(1 + \epsilon)|K^0\rangle + (1 - \epsilon)|\bar{K}^0\rangle] \quad (7.48b)$$

If T invariance holds, then $(\Gamma + iM)_{12} = (\Gamma + iM)_{21}$. Thus, from (7.36), $e^{i\phi} = e^{-i\phi} = 1$, $\phi = 0$, and

$$\frac{1 - \epsilon_1}{1 + \epsilon_1} = -\tan\frac{\theta}{2} = \frac{1 + \epsilon_2}{1 - \epsilon_2}$$

or

$$\epsilon_1 = -\epsilon_2, \quad \epsilon = 0, \quad \delta = \epsilon_1 = -\epsilon_2 \qquad T \text{ invariance} \quad (7.49)$$

Consequently, T invariance implies

$$|K_S^0\rangle = [2(1 + |\delta|^2)]^{-1/2}[(1 + \delta)|K^0\rangle - (1 - \delta)|\bar{K}^0\rangle] \quad (7.50a)$$
$$|K_L^0\rangle = [2(1 + |\delta|^2)]^{-1/2}[(1 - \delta)|K^0\rangle + (1 + \delta)|\bar{K}^0\rangle] \quad (7.50b)$$

Obviously, if T invariance and CPT invariance both hold, then CP invariance is also valid and $\epsilon = \delta = 0$. Of course, we know the CP violation is an extremely small effect, so $|\epsilon|$, $|\delta|$, $|\epsilon_1|$, and $|\epsilon_2|$ are all very small compared with unity. It follows that $\phi \ll 1$ and that $\theta \simeq -\pi/2$, even without the assumptions of CPT or T invariance.

The existence of nonzero ϵ_1 and ϵ_2 implies that $|K_S^0\rangle$ and $|K_L^0\rangle$ are not orthogonal states. From (7.46a,b) we have

$$\langle K_S^0|K_L^0\rangle = \frac{(1 + \epsilon_1^*)(1 + \epsilon_2) - (1 - \epsilon_1)^*(1 - \epsilon_2)}{2(1 + |\epsilon_1|^2)^{1/2}(1 + |\epsilon_2|^2)^{1/2}} \quad (7.51)$$

Neglecting $|\epsilon_1|^2$, $|\epsilon_2|^2$, and other second-order small quantities, this reduces to

$$\langle K_S^0|K_L^0\rangle \simeq \epsilon_1^* + \epsilon_2 = 2(\text{Re }\epsilon - i\text{ Im }\delta) \quad (7.52)$$

Therefore, if CPT invariance holds, $\langle K_S^0|K_L^0\rangle$ is real:

$$\langle K_S^0|K_L^0\rangle = 2\text{ Re }\epsilon \qquad CPT \text{ invariance} \quad (7.53)$$

while if T invariance holds, $\langle K_S^0|K_L^0\rangle$ is imaginary:

$$\langle K_S^0|K_L^0\rangle = -2i\text{ Im }\delta \qquad T \text{ invariance} \quad (7.54)$$

To summarize the results of the preceding two sections, CP violation in neutral K decay may be parameterized by the mixing of CP eigenstates $|K_1^0\rangle$ and $|K_2^0\rangle$:

$$|K_S^0\rangle = (|K_1\rangle + \epsilon_1|K_2\rangle)/(1 + |\epsilon_1|^2)^{1/2} \qquad (7.55)$$

$$|K_L^0\rangle = (|K_2\rangle + \epsilon_2|K_1\rangle)/(1 + |\epsilon_2|^2)^{1/2} \qquad (7.56)$$

where ϵ_1 and ϵ_2 are complex numbers. If CP violation is accompanied by T violation so that CPT is conserved, then $\epsilon_1 = \epsilon_2 \equiv \epsilon$. However, if CPT is violated and T is conserved, then $\epsilon_1 = -\epsilon_2 \equiv \delta$.

7.5 Transition amplitudes for $K_S^0 \to \pi\pi$ and $K_L^0 \to \pi\pi$

We now calculate the K_S and K_L decay modes to two pions, assuming CPT invariance is valid. The final-state pions are bosons, so the total wavefunction must be symmetric with respect to particle interchange. The isospin of a pion is $I = 1$ and the final state can have $I = 0$ or 2, $I_3 = 0$. There are four amplitudes:

$$\langle \pi\pi, I = 0|H_{wk}|K_S\rangle, \qquad \langle \pi\pi, I = 2|H_{wk}|K_S\rangle$$
$$\langle \pi\pi, I = 0|H_{wk}|K_L\rangle, \qquad \langle \pi\pi, I = 2|H_{wk}|K_L\rangle$$

which may be rewritten in terms of the physical pion states $\pi^+\pi^-$ and $\pi^0\pi^0$ by Clebsch–Gordan coefficients:

$$\langle \pi^+\pi^-| = \sqrt{\tfrac{1}{3}}\langle \pi\pi, I = 2| + \sqrt{\tfrac{2}{3}}\langle \pi\pi, I = 0|$$
$$\langle \pi^0\pi^0| = \sqrt{\tfrac{2}{3}}\langle \pi\pi, I = 2| - \sqrt{\tfrac{1}{3}}\langle \pi\pi, I = 0|$$

where $\pi^+\pi^-$ is understood to mean $(\pi_1^+\pi_2^- + \pi_1^-\pi_2^+)/\sqrt{2}$. The pions undergo final-state phase shifts $e^{i\delta_0}$ and $e^{i\delta_2}$ for $I = 0$ and 2, respectively:

$$\langle \pi^+\pi^-| = \sqrt{\tfrac{1}{3}}e^{i\delta_2}\langle \pi\pi, I = 2| + \sqrt{\tfrac{2}{3}}e^{i\delta_0}\langle \pi\pi, I = 0| \qquad (7.57a)$$

$$\langle \pi^0\pi^0| = \sqrt{\tfrac{2}{3}}e^{i\delta_2}\langle \pi\pi, I = 2| - \sqrt{\tfrac{1}{3}}e^{i\delta_0}\langle \pi\pi, I = 0| \qquad (7.57b)$$

The amplitudes for decay are defined as

$$A_0 = \langle \pi\pi, I = 0|H_{wk}|K^0\rangle \qquad (7.58a)$$

$$A_2 = \langle \pi\pi, I = 2|H_{wk}|K^0\rangle \qquad (7.58b)$$

And similar $\bar{K}^0$ amplitudes may be derived from symmetry. Under CPT,

$$|K_0\rangle \to -\langle \bar{K}_0| \qquad (7.59a)$$

$$\langle \pi\pi, I = 0| \to |\pi\pi, I = 0\rangle \qquad (7.59b)$$

$$\langle \pi\pi, I = 2| \to |\pi\pi, I = 2\rangle \qquad (7.59c)$$

and therefore if H_{wk} is *invariant under CPT*,

$$\langle \pi\pi, I = 0|H_{wk}|\bar{K}^0\rangle = -A_0^* \qquad (7.60a)$$

7.5 Transition amplitudes for $K_S \to \pi\pi$ and $K_L \to \pi\pi$

$$\langle \pi\pi, I = 2|H_{\text{wk}}|\bar{K}^0\rangle = -A_2^* \tag{7.60b}$$

An arbitrary overall phase may be eliminated by taking A_0 real.

The initial state of an actual kaon beam is some superposition of K^0 and $\bar{K}^0$, which both have $I = \tfrac{1}{2}$. Therefore transitions represented by A_2 have $|\Delta I| = \tfrac{3}{2}$, in violation of the $|\Delta I| = \tfrac{1}{2}$ rule (see Section 6.3). Experimentally, $|\Delta I| = \tfrac{3}{2}$ transitions are suppressed by a factor $\sim \tfrac{1}{20}$.

Using (7.48a,b) we may now express the observed transitions in terms of A_0, A_2, and the CP-violating, CPT-conserving parameter ϵ:

$$\langle \pi^+\pi^-|H_{\text{wk}}|K_S^0\rangle = 3^{-1/2}[2(1 + |\epsilon|^2)]^{-1/2}$$
$$\times [(A_2 + A_2^*)e^{i\delta_2} + (4/\sqrt{2})A_0 e^{i\delta_0} + (A_2 - A_2^*)e^{i\delta_2}\epsilon] \tag{7.61}$$

$$\langle \pi^0\pi^0|H_{\text{wk}}|K_S^0\rangle = \sqrt{\tfrac{2}{3}}[2(1 + |\epsilon|^2)]^{-1/2}$$
$$\times [(A_2 + A_2^*)e^{i\delta_2} - \sqrt{2}A_0 e^{i\delta_0} + (A_2 - A_2^*)e^{i\delta_2}\epsilon] \tag{7.62}$$

$$\langle \pi^+\pi^-|H_{\text{wk}}|K_L^0\rangle = 3^{-1/2}[2(1 + |\epsilon|^2)]^{-1/2}$$
$$\times \{(A_2 - A_2^*)e^{i\delta_2} + \epsilon[(A_2 + A_2^*)e^{i\delta_2} + (4/\sqrt{2})A_0 e^{i\delta_0}]\} \tag{7.63}$$

$$\langle \pi^0\pi^0|H_{\text{wk}}|K_L^0\rangle = \sqrt{\tfrac{2}{3}}[2(1 + |\epsilon|^2)]^{-1/2}$$
$$\times \{(A_2 - A_2^*)e^{i\delta_2} + \epsilon[(A_2 + A_2^*)e^{i\delta_2} - \sqrt{2}A_0 e^{i\delta_0}]\} \tag{7.64}$$

The experimental observables are the ratios

$$\eta^{+-} = \frac{\langle \pi^+\pi^-|H_{\text{wk}}|K_L^0\rangle}{\langle \pi^+\pi^-|H_{\text{wk}}|K_S^0\rangle} \tag{7.65}$$

$$\eta^{00} = \frac{\langle \pi^0\pi^0|H_{\text{wk}}|K_L^0\rangle}{\langle \pi^0\pi^0|H_{\text{wk}}|K_S^0\rangle} \tag{7.66}$$

Neglecting terms that are second order in the small quantities ϵ and $|A_2|$, we find

$$\eta^{+-} \simeq \epsilon + \epsilon' \tag{7.67}$$
$$\eta^{00} \simeq \epsilon - 2\epsilon' \tag{7.68}$$

where

$$\epsilon' = \frac{1}{\sqrt{2}} \text{Im}\left(\frac{A_2}{A_0}\right) e^{i(\pi/2 + \delta_2 - \delta_0)} \tag{7.69}$$

The quantity ϵ' is a new CP- and T-violating parameter, which did not appear in the K^0-$\bar{K}^0$ analysis. If there is a phase difference between A_0 and A_2, η^{+-} and η^{00} can be nonzero even without K_L-K_S mixing ($\epsilon = 0$), owing to the phase difference between $I = 0$ and $I = 2$ decay modes.

7.6 Measurement of η^{+-} and η^{00}

The amplitudes and phases of η^{+-} and η^{00} have been measured with increasing precision since 1964. We shall describe a recent experiment of Christenson et al (1979) who independently determined all four quantities $|\eta^{+-}|$, $|\eta^{00}|$, ϕ^{+-}, and ϕ^{00} in a single apparatus.

The experiment was performed at the Brookhaven AGS proton accelerator, where a beam of kaons was produced by sending high-energy (28.5-GeV) protons into a 3.5-in. platinum target (see Figure 7.6). A brass collimator in a transverse magnetic field defined the angular divergence of the neutral beam and swept away charged particles. The kaons, along with a contamination of neutrons and some neutral hyperons, drifted through a helium-filled region where some decayed via $K^0_{L,S} \to 2\pi^0 \to 4\gamma$. The γ rays were converted to e^+e^- pairs in lead sheets. These were then detected in proportional wire chambers (PWC), allowing later computer reconstruction of the decay vertex. The total energy of all γs was measured in a segmented lead-glass calorimeter. In the measurement of η^{00}, the principal background came from the CP-conserving three-pion decay $K_L^0 \to \pi^0\pi^0\pi^0$. Layers of lead and scintillator around the fiducial region were designed to veto events with extra γ rays.

Measurements of η^{+-} were made in the same detector by energizing

Figure 7.6. Apparatus of Christenson et al. (79), where N and C are scintillators used to trigger on desired events. For example, a $K_L \to \pi^0\pi^0$ decay candidate was indicated by the absence of a signal in N but the presence of signals after the lead sheets, which convert γ's from π^0 decays to e^+e^- pairs. PWC: proportional wire chamber; 72D18: spectrometer magnet (not used for neutrals). (Reprinted with permission.)

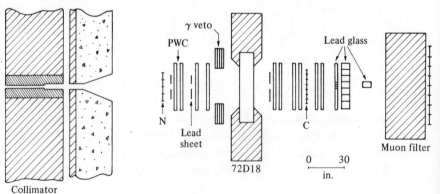

7.6 Measurements of η^{+-} and η^{00}

the spectrometer magnet labeled 72D18 in the figure. The pions were then momentum-analyzed in the PWC array. A large background caused by contamination decays $\Lambda^0 \to p\pi^-$ was minimized by discarding events with a fast proton and, in later analysis, by eliminating events where the reconstructed mass of the decaying particle was close to $m_\Lambda = 1115$ MeV/c^2.

The decay intensity versus proper time is shown in Figure 7.7. Interference shows up clearly as a deviation from the exponential decay law. Its form may be derived from the time development of K_L^0 and K_S^0 states in the beam. The high-energy proton target produces mostly K^0s, along with a smaller number of $\bar{K}^0$s, which are made in hadronic showers. The kaons are incoherent, and the beam may be represented by the state

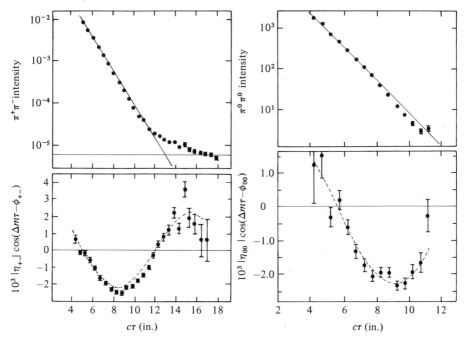

Figure 7.7. Intensity of $\pi^+\pi^-$ and $\pi^0\pi^0$ as a function of proper time τ. The exponential decay is modified by a sinusoidal term shown in the lower half of the diagrams. The curve through the lower data is a best fit to $|\eta|$ and ϕ. (From Christenson et al. 79. Reprinted with permission.)

$$|\psi\rangle = N_K|K_0\rangle + N_{\bar{K}}|\bar{K}^0\rangle$$

$$= N_K \frac{1}{\sqrt{2}} (1 - \epsilon)[|K_S\rangle e^{-(im_S + \gamma_S/2)\tau} + |K_L\rangle e^{-(im_L + \gamma_L/2)\tau}]$$

$$+ N_{\bar{K}} \frac{1}{\sqrt{2}} (1 + \epsilon)[|K_L\rangle e^{-(im_L + \gamma_L/2)\tau} - |K_S\rangle e^{-(im_S + \gamma_S/2)\tau}] \quad (7.70)$$

where $N_{K,\bar{K}}$ are the fractions of K and $\bar{K}$ particles produced. The $\pi\pi$ decay amplitude is

$$\langle \pi\pi|H_{wk}|\psi(\tau)\rangle = \frac{1}{\sqrt{2}} \langle \pi\pi|H_{wk}|K_S\rangle [(1 - \epsilon)N_K(e^{-(im_S + \gamma_S/2)\tau}$$

$$+ \eta e^{-(im_L + \gamma_L/2)\tau})$$

$$+ (1 + \epsilon)N_{\bar{K}}(\eta e^{-(im_L + \gamma_L/2)\tau} - e^{-(im_S + \gamma_S/2)\tau})] \quad (7.71)$$

Squaring this, we find the decay rate for K^0 and $\bar{K}^0$:

$$\Gamma(\pi\pi) = |\langle \pi\pi|H_{wk}|K_S\rangle|^2 R(t)(N_K + N_{\bar{K}}) \quad (7.72)$$

where

$$R(t) = |\eta|^2 e^{-\gamma_L \tau} + e^{-\gamma_S \tau} + \frac{2N_K - N_{\bar{K}}}{N_K + N_{\bar{K}}} |\eta| e^{-(\gamma_L + \gamma_S)\tau/2} \cos(\Delta m t + \phi) \quad (7.73)$$

The cosinusoidal variation may be used to determine ϕ, since Δm is known to high precision from other experiments (Section 7.7). From the data shown in Figure 7.5, Christenson et al. found these results:

$$\begin{aligned} |\eta^{00}| &= (2.33 \pm 0.18) \times 10^{-3}, & \phi^{00} &= 55.7° \pm 5.8° \\ |\eta^{+-}| &= (2.27 \pm 0.12) \times 10^{-3}, & \phi^{+-} &= 41.7° \pm 3.5° \end{aligned} \quad (7.74)$$

Cancellation of systematic errors common to both experiments reduces the uncertainty in the ratio η^{00}/η^{+-}:

$$|\eta^{00}/\eta^{+-}| = 1.00 \pm 0.09, \quad \Delta\phi \equiv \phi^{00} - \phi^{+-} = 12.6° \pm 6.2° \quad (7.75)$$

The current world average for η^{00} is heavily influenced by this experiment, which contains most of the known and analyzed $K_L \to \pi^0\pi^0$ decays. The averages are (Kelly et al. 80)

$$\left|\frac{\eta^{00}}{\eta^{+-}}\right| = 1.023 \pm 0.26, \quad \phi^{+-} = 44.6° \pm 1.2°,$$

$$\Delta\phi = 9.8° \pm 5.3°, \quad \phi^{00} = 54.5° \pm 5.3° \quad (7.76)$$

7.7 Regeneration and the $K_L - K_S$ mass difference

In Section 7.2 we briefly described the phenomenon of regeneration, where a pure K_L^0 beam is converted to a mixture of K_L^0 and K_S^0 as it passes through matter. We shall now consider this in greater detail. Let $|\psi\rangle$ denote a two-component neutral kaon state, as in (7.18).

7.7 Regeneration and the $K_L - K_S$ mass difference

The change in $|\psi\rangle$ with respect to proper time τ as the beam propagates through matter has vacuum and nuclear components:

$$\frac{d\psi}{d\tau} = \frac{d\psi}{d\tau}\bigg|_{\text{vac}} + \frac{d\psi}{d\tau}\bigg|_{\text{nuc}} = \frac{1}{\beta\gamma}\frac{d\psi}{dz} \tag{7.77}$$

where β and γ are the Lorentz transformation coefficients v/c and $(1 - v^2/c^2)^{-1/2}$, respectively. The indexes of refraction n and $\bar{n}$ for K^0 and $\bar{K}^0$ can be expressed in terms of the forward-scattering amplitudes $f(0)$ and $\bar{f}(0)$ as

$$n = 1 + (2\pi N/k^2)f(0) \tag{7.78a}$$
$$\bar{n} = 1 + (2\pi N/k^2)\bar{f}(0) \tag{7.78b}$$

where N is the number of scattering centers per unit volume and k the wave number. We have

$$\frac{d\psi}{dz}\bigg|_{\text{nuc}} = ik\begin{pmatrix} n-1 & 0 \\ 0 & \bar{n}-1 \end{pmatrix}\psi = \frac{2\pi i N}{k}\begin{pmatrix} f(0) & 0 \\ 0 & \bar{f}(0) \end{pmatrix}\psi \tag{7.79}$$

or, in units where $c = 1$,

$$\frac{d\psi}{d\tau}\bigg|_{\text{nuc}} = \frac{2\pi Nvi}{k(1-v^2)^{1/2}}\begin{pmatrix} f(0) & 0 \\ 0 & \bar{f}(0) \end{pmatrix}\psi \tag{7.80}$$

Now

$$\frac{d\psi}{d\tau}\bigg|_{\text{vac}} = -(\Gamma + iM)\psi$$

We define new matrices Γ' and M' by

$$\frac{d\psi}{d\tau} = -(\Gamma' + iM')\psi$$
$$= -(\Gamma + iM)\psi + \frac{2\pi Nvi}{k(1-v^2)^{1/2}}\begin{pmatrix} f & 0 \\ 0 & \bar{f} \end{pmatrix}\psi \tag{7.81}$$

or

$$\Gamma' + iM' = \Gamma + iM - i\alpha_0\begin{pmatrix} f & 0 \\ 0 & \bar{f} \end{pmatrix}$$

where

$$\alpha_0 = 2\pi Nv/k(1-v^2)^{1/2}$$

From now on we assume *CPT* invariance but not *CP* invariance. In this case,

$$\Gamma + iM = D + iE_1\sigma_1 + iE_2\sigma_2 \tag{7.82}$$

It is then straightforward to diagonalize the matrix $\Gamma' + iM'$ and find its eigenvalues and eigenvectors. Defining these by

$$(\Gamma' + iM')|K'_S\rangle \equiv \lambda'_S|K'_S\rangle \equiv \left(im'_S + \frac{\gamma'_S}{2}\right)|K'_S\rangle$$

$$(\Gamma' + iM')|K'_L\rangle \equiv \lambda'_L|K'_L\rangle \equiv \left(im'_L + \frac{\gamma'_L}{2}\right)|K'_L\rangle$$

we find that

$$|K'_S\rangle = [2(1 - |\epsilon'_1|^2)^{-1/2}][(1 + \epsilon'_1)|K^0\rangle - (1 - \epsilon'_1)|\bar{K}^0\rangle] \quad (7.83a)$$

and

$$|K'_L\rangle = [2(1 + |\epsilon'_2|^2)^{-1/2}][(1 + \epsilon'_2)|K^0\rangle + (1 - \epsilon'_2)|\bar{K}^0\rangle] \quad (7.83b)$$

where

$$\epsilon'_1 = \epsilon - \frac{i\pi N\Lambda}{k}\frac{f - \bar{f}}{\frac{1}{2} + i\mu} \quad (7.84)$$

$$\epsilon'_2 = \epsilon + \frac{i\pi N\Lambda}{k}\frac{f - \bar{f}}{\frac{1}{2} + i\mu} \quad (7.85)$$

$$\mu = \frac{m'_S - m'_L}{\gamma'_S - \gamma'_L} \quad (7.86)$$

$$\Lambda = \frac{v}{(1 - v^2)^{1/2}}\frac{1}{(\gamma'_S - \gamma'_L)} \quad (7.87)$$

In (7.83)–(7.87) all quadratic terms in ϵ'_1, ϵ'_2, and $\alpha_0(f \pm \bar{f})$ are neglected. Of course, $\gamma'_S \gg \gamma'_L$ since γ'_S and γ'_L and not very different from their respective values in vacuum. Therefore, to a good approximation, Λ is simply the mean decay length of the short-lived kaon in matter.

Let us express the states $|K'_S\rangle$ and $|K'_L\rangle$ of (7.83a,b) in terms of the vacuum states $|K_S\rangle$ and $|K_L\rangle$. Neglecting quadratic terms in ϵ, ϵ'_1, and ϵ'_2, we have

$$|K'_S\rangle \simeq \frac{1}{\sqrt{2}}[(1 + \epsilon - r)|K^0\rangle - (1 - \epsilon + r)|\bar{K}^0\rangle]$$

$$= \frac{1}{\sqrt{2}}[(1 + \epsilon)|K^0\rangle - (1 - \epsilon)|\bar{K}^0\rangle - r(|K^0\rangle + |\bar{K}^0\rangle)] \quad (7.88)$$

$$\simeq |K_S\rangle - r|K_L\rangle$$

and similarly

$$|K'_L\rangle = |K_L\rangle + r|K_S\rangle \quad (7.89)$$

where

$$r(k) = -i\pi N\Lambda[f(0) - \bar{f}(0)]/k(i\mu + \tfrac{1}{2}) \quad (7.90)$$

The *regeneration parameter r* characterizes the regenerating power of a given medium for given k. It is important to notice that $r(k)$ is inde-

7.7 Regeneration and the $K_L - K_S$ mass difference

pendent of the thickness of the regenerator. In all real cases, $|r| \ll 1$ and terms quadratic in r may be neglected when analyzing regeneration effects.

In almost all regenerators, $r \gg \epsilon$, so that CP violation may be neglected. However, if $r \simeq \epsilon$, CP violation interferes with regeneration. This was first observed by Fitch et al. (65), who thereby demonstrated conclusively that CP symmetry is indeed violated. Prior to that time, it seemed possible to entertain alternative hypotheses for $K_L{}^0 \to \pi\pi$ that conserve CP. For example, it was imagined that, along with the two pions, a third particle was emitted with very small mass and energy or that one of the "pions" in $K_L{}^0 \to \pi\pi$ was a new type of particle with positive intrinsic parity. Such hypotheses were not immediately dismissed, perhaps because they had an important precedent: Pauli proposed neutrinos for similar reasons.

The Fitch experiment clearly showed that ϵ and r interfere and, therefore, that final states from K_S or K_L are identical. This also provided an absolute means to distinguish matter from antimatter. In general, almost any experiment performed in a world made of antimatter will give the same result as an equivalent experiment performed in a matter-filled world, provided we invert the definitions of left- and right-handedness to account for weak parity violation. However, in the antiworld, the relative signs of ϵ and r reverse, so that the total rate for the two-pion decay mode has an interference term with the opposite sign. The difference is shown in Figure 7.8 along with the results of Fitch et al. (65).

Regeneration can be used to measure the $K_L - K_S$ mass splitting Δm. We describe a technique called the *gap method*, which utilizes the interference between a K_L beam and a regenerated K_S beam when both pass through a second regenerator (see Figure 7.9).

Consider the regenerators R_1 and R_2 of the same material with thicknesses L_1 and L_2, respectively. These are separated by an air gap of variable length G. Let us analyze the wavefunction ψ at various points along the beam, retaining only terms of first order in the regeneration parameter r.

(a) Before entering R_1, the system is in a pure $K_L{}^0$ state,
$$\psi = K_L{}^0 = K_L' - rK_S' \tag{7.91}$$

(b) After a distance $L_1 = vl_1/(1 - v^2)^{1/2}$ in regenerator R_1,
$$\begin{aligned}\psi &= K_L' \exp(-\lambda_L' l_1) - rK_S' \exp(-\lambda_S' l_1) \\ &= (K_L + rK_S)\exp(-\lambda_L' l_1) - r(K_S - rK_L)\exp(-\lambda_S' l_1) \\ &\simeq K_L \exp(-\lambda_L' l_1) + rK_S[\exp(-\lambda_L' l_1) - \exp(-\lambda_S' l_1)]\end{aligned} \tag{7.92}$$

(c) After a gap distance $G = vg/(1 - v^2)^{1/2}$

$$\begin{aligned}\psi &= K_L \exp(-\lambda'_L l_1) \exp(-\lambda_L g) \\ &\quad + r K_S [\exp(-\lambda'_L l_1) - \exp(-\lambda'_S l_1)] \exp(-\lambda_S g) \\ &\simeq K'_L \exp(-\lambda'_L l_1) \exp(-\lambda_L g) \\ &\quad + r K'_S \{\exp(-\lambda_S g)[\exp(-\lambda'_L l_1) - \exp(-\lambda'_S l_1)] \\ &\quad - \exp(-\lambda_L g) \exp(-\lambda'_L l_1)\}\end{aligned} \quad (7.93)$$

(d) After a distance $L_2 = v l_2 (1 - v^2)^{1/2}$ in regenerator R_2,

$$\begin{aligned}\psi &= K'_L \exp(-\lambda'_L l_1) \exp(-\lambda_L g) \exp(-\lambda'_L l_2) + r K'_S \exp(-\lambda'_S l_2) \\ &\quad \times \{\exp(-\lambda_S g)[\exp(-\lambda'_L l_1) - \exp(-\lambda'_S l_1)] - \exp(-\lambda_L g) \\ &\quad \times \exp(-\lambda'_L l_1)]\end{aligned}$$

Figure 7.8. Plot of $|\epsilon + r|$ vs. r for the experiment of Fitch et al. (65), showing the predictions for total constructive (solid line) or destructive (dashed line) interference. (Reprinted with permission.)

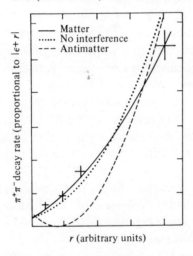

Figure 7.9. Gap method for determining Δm.

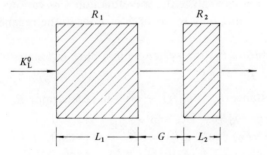

7.7 Regeneration and the $K_L - K_S$ mass difference

or

$$\psi = A_0 e^{-\lambda_L g}\{K_L + rK_S[B_0 + C_0 e^{-(\lambda_S - \lambda_L)g}]\} \qquad (7.94)$$

where A_0, B_0, and C_0 are all constants:

$$A_0 = \exp[-(\lambda'_L l_1 + \lambda'_L l_2)] \qquad (7.95)$$
$$B_0 = 1 - \exp[(\lambda'_L - \lambda'_S)l_2] \qquad (7.96)$$
$$C_0 = \exp[(\lambda'_L - \lambda'_S)l_2]\{1 - \exp[(\lambda'_L - \lambda'_S)l_1]\} \qquad (7.97)$$

It is now easy to see why a measurement of the $\pi^+\pi^-$ decay rate depends sensitively on g and leads to a determination of the mass difference Δm. Since $K_L \to \pi^+\pi^-$ violates CP invariance, the overwhelming contribution to the $\pi^+\pi^-$ decay rate comes from the K_S term on the right-hand side of (7.94), which is proportional to $B_0 + C_0 e^{-(\lambda_S - \lambda_L)g}$. The complex numbers B_0 and C_0 are represented in Figure 7.10, and for zero gap size $g = 0$, the K_S decay amplitude is proportional to the vector sum of B_0 and C_0. As the gap size increases, the vector $C_0 e^{-(\lambda_S - \lambda_L)g}$ decreases in length at a rate determined by $(\gamma_S - \gamma_L)$ and rotates in a sense determined by the sign of the mass difference and at a rate determined by its magnitude. (Note that for the appropriate choice of lengths L_1 and L_2, B_0 and C_0 are such that for a convenient gap length g the K_S amplitude is reduced to zero. Thus one finds a sharp minimum in the $\pi^+\pi^-$ decay rate as a function of increasing gap size. The method is obviously relatively insensitive to the magnitude and phase of r, and if CP were conserved, r could be disregarded.)

Although several experiments have used regenerators with a movable gap, it is also possible to fix the laboratory gap G and use the momentum spread of the kaon beam to vary g. Using this technique, an experiment performed at CERN has found (Geweniger et al. 74)

Figure 7.10. B_0 and C_0 from (7.95)–(7.96), represented in the complex plane. The sharp minimum in the vector sum $|B_0 + C_0|$ can be used to determine Δm.

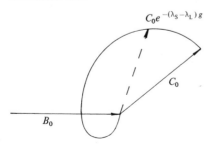

$$\Delta m = (0.534 \pm 0.003) \times 10^{10} \; \hbar c^{-2} \; \text{sec}^{-1} \tag{7.98}$$

The result depends on the K_S lifetime, which must be independently measured. The value used was

$$\gamma_S = 1.20 \times 10^{10} \; \text{sec}^{-1} \tag{7.99}$$

Another sensitive technique that has been used to measure Δm is the interference in the charge asymmetry of $K_L \to \pi^{\pm} l^{\mp} \overset{(-)}{\nu}$, which will be discussed in the following section.

7.8 CP violation in K_{l3} decays

In Section 4.13 we considered the rate $R^{\pm}$ for $K^0 \to \pi^{\mp} l^{\pm} \nu$, assuming CP conservation and found

$$R^{\pm}(t) = |1 + x|^2 e^{-\gamma_S \tau} + |1 - x|^2 e^{-\gamma_L \tau} \pm 2(1 - |x|^2)$$
$$\times e^{-(\gamma_S + \gamma_L)\tau/2} \cos \Delta m \tau - 4 \, \text{Im}(x) e^{-(\gamma_S + \gamma_L)\tau/2} \sin \Delta m \tau, \tag{7.100}$$

where x is the ratio of $\Delta S = -\Delta Q$ to $\Delta S = \Delta Q$ amplitudes. The charge asymmetry is defined as

$$\delta_c \equiv \frac{R^+ - R^-}{R^+ + R^-} \tag{7.101}$$

Direct substitution yields (for $\tau \gg 1/\gamma_S$)

$$\delta_c \equiv \frac{2(1 - |x|^2)}{|1 - x|^2} e^{-(\gamma_S + \gamma_L)\tau/2} \cos \Delta mt \tag{7.102}$$

To include the effects of CP violation, we repeat the steps of Section 4.13, assuming (7.48a,b) for K_L^0 and K_S^0. Keeping terms to order ϵ only, (7.102) is modified by the addition of another term

$$\delta_c = \frac{2(1 - |x|^2)}{|1 - x|^2} [\text{Re}(\epsilon) + A e^{-(\gamma_S + \gamma_L)\tau/2} \cos \Delta mt] \tag{7.103}$$

where A is the dilution factor

$$A = \frac{N_{K^0} - N_{\bar{K}^0}}{N_{K^0} + N_{\bar{K}^0}} \tag{7.104}$$

Figure 7.11 shows the results of a measurement of δ_c carried out at CERN (Gjesdal et al. 74). For proper time $\tau < 10^{-10}$ sec, the second term in (7.103) dominates. A measurement of Δm from this yields

$$\Delta m = (0.533 \pm 0.04) \times 10^{10} \; \hbar c^{-2} \; \text{sec}^{-1}, \tag{7.105}$$

comparable in magnitude and precision to Δm measurements made with the gap method of Section 7.7.

For $\tau \gg 10^{-10}$ sec, only the first term is important. This experiment then yields

7.9 CPT conservation in K^0 decay

$$\frac{2(1+|x|^2)}{|1-x|^2} \operatorname{Re}(\epsilon) = \begin{cases} (3.41 \pm 0.18) \times 10^{-3} & \text{for } K_L \to \pi^{\pm} e^{\mp}(\bar{\nu})_e \\ (3.13 \pm 0.29) \times 10^{-3} & \text{for } K_L \to \pi^{-} \mu^{\mp}(\bar{\nu})_\mu \end{cases}$$
(7.106)

The $(\Delta S = -\Delta Q)/(\Delta S = \Delta Q)$ fraction x can be determined separately. A world average then gives

$$\operatorname{Re}(\epsilon) = (1.621 \pm 0.088) \times 10^{-3} \tag{7.107}$$

Figure 7.11. Charge asymmetry in the decay $K^0 \to \pi^{\pm} e^{\pm}(\bar{\nu})$ as a function of proper time. The curve is a fit to (7.103). Both the K_L–K_S interference and the CP-violating offset are clearly shown. (From Gjesdal et al. 74. Reprinted with permission.)

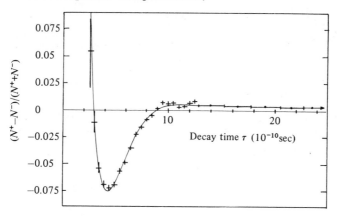

7.9 CPT conservation in K^0 decay

The equation of continuity must be valid for an arbitrary superposition of K_L^0 and K_S^0. Thus, we may rewrite (7.30) as follows:

$$-\frac{d}{dt}\langle K_S^0|K_L^0\rangle = 2\pi \sum_F \rho(F)\langle K_S^0|H_{wk}|F\rangle\langle F|H_{wk}|K_L^0\rangle \tag{7.108}$$

$$= \left[i(m_L - m_S) + \frac{\gamma_S + \gamma_L}{2}\right]\langle K_S^0|K_L^0\rangle \tag{7.109}$$

The right side of (7.108) may be simplified by consideration of possible final states F. The K_S^0 decays occur principally to two pions. All other K_S^0 decay modes and all K_L^0 decays are slower by at least three orders of magnitude. Furthermore, $(\pi\pi, I = 2)$ final states are suppressed according to the $|\Delta I| = \frac{1}{2}$ rule by a factor of at least 20 in the amplitude.

The only other states F that might contribute to this sum are $F = (\pi\pi\pi)$ or $(\pi l \nu)$, the main channels of K_L^0 decay. Current experiments limit all contributions to the sum other than $(\pi\pi, I = 0)$ to 10 percent or less (Schubert et al. 70). We may write

$$2\pi \sum_F \rho(F)\langle K_S^0|H_{wk}|F\rangle\langle F|H_{wk}|K_L^0\rangle$$
$$= 2\pi\rho(\pi\pi, I = 0)\langle K_S^0|H_{wk}|\pi\pi, I = 0\rangle\langle \pi\pi, I = 0|H_W|K_L^0\rangle$$
$$= 2\pi[\eta^{+-}\rho(\pi^+\pi^-)|\langle K_S^0|H_{wk}|\pi^+\pi^-, I = 0\rangle|^2 \quad (7.110)$$
$$+ \eta^{00}\rho(\pi^0\pi^0)|\langle K_S^0|H_{wk}|\pi^0\pi^0, I = 0\rangle|^2]$$
$$= [\eta^{+-}\gamma_S^{+-} + \eta^{00}\gamma_S^{00}]$$

where γ_S^{+-} and γ_S^{00} are the partial rates for $K_S^0 \to \pi^+\pi^-$ and $\pi^0\pi^0$, respectively. Combining (7.108)–(7.110), we obtain

$$\langle K_S^0|K_L^0\rangle = 2\left[\frac{\eta^{+-}\gamma_S^{+-} + \eta^{00}\gamma_S^{00}}{2i\,\Delta m + \gamma_S + \gamma_L}\right] \quad (7.111)$$

According to (7.52):

$$\langle K_S^0|K_L^0\rangle = 2\,\mathrm{Re}(\epsilon) - 2i\,\mathrm{Im}(\delta) \quad (7.112)$$

where $\epsilon(\delta)$ is zero if $T(CPT)$ is conserved. Equating (7.111) and (7.112), we find that *CPT* is conserved if the relative phase of the decay amplitude for K_L^0 and K_S^0 (represented by η^{+-} and η^{00}) is equal to the relative *propagation* phase $\phi_{K_L} - \phi_{K_S}$. In that case, the right-hand side of (7.111) is real.

Substituting the experimental values for the η's, γ's, and Δm, we find

$$\langle K_S^0|K_L^0\rangle = (3.30 \pm 0.16) \times 10^{-3}\,\exp i(3 \pm 2)° \quad (7.113)$$

Therefore, the *CPT*-violating interaction strength is at most a few percent of the *CPT*-conserving, *T*-violating interaction. This is one of the most stringent tests of *CPT* invariance in physics, and we stress that the argument rests on very few assumptions. We have used only the continuity equation, which depends on conservation of probability, the Hermiticity of the mass and decay Hamiltonians, and the principle of superposition.

Combining (7.112) and (7.113), we derive

$$\mathrm{Re}(\epsilon) = (1.65 \pm 0.08) \times 10^{-3} \quad (7.114)$$

This is in excellent agreement with the value derived from K_{l3} charge asymmetry (Section 7.8), which was arrived at altogether independently.

7.10 The phase of ϵ and the magnitude of Γ_{12} and M_{12}

Combining (7.111), (7.67), and (7.68), we obtain

$$\mathrm{Re}(\epsilon) = \frac{(\epsilon + \epsilon')\gamma_S^{+-} + (\epsilon - 2\epsilon')\gamma_S^{00}}{2i\,\Delta m + \gamma_S + \gamma_L} \quad (7.115)$$

7.11 *The phase of* ϵ'

We know that $\epsilon' \ll \epsilon$ because of the approximate validity of the $|\Delta I| = \frac{1}{2}$ rule. Furthermore, it so happens that $\gamma_S^{+-} \simeq 2\gamma_S^{00}$, so that ϵ' cancels out of (7.115). We therefore ignore ϵ' and obtain

$$\text{Re}(\epsilon) \simeq \frac{\epsilon \gamma_S}{2i\,\Delta m + \gamma_S + \gamma_L} \qquad (7.116)$$

Since $\gamma_L \ll \gamma_S$, this yields

$$\phi_\epsilon \simeq \tan^{-1}(2\,\Delta m/\gamma_S) = (43.7 \pm 0.2)° \qquad (7.117)$$

However, it is easy to show from (7.36) and (7.38a,b) that

$$\epsilon \simeq \frac{(\Gamma_{12}^* - \Gamma_{12}) + i(M_{12}^* - M_{12})}{(\gamma_S - \gamma_L) + 2i(m_S - m_L)} \qquad (7.118)$$

Therefore we conclude that $\text{Im}\,\Gamma_{12} \ll \text{Im}\,M_{12}$. This shows that ϵ arises primarily from mixing in the mass matrix, not the decay matrix. The same conclusion may be drawn from an explicit consideration of the terms in the sum for Γ_{12} [(7.31c)]. The experimental limit arrived at in this way is

$$\tan^{-1}(\text{Im}\,\Gamma_{12}/2\,\text{Im}\,M_{12}) < 8° \qquad (7.119)$$

7.11 The phase of ϵ'

If the $I = 2$ and $I = 0$ amplitudes for $K^0 \to 2\pi$ are out of phase, CP is violated and the size of the violation is given by the parameter ϵ', as discussed in Section 7.5. From (7.67) and (7.68), we find:

$$\epsilon' = \tfrac{1}{3}(\eta^{+-} - \eta^{00}) = \tfrac{1}{3}\eta^{+-}(1 - \eta^{00}/\eta^{+-}) \qquad (7.120)$$

From the current experimental values, we have

$$\begin{aligned}\text{Re}(1 - \eta^{00}/\eta^{+-}) &= 0.00 \pm 0.06 \\ \text{Im}(1 - \eta^{00}/\eta^{+-}) &= 0.17 \pm 0.09\end{aligned} \qquad (7.121)$$

The experimental evidence does not conclusively rule out $\epsilon' = 0$. The phase of ϵ' is given by (7.69) as:

$$\phi'_\epsilon = \pi/2 + \delta_2 - \delta_0 \qquad (7.122)$$

This has been determined by scattering experiments such as $\pi^- p \to n\pi^+\pi^-$ or $\pi^- p \to \Delta^{++}\pi^-\pi^-$, which yield (Kleinknecht 76)

$$\phi'_\epsilon = (37 \pm 5)° \qquad (7.123)$$

which lies outside the values allowed by (7.120) and (7.121), but by less than two standard deviations.

7.12 Summary of CP-violation experimental data

The experimental data for CP violation are summarized in Table 7.2. There is strong support for the conclusion that $\eta^{+-} = \eta^{00} = \epsilon$, $\epsilon' = 0$. One contradictory piece of evidence exists: $\phi^{00} - \phi^{+-} = (9.8 \pm 5.4)°$. This comes from a single experiment (Christenson et al. 1979) and has not yet been confirmed.

Table 7.2. CP *violation by experiment*[a]

Value	Source
$|\eta^{+-}| = (2.273 \pm 0.022) \times 10^{-3}$	$K_{\pi 2}^0$ decay
$|\eta^{00}| = (2.325 \pm 0.082) \times 10^{-3}$	$K_{\pi 2}^0$ decay
$\phi^{+-} = (44.6 \pm 1.2)°$	$K_{\pi 2}^0$ decay
$\phi^{00} = (54.5 \pm 5.3)°$	$K_{\pi 2}^0$ decay
$\phi_\epsilon = (43.7 \pm 0.2)°$	$\tan^{-1}(2\,\Delta m/\gamma_S)$
$\mathrm{Re}(\epsilon) = (1.621 \pm 0.088) \times 10^{-3}$	K_{l3}^0 decay
$|\epsilon| = (2.25 \pm 0.13) \times 10^{-3}$	From above two lines
$\phi'_\epsilon = (37 \pm 5)°$	$\pi\pi$ phase shifts
$|\epsilon'| = (0.17 \pm 0.10)|\eta^{+-}|$	From $\epsilon' = \frac{1}{3}(\eta^{+-} - \eta^{00})$

[a] Connecting relations: $\eta^{+-} = \epsilon + \epsilon'$ and $\eta^{00} = \epsilon - 2\epsilon'$.

7.13 Origins of CP violation

Many speculations have been offered since 1964 on the possible origins of CP violation in K_L^0 decay. Despite the high precision of the experimental results just summarized, few models have been ruled out. One difficulty in testing models is that CP violation has been seen only in the K^0–$\bar{K}^0$ system. Future measurements of the charmed meson system D^0–$\bar{D}^0$, and especially the bottom quark mesons B^0 and $\bar{B}^0$, may shed some light on the matter, but the experimental problems are formidable. For now, all we can do is summarize the theoretical problem.

CP violation in K_L^0 decay may be caused by one (or more) of the following:

  (i) a small T or C violation ($\sim 10^{-3}$) in the strong interaction (millistrong);
 (ii) a large T or C violation (~ 0.1 to 0.01) in the hadronic electromagnetic interaction;
(iii) a 10^{-3} T violation in the weak interaction (milliweak);
(iv) a $\Delta S = 2$ interaction, which is "superweak," i.e., $\sim 10^{-10} G_F$;
 (v) a CP-violating mixing, similar to Cabibbo mixing, between the

7.13 Origins of CP violation

s quark and the heavier quarks, c, b, and the hypothetical t; and

(vi) the existence of extra scalar (Higgs) bosons in nature, which violate CP.

7.13.1 Millistrong interaction

Here we presume that $K_L^0 \to 2\pi$ occurs via a second-order process, with amplitude

$$A \simeq \sum_F \frac{\langle K_L^0|H_{wk}|F\rangle\langle F|gH_S|2\pi\rangle}{m_K - E_F} \qquad (7.124)$$

where H_{wk} is the ordinary $|\Delta S| = 1$ CP-conserving weak interaction and F a hadronic state with quantum numbers $B = 0$, $S = 0$, $CP = -1$, for example, three pions. The number g is the CP-violating fraction of the strong Hamiltonian. To account for the observed rate for $K_L^0 \to 2\pi$, we must have

$$G_F g m_\pi/(m_K - E_F) \simeq 2 \times 10^{-3} \; G_F \qquad (7.125)$$

where we assume a typical size for H_S. Since the denominator is $\sim m_\pi$, we have $g \simeq 10^{-3}$, hence the name *millistrong*.

Now, we know that CPT is conserved in $K_L^0 \to 2\pi$, and it may also be established that parity is conserved in the strong and electromagnetic interaction at the level of 10^{-5}; otherwise parity violation in nuclear forces would be much larger than the observed size, which is roughly compatible with the weak interaction (see Chapter 9). It therefore follows that we require C and T violation in the strong interaction of order 10^{-3}.

T invariance is tested directly by comparison of differential cross sections for certain nuclear reactions and their inverse, for example,

$$d + {}^{24}\text{Mg} \rightleftarrows p + {}^{25}\text{Mg}$$
$$\alpha + {}^{24}\text{Mg} \rightleftarrows p + {}^{27}\text{Al}$$
$$d + {}^{16}\text{O} \rightleftarrows \alpha + {}^{14}\text{N}$$

These experiments show that T-odd amplitudes must be less than 3×10^{-3} of corresponding T-even amplitudes.

C invariance in the strong interaction has been tested by comparing the angular distributions for charge conjugate final states in $p\bar{p}$ annihilations in a hydrogen bubble chamber (Pais 59, Dobrzynski et al. 66):

$$p\bar{p} \to \bar{K}^0 K^+ \pi^- \qquad \text{vs.} \qquad p\bar{p} \to K^0 K^- \pi^+$$

or

$$p\bar{p} \to \bar{K}^0 K^+ \pi^0 \pi^- \qquad \text{vs.} \qquad p\bar{p} \to K^0 K^- \pi^0 \pi^+$$

Results confirm C invariance to 1 percent, and therefore from current experimental limits, one cannot rule out a millistrong interaction.

7.13.2 CP-violating electromagnetic interaction

The idea is the same as in the millistrong theory, except that the transition $F \to 2\pi$ is assumed to be electromagnetic. The strength of the required interaction, $\sim 10^{-3}$, is close to $\alpha/2\pi$. The theory requires a fairly strong CP violation, perhaps of order unity in electromagnetic decays and, for this reason, can be all but ruled out by existing experimental evidence, a status not enjoyed by the other theories discussed here.

Parity conservation is known to hold to very high precision in electromagnetic processes. Both C- and P-violating electromagnetic amplitudes are $< 10^{-12}$ of C- and P-conserving amplitudes in atoms, as tested by experiments in which searches are made for weak neutral currents (Chapter 9), and T and P violations are known to be still smaller. These would give rise to a permanent electric dipole moment (edm) in the neutron, electron, and proton. Current experimental limits are shown in Table 7.3. A large $[O(1)]$ C- and T-violating EM interaction coupled to a P- and C-violating weak interaction might give rise to an effective P and T violation, which is small only because of G_F. A rough estimate for the neutron edm would then be

$$\mu \simeq e \times \lambda_n \times G_F m_n^2 = (e\hbar/m_n c) G_F m_n^2 \simeq 10^{-19} \, e \quad \text{cm} \qquad (7.126)$$

More sophisticated calculations, which invoke C and T violation in the electromagnetic interaction to explain $K_L^0 \to 2\pi$, yield a neutron edm of order 10^{-23} e-cm, but these are now ruled out by experiment.

Searches for T and C violations in the electromagnetic interaction have been carried out in several other processes – and no violations have been found – at a level that seems to rule out EM-induced CP violation. These include:

Table 7.3. *Electric dipole moments*

Particle	edm (e-cm)	Reference
n	$<6 \times 10^{-25}$	Altarev et al. (81)
p	$<7 \times 10^{-21}$	Harrison et al. (69)
e	$<3 \times 10^{-24}$	Weisskopf et al. (68)

7.13 Origins of CP violation

(a) *Charge asymmetry in η decay.* It was predicted that C violation in the EM interaction would produce a charge asymmetry of order 5 percent in $\eta \to \pi^+\pi^-\gamma$ and $\eta \to \pi^+\pi^-\pi^0$ (Lee 65). Current limits are well below this. One charge asymmetry that may be observed is

$$\Delta = (N_+ - N_-)/(N_+ + N_-) \tag{7.127}$$

where $N_+(N_-)$ refer to the number of η decays in which the $\pi^+(\pi^-)$ possess more momentum than $\pi^-(\pi^+)$. For $\eta \to \pi^+\pi^-\pi^0$, the current limit is $\Delta = (1.2 \pm 1.7) \times 10^{-3}$ (Kelly et al. 80). For $\eta \to \pi^+\pi^-\gamma$, one experiment (Jane 74) yields $\Delta = (1.2 \pm 0.6) \times 10^{-2}$, but this has not been confirmed. The world average is $\Delta = (9 \pm 4) \times 10^{-3}$.

(b) *Detailed balance in photonuclear reactions.* Here one compares the relative magnitude and phase of a photonuclear reaction with its inverse. Experiments have been performed on $\gamma d \rightleftarrows pn$, $\pi^- p \rightleftarrows \gamma n$, and $\gamma\,^3$He $\rightleftarrows pd$. No T violation has been seen at the level of a few percent (reviewed in Kleinknecht 76).

(c) *Nuclear transitions.* Time-reversal invariance requires that the relative phase between two competing amplitudes be zero or π in an electromagnetic transition between two nuclear levels (for example, an E2, M1 mixed transition). Mössbauer observations of such "mixed" γ rays from ^{99}Ru (Kistner 67) and ^{193}Au (Atac et al. 68) indicate that a T-odd amplitude, if any, must be less than about 3×10^{-3} of the corresponding T-even amplitude.

(d) *Inelastic scattering.* Here one makes use of high-energy inelastic EM reactions with polarized proton targets, for example,

$$e + p \to \Delta^+ + e$$

to search for a correlation of the form

$$\sigma_p \cdot \mathbf{p}_{ei} \times \mathbf{p}_{ef}$$

A SLAC experiment finds no T violation $[(-0.3 \pm 1.3)\%]$ (Rock et al. 70).

7.13.3 CP violation in the weak interaction

Some theories propose that $K_L^0 \to 2\pi$ is a first-order weak process. In that case, the weak interaction must violate CP at the level of 10^{-3} (milliweak). This has been tested in β decay of free neutrons and ^{19}Ne, hyperon decays ($\Lambda^0 \to p\pi^-$), and $K_{\mu 3}$ decay. In neutron and ^{19}Ne decay, one searches for a T-violating correlation $\sigma_N \cdot \mathbf{p}_e \times \mathbf{p}_\nu$

(see Section 5.3). No evidence for T violation is found at a level of precision of one part in 10^3.

Observation of an imaginary part of the quantity x in $K_{\mu3}$ decay (Section 7.8) would be evidence for simultaneous violation of T and $\Delta S = \Delta Q$. No such violation is found:

$$\text{Im}(x) = (-1.38 \pm 2.4) \times 10^{-2}$$

T violation would also be indicated by a nonvanishing muon polarization transverse to the π-μ decay plane. The correlation is $\boldsymbol{\sigma}_\mu \cdot \mathbf{p}_\mu \times \mathbf{p}_\pi$. Here the result can be expressed in terms of $\text{Im}(\xi)$ (see Section 6.1). One finds

$$\text{Im}(\xi) = -0.016 \pm 0.025 \tag{7.128}$$

(Campbell et al. 81).

Finally, one can look for a T violation in the decay $\Lambda^0 \to p\pi^-$, proportional to $\boldsymbol{\sigma}_p \times \boldsymbol{\sigma}_\Lambda \cdot \mathbf{p}_p$. This is sensitive to the phase between s- and p-wave decay amplitudes. No phase difference is observed in excess of the expected (and separately measured) final-state interaction phase shift of $(7 \pm 1)°$ (see Section 6.5).

7.13.4 Superweak CP violation

This theory postulates a $\Delta S = 2$ interaction ($K^0 \to \bar{K}^0$), which, in the (K_L, K_S) basis, produces $K_L^0 \to K_S^0$ (Wolfenstein 64). Then $K_L^0 \to 2\pi$ proceeds via the intermediate state K_S^0, which decays weakly to two pions. The amplitude is

$$A \simeq \frac{\langle \pi\pi | H_{\text{wk}} | K_S^0 \rangle \langle K_S^0 | H_{\text{Swk}} | K_L^0 \rangle}{\Delta m} \tag{7.129}$$

which must be equated to

$$A \simeq 10^{-3} \langle \pi\pi | H_{\text{wk}} | K_S^0 \rangle \tag{7.130}$$

in order to account for CP violation. Inserting the experimental value of Δm, we find that only an extraordinarily minute coupling H_{Swk} is required:

$$H_{\text{Swk}} \simeq 2 \times 10^{-10} H_{\text{wk}} \tag{7.131}$$

The reason is that K_L^0 and K_S^0 are nearly degenerate in mass, so that even a tiny perturbation can cause a substantial effect. If this is the cause, however, it may not manifest itself anywhere else in nature.

In the superweak model, CP violation arises entirely from the K^0–$\bar{K}^0$ mass matrix, and ϵ' is predicted to be zero. Therefore we have

$$\eta^{+-} = \eta^{00} = \epsilon \tag{7.132}$$

and

7.13 Origins of CP violation

$$\phi^{+-} = \phi^{00} = \tan^{-1}(2\,\Delta m/\gamma_S) \qquad (7.133)$$

Of course, these predictions are supported by experiment. However, ϵ' is also expected to be small in other models, since the $I = 2$ final states that give rise to ϵ' are suppressed by the $|\Delta I| = \tfrac{1}{2}$ rule. Finally, some authors have attempted to incorporate the superweak model into unified gauge theories (see, for example, Mohapatra et al. 75).

7.13.5 CP violation in gauge theories

In recent years, attention has been focused on attempts to explain CP violation in terms of the standard model or extensions of it. Many ideas have been proposed, with two principal schemes most widely discussed. In one of these, originally suggested by Kobayashi and Maskawa (73), CP violation comes about by means of quantity δ in the six-quark mixing matrix (4.70). The model can be made to reproduce the superweak predictions for K_L^0 decay and has the advantage that no particles are needed, other than those already appearing in the standard model. The second major alternative postulates new Higgs bosons (in addition to the one required to impart mass to W and Z). In such theories, CP violation is generated without specific appeal to heavy quarks.

(a) *CP violation in the Kobayashi–Maskawa (KM) model.* Let us recall the six-quark scheme of Section 4.3, according to which the charged hadronic weak current takes the form:

$$J_\lambda^\dagger = (\bar{u}, \bar{c}, \bar{t})\gamma_\lambda(1 - \gamma_5)$$

$$\times \begin{pmatrix} c_1 & s_1 c_3 & s_1 s_3 \\ -s_1 c_2 & c_1 c_2 c_3 - s_2 s_3 e^{i\delta} & c_1 c_2 c_3 + s_2 c_3 e^{i\delta} \\ -s_1 s_2 & c_1 s_2 c_3 + c_2 s_3 e^{i\delta} & c_1 s_2 s_3 - c_2 c_3 e^{i\delta} \end{pmatrix} \begin{pmatrix} d \\ s \\ b \end{pmatrix} \qquad (7.134)$$

We shall now explain how δ contributes to CP violation by means of several mechanisms, which are shown in Figures 7.12 and 7.13 (Ellis et al. 76b). Figure 7.12 is the same as Figure 7.5, except for the inclusion of the t quark in addition to the c and u quarks. Therefore, as shown in Section 7.3, contributions are made to the off-diagonal matrix element M_{12}. However, whereas $\text{Re}(M_{12})$ is proportional to $\Delta m = m_L - m_S$, $\text{Im}(M_{12})$ contributes to the CP-violating parameter ϵ [(7.118)]. Figure 7.13 is related to $\text{Im}\,\Gamma_{12}$, which also contributes to ϵ. From the experimental results, we know that $\text{Im}\,\Gamma_{12} \ll \text{Im}\,M_{12}$; therefore the amplitude corresponding to Figure 7.13 must be greatly suppressed, and this is a feature that should appear in our explanation.

First, we ignore $\text{Im}\,\Gamma_{12}$ and assume that $|\epsilon| \simeq \text{Im}\,M_{12}/\Delta m$. Our task is then to calculate this ratio from Figure 7.12. First, we reduce that figure to the four-fermion diagrams of Figure 7.14 in the limit $m_W \to \infty$. The amplitude corresponding to Figure 7.14a is

$$\mathcal{M} = \frac{G_F^2}{2} \int \frac{d^4k}{(2\pi)^4} \sum_{i,j} a_i a_j \bar{u}_d \gamma_\lambda (1 - \gamma_5) \frac{\not{k} + m_i}{k^2 - m_i^2} \gamma^\lambda (1 - \gamma_5) u_s$$

$$\times \bar{v}_d \gamma_\sigma (1 - \gamma_5) \frac{\not{k} + m_j}{k^2 - m_j^2} \gamma^\sigma (1 - \gamma_5) v_s \qquad (7.135)$$

where $i,j = u, c, t$ and where $a_{i,j}$ are the K–M matrix elements in (7.134) for transitions of the form $s \to i \to d$:

$$a_1 = +c_1 s_1 c_3 \qquad (7.136)$$
$$a_2 = -s_1 c_2 (c_1 c_2 c_3 - s_2 s_3 e^{i\delta}) \qquad (7.137)$$
$$a_3 = -s_1 s_2 (c_1 s_2 c_3 + c_2 s_3 e^{i\delta}) \qquad (7.138)$$

The expression corresponding to Figure 7.14b is identical. To obtain ϵ, we separate the integral into real and imaginary parts and evaluate

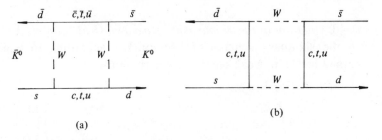

Figure 7.12. Second-order contributions to $K^0 \to \bar{K}^0$ as in Figure 7.5, except that the t quark is now included.

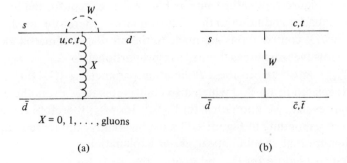

Figure 7.13. First-order contributions to $K^0 \to \bar{K}^0$.

7.13 Origins of CP violation

them by means of contour integration, assuming $m_{c,t}^2 \gg m_u^2 \simeq 0$. Thus we find:

$$\text{Im}(M_{12})/\text{Re}(M_{12}) = 2s_2 c_2 s_3 \sin \delta \, P(\theta_2, \eta) \quad (7.139)$$

where

$$\eta = m_c^2/m_t^2$$

and

$$P(\theta_2, \eta) = \frac{s_2^2\{1 + [\eta/(1-\eta)] \ln \eta\} - c_2^2\{\eta - [\eta/(1-\eta)] \ln \eta\}}{c_1 c_3 \{c_2^4 \eta + s_2^2 - 2s_2^2 c_2^2 [\eta/1-\eta)] \ln \eta\}} \quad (7.140)$$

Unfortunately, the various parameters that enter into these expressions are not well known, so a firm value of ratio (7.139) cannot be obtained. All that can be said is that this model yields CP violation with approximately the right magnitude (Ellis et al. 76b, Shrock and Treiman 79).

Next we turn to the amplitudes associated with Figure 7.13. Those corresponding to c and t intermediate states tend to cancel, since

$$\text{Im } \mathcal{M}(s \to c \to d) \propto +s_1 s_2 s_3 c_2 \sin \delta \quad (7.141)$$

and

$$\text{Im } \mathcal{M}(s \to t \to d) \propto -s_1 s_2 s_3 c_2 \sin \delta. \quad (7.142)$$

The cancellation is imperfect, however, because of the $t - c$ mass dif-

Figure 7.14. Four-fermion limits of Figure 7.12, as $m_W \to \infty$.

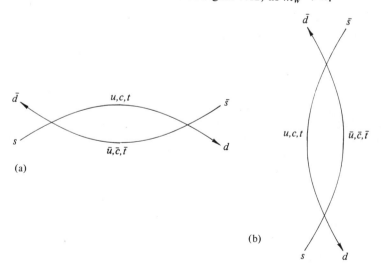

ference, which must be quite large: $m_t > 10 m_c$. Figure 7.13a is a penguin diagram, and as we know from Section 6.6, Feynman graphs of this type have been proposed as the dominant contribution to non-leptonic decay amplitudes and may thus explain the $|\Delta I| = \tfrac{1}{2}$ rule. Assuming this is true, the CP-violation contribution of Figure 7.13a is

$$\left|\frac{\text{Im(penguin diagram)}}{\langle \pi\pi, I = 0|H_W|K_0\rangle}\right| = \left|\frac{\text{Im(penguin diagram)}}{\text{Re(penguin diagram)}}\right| \equiv \epsilon_{\text{penguin}} \qquad (7.143)$$

The actual calculation is quite complicated (see Guberina and Peccei 80), but we can see the general features of the answer by making the approximation $m_W \gg m_{c,t} \gg m_u$. Then Figure 7.13a becomes Figure 7.15, which is logarithmically divergent with the W mass m_W as the relevant cut off.

$$\mathcal{M} \propto G_F \alpha_S \left(a_1 \ln \frac{m_W^2}{k^2} + a_2 \ln \frac{m_W^2}{m_c^2} + a_3 \ln \frac{m_W^2}{m_t^2}\right) \qquad (7.144)$$

where k is the gluon momentum transfer and we make use of the fact that $m_u \ll k \simeq m_{K^0}$, but $m_{c,t} \gg k$. The desired ratio is then (Gilman and Wisc 79) approximately

$$|\epsilon_{\text{penguin}}| \simeq s_2 c_2 s_3 \sin \delta \frac{\ln(m_t^2/k^2) - \ln(m_c^2/k^2)}{c_2^2 \ln(m_c^2/k^2) + s_2^2 \ln(m_t^2/k^2)} \qquad (7.145)$$

Figure 7.15. "Penguin" diagram contribution to CP violation.

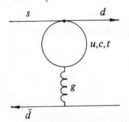

Figure 7.14b requires that $c\bar{c}$ or $t\bar{t}$ annihilate into gluons in the final state. This type of process is suppressed by Zweig's rule, which forbids final-state hadrons whose quark lines were all created inside the diagram. The suppression is caused by the dynamics of the strong interaction, details of which are not understood. Nevertheless we expect the rule to hold here and to reduce the amplitude in Figure 7.14b by a factor of 0.1 to 0.01.

Both of these diagrams involve an effective $s \to d$ transition, which is strictly $|\Delta I| = \tfrac{1}{2}$. In the KM model, there is no direct CP-

7.13 Origins of CP violation

nonconserving contribution to $|\Delta I| = \frac{3}{2}$ amplitudes A_2. However, imaginary contributions to the $I = 0$ decay amplitude A_0 will cause a phase shift between A_0 and A_2 and contribute to ϵ', defined in (7.69). Since we defined A_0 to be real in (7.58a,b), we must now transform the phases of $|K^0\rangle$ and $|\bar{K}^0\rangle$. If the first-order diagrams contribute an imaginary fraction to A of

$$A_0 = A_0^{\delta=0} + i\epsilon_r A_0^{\delta=0} = A_0 e^{i\xi} \tag{7.146}$$

where $\epsilon_r = \epsilon_{\text{penguin}} +$ other first-order contributions, then we must transform A_0 by

$$A_0 \to A_0^{\delta=0} \tag{7.147}$$

At the same time, we must transform

$$\epsilon \to \epsilon + 2\xi \tag{7.148}$$

$$\epsilon' \to \frac{1}{\sqrt{2}} |\xi| \left|\frac{A_2}{A_0}\right| \tag{7.149}$$

$$\left|\frac{\epsilon'}{\epsilon}\right| \simeq \frac{1}{20} \left|\frac{2\xi}{\epsilon}\right| \tag{7.150}$$

Finally, let us consider the neutron electric dipole moment according to this model of CP violation. It is easy to show that no contribution to the edm can arise at the one-loop level. Ellis, Gaillard, and Nanopoulos (76b) considered various nonvanishing diagrams at the two-loop level but did not recognize that the total contribution of these diagrams is zero, as was later shown by Shabalin (79). Thus it turned out that at the two-loop level there is no contribution to the neutron edm. Subsequently, however, Deshpande, Eilam, and Spence (82) demonstrated that a different contribution of order G_F^2 exists (see also Ellis and Gaillard 79 and Shabalin 80). On the other hand, this contribution is extremely small, of order 10^{-32} e-cm and not observable with present or foreseeable experimental methods.

(b) CP *violation with extra Higgs bosons.* In models of this type (see, for example, Weinberg 76b, Anselm and Dyakonov 78, Deshpande and Ma 78), it is assumed that there exist additional $SU(2)$ complex scalar doublets. Since $W^{\pm}$ and Z^0 already have mass because of the original scalars, the new scalar fields (charged as well as neutral) cannot be eliminated by a gauge transformation. CP violation may be introduced in a variety of ways here: in the Yukawa couplings to fermions or in the scalar self-couplings. The Yukawa couplings of the quarks to these new fields include terms corresponding to charged Higgs exchange:

$$A(\bar{N}_R m_n U P_L) \cdot (\bar{P}_R m_P U N_L) + \text{h.c.} \tag{7.151}$$

where the constant A may be complex. This yields CP violation, and leads to estimates of the neutron edm of order 10^{-23} e-cm, which is already contradicted by experiment. The estimates can be revised downward somewhat, but if experiment shows that the neutron edm is less than 10^{-25} e-cm, the Higgs models of CP violation would be in serious difficulty.

We shall conclude this section with a very brief mention of several other theoretical possibilities. In left–right symmetric models such as those discussed in Section 2.12, one may arrange the left- and right-handed coupling coefficients to be relatively complex, which results in CP violation (see, e.g., Mohapatra et al. 75). The size of the effect can be adjusted by choosing the ratio m_{WR}/m_{WL}. In general such models lead to fairly large neutron edm estimates. Some gauge theories introduce CP violation into flavor-changing *neutral* currents by proposing the existence of extra heavy neutral vector or scalar bosons. For example (Lee 73, Sikivie 76), if we couple s and d quarks to new fields Z' or h' by

$$gZ'_\mu[\bar{d}\gamma^\mu(c + \gamma_5 c')s] \tag{7.152}$$

or

$$gh'[\bar{d}(c + \gamma_5 c')s] \tag{7.153}$$

where c and c' are complex, then the K^0–$\bar{K}^0$ mass matrix will acquire off-diagonal terms with imaginary components:

$$\text{Im } M_{12} \simeq \frac{g^2}{m^2_{(Z',h')}} \text{Im}(c^2 + c'^2) m_K^3 \tag{7.154}$$

From

$$|\epsilon| \simeq \text{Im } M_{12}/2 \, \Delta m \simeq 10^{-3} \tag{7.155}$$

we deduce

$$(m_Z^2/m_{Z',h'}^2) \text{Im}(c^2 + c'^2) \simeq 10^{-10} \tag{7.156}$$

so the interaction is superweak. Such theories have small neutron edms but can lead to large CP-violating effects in D^0–$\bar{D}^0$ or B^0–$\bar{B}^0$.

As a final example, we mention the possibility that in quantum chromodynamics there exists a CP-odd Lagrangian term

$$(\theta/32\pi^2)\epsilon_{\mu\nu\rho\sigma}F^{a,\mu\nu}F^{a,\rho\sigma} \tag{7.157}$$

where ϵ is the completely antisymmetric tensor, F^a the Yang–Mills field tensor, and θ an arbitrary parameter that must be very small ($<10^{-8}$) for consistency with the neutron edm limit. A new type of pseu-

7.14 CP violation in heavy-quarked mesons D^0, B^0, T^0

doscalar particle, the *axion*, was suggested to eliminate this term altogether, but it has not been observed, and some theorists have speculated that CP violation arises from this source.

7.14 CP violation in heavy-quarked mesons D^0, B^0, T^0

The K^0–$\bar{K}^0$ system may never provide enough experimental information to distinguish among the various theories offered to explain CP violation. A search for alternative systems in which the effect might be observed leads us to consider other hadron state pairs having opposite CP eigenvalues which are close in mass. These include neutral pseudoscalar mesons listed in Table 7.4.

Mass mixing and CP violation in heavy-meson systems may be described by a generalization of the formalism developed for the K mesons. Thus, we write the Hamiltonian for the M^0–$\bar{M}^0$ system as $H = M - i\Gamma$, where M and Γ are Hermitian. Assuming CPT invariance, H takes the form

$$H = \begin{pmatrix} M_{11} - i\Gamma_{11} & M_{12} - i\Gamma_{12} \\ M_{12}^* - i\Gamma_{12}^* & M_{11} - i\Gamma_{11} \end{pmatrix} \quad (7.158)$$

Just as before we may write expressions for the states of definite mass and lifetime, M_S^0 and M_L^0, in terms of the states M^0 and $\bar{M}^0$:

$$\begin{pmatrix} M_S^0 \\ M_L^0 \end{pmatrix} = [2(1 + |\epsilon|^2)]^{-1/2} \begin{pmatrix} 1 + \epsilon & \epsilon - 1 \\ 1 + \epsilon & 1 - \epsilon \end{pmatrix} \begin{pmatrix} M^0 \\ \bar{M}^0 \end{pmatrix} \quad (7.159)$$

However, the subscripts S and L no longer necessarily mean short and long, respectively, for we may have $\gamma_S < \gamma_L$ in some of the cases in Table 7.4. In fact, all of the mesons in Table 7.4 decay so quickly that it is quite impossible to study mass-mixing effects in neutral beams, as was feasible for K^0–$\bar{K}^0$. All that can be observed is the integral over the decay history of all the particles. Thus we consider a pure meson state M^0 at time $t = 0$, which evolves in time. We then define the integrated mass-mixing fraction by the formula:

Table 7.4. *Heavy neutral pseudoscalar mesons*

Meson	$q\bar{q}$ bound state	Mass (GeV/c^2)
D^0–$\bar{D}^0$	$c\bar{u}$–$\bar{c}u$	1.863
B_d^0–$\bar{B}_d^0$	$b\bar{d}$–$\bar{b}d$	5.2 ?
B_s^0–$\bar{B}_s^0$	$b\bar{s}$–$\bar{b}s$	5.5 ?
T_u^0–$\bar{T}_u^0$	$t\bar{u}$–$\bar{t}u$	>15
T_c^0–$\bar{T}_c^0$	$t\bar{c}$–$\bar{t}c$	>15

$$\rho = \frac{\int_0^\infty \langle \overline{M}^0 | \psi(t) \rangle \, dt}{\int_0^\infty \langle M^0 | \psi(t) \rangle \, dt}$$

$$= \frac{4(\Delta m/\gamma)^2 + (\Delta\gamma/\gamma)^2}{2 + 4(\Delta m/\gamma)^2 - (\Delta\gamma/\gamma)^2} \tag{7.160}$$

where the second expression on the right-hand side is easily derived and where Δm and $\Delta\gamma$ are given by the expressions:

$$\Delta\gamma = \gamma_S - \gamma_L = 2\,\mathrm{Re}[(M_{12} - i\Gamma_{12})(M_{12}^* - i\Gamma_{12}^*)]^{1/2} \tag{7.161}$$

$$\Delta m = m_S - m_L = -2\,\mathrm{Im}[(M_{12} - i\Gamma_{12})(M_{12}^* - i\Gamma_{12}^*)]^{1/2} \tag{7.162}$$

In the case of kaons, $\Delta\gamma \simeq \gamma \simeq \Delta m$, which means that $\rho_K \simeq 1$ and the mixing is essentially complete. However, in a heavy meson system, the huge increase in phase space makes γ quite enormous, and therefore we anticipate that $\Delta m \ll \gamma$ and $\Delta\gamma \ll \gamma$, thus $\rho \ll 1$.

We next attempt to estimate the various mass differences for the pairs listed in Table 7.4. In the case of neutral kaons, we have shown that

$$\Delta m \simeq \frac{G_F^2}{4\pi^2} f_K^2 m_c^2 m_K \sin^2\theta_C \cos^2\theta_C \tag{7.163}$$

An extension of these arguments to the mesons of Table 7.4 leads to the estimates:

$$\Delta m(D^0) \propto (m_s^2 - m_d^2) s_1^2 c_1^2 m_D \tag{7.164}$$

$$\Delta m(B_d^0) \propto (m_t^2 + m_c^2) s_1^2 s_2^2 m(B_d^0) \cos 2\delta \tag{7.165}$$

$$\Delta m(B_s^0) \propto (m_t^2 + m_c^2)(s_3 + s_2 c_\delta)^2 m(B_s^0) \tag{7.166}$$

The ratio $\Delta m_D/\Delta m_K$ can be estimated if we assume $f_D \simeq f_K$, in which case we find

$$\frac{\Delta m_D}{\Delta m_K} \simeq \frac{f_D^2}{f_K^2} \frac{m_D}{m_K} \frac{m_s^2}{m_c^2} \simeq 0.04 \tag{7.167}$$

which yields

$$\Delta m_D/\gamma_D \simeq 10^{-3} \tag{7.168}$$

The rate difference $\Delta\gamma$ is also small for heavy neutral meson systems. It acquires contributions from those final states that are CP eigenstates, that is, have zero net flavor when summed over the quarks in all of the final-state hadrons, and these can be reached by only one of the two CP eigenstates M_1^0 or M_2^0. The main decay mode of K_S^0 is a $CP = +1$ eigenstate, so $(\Delta\gamma/\gamma)_K \simeq 1$. By contrast, pure CP decay

7.14 CP violation in heavy-quarked mesons D^0, B^0, T^0

modes of heavier mesons are a small fraction of the total. The lone exception may be B_s^0, which should have a relatively large decay probability to CP eigenstates $F^{\pm}$, $D^{\pm}$, and $D^0\bar{D}^0$. In D and T decays, the pure CP final states are suppressed by mixing angles, and in all heavy meson decays, there are more quark flavor combinations that dilute the ratio.

For example, in D^0 decay, CP final states must have zero total strangeness, and these decays are Cabibbo-suppressed. The current experimental values are

$$\frac{D^0 \to \pi^+\pi^-}{D^0 \to \text{anything}} = (5.9 \pm 3.2) \times 10^{-4} \quad (7.169)$$

$$\frac{D^0 \to K^+K^-}{D^0 \to \text{anything}} = (2.0 \pm 0.8) \times 10^{-3} \quad (7.170)$$

which imply $\Delta\gamma_D/\gamma_D \simeq 2 \times 10^{-3}$.

These results show that there is very little mass mixing in the D^0 system:

$$\rho_D \leq 10^{-4} \quad (7.171)$$

Values for $\Delta m(B_d^0)$ and $\Delta m(B_s^0)$ are even less certain, since we do not know θ_2, θ_3, δ, or m_t. For $m_t < 30$ GeV/c^2, assuming favorable numbers for mixing angles, $\Delta m(B_d^0)$ may be larger than Δm_K by approximately a factor of 10^2 to 10^3. On the other hand, γ_B will be larger than γ_K:

$$\frac{\gamma_B}{\gamma_K} \simeq \frac{m_B^5}{m_K^5} \times [\text{ratios of mixing angles}, \mathcal{O}(1)] \sim 10^5 \quad (7.172)$$

so at best we can hope that

$$(\Delta m/\gamma)_{B_d^0} \simeq 10^{-2} \quad (7.173)$$

The fractional lifetime difference $(\Delta\gamma/\gamma)_B$ may be somewhat larger than for D mesons because, in B_d^0 decay, all the channels are "Cabibbo-suppressed" in the sense that they involve *family*-changing ($b \to c$ or $b \to u$) transitions.

In heavy meson systems, mass mixing causes charge asymmetries in semileptonic decay rates. For example, if a D^0 meson is produced at $t = 0$, it will evolve into a superposition of D^0 and $\bar{D}^0$ in a way that is dependent on m_D. Because the $c \to s, d$ decay satisfies the $\Delta C = \Delta Q$ rule, the D^0 component contributes to final states of the form ($l^+\nu_l$ + hadrons) and the $\bar{D}^0$ component results in ($l^-\bar{\nu}_l$ + hadrons). The relative intensity of these two channels can be characterized by the parameter ρ. Specifically, if we define $N^{\pm\pm}$ as the number of events produced in e^+e^- collisions of the form

286   7 Neutral K mesons and CP violation

$$e^+e^- \to \psi'' \to D^0\bar{D}^0$$
$$\phantom{e^+e^- \to \psi'' \to D^0\bar{D}^0} \to l^\pm(\bar{\nu}) + \text{hadrons}$$
$$\phantom{e^+e^- \to \psi'' \to D^0\bar{D}^0} \to l^\pm(\bar{\nu}) + \text{hadrons} \qquad (7.174)$$

then there are three different types of events: N^{--} (two charged leptons), N^{++} (two charged antileptons), and N^{+-} (one of each). It can be shown that

$$(N^{++} + N^{--})/N^{+-} = 2\rho_D/(1 + \rho_D^2) \qquad (7.175)$$

Measurements of kaon pairs produced in D^0 decay yield the limit (Goldhaber 77):

$$(N^{++} + N^{--})/N^{+-} < 16 \times 10^{-2} \qquad (7.176)$$

which implies

$$\rho_D < 0.08$$

This limit is well above the theoretical estimate but already small enough to ensure that CP-violating effects will be very difficult to observe.

Just as the sum of N^{++} and N^{--} gives the mass-mixing fraction, the difference is a measure of CP violation. It can be shown (Pais and Treiman 75) that

$$\frac{N^{++} - N^{--}}{N^{++} + N^{--}} = \frac{-2\,\text{Im}(\Gamma_{12}/M_{12})}{1 + |\Gamma_{12}/M_{12}|^2} \simeq 4\,\text{Re}(\epsilon) \qquad (7.177)$$

The parameter ϵ may be estimated in the Kobayashi–Maskawa model by means of the techniques of Subsection 7.13.5. One finds

$$|\epsilon|_K \propto s_2 s_3 s_\delta \simeq 10^{-3} \qquad (7.178)$$

$$|\epsilon|_D \propto s_2 s_3 s_\delta \simeq 10^{-3} \qquad (7.179)$$

$$\left.\begin{array}{l}|\epsilon|_{B_d} \propto \tan 2\delta \\[4pt] |\epsilon|_{B_s} \propto \dfrac{s_2}{s_3 + s_2 c_\delta} \tan 2\delta\end{array}\right\} \gg 10^{-3} \text{ (possibly)} \qquad \begin{array}{l}(7.180)\\[20pt](7.181)\end{array}$$

This would suggest that the B_d and B_s systems offer the best possibilities for observing CP violation in the charge asymmetry (7.177). On the other hand, it has been shown (Hagelin and Wise 80) that for B_d and B_s the leading terms in M_{12} and Γ_{12} have the same sign; consequently, the charge asymmetry should be small. Measurements of this quantity would be further complicated by leptons from subsequent decays, such as

$$B^0 \to l^-\bar{\nu}D^+$$
$$\phantom{B^0 \to l^-\bar{\nu}D^+} \to l^+ + \cdots$$

The solution to these problems may be so difficult that CP violation is likely to remain an oddity of the K-meson system for many years to come.

Problems

7.1 Verify (7.118).

7.2 Consider the integral

$$I_{\mu\nu} = \int \frac{d^4k}{(2\pi)^4} \frac{k_\mu k_\nu}{(k^2 - m_W^2)^2(k^2 - m_u^2)^2} \qquad (1)$$

which enters into the calculation of $\Delta m = m_L - m_S$. It is a "Feynman integral," of a type that appears frequently in quantum electrodynamics.

(a) By contour integration, show that

$$\int \frac{d^4k}{(2\pi)^4} \frac{1}{(k^2 - c)^3} = \frac{1}{32\pi^2 ic} \qquad (2)$$

where c is a constant.

(b) Make the substitutions $k_\mu \to k_\mu - p_\mu$ and $c \to c + p_\mu p^\mu$, where p_μ is a constant 4-vector, to show that

$$\int \frac{d^4k}{(2\pi)^4} \frac{k_\mu}{(k^2 - 2k\cdot p - c)^3} = \frac{p_\mu}{32\pi^2 i(c - p^2)} \qquad (3)$$

(c) Differentiate both sides of (3) to obtain:

$$\int \frac{d^4k}{(2\pi)^4} \frac{k_\mu k_\nu}{(k^2 - c)^4} = \frac{g_{\mu\nu}}{192\pi^2 ic} \qquad (4)$$

(d) Verify the identity

$$\frac{1}{ab} = \int_0^1 \frac{dx}{[ax + b(1-x)]^2}$$

then differentiate with respect to a and b to obtain the result

$$I_{\mu\nu} = 6 \int_0^1 x(1-x)\, dx \int \frac{d^4k}{(2\pi)^4} \frac{k_\mu k_\nu}{[k^2 + (m_u^2 - m_W^2)x - m_u^2]^4}$$

which, together with (4), yields (7.27).

7.3 How does the inclusion of color modify result (7.29)?

7.4 Derive (7.164)–(7.166) by the methods of Section 7.3.

7.5 Derive the Pais–Treiman relation (7.177), and, similarly, obtain (7.175).

7.6 In Section 7.7, we found that removal of different amounts of the K^0 and $\bar{K}^0$ components of a K_L^0 beam leads to coherent production of K_S^0 in the remaining beam. K_S^0 may also be produced in the scattered portion of the beam through "diffractive regeneration." Equation (7.92) describes the composition of the unscattered beam as a function of distance l_1 through the absorber. Show that, for thin absorbers (thickness x), the probability for coherent K_S^0 production in the forward direction is

$$n \simeq \frac{\pi^2 N^2}{k^2} |f(0) - \bar{f}(0)|^2 x^2$$

Now consider the scattered beam. Show that the probability for producing a K_S^0 in an absorber of thickness dx, scattered into solid angle $d\Omega$, is

$$dn \simeq \tfrac{1}{4}|f(\theta) - \bar{f}(\theta)|^2 N\, dx\, d\Omega$$

Finally, consider a K_S^0 detector placed downstream from the regenerator with finite angular resolution $\delta\Omega$. Show that the scattered K_S^0 component contributes a fraction $k^2\delta\Omega/4\pi^2 Nx$ to the total K_S^0 signal at $\theta = 0$. (See Good 58.)

8
High-energy neutrino–nucleon collisions

8.1 Introduction

During the past decade, progress in weak interactions and, indeed, in strong interactions has come about in very large part because of high-energy neutrino-beam scattering experiments at proton accelerators. Neutrinos have several extremely important advantages as projectiles. First, they do not suffer strong or electromagnetic interactions, which would otherwise obscure weak effects in interactions with nucleons. Second, neutrino scattering cross sections grow linearly with energy in the laboratory frame, and notwithstanding the fact of weak coupling, large numbers of events can be accumulated at high energy. Thus, relatively rare processes, such as charmed-quark production, can be studied in detail. Third, like electrons or muons, high-energy neutrinos have short de Broglie wavelengths ($\sim 10^{-15}$ cm at 20 GeV) and may be employed to probe the interior of nucleons at short distances. Finally, neutrinos are always polarized. This is useful for a variety of reasons, in particular, because in charged-current interactions the left-handed ν scatters preferentially from quarks, whereas the right-handed $\bar{\nu}$ scatters mainly from antiquarks. Thus, the composition of the $q\bar{q}$ sea in a nucleon can be investigated.

In this chapter we shall discuss experimental methods of neutrino-beam production and detection, elastic charged and neutral νN scattering, deep inelastic charged νN reactions, charm production in neutrino–nucleon collisions, and measurements of fundamental electroweak parameters in neutral νN collisions. The quark–parton model is of central importance in deep inelastic charged and neutral reactions and we shall develop it in some detail. The principal consequence of this model is the prediction of scaling, which is verified reasonably well by experimental data. There are, however, small deviations from scaling, which are attributed in part to residual quark–gluon interactions and are treated by QCD. This subject has in fact become of major importance in neutrino physics, but it lies outside of our domain of interest, and we shall discuss it only very briefly.

Three major areas of neutrino physics appear elsewhere in this book. Chapter 3 contains a discussion of neutrino–electron scattering; Chapter 10 is devoted to the properties of neutrinos per se and includes a discussion of neutrino oscillations; and Chapter 11 is devoted to neutrino astrophysics.

8.2 Experimental methods

8.2.1 Neutrino beams

High-energy neutrino beams are generated from the in-flight decays $\pi_{\mu 2}$, $K_{\mu 2}$, and K_{e3} of highly relativistic pions and kaons produced in a proton synchrotron target. The collision products are focused by a variety of means to select the desired meson type and momentum. Then they enter an evacuated tunnel several meters long, where, because of relativistic time dilation, only ~5 to 20 percent of the K's and π's decay. Following this is an absorber for the remaining hadrons and muons, consisting of steel, rock, or earth, as much as 1 km thick. Neutrinos emerge from this shield together with a small background of hadrons and muons produced in the absorber.

The neutrino energy spectrum is determined by the mesons that enter the decay tunnel. Three types of focusing are used to shape the spectrum. They result in (a) narrowband, or dichromatic, beams, (b) wideband beams, and (c) highband beams. The first type employs a series of dipole and quadrupole magnets to select and focus pions and kaons with the desired charge and momentum. The energy E_ν of a neutrino produced in $\pi_{\mu 2}$ or $K_{\mu 2}$ decay is sharply defined in the meson rest frame and is a function of decay angle θ in the laboratory frame:

$$E_\nu^{\pi,K} = E_{\pi,K}(1 - m_\mu^2/m_{\pi,K}^2)(1 + \gamma^2\theta^2)^{-1} \qquad (8.1)$$

The energy spread $\Delta E_\nu/E_\nu$ for neutrinos that enter the detector at a distance R from the meson beam axis depends on the meson momentum distribution and the relative length of the decay tunnel and absorber. For a monoenergetic beam, the energy spread is

$$\frac{\Delta E_\nu}{E_\nu} \simeq \frac{1}{1 + \gamma^2 R/(L_A + L_D)} - \frac{1}{1 + \gamma^2 R/L_A} \qquad (8.2)$$

(see Figure 8.1). The actual energy dependence with R is shown in Figure 8.2. Pions and kaons have different decay energies, hence the beam is dichromatic, that is, has two momentum peaks.

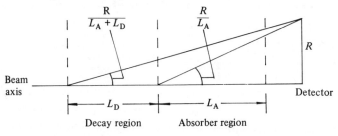

Figure 8.1. Relationship of distance R to lengths of decay and absorber regions.

Wideband beam focusing maximizes the total neutrino flux regardless of momentum, by collimating the target secondaries. This is done with aluminum sheet conductors in the shape of a horn or cone placed around the beam and pulsed with 0.25 to 0.5×10^6 A of current when the protons strike the target. The resulting magnetic field lines describe concentric circles around the beam axis. Thus secondaries of one

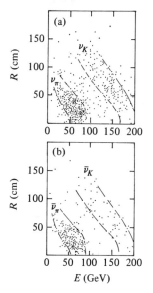

Figure 8.2. Neutrino energy vs. distance R from the beam axis for (a) muon neutrinos and (b) antineutrinos in the CERN SPS narrowband neutrino beam.

292   8 High-energy neutrino–nucleon collisions

charge are focused and those of opposite sign are defocused. Much higher fluxes are possible for the wideband than for the narrowband system but the spectrum is peaked at lower energy.

Finally, the highband beam system consists of quadrupole focusing magnets downstream from the target, employed to select mesons of high energy. The result is a suppression of the low-energy peak of the wideband beam and an enhancement of the high-energy flux (see Figures 8.3a and 8.3b).

8.2.2   Neutrino detectors

The very small cross sections for neutrino interaction with matter $[\sigma \simeq 10^{-38} \text{ cm}^2 \times E(\text{GeV})]$ require that detectors be very massive. An ideal detector should also satisfy the following requirements:

(i) good energy resolution for scattered particles, since the neutrino energy is not well known;
(ii) tracking and particle identification for the final muon in charged-current ν_μ and $\bar{\nu}_\mu$ scattering;
(iii) the ability to identify neutral current events, where the final lepton always goes undetected;
(iv) simple target composition for the study of isospin dependence, and

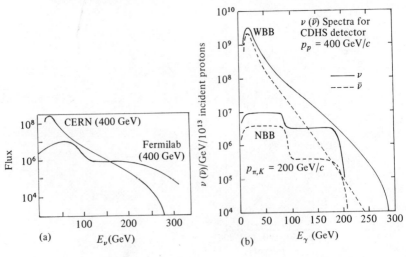

Figure 8.3. (a) Comparison of CERN wideband SPS beam and Fermilab highband beam (quad triplet). (b) WBB: wideband beam; NBB: narrowband beam.

8.2 Experimental methods

(v) sophisticated tracking ability for detection of charmed particles and other particular final states.

No single detector satisfies all these criteria and several different designs have been developed and optimized for various purposes (see Table 8.1). Indeed, several different experiments may be placed in a single beam line and run simultaneously because of the small cross sections for neutrino absorption.

Counter experiments consist of a target, a calorimeter, and a muon spectrometer. For example, the CERN–Dortmund–Heidelberg–Saclay collaboration (CDHS) employs a 1,400-metric-ton apparatus consisting of 5- and 15-cm-thick magnetized iron plates sandwiched together with drift chambers and plastic scintillators, producing a stack 23 m long, which acts as a target, calorimeter, and spectrometer all in one (see Figure 8.4). Another detector, built by the CERN–Hamburg–Amsterdam–Rome–Moscow (CHARM) collaboration, uses a nonmagnetic isoscalar target (150 metric tons of marble: $CaCO_3$) sandwiched between scintillators and proportional chambers, followed by toroidal iron magnets employed to measure the muon momentum

Table 8.1. *Neutrino detectors*

Acronym	Group	Method	Location
CDHS	CERN–Dortmund–Heidelberg–Saclay	Counter	CERN
CHARM	CERN–Hamburg–Amsterdam–Rome–Moscow	Counter	CERN
CFRR	Caltech–Fermilab–Rochester–Rockefeller	Counter	Fermilab
HPWF, also HPWFRO	Harvard–Pennsylvania–Wisconsin–Fermilab–Rutgers–Ohio St.	Counter	Fermilab
BEBC	Big European Bubble Chamber used by many groups, e.g.: Aachen–Bonn–CERN–Demokritos–London–Oxford–Saclay	Bubble chamber	CERN
ABCDLOS			
GGM	Gargamelle (many groups)	Bubble chamber	CERN
15′ FNAL	Many groups	15′ bubble chamber	Fermilab

294   8 High-energy neutrino–nucleon collisions

(see Figure 8.5). The resolution in such detectors is quite poor: Muons suffer multiple Coulomb scattering in the magnetized iron, which limits momentum determinations to 10 to 15 percent. Furthermore, the calorimeters can only resolve hadronic energy to 10 to 15 percent, and since the shower must be contained in the detector for measurement of E_ν, the active volume is much smaller than the total volume.

Momentum resolution can be improved with bubble chambers, and several have been used with active masses of 1 to 10 tons. For example, the BEBC bubble chamber at CERN consists of a 17-m³ fiducial volume, which can be filled with a Ne–H$_2$ mixture. The chamber is placed in a 35-kg field produced by a superconducting magnet, which makes possible a momentum resolution $\Delta p/p$ of ~4 percent. The en-

Figure 8.4. CDHS detector. (From deGroot et al. 1979. Reprinted with permission.)

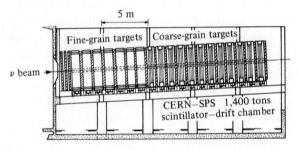

Figure 8.5. CHARM detector. (From Diddens et al. 1980. Reprinted with permission.)

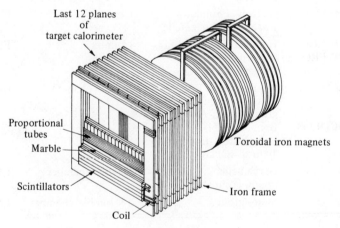

ergy resolution is, by contrast, quite poor, since many neutrals escape detection, and as a result, about 20 percent of the total hadronic energy is lost. Although corrections can be applied, the latter lead to systematic errors that are difficult to evaluate, and thus the ultimate precision is quite limited.

A major problem in high-energy neutrino physics is the detection of short-lived charmed hadrons that are produced in νN collisions. Conventional bubble-chamber techniques, adequate for identifying relatively long-lived hadrons, such as K and Σ^+, are of no use for a charmed hadron, where a typical track length of a few hundred micrometers is scarcely larger than the size of individual bubbles. Specialized bubble chambers with smaller bubbles, such as the Little European Bubble Chamber LEBC (Allison et al. 80), offer one solution.[1] However, photographic emulsion is by far the best track medium, since the track element, a photographic grain, is typically 0.5 μm in diameter. This excellent spatial resolution can be augmented by counters or conventional bubble chambers placed behind the emulsion stack to provide final-state particle identification and momentum determination. It is now possible with such hybrid systems to study charmed hadronic decays with some precision.

8.3 Charged-current neutrino–nucleon elastic scattering

We begin our discussion of the theory of neutrino–nucleon interactions with consideration of the charged-current elastic reactions:

$$\bar{\nu}_\mu p \rightarrow \mu^+ n \tag{8.3}$$

and

$$\nu_\mu n \rightarrow \mu^- p \tag{8.4}$$

(see Figure 8.6). These reactions are important because they provide an opportunity, at least in principle, to study the charged-current hadronic form factors at high energy and fairly large momentum transfer (test of CVC) and also because they may be used to normalize neutral-current (NC) neutrino scattering cross sections.

We denote the 4-momenta for ν, μ, N, and N' by $k = (\omega, \mathbf{k})$, $k' = (\omega', \mathbf{k}')$, $p = (E, \mathbf{p})$, and $p' = (E', \mathbf{p}')$, respectively. Also $q = k - k'$ is

---

[1] One novel approach is the British Infinitesimal Bubble Chamber (BIBC), which employs holography to improve photographic emulsion without loss of depth of field.

Figure 8.6. Charged-current elastic reactions.

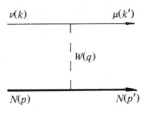

the 4-momentum transfer. We also define $Q^2 = -q^2 > 0$. In all experiments to date, $Q^2 \ll m_W^2$, so the amplitude for reaction (8.4) may be written as

$$\mathcal{M} = \frac{G_F}{\sqrt{2}} \bar{u}_p(p',s')[f_1(q^2)\gamma^\alpha + if_2(q^2)\sigma^{\alpha\beta}q_\beta - g_1(q^2)\gamma^\alpha\gamma^5$$
$$- g_3(q^2)q^\alpha\gamma_5]u_n(p,s)\bar{u}_\mu\gamma_\alpha(1-\gamma_5)u_\nu \quad (8.5)$$

In (8.5) and henceforth we shall ignore second-class terms, and the form factors are taken to be real in accordance with T invariance. As noted in Section 4.9, the CVC hypothesis implies [(4.130) and (4.131)] that

$$f_1(q^2) = C^{(1)}(q^2) \quad (8.6)$$
$$f_2(q^2) = M^{(1)}(q^2) \quad (8.7)$$

where $C^{(1)}$ and $M^{(1)}$ are the isovector form factors appearing in the electromagnetic matrix element of the nucleon [(4.128)]. As for the axial vector form factors in (8.5), neutron β decay tells us that

$$g_1(0) = 1.25$$

and the q^2 dependence of g_1 must be determined by neutrino scattering, whereas the g_3 term can be neglected if the final lepton is highly relativisitic (see Problem 8.1).

To understand the q^2 dependence better, let us consider briefly the electromagnetic case once again. Starting from formula (4.128), one can derive the cross section for unpolarized electron–proton scattering by photon exchange (the so-called Rosenbluth formula):

$$\frac{d\sigma}{d\Omega} = \left(\frac{\alpha^2}{4\omega^2 \sin^4(\theta/2)} \cos^2\frac{\theta}{2}\right) \times \frac{1}{1 + 2\omega/m_p \sin^2(\theta/2)}$$
$$\times \left[(C_p^2 - M_p^2 q^2) - \frac{q^2}{2m_p^2}(C_p + 2m_p M_p)^2 \tan^2\frac{\theta}{2}\right] \quad (8.8)$$

8.3 Charged-current ν–N elastic scattering

If the proton were a Dirac particle, we would have $C_p = 1$, $M_p = 0$. The factors that actually appear in the first and second terms of (8.8) account for the finite charge and magnetic moment distribution of the real proton. It is therefore natural to define the proton electric and magnetic form factors by

electric: $\quad G_E^p(q^2) = C_p(q^2) + \dfrac{q^2}{2m_p} M_p(q^2); \quad G_E^p(0) = 1 \quad$ (8.9)

magnetic: $\quad G_M^p(q^2) = C_p(q^2) + 2m_p M_p(q^2); \quad G_M^p(0) = 2.79 \quad$ (8.10)

Similarly, for the neutron, one has

$$G_E^n(q^2) = C_n(q^2) + \frac{q^2}{2m_p} M_n(q^2); \quad G_E^n(0) = 0 \quad (8.11)$$

$$G_M^n(q^2) = C_n(q^2) + 2m_p M_n(q^2); \quad G_M^n(0) = -1.91 \quad (8.12)$$

Electron scattering experiments reveal that G_E^p, G_M^p, and G_M^n have the same q^2 dependence:

$$\frac{G_E^p(q^2)}{G_E^p(0)} = \frac{G_M^p(q^2)}{G_M^p(0)} = \frac{G_M^n(q^2)}{G_M^n(0)} = \left[1 + \frac{Q^2}{m_V^2}\right]^{-2} \quad (8.13)$$

with $m_V = 0.84$ GeV/c^2. Also, one finds $G_E^n(q^2) = 0$. Using these results and CVC [(4.130) and (4.131)], we easily obtain

$$f_1(q^2) = \frac{(G_E^p - G_E^n) + (Q^2/4m_p^2)(G_M^p - G_M^n)}{1 + Q^2/4m_p^2}$$

$$= \frac{[1 + 4.70 Q^2/4m_p^2]}{(1 + Q^2/4m_p^2)(1 + Q^2/m_V^2)^2} \quad (8.14)$$

and

$$f_2(q^2) = \frac{3.70}{(1 + Q^2/4m_p^2)(1 + Q^2/m_V^2)^2} \quad (8.15)$$

The axial vector form factor is presumed to obey the "dipole" formula:

$$g_1(q^2) = \frac{1.25}{(1 + Q^2/m_A^2)^2} \quad (8.16)$$

and m_A is extracted from the q^2 dependence of the experimental cross section, assuming forms (8.14) and (8.15) for $f_1(q^2)$ and $f_2(q^2)$, respectively, as given by CVC. We shall return to this point later.

The differential cross section for $\nu_\mu n \to \mu^- p$ and $\bar\nu_\mu p \to \mu^+ n$ may be found from (8.5) by a straightforward, if tedious, calculation. Here we merely present the result, averaged over nucleon spins in the high-energy limit, where m_μ may be neglected:

$$\frac{d\sigma_{\text{elas}}}{dQ^2} = \frac{G_F^2}{2\pi} \cos^2\theta_1 \left[f_1^2(q^2) \left(1 + \frac{q^2}{2m_p\omega} + \frac{q^2}{4\omega^2} + \frac{q^4}{8m_p^2\omega^2} \right) \right.$$

$$+ f_2^2(q^2) \left(-q^2 - \frac{q^4}{2m_p\omega} + \frac{q^4}{4\omega^2} \right)$$

$$+ g_1^2(q^2) \left(1 + \frac{q^2}{2m_p\omega} - \frac{q^2}{4\omega^2} + \frac{q^4}{8m_p^2\omega^2} \right) + f_1(q^2) f_2(q^2) \frac{q^4}{2m_p\omega^2}$$

$$\mp f_1(q^2) g_1(q^2) \left(\frac{q^2}{m_p\omega} + \frac{q^4}{4m_p^2\omega^2} \right)$$

$$\left. \mp f_2(q^2) g_1(q^2) \left(\frac{2q^2}{\omega} + \frac{q^4}{4m_p\omega^2} \right) \right] \qquad (8.17)$$

In this formula $\mp$ refer to ν and $\bar{\nu}$, respectively.

In the limit $q^2 \to 0$ (forward scattering), (8.17) takes the simple form:

$$\left. \frac{d\sigma}{dQ^2} \right|_{q^2=0} = \frac{G_F^2}{2\pi} \cos^2\theta_1 [f_1^2(0) + g_1^2(0)] = 2.59 \frac{G_F^2}{2\pi} \cos^2\theta_1 \qquad (8.18)$$

which is independent of neutrino energy. Another case of interest is the high-energy limit of (8.17), where $\omega \to \infty$:

$$\left. \frac{d\sigma}{dQ^2} \right|_{\omega \to \infty} = \frac{G_F^2 \cos^2\theta_1}{2\pi} [f_1^2(q^2) + g_1^2(q^2) - q^2 f_2^2(q^2)] \qquad (8.19)$$

Let us now consider the experimental situation very briefly. Assuming CVC, the only unknown in $d\sigma/dQ^2$ is m_A, and in principle, careful measurements should reveal the latter quantity. In practice, however, this is extremely difficult and requires a detailed knowledge of target nuclear physics. The most straightforward cases are the reactions

$$\bar{\nu}_\mu p \to \mu^+ n \qquad (8.20)$$

$$\nu_\mu D \to \mu^- pp \qquad (8.21)$$

Reaction (8.20) would be preferred for simplicity, but (8.21) has been studied more thoroughly because of the availability of higher fluxes of ν_μ. For example, a deuterium bubble-chamber experiment at Argonne Laboratory recorded 166 events of type (8.21) and established (Mann et al. 73) that

$$m_A = 0.95 \pm 0.12 \quad \text{GeV}/c^2 \qquad (8.22)$$

The average experimental values preferred recently are $m_A = 0.9$ (Hung and Sakurai 81) and $m_A = 1.00 \pm 0.05$ (Kim et al. 81). The quantity m_A has also been measured indirectly in the electroproduction reaction $ep \to en\pi^+$, where the electromagnetic cross section can be related to $g_1(q^2)$ by current algebra. These experiments yield somewhat higher values for m_A (Gourdin 74).

8.4 Neutral-current ν–N elastic scattering

Ultimately, very detailed analyses of νN charged-current elastic scattering might help to confirm the predictions of CVC at large Q^2. Unfortunately, low statistics and nuclear target uncertainties presently prohibit such refinements, and these reactions are used for the most part to normalize the neutral-current elastic-scattering form factors in tests of the standard model.

8.4 Neutral–current neutrino–nucleon elastic scattering

As mentioned previously, detailed study of NC neutrino–nucleon reactions is of critical importance for determining the neutral-current coupling parameters. The simplest of these reactions is elastic scattering, and in Section 4.12, we presented the matrix elements according to the standard model. However, it is desirable to write these quantities in a more general form in order to compare that model with alternative descriptions. For this purpose, we continue to assume V–A couplings, and ignore the possibility of S, T, and P contributions. Our starting point is therefore a general effective neutral weak Lagrangian for neutrino–quark interactions (Hung and Sakurai 81):

$$\mathscr{L}_{NC}^{\nu q} = \bar{u}_\nu \gamma_\lambda (1 - \gamma_5) u_\nu \{\tfrac{1}{2}[\bar{U}\gamma^\lambda(\alpha + \beta\gamma_5)U - \bar{D}\gamma^\lambda(\alpha + \beta\gamma_5)D]$$
$$+ \tfrac{1}{2}[\bar{U}\gamma^\lambda(\gamma + \delta\gamma_5)U + \bar{D}\gamma^\lambda(\gamma + \delta\gamma_5)D]$$
$$+ \bar{S}\gamma^\lambda(\gamma' + \delta'\gamma_5)S + \cdots\} \quad (8.23)$$

Here the coefficients α, β, γ, δ, ... represent the following types of terms: α, isovector–vector; β, isovector–axial vector; γ, isoscalar–vector; δ, isoscalar–axial vector; γ', strange–vector; and δ', strange–axial vector. Also, in (8.23), the ellipses represent additional terms from heavy quarks c, t, and b, which should make only a very small contribution. According to the GIM mechanism, c and s quarks must form an isospin doublet with the same neutral weak couplings as u and d. This implies:

$$\gamma' = \tfrac{1}{2}(\gamma - \delta) \quad (8.24)$$

and

$$\delta' = \tfrac{1}{2}(\delta - \beta) \quad (8.25)$$

Also, we note for future reference that the standard model yields

$$\alpha = 1 - 2\sin^2\theta_W \quad (8.26)$$
$$\beta = 1 \quad (8.27)$$
$$\gamma = -\tfrac{2}{3}\sin^2\theta_W \quad (8.28)$$
$$\delta = 0 \quad (8.29)$$

8 High-energy neutrino–nucleon collisions

The matrix element of the neutral weak hadronic current between nucleon states is given by

$$\langle N|J_z^\lambda|N\rangle = \bar{u}_N(p', s')\{[f_1^{(1)}(q^2)\gamma^\lambda + if_2^{(1)}(q^2)\sigma^{\lambda\nu}q_\nu \\ - g_1^{(1)}(q^2)\gamma^\lambda\gamma_5 - g_3^{(1)}(q^2)q^\lambda\gamma_5]\tau_3 \\ + \tfrac{1}{2}[f_1^{(0)}(q^2)\gamma^\lambda + if_2^{(0)}(q^2)\sigma^{\lambda\nu}q_\nu \\ - g_1^{(0)}(q^2)\gamma^\lambda\gamma_5 - g_3^{(0)}(q^2)q^\lambda\gamma_5]\}u_N(p, s) \quad (8.30)$$

where superscripts (1) and (0) refer, as before, to isovector and isoscalar components. Let us relate the various form factors in (8.30) to the coefficients α, β, ... appearing in (8.23), as follows:

(i) The terms in $g_3^{(0)}(q^2)$ and $g_3^{(1)}(q^2)$ make no contribution in ν–N scattering because $m_\nu = 0$, and they may be ignored.

(ii) The vector weak couplings are unrenormalized by strong interaction. Therefore:

$$\langle N|\tfrac{1}{2}\alpha(\bar{U}\gamma^\lambda U - \bar{D}\gamma^\lambda D|N\rangle = \bar{u}_N f_1^{(1)}(0)\gamma^\lambda u_N \quad (8.31)$$

$$\langle N|\tfrac{1}{2}\gamma(\bar{U}\gamma^\lambda U + \bar{D}\gamma^\lambda D|N\rangle = \bar{u}_N f_1^{(0)}(0)\gamma^\lambda u_N \quad (8.32)$$

To evaluate the matrix elements on the left-hand side, we employ the color-symmetric $SU(6)$ quark spin functions for p [(4.48)] and n [(4.49)] to obtain

$$f_1^{(1)}(0) = \alpha, \quad f_1^{(0)}(0) = 3\gamma \quad (8.33)$$

(iii) The vector form factors $f_2^{(1),(0)}$ are not represented in any simple way by linear combinations of α, β, ... since they represent anomalous magnetic properties of the nucleon not directly derivable from the fundamental fermions.

(iv) In order to arrive at expressions for $g_1^{(1)}(0)$ and $g_1^{(0)}(0)$, we may employ $SU(3)_f$ and a generalized form of Cabibbo's hypothesis, according to which *all* hadronic axial weak currents (including the neutral currents) transform as members of an $SU(3)_f$ axial octet g_i. Thus the isovector–axial vector neutral current is associated with g_3, and the isoscalar–axial vector current is associated with g_8. Repeating essentially the same steps as in Section 6.4.2, we arrive at

$$g_1^{(1)}(0) = \tfrac{1}{2}\beta(F + D) \quad (8.34)$$

$$g_1^{(0)}(0) = \tfrac{1}{2}\delta(3F - D) \quad (8.35)$$

and thus

$$\frac{g_1^{(0)}(0)}{g_1^{(1)}(0)} = \left(\frac{3F - D}{F + D}\right)\frac{\delta}{\beta} = (0.40 \pm 0.04)\frac{\delta}{\beta} \quad (8.36)$$

where the numerical value of the ratio $(3F - D)/(F + D)$ is obtained from hyperon semileptonic decays.

8.4 Neutral-current ν–N elastic scattering

The quantities $g_1^{(1)}(0)$ and $g_1^{(0)}(0)$ may also be obtained from the color-symmetric $SU(6)$ quark model. By means of (4.48) and (4.49) and using the same steps that lead to (6.71), we obtain

$$\frac{g_1^{(0)}(0)}{g_1^{(1)}(0)} = \frac{\left\langle p \left| \sum_{\text{quarks}} \bar{Q}\gamma^\mu\gamma^5 Q \right| p \right\rangle}{2\left\langle p \left| \sum_{\text{quarks}} \bar{Q}\gamma^\mu\gamma^5\tau_3 Q \right| p \right\rangle} = \frac{3}{5}\frac{\delta}{\beta} \qquad (8.37)$$

In addition, we know from neutron β decay that $F + D = 1.25$, thus that[2]

$$g_1^{(1)}(0) = 0.63\beta \qquad (8.38)$$

Combining (8.36) and (8.38), we obtain $g_1^{(0)}(0) = 0.25\delta$, whereas (8.37) and (8.38) would predict $g_1^{(0)}(0) = 0.38\delta$. Further discussion is given by Hung (78), who presents a third method for evaluating $g_1^{(0)}(0)$.

(v) Finally, in electron-nucleon scattering, the isovector and isoscalar form factors have the same q^2 dependence, and we shall therefore assume this to be true for neutral weak interactions as well. Thus,

$$\frac{f_1^{(1)}(q^2)}{f_1^{(1)}(0)} = \frac{f_1^{(0)}(q^2)}{f_1^{(0)}(0)} = \frac{C^{(1)}(q^2)}{C^{(1)}(0)} \qquad (8.39)$$

$$\frac{f_2^{(1)}(q^2)}{f_2^{(1)}(0)} = \frac{f_2^{(0)}(q^2)}{f_2^{(0)}(0)} = \frac{M^{(1)}(q^2)}{M^{(1)}(0)} \qquad (8.40)$$

We also assume that all axial form factors for charged and neutral currents have the same q^2 dependence:

$$\frac{g_1^{(1)}(q^2)}{g_1^{(1)}(0)} = \frac{g_1^{(0)}(q^2)}{g_1^{(0)}(0)} = \frac{g_1(q^2)}{g_1(0)} = \frac{1}{(1 - (q^2/m_A^2))^2} \qquad (8.41)$$

The cross section for neutral-current (NC) νN elastic scattering is obtained directly from the charged-current cross section (8.17) with the following substitutions:

$$F(q^2) \rightarrow \begin{cases} \frac{1}{2}[F^{(1)}(q^2) + F^{(0)}(q^2)] & \text{for protons} \\ \frac{1}{2}[F^{(1)}(q^2) - F^{(0)}(q^2)] & \text{for neutrons} \end{cases}$$

with $F = f_1, f_2$, and g_1. Comparison of the resulting NC cross section with actual experimental results would require detailed knowledge of

---

[2] Some authors argue that $g_1^{(1)}(0)$ may be slightly larger than the result (8.38) because of the effects of heavy quarks.

the neutrino spectrum and q^2 dependence of detector efficiency. It is therefore more practical to determine the ratios

$$R_{\nu p}^{el} = \frac{\sigma(\nu_\mu p \to \nu_\mu p)}{\sigma(\nu_\mu n \to \mu^- p)} \quad (8.42)$$

and

$$R_{\bar\nu p}^{el} = \frac{\sigma(\bar\nu_\mu p \to \bar\nu_\mu p)}{\sigma(\bar\nu_\mu p \to \mu^+ n)} \quad (8.43)$$

At $q^2 = 0$, one expects these ratios to be equal and to be given by the formula

$$R_{\bar\nu p}^{el} = R_{\nu p}^{el} = \frac{1}{4\cos^2\theta_1}\left[\frac{(1.25)^2 + (1 - 4\sin^2\theta_W)}{(1.25)^2 + 1}\right] \quad (8.44)$$

However, no experiment can be performed at $q^2 = 0$ (where the final proton has zero momentum and is undetectable); thus the q^2 dependence of the form factors must be taken into account and an extrapolation to $q^2 = 0$ carried out. In fact, experiments at finite q^2 have an advantage: in principle, one can observe the V–A interference terms, which vanish at $q^2 = 0$ [see (8.17)].

Neutral-current νN elastic-scattering experiments have been carried out in the Gargamelle bubble chamber at CERN and in counter/calorimeter experiments at Brookhaven and CERN. The highest statistics and lowest background have been achieved by a Harvard–Brookhaven–Pennsylvania collaboration (Entenberg et al. 1979) in which 217 $\nu p \to \nu p$ events and 82 $\bar\nu p \to \bar\nu p$ events were observed with the wideband neutrino beam at the Brookhaven AGS. Their detector consisted of 30 tons of liquid scintillator subdivided into cells that were individually viewed by photomultiplier tubes. Thus it was possible to reconstruct the track of the scattered proton and measure its total kinetic energy. The primary sources of background in such experiments came from neutrons produced in the walls of the detector, which can enter the viewing region and scatter protons, and NC single-pion productions such as $\nu p \to \nu n\pi^+$ or $\nu p \to \nu p\pi^0$, where one of the final-state hadrons has too little kinetic energy to be detected. The latter background can be reduced by observing the decays of the pions in delayed coincidence. The neutron background was eliminated by shielding, and the experimenters were able to show that the neutron contamination was small because true events were distributed uniformly in the scintillator, rather than peaked near the walls. The observed events ranged in Q^2 from 0.4 GeV^2 to 0.9 GeV^2, and when this was folded into the cross section together with the ν spectrum, the ratios obtained were

$$R_{\nu p}^{el} = 0.11 \pm 0.02 \tag{8.45}$$

$$R_{\bar{\nu} p}^{el} = 0.19 \pm 0.05 \tag{8.46}$$

$$\frac{\sigma(\nu p \to \nu p)}{\sigma(\bar{\nu} p \to \bar{\nu} p)} = 0.53 \pm 0.17 \tag{8.47}$$

The experimentally determined form factors were extrapolated to $q^2 = 0$, with the results:

$$G_E(\nu p \to \nu p, q^2 = 0) = f_{1p}(0) \qquad = 0.5 \begin{cases} +0.05 \\ -0.5 \end{cases} \tag{8.48}$$

$$G_M(\nu p \to \nu p, q^2 = 0) = f_{1p}(0) + 2mf_{2p}(0) = 1.0 \begin{cases} +0.35 \\ -0.04 \end{cases} \tag{8.49}$$

$$G_A = g_{1p}(0) \qquad = 0.5 \begin{cases} +0.2 \\ -0.15 \end{cases} \tag{8.50}$$

Although these results contain large uncertainties, they are in agreement with predictions of the standard model and have been instrumental in establishing its validity, as we shall see in Section 8.7.

8.5 Deep inelastic charged-current neutrino–nucleon interactions

8.5.1 Cross sections according to the quark–parton model

Let us now consider deep inelastic collisions:

$$\nu(\text{or } \bar{\nu}) + N \to \mu + X \tag{8.51}$$

where X stands for highly excited hadronic matter. As before, the 4-momenta of ν, μ, N, and X are $k = (\omega, \mathbf{k})$, $k' = (\omega', \mathbf{k}')$, $p = (E, \mathbf{p})$, and $p' = (E', \mathbf{p}')$, where $q = p' - p$ is the 4-momentum transfer. In the laboratory frame, $\mathbf{p} = 0$, and $E = m_N =$ nucleon mass. It is convenient to define $\nu = \omega - \omega'$, the energy transferred to the hadrons. It is easy to see that, in general,

$$m_N \nu = p \cdot q$$

Reaction (8.51) is a "deep inelastic" collision when $Q^2 \gg m_N^2$, $\nu \gg m_N$.

In the quark–parton model, the nucleon is assumed to consist of pointlike "partons": three fractionally charged valence quarks and an ocean of $q\bar{q}$ pairs. We consider the nucleon in a Lorentz frame where its 4-momentum p (and the momentum of its constituent partons) is much larger than the typical transverse momentum associated with the strong interaction between quarks. This is called the "infinite momentum frame" (Bjorken and Paschos 69). In this approximation the partons move in parallel trajectories, each carrying a fraction of the total momentum p. Let one quark with momentum ξp interact with a neutrino, thus absorbing 4-momentum q (Figure 8.7). In the present approximation, the other quarks are mere "spectators." The final momentum of the affected quark is $\xi p + q$, and

$$(\xi p + q)^2 = m^2$$

where m is the quark mass. Since $Q^2 \gg m_N^2 > m^2$, we have

$$(\xi p + q)^2 \simeq q^2 + 2\xi p \cdot q \simeq 0$$

or

$$\xi \simeq \frac{q^2}{-2p \cdot q} = \frac{Q^2}{2m_N \nu} \equiv x \qquad (8.52)$$

where x is defined by the last equality. Formula (8.52) is very important because it expresses the fractional momentum ξ in terms of Q^2 and ν, which are measurable quantities.

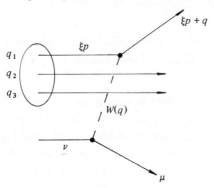

Figure 8.7. Diagram illustrating deep inelastic scattering according to the quark–parton model (see text for explanation).

Now, a quark with fractional momentum $\xi = x$ in the infinite momentum frame has an effective mass xm_N in the laboratory frame. We can easily find the differential cross section for charged-current scattering of ν or $\bar{\nu}$ from such a quark merely by borrowing the results obtained for neutrino–electron scattering from Chapter 3 (3.141) and applying obvious modifications. We thus obtain

8.5 Deep inelastic charged-current ν–N interactions

$$d\sigma(\nu q)/dy = 2G_F^2\rho^2 m_N x\omega_L/\pi \qquad (8.53a)$$
$$d\sigma(\bar{\nu}q)/dy = 2G_F^2\rho^2 m_N x\omega_L(1-y)^2/\pi \qquad (8.53b)$$

where ρ is a suitable Cabibbo factor, ω_L the neutrino energy in the laboratory frame, and $y = \nu/\omega_L$. A sum over all fermions in the nucleon then yields

$$\partial^2\sigma_{CC}(\nu N)/\partial y\,\partial x = 2G_F^2\rho^2 m_N\omega_L/\pi \times [Q(x) + \bar{Q}(x)(1-y)^2] \qquad (8.54)$$
$$\partial^2\sigma_{CC}(\bar{\nu}N)/\partial y\,\partial x = 2G_F^2\rho^2 m_N\omega_L/\pi \times [\bar{Q}(x) + Q(x)(1-y)^2] \qquad (8.55)$$

where $Q(x)$ and $\bar{Q}(x)$ are the quark and antiquark distribution functions, respectively.[3]

The factors

$$D^\nu(x) = \rho^2[Q(x) + \bar{Q}(x)(1-y)^2] \qquad (8.56)$$
$$D^{\bar{\nu}}(x) = \rho^2[\bar{Q}(x) + Q(x)(1-y)^2] \qquad (8.57)$$

require some explanation. For CM energies below the charm threshold but high enough so that the s-quark mass can be neglected, the possible transitions are

$$\begin{array}{ll}
\bar{\nu}u \to \mu^+ d, & \nu\bar{u} \to \mu^- \bar{d} \\
\nu d \to \mu^- u, & \bar{\nu}\bar{d} \to \mu^+ \bar{u} \\
\bar{\nu}u \to \mu^+ s, & \nu\bar{u} \to \mu^- \bar{s} \\
\nu s \to \mu^- u, & \bar{\nu}\bar{s} \to \mu^+ \bar{u}
\end{array} \qquad (8.58)$$

Thus, neglecting mixing angles θ_2 and θ_3, we have

$$D^\nu(x) = \bar{Q}_u(x)(1-y)^2 + Q_d(x)\cos^2\theta_1 + Q_s(x)\sin^2\theta_1 \qquad (8.59)$$
$$D^{\bar{\nu}}(x) = Q_u(x)(1-y)^2 + \bar{Q}_d(x)\cos^2\theta_1 + \bar{Q}_s(x)\sin^2\theta_1 \qquad (8.60)$$

In an isoscalar target, $Q_u(x) = Q_d(x)$, $\bar{Q}_u(x) = \bar{Q}_d(x)$, and $\bar{Q}_s(x) = Q_s(x)$. Thus for such a target, (8.59) becomes

$$D^\nu(x)_{iso} = \tfrac{1}{2}[\bar{Q}_u(x) + \bar{Q}_d(x)](1-y)^2 + \tfrac{1}{2}[Q_u(x) + Q_d(x)]\cos^2\theta_1$$
$$+ \tfrac{1}{2}[Q_s(x) + \bar{Q}_s(x)]\sin^2\theta_1 \qquad (8.61)$$

and similarly for $D^{\bar{\nu}}(x)_{iso}$. On the other hand, at energies well above the threshold for charm production, $D^{\nu,\bar{\nu}}(x)$ assumes the forms of (8.54) and (8.55) with $\rho^2 = \tfrac{1}{2}$ for isoscalar targets, since only one-half the quarks participate in charged-current νN interactions.

---

[3] The notation, commonly used, is quite unfortunate: q is the momentum transfer and $Q^2 = -q^2$. But q also stands for quark and $Q(x)$ for a quark x distribution. Most of the time this does not lead to confusion.

8.5.2 Structure functions

To proceed further we must formulate the scattering cross sections in terms of structure functions. Let us introduce them by considering deep inelastic electron–nucleon scattering by single-photon exchange (Figure 8.8). Here, the differential cross section is given by the standard formula:

$$d\sigma^{eN} = \frac{(2\pi)^4\, \delta^4(k + p - k' - p')}{[(k \cdot p)^2 - m_N^2 m_e^2]^{1/2}} \frac{m_e}{\omega'} \frac{m'}{E'} \frac{d^3\mathbf{p}'\, d^3\mathbf{k}'}{(2\pi)^6} |\overline{\mathcal{M}}|^2 \qquad (8.62)$$

where m' is the invariant final hadronic mass $m' = (p' \cdot p')^{1/2}$ and $|\overline{\mathcal{M}}|^2$ the square of the matrix element, summed over final and averaged over initial spins. In the laboratory frame, (8.62) becomes, in the limit $m_e \ll \omega_L$,

$$d\sigma^{eN} = \frac{1}{(2\pi)^2} \frac{\delta^4(k + p - k' - p')}{\omega_L \omega_L' m_N E'} m_e m' \, d^3\mathbf{p}'\, \omega_L'^2 \, d\omega_L'\, d\Omega_e' \, |\overline{\mathcal{M}}|^2$$

Integration over the unobserved $\mathbf{p}'$ yields:

$$d\sigma^{eN} = \frac{1}{(2\pi)^2} \frac{\delta(\omega_L + m_N - \omega_L' - E') m_e m' \omega_L' \, d\omega_L' \, d\Omega_e' \, |\overline{\mathcal{M}}|^2}{\omega_L m_N E'} \qquad (8.63)$$

Figure 8.8. Deep inelastic scattering by single-photon exchange.

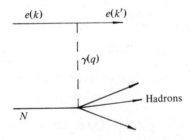

For single-photon exchange, $|\overline{\mathcal{M}}|^2$ must take the form:

$$|\overline{\mathcal{M}}|^2 = \tfrac{1}{4} \sum_{\text{spins}} L_\lambda L_\sigma^* W^{\lambda\sigma}/q^4 \qquad (8.64)$$

where $L_\lambda = (4\pi\alpha)^{1/2} \bar{u}_e(k') \gamma_\lambda u_e(k)$ is the lepton vertex factor, q^{-4} the square of the photon propagator, and $W^{\lambda\sigma}$ the tensor describing the hadronic vertex. Quite generally, $W^{\lambda\sigma}$ must be a function of the kinematic invariants q^2 and $q \cdot p$ or x, and it can be written as follows (Drell and Walecka 64):

$$\sum_{N\text{ spin}} W^{\lambda\sigma} = \frac{4\pi\alpha}{m_N m'} [A g^{\lambda\sigma} + B q^\lambda q^\sigma + C p^\lambda p^\sigma + D(q^\lambda p^\sigma + p^\lambda q^\sigma)$$

$$+ E(q^\lambda p^\sigma - p^\lambda q^\sigma)] \qquad (8.65)$$

8.5 Deep inelastic charged-current ν–N interactions

where $A, B, \ldots, E$ are functions of q^2 and $q \cdot p$. Electromagnetic current conservation implies $q_\lambda W^{\lambda\sigma} = q_\sigma W^{\lambda\sigma} = 0$, and therefore that

$$E = 0$$

$$C = \frac{q^2}{(q \cdot p)^2}(A + Bq^2)$$

and

$$D = -\frac{1}{p \cdot q}(A + Bq^2)$$

Thus we obtain:

$$\sum_{N \text{ spin}} W^{\lambda\sigma} = \frac{8\pi\alpha}{m_N m'}\left[a\left(p^\lambda - \frac{q \cdot p q^\lambda}{q^2}\right)\left(p^\sigma - \frac{q \cdot p q^\sigma}{q^2}\right)\right.$$

$$\left. - bm_N^2\left(g^{\lambda\sigma} - \frac{q^\lambda q^\sigma}{q^2}\right)\right] \quad (8.66)$$

where a and b are new functions of q^2 and $q \cdot p$. The leptonic factor reduces by standard techniques to

$$\sum_{e \text{ spin}} L_\lambda L_\sigma^* = \frac{4\pi\alpha}{m_e^2}[k'_\lambda k_\sigma + k'_\sigma k_\lambda + (m_e^2 - k' \cdot k)g_{\lambda\sigma}] \quad (8.67)$$

Thus employing (8.66) and (8.67) and taking the limit $m_e \ll \omega_L$, we find that, in the laboratory frame, (8.64) becomes

$$|\overline{\mathcal{M}}|^2 = \frac{16\pi^2 m_N \alpha^2 \omega'_L \omega_L}{m_e^2 m' q^4}\left(a \cos^2\frac{\theta}{2} + 2b \sin^2\frac{\theta}{2}\right) \quad (8.68)$$

Functions a and b must be determined by experiment, but we can gain insight into them by comparing (8.68) with the formula for eN elastic scattering (8.8). Thus $|\overline{\mathcal{M}}|^2$ of (8.68) reduces to $|\overline{\mathcal{M}}|^2_{\text{elas}}$ when

$$a = C_N^2 - M_N^2 q^2 \quad (8.69)$$

and

$$b = (Q^2/4m_N^2)(C_N + 2m_N M_N)^2 \quad (8.70)$$

To complete our formulation of the inelastic cross section, we make the change of variable:

$$d(Q^2) = \omega_L \omega'_L \, d\Omega'/\pi \quad (8.71)$$

and obtain from (8.63) and (8.68) the result:

$$\frac{\partial^2 \sigma_{\text{inelas}}(e, N)}{\partial Q^2 \, \partial \nu} = \frac{4\pi\alpha^2 \omega'_L}{q^4 \omega_L}\cos^2\frac{\theta}{2}\left[W_2(q^2, \nu) + 2\tan^2\frac{\theta}{2} W_1(q^2, \nu)\right] \quad (8.72)$$

where

$$W_2(q^2, \nu) = (am_N/E')\delta(\omega_L + m_N - \omega'_L - E') \tag{8.73}$$

and

$$W_1(q^2, \nu) = (bm_N/E')\delta(\omega_L + m_N - \omega'_L - E') \tag{8.74}$$

The important formula (8.72) was derived on the basis of two simple assumptions: that the interaction is mediated by a single vector boson and that the vector current is conserved. However, according to CVC, this is also true of the vector part of the ν–N interaction. Thus we may immediately extend (8.72) to describe ν–N scattering by single W exchange:

$$\frac{\partial^2 \sigma_{\text{inelas}}(\nu, N)}{\partial Q^2 \, \partial \nu} = \frac{\omega'_L}{\omega_L} \frac{G_F^2}{2\pi} \left\{ \mathcal{W}_2 \cos^2 \frac{\theta}{2} + 2\mathcal{W}_1 \sin^2 \frac{\theta}{2} + \mathcal{A} \right\} \tag{8.75}$$

where $\mathcal{W}_{1,2}$ are new structure functions analogous to (but not equal to) $W_{1,2}$, respectively, of (8.72). As for the third term on the right-hand side of (8.75), it may be shown (Pais 71) that, if T invariance holds and there are no second-class currents, $\mathcal{A}$ takes the form:

$$\mathcal{A}^{\nu, \bar{\nu}} = \mp \frac{\omega_L + \omega'_L}{m_N} \mathcal{W}_3^{\nu, \bar{\nu}} \sin^2 \frac{\theta}{2} \tag{8.76}$$

where $\mathcal{W}_3$ is a third structure function arising from interference between vector and axial currents and the $\mp$ sign in (8.76) applies for νN and $\bar{\nu}N$ scattering respectively.

Actually, it is most useful to define new form factors by the relations:

$$F_1^{\nu, \bar{\nu}}(x, q^2) = \mathcal{W}_1^{\nu, \bar{\nu}}(q^2, \nu) \tag{8.77}$$

$$F_2^{\nu, \bar{\nu}}(x, q^2) = \nu \mathcal{W}_2^{\nu, \bar{\nu}}(q^2, \nu)/m_N \tag{8.78}$$

$$F_3^{\nu, \bar{\nu}}(x, q^2) = \nu \mathcal{W}_3^{\nu, \bar{\nu}}(q^2, \nu)/m_N \tag{8.79}$$

Making these substitutions in (8.76) and (8.75) and changing variables from Q^2 and ν to x and y, we obtain:

$$\frac{\partial^2 \sigma^{\nu N, \bar{\nu} N}}{\partial x \, \partial y} = \frac{G_F^2 m_N \omega_L}{\pi} \left[\left(1 - y - \frac{m_N xy}{2\omega_L}\right) F_2^{\nu, \bar{\nu}}(x, q^2) + y^2 x F_1^{\nu, \bar{\nu}}(x, q^2) \right.$$
$$\left. \mp y\left(1 - \frac{y}{2}\right) x F_3^{\nu, \bar{\nu}}(x, q^2) \right] \tag{8.80}$$

or, ignoring the small term in $m_N xy/2\omega_L$,

$$\frac{\partial^2 \sigma^{\nu N, \bar{\nu} N}}{\partial x \, \partial y} = \frac{G_F^2 m_N \omega_L}{\pi} \left[\left(x F_1 \mp \frac{x F_3}{2}\right) \right.$$
$$\left. + (F_2 - 2xF_1)(1 - y) + \left(x F_1 \pm \frac{x F_3}{2}\right)(1 - y)^2 \right] \tag{8.81}$$

where we have expressed the differential cross section in powers of $1 - y$ in the last formula.

8.5 Deep inelastic charged-current ν–N interactions

A priori, there are 12 form factors in (8.80) or (8.81): 3 each for νp, νn, $\bar{\nu} p$, and $\bar{\nu} n$ scattering. However, if we assume charge symmetry in strong interactions, it follows that

$$F_i^{\nu n} = F_i^{\bar{\nu} p} \quad \text{and} \quad F_i^{\bar{\nu} n} = F_i^{\nu p}, \quad i = 1,2,3 \tag{8.82}$$

The remaining six are reduced to three if we restrict consideration to targets where there are approximately equal numbers of protons and neutrons, such as the isoscalar marble target in the CHARM detector. This is valid, of course, only if the $q\bar{q}$ sea and Cabibbo-suppressed weak couplings are neglected, since the s–c quark mass difference breaks charge symmetry.

8.5.3 Predictions of the quark–parton model

We are now ready to compare the general differential cross section (8.80) or (8.81) with the corresponding formulas (8.54) or (8.55) of the quark–parton model. A number of important results are thus obtained in this comparison, as follows:

(a) Since (8.54) and (8.55) are valid in the limit $Q^2 \gg m_N^2$ and $\nu \gg m_N$, the form factors $F_i(x, q^2)$ must be independent of Q^2 in this "scaling" limit:

$$F_i(x, q^2) \to F_i(x) \tag{8.83}$$

This was first conjectured by Bjorken and Paschos (1969) and was shortly thereafter confirmed in electron–proton scattering at SLAC, to a precision of about 10 percent. As we have seen, scaling follows from the assumption that, at large Q^2, the hadron appears to be composed of pointlike "free" quark–partons.

(b) From (8.80) and (8.83) neglecting the term $m_N xy/2\omega_L$, scaling implies that the total cross section for charged-current ν–N or $\bar{\nu}$–N scattering must rise linearly with ω_L, since x and y are dimensionless parameters. Of course, this conclusion breaks down when the CM energy approaches m_W, but in all neutrino experiments so far, one has $E_{CM} < 30$ GeV $< m_W$, and the linear increase is confirmed within errors. The average of several experiments with values of ω_L ranging up to 260 GeV is (Kelly et al. 80):

$$\sigma^{\nu N}/\omega_L = (0.63 \pm 0.02) \times 10^{-38} \text{ cm}^2/\text{GeV} \tag{8.84}$$

$$\sigma^{\bar{\nu} N}/\omega_L = (0.30 \pm 0.01) \times 10^{-38} \text{ cm}^2/\text{GeV} \tag{8.85}$$

(see Figure 8.9).

(c) Comparison of (8.81) with (8.54) or (8.55) yields the following quark–parton model predictions:

$$F_L(x) \equiv F_2(x) - 2xF_1(x) = 0 \tag{8.86}$$

$$xQ(x) = xF_1(x) + \tfrac{1}{2}xF_3(x) \tag{8.87}$$

$$x\bar{Q}(x) = xF_1(x) - \tfrac{1}{2}xF_3(x) \tag{8.88}$$

Equation (8.86) is called the Callan–Gross relation. Within the quark–parton approximation, it holds only for spin-$\tfrac{1}{2}$ partons, since other spin assignments would lead to terms proportional to $1 - y$ in (8.54) or (8.55).[4] More generally, the "longitudinal structure function" $F_L(x, q^2)$ can be nonzero because of scale-breaking (QCD) corrections. It actually accounts for less than 20 percent of the total cross section experimentally and for most purposes we may ignore it.

Structure functions F_2 and F_3 may be separated by writing the sum and difference of (8.80) for ν and $\bar{\nu}$:

$$\frac{\partial^2}{\partial x\, \partial y}(\sigma_{CC}^{\nu N} + \sigma_{CC}^{\bar{\nu} N}) = \frac{G_F^2 m_N \omega_L}{\pi}\{F_2(x)[1 + (1-y)^2] - F_L(x)y^2\} \tag{8.89}$$

---

[4] Here we assume that the weak interaction contains only V–A couplings. S, T, or P would all contribute to $1 - y$ terms even for spin $\tfrac{1}{2}$.

Figure 8.9. Neutrino–nucleon and antineutrino–nucleon total cross section for isoscalar targets. [Source: Kelly et al. (80) and ●, Barish et al. (79a); ○, Barish et al. (77); ▽, de Groot et al. (79); △, Ciampolillo et al. (79); □, Colley et al. (79); ◇, Barish et al. (79b); and ■, Asratyan et al. (78).]

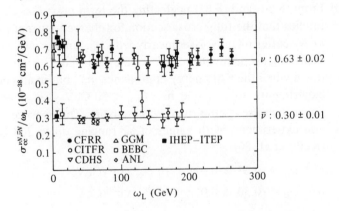

8.5 Deep inelastic charged-current ν–N interactions

$$\frac{\partial^2}{\partial x\, \partial y} (\sigma_{CC}^{\nu N} - \sigma_{CC}^{\bar\nu N}) = \frac{G_F^2 m_N \omega_L}{\pi} \{xF_3(x)[1 - (1-y)^2]\} \tag{8.90}$$

Also from (8.86) to (8.88) we obtain

$$\frac{1}{x} F_2(x) = Q(x) + \bar Q(x) + \frac{F_L(x)}{x} \simeq Q(x) + \bar Q(x) \tag{8.91}$$

and

$$F_3(x) = Q(x) - \bar Q(x) \tag{8.92}$$

Equations (8.89) to (8.92) suggest ways in which neutrino scattering may be used to verify the quark–parton model. For example, from measurements of F_2, it is possible to show that quarks have fractional charge, by comparing (8.89) to the analogous differential cross section for electron–nucleon scattering. The latter is identical to (8.89) except for the substitutions

$$G_F^2 m_N \omega_L / \pi \rightarrow 8\pi\alpha^2/q^4$$

and $F_3(x) \rightarrow 0$. Thus

$$\frac{\partial^2 \sigma^{eN}}{\partial x\, \partial y} = \frac{8\pi\alpha^2}{q^4} \{F_2^{eN}(x)[1 + (1-y)^2] - F_L^{eN}(x)y^2\} \tag{8.93}$$

Furthermore, the quark distribution functions for eN scattering differ from those of (8.87) and (8.88) only by the change

$$Q(x) \rightarrow [\text{quark charge}/e]^2 Q(x) = \begin{cases} \frac{4}{9} Q_i(x) & \text{for } i = u, c, t \\ \frac{1}{9} Q_i(x) & \text{for } i = d, s, b \end{cases}$$

Therefore the ratio $F_2^{eN}/F_2^{\nu N}$ is just the mean square quark charge:

$$\frac{F_2^{eN}}{F_2^{\nu N}} = \frac{\frac{4}{9}(Q_u + \bar Q_u) + \frac{1}{9}(Q_d + \bar Q_d)}{\bar Q + Q} = \frac{5}{18} \tag{8.94}$$

for isoscalar targets. If quarks had integer charges, this ratio would be approximately $\frac{1}{2}$. Experiments confirm the fractional charge hypothesis (Figure 8.10).

According to (8.92), $F_3^{\bar\nu N}$ is equal to the difference between quark and antiquark concentrations. Since the sea must have equal numbers of each, we arrive at the Gross–Llewellyn–Smith sum rule in the limit $Q^2 \rightarrow \infty$:

$$I = \int_0^1 F_3^{\nu N}(x)\, dx = \text{number of valence quarks} = 3 \tag{8.95}$$

This is well confirmed by experiment: $I = 3.2 \pm 0.5$ (de Groot et al. 79) and $I = 2.8 \pm 0.6$ (Benvenuti et al. 79).

Antiquarks (which belong entirely to the sea) can be distinguished from quarks by their different y dependence as follows. First, we suppose that there were no antiquarks. Then we would have $\partial \sigma^{\nu N}/\partial y =$ constant and $\partial \sigma^{\bar\nu N}/\partial y \propto (1 - y)^2 \to 0$ at $y = 1$. However, Figure 8.11 shows how $\partial \sigma^{\nu N}/\partial y$ and $\partial \sigma^{\bar\nu N}/\partial y$ actually vary with y. These results reveal that antiquarks make up about 10 to 15 percent of the partons. Also, the ratio of total cross sections $\sigma^{\nu N}/\sigma^{\bar\nu N}$ deviates from the naive quark model prediction of $\frac{1}{3}$ because of the $q\bar q$ pairs in the sea. One obtains the experimental ratios:

$$\sigma_{CC}^{\nu N}/\sigma_{CC}^{\bar\nu N} = 0.38 \pm 0.025 \qquad \text{Gargamelle,} \quad 2 < \omega_L < 10 \text{ GeV} \tag{8.96}$$

and

$$\sigma_{CC}^{\nu N}/\sigma_{CC}^{\bar\nu N} = 0.48 \pm 0.03 \qquad \text{(CDHS and CTFR, 30–200 GeV)} \tag{8.97}$$

(from, respectively, Eichten et al. 73; Barish et al. 77 and de Groot et al. 79).

The slight energy dependence here is a combination of two effects: first, there are scaling violations attributable to the strong interaction; second, the composition of the $q\bar q$ sea changes at high energies because of the threshold for charm-quark production.

The x distributions for quarks and antiquarks are quite different, as shown in Figure 8.12, which shows $d\sigma/dx$ for high values of y, where neutrinos scatter only from quarks, and antineutrinos from antiquarks. Again, this is in accord with the naive quark model, which states that the valence quarks ought to carry a greater fraction of the total nuclear momentum.

Finally, the total cross section is a measure of the total momentum

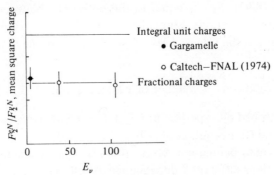

Figure 8.10. Ratio of electromagnetic to νN structure functions F_2, equal to the mean square parton charge. (From Sciulli 80.)

Figure 8.11. y dependence of cross section for νN and $\bar{\nu}N$ scattering, showing portion attributable to sea antiquarks. Note that antiquarks make up a different fraction of the total at low y (measured by νN scattering) than at high y (from $\bar{\nu}N$ scattering). (From de Groot et al. 79.) (Reprinted with permission.)

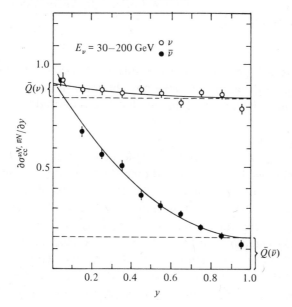

Figure 8.12. x distributions for νN and $\bar{\nu}N$ at large y. In the quark model, the νN plot is the quark distribution $Q(x)$ and the $\bar{\nu}N$ plot is the antiquark distribution $\bar{Q}(x)$ (CDHS data). (From de Groot, et al. 79. Reprinted with permission.)

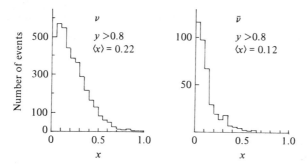

carried by all the quarks. One might expect that this is simply equal to the momentum of the nucleon, that is,

$$\int x[Q(x) + \bar{Q}(x)]\, dx = 1 \tag{8.98}$$

in which case, neglecting a slight correction for the lower $\bar{q}$ cross section, we would simply have

$$\sigma_{CC}^{\nu N} = G^2 m_N \omega_L / \pi = 1.3 \times 10^{-38} \quad \text{cm}^2 \ (\omega_L \text{ in GeV}) \tag{8.99}$$

However, the actual cross section for neutrinos is smaller by about a factor of 2, indicating that the partons carry about one-half the total nucleon momentum. It is presumed that the balance is carried by gluons.

8.5.4 Scaling violations

We have just seen that the simple quark–parton model accounts remarkably well for scaling and other features of deep inelastic lepton–hadron scattering. However, the experimental data deviate slightly from the ideal behavior of this model, providing clues about the structure of the nucleon and the nature of the strong force. It is thought that these scaling violations are caused partly by residual quark–gluon interactions, which are characterized by the effective coupling constant $\alpha_S(Q^2)$ of QCD [see (6.109)]. Some features of the violations also arise simply because we have made a choice of scaling variables that is only approximately appropriate. In the following brief discussion, we shall limit ourselves to simple qualitative remarks concerning these effects. The reader interested in more details should consult recent reviews such as Mueller (81) or Reya (81).

When probed by a neutrino or charged lepton that scatters with a certain value of Q^2, a quark q may appear to have momentum fraction x. However, at higher Q^2, the quark will be resolved into a gluon plus a quark with smaller x' (Figure 8.13a). Thus we expect the quark distri-

Figure 8.13. Quark–gluon interactions that lead to deviations from scaling.

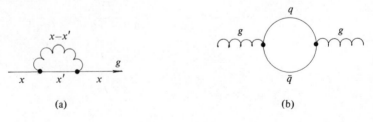

Figure 8.14. (a) $F_2(x)$ in different x bins as a function of $\ln(Q^2)$; (b) $xF_3(x)$ in different x bins as a function of $\ln(Q^2)$. (From de Groot et al. 79. Reprinted with permission.)

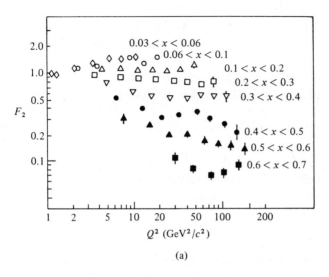

(a)

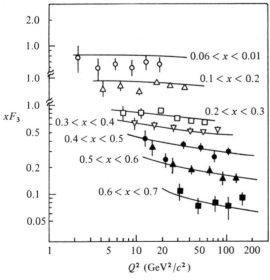

(b)

bution functions $F_2(x)$ and $xF_3(x)$ to decrease with increasing Q^2 at large x, in violation of the scaling prediction. Also, there are many $q\bar{q}$ pairs with fairly small x in the nucleon, formed by gluon pair production (Figure 8.13b). At larger values of Q^2, more of these pairs can be probed, so we expect the structure functions to increase with increasing Q^2 at low x, again in violation of scaling. These effects are seen in the CDHS data (Figure 8.14).

One may characterize the QCD effects in first order in terms of certain moments of the quark distribution defined as

$$M_N(Q^2) = \int_0^1 x^{N-1} Q(x, Q^2)\, dx \tag{8.100}$$

It can be shown that in first-order QCD these quantities vary as the logarithm of Q^2 raised to a certain negative power. They are therefore called *logarithmic scaling violations*.

When the variable

$$x = Q^2/2m_N\nu$$

was defined as the momentum fraction ξ of the quark, we assumed that in the infinite momentum frame both m_N and the quark mass were negligible. The failure of this approximation leads to scaling violations proportional to $1/Q^2$. The actual momentum fraction ξ is defined by

$$(\xi p + q)^2 = m^2$$

Solving for ξ, we find:

$$\xi = \frac{-\nu + (q^2 + \nu^2 + m^2)^{1/2}}{m_N} = x\left(1 + \frac{m_N^2 x^2 - m^2}{Q^2} + \cdots\right) \tag{8.101}$$

Therefore the approximation $\xi = x$ causes an error in ξ of order $m_N^2 x^2/Q^2$. Also, charm production results in scaling violation attributable to the nonnegligible c-quark mass, which is of order m_c^2/Q^2. It is desirable to separate these mass effects from the logarithmic scaling violations of first-order QCD, and therefore the structure functions are often expressed in terms of the *Nachtman variable*:

$$\xi_N = \frac{2x}{1 + (1 - 4m_N^2 x^2/Q^2)^{1/2}} \simeq x\left(1 + \frac{m_N^2 x^2}{Q^2}\right) \tag{8.102}$$

which is a better approximation to the true momentum fraction ξ than x. These effects should be most important at small Q^2. Unfortunately, first-order QCD calculations also break down at small Q^2, because $\alpha_S(Q^2)$ becomes large. The ambiguities of higher-order QCD corrections and other low-Q^2-scale breaking phenomena remain to be resolved.

8.6 Charmed-particle production in neutrino beams

The standard model provides a very simple picture of charmed particle production by ν–N and $\bar{\nu}$–N collisions. We neglect t and b quarks and employ the GIM scheme to find the following production ratios for neutrinos:

$$d\sigma/dy\ (\nu_\mu d \to \mu^- u) \propto \cos^2 \theta_1 \tag{8.103}$$

$$d\sigma/dy\ (\nu_\mu d \to \mu^- c) \propto \sin^2 \theta_1 \tag{8.104}$$

$$d\sigma/dy\ (\nu_\mu s \to \mu^- c) \propto \epsilon \cos^2 \theta_1 \tag{8.105}$$

where ϵ is the ratio of s-type quarks (entirely in the sea) to valence and sea-type d quarks. Since $\sin^2 \theta_1 \simeq 0.05$ and ϵ turns out to be about 0.03, we expect charm production in about 8 percent of the ν–N reactions. For $\bar{\nu}$–N, the situation is somewhat more favorable:

$$d\sigma/dy\ (\bar{\nu} u \to \mu^+ d) \propto \cos^2 \theta_1 (1-y)^2 \tag{8.106}$$

$$d\sigma/dy\ (\bar{\nu} \bar{s} \to \mu^+ \bar{c}) \propto \epsilon \cos^2 \theta_1 \tag{8.107}$$

The $(1-y^2)$ suppression in (8.106) means that anticharm may be produced in up to 10 percent of $\bar{\nu}$–N reactions. In addition, although the following reactions are forbidden to first order:

$$\nu q \not\to \bar{c} + \text{anything} \tag{8.108}$$

$$\bar{\nu} q \not\to c + \text{anything} \tag{8.109}$$

they can occur in secondary processes such as $c\bar{c}$ pair production.

Clearly, then, neutrino beams offer very attractive possibilities for the study of charmed particles, provided a means can be found for identifying the latter. As noted in Section 8.2, charmed-particle lifetimes are so short (10^{-12}–10^{-13} sec) that photographic emulsion and/or bubble-chamber techniques must be employed to observe these particles directly. In order to achieve good statistics, however, it is necessary to rely on counters, in which case the evidence for charmed particles is indirect and comes from their semileptonic decay modes.

As noted in Chapter 6, these are expected to have branching ratios $B \simeq 10$–20% for electrons and for muons. The relevant transitions are as follows, with the lepton production rates given on the right-hand side:

$$\begin{array}{l} \nu d \to \mu^- c \\ \hookrightarrow l^+ \nu_l s \end{array} \quad \propto B \sin^2 \theta_C \tag{8.110}$$

$$\begin{array}{l} \nu s \bar{s}_T \to \mu^- c \bar{s}_T \\ \phantom{\nu s \bar{s}_T \to \mu^- c}\hookrightarrow l^+ \nu_l s \end{array} \quad \propto B\epsilon \tag{8.111}$$

$$\begin{array}{l} \bar{\nu} \bar{s} s_T \to \mu^+ \bar{c} s_T \\ \phantom{\bar{\nu} \bar{s} s_T \to \mu^+ \bar{c}}\hookrightarrow l^- \bar{\nu}_l \bar{s} \end{array} \quad \propto 3B\epsilon \tag{8.112}$$

where the subscript T denotes strange quarks in the sea that are stranded in the target when their charge-conjugate quark is scattered. We are thus led to the following general predictions:

(a) Strange mesons should accompany charm production;
(b) The two leptons produced should have opposite sign.

In dimuon events, the muon with the higher momentum comes from the primary νW vertex, and the other one from charm decay. Thus, according to the model, it is easy to determine whether c or $\bar{c}$ was produced, simply from the charge of the faster muon. This also means that x and y can be measured without ambiguity as to which lepton came from the neutrino beam, and the differential cross section may be determined. The following general features for $d\sigma/dy$ are predicted by the model and confirmed in experiments.

(i) The y distribution is flat, for both ν and $\bar{\nu}$, since neutrinos produce charm from quarks, whereas antineutrinos produce antineutrinos produce antineutrinos produce anticharm from antiquarks. This is in contrast to the $(1-y)^2$ dependence of the inclusive $\bar{\nu}$ cross section.

(ii) The $\bar{\nu}$ cross section has a lower average x than the inclusive cross section, since all $\bar{\nu}$ charmed interactions involve sea quarks only.

(iii) The relative angle between the two muon momentum projections in the plane normal to the neutrino beam axis (the azimuthal angle) is peaked at 180°. This follows from the fact that the transverse momentum of the initial quark is negligible, so that the fast muon and the c quark are emitted back to back in the transverse plane.

Figure 8.15. y distributions for neutrino and antineutrino dimuon events. The solid curves indicate the acceptance of the detector. If the data follow the curve, the cross sections have no y dependence. The dotted curve shows the expected shape of a cross section that has a $(1-y)^2$ dependence. (From Willutzki 79.)

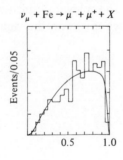
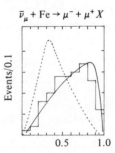

8.6 Charmed-particle production in neutrino beams

(iv) Finally, the direction of the hadronic shower is approximately the direction of the scattered charmed quark. The transverse momentum of the slower muon with respect to this direction should be small, representing the average momentum of the semileptonic decay.

Dimuon data from the CDHS detector are shown in Figures 8.15–8.18 and confirm all of the preceding predictions. Thus we may be fairly cer-

Figure 8.16. Dimuon x distributions. The antineutrino cross section is typical of sea quarks, which is expected according to the GIM scheme for charm production. Curves are fits to the data of the function $(1 - x)^\alpha$. (From Willutzki 79.)

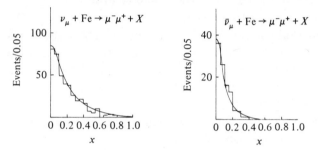

Figure 8.17. Azimuthal angle between scattered dimuons, showing the expected peak at 180°. (From Holder et al. 77. Reprinted with permission.)

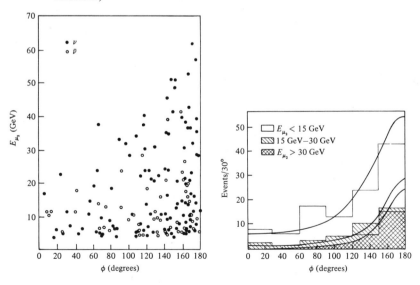

320   8 High-energy neutrino-nucleon collisions

tain that charm is responsible for dimuon production in neutrino beams.

In about 1 dimuon event in 10, the leptons have the same sign, and trimuon events have also been observed. Many of these can be ascribed to various backgrounds, such as π and K decays, but some evidence exists for a "prompt" signal, that is, one that has the characteristics of heavy-quark decay. Several exotic explanations have been proposed to explain these events, such as new heavy leptons or b quark production, but a more natural explanation is $c\bar{c}$ pair creation and present data seem to favor this hypothesis (Smith 79).

8.7   Tests of the standard model with neutral neutrino–nucleon reactions

Neutral-current neutrino–nucleon reactions have played an essential role in the study of electroweak interactions and have been crucial in establishing the standard model. In addition to the elastic reactions already considered in Section 8.4, there are the following types of inelastic processes:

(i) deep inelastic "inclusive" scattering ($\nu N \to \nu x$) from proton, neutron, and isoscalar targets;

Figure 8.18. Transverse momentum of the *slower* scattered dimuon with respect to the hadronic shower axis. According to the model of charm production, this should be approximately the momentum of the decay muon in the charmed-particle rest frame. (From Holder et al. 1977. Reprinted with permission.)

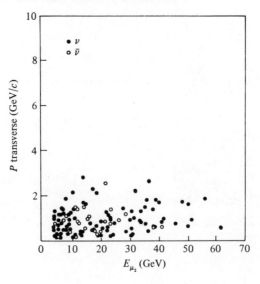

8.7 Tests of the standard model

(ii) "semi-inclusive" scattering ($\nu N \to \nu \pi x$) in which a pion is observed in the final state;

(iii) exclusive reactions, for example, $\nu N \to \nu \pi N'$ and $\nu N \to \nu \Delta$; and

(iv) finally, the low-energy energy reaction $\bar{\nu}_e D \to \bar{\nu}_e p n$ generated with reactor antineutrinos.

We shall now consider how each of these contributes to our general understanding of the neutral weak interaction.

8.7.1 Deep inelastic neutral-current scattering

(*a*) *Cross sections.* In the following, scale-breaking effects will be ignored (or else corrections applied from CC experiments) and our analysis will be based on the quark–parton model. In this approximation, the NC differential cross sections for ν–N and $\bar{\nu}$–N scattering are functions of $x = Q^2/2m_N \nu$ and $y = \nu/\omega$ only. We shall also continue to assume V, A couplings; S, T, and P couplings might also enter, but in principle, these can be distinguished by their y dependence:[5]

$$\partial \sigma / \partial y \; (V, A) \propto 1, (1 - y)^2$$
$$\partial \sigma / \partial y \; (S, P) \propto y^2$$
$$\partial \sigma / \partial y \; (T) \propto (2 - y)^2$$

if one assumes that partons have spin $\tfrac{1}{2}$.

For a target nucleon composed of u, d, and s quarks, the NC differential cross sections may then be derived by following the same steps that lead to the CC cross sections (8.54) and (8.55). One thus obtains:

$$\frac{\partial \sigma_{NC}^{\nu N}}{\partial x \, \partial y} = \frac{2 G_F^2 m_N \omega x}{\pi} \{ |u_L|^2 [Q_u(x) + \bar{Q}_u(x)(1-y)^2] $$
$$+ |u_R|^2 [\bar{Q}_u(x) + Q_u(x)(1-y)^2]$$
$$+ |d_L|^2 [Q_d(x) + Q_s(x) + (\bar{Q}_d(x) + \bar{Q}_s(x))(1-y)^2]$$
$$+ |d_R|^2 [\bar{Q}_d(x) + \bar{Q}_s(x) + (Q_d(x) + Q_s(x))(1-y)^2] \} \quad (8.113)$$

$$\frac{\partial \sigma_{NC}^{\bar{\nu} N}}{\partial x \, \partial y} = \frac{2 G_F^2 m_N \omega x}{\pi} \{ |u_R|^2 [Q_u(x) + \bar{Q}_u(x)(1-y)^2]$$
$$+ |u_L|^2 [\bar{Q}_u(x) + Q_u(x)(1-y)^2]$$
$$+ |d_R|^2 [Q_d(x) + Q_s(x) + (\bar{Q}_d(x) + \bar{Q}_s(x))(1-y^2)]$$
$$+ |d_L|^2 [\bar{Q}_d(x) + \bar{Q}_s(x) + (Q_d(x) + Q_s(x))(1-y^2)] \} \quad (8.114)$$

---

[5] As we shall see later, actual experimental limits on S–P couplings from y dependence are none too stringent.

where $u_{L,R}$ and $d_{L,R}$ are left- and right-handed couplings to up and down quarks, respectively.

These differ from the charged-current cross sections (8.54) and (8.55) in two important respects: first, there is no dependence on the Cabibbo angles; second, the ν–N and $\bar{\nu}$–N cross sections each involve u and d quark distribution functions.

The NC differential cross sections are extremely difficult to measure. In the CC case, one observes the final-state muon momentum and angle and the total energy of the hadronic shower. Thus it is possible to determine ω, y, and x. In neutral interactions, one may observe only the hadronic shower. The energy of the incident neutrino must be determined from the spectrum of the neutrino beam measured in CC events. Usually a dichromatic beam is used in which the energy at a given radial distance from the beam axis has a spread of 10 to 20 percent. This information, plus the shower energy, determines y. It is much more difficult to determine x, which depends on momentum transfer and therefore requires knowledge of the shower direction. Although some x distributions have been measured (Mess 79), in most experiments one integrates over this variable.

The y distributions have been used to set limits on S or P couplings by fitting the differential cross sections to

$$d\sigma_{NC}/dy = h_1 + h_2(1-y)^2 + h_3 y^2 \tag{8.115}$$

One finds

$$h_3/(h_1 + h_2) < \begin{cases} 0.16 & \text{from CDHS} \\ 0.12 & \text{from BEBC} \end{cases} \tag{8.116}$$
$$\tag{8.117}$$

(respectively, Holder et al. 77 and Deden et al. 79). Unfortunately these results yield rather poor limits on scalar and pseudoscalar couplings to up and down quarks $u_{S,P}$, $d_{S,P}$ (Deden et al. 79):

$$\frac{|u_S|^2 + |u_P|^2 + |d_S|^2 + |d_P|^2}{|u_L|^2 + |u_R|^2 + |d_L|^2 + |d_R|^2} < 0.50 \tag{8.118}$$

There is also a small contribution to h_3 from scaling violations, which gives rise to finite $F_L(x)$ in charged currents. Thus it is doubtful that the previous limit can be improved substantially.

(b) *General results for isoscalar targets.* In the difference

$$\frac{\partial^2 \sigma_{NC}^{\nu N}}{\partial x \, \partial y} - \frac{\partial^2 \sigma_{NC}^{\bar{\nu} N}}{\partial x \, \partial y}$$

the sea–quark contributions cancel. The remaining distributions refer

8.7 Tests of the standard model

only to valence quarks, and strong isospin invariance requires that $Q_u^{\text{val.}}(x) = Q_d^{\text{val.}}(x)$ for isoscalar targets. Thus we obtain

$$\frac{\partial^2 \sigma_{\text{NC}}^{\nu N}}{\partial x \, \partial y} - \frac{\partial^2 \sigma_{\text{NC}}^{\bar{\nu} N}}{\partial x \, \partial y} = \frac{G_F^2 m_N \omega x}{\pi} Q^{\text{val}}(x)$$
$$\times [|u_L|^2 + |d_L|^2 - |u_R|^2 - |d_R|^2][1 - (1-y)^2] \quad (8.119)$$

This can be compared with the difference in CC cross sections:

$$\frac{\partial^2 \sigma_{\text{CC}}^{\nu N}}{\partial x \, \partial y} - \frac{\partial^2 \sigma_{\text{CC}}^{\bar{\nu} N}}{\partial x \, \partial y} = \frac{G_F^2 m_N \omega x}{\pi} Q(x)[1 - (1-y)^2] \quad (8.120)$$

The ratio of (8.119) and (8.120) gives the Paschos–Wolfenstein relation (73):

$$\frac{(\partial \sigma^{\nu N} - \partial \sigma^{\bar{\nu} N})_{\text{NC}}}{(\partial \sigma^{\nu N} - \partial \sigma^{\bar{\nu} N})_{\text{CC}}} = |u_L|^2 + |d_L|^2 - |u_R|^2 - |d_R|^2$$
$$= \tfrac{1}{2} - \sin^2 \theta_W \quad (8.121)$$

where the last equality holds in the standard model. Formula (8.121) should be valid independent of the details of hadronic structure. It has been used to determine the Weinberg angle:

$$\sin^2 \theta_W = 0.230 \pm 0.023 \quad \text{(CHARM)} \quad (8.122)$$

The ratios of NC to CC total cross sections for an isoscalar target are given by

$$R_\nu = \sigma_{\text{NC}}^{\nu N}/\sigma_{\text{CC}}^{\nu N} = |u_L|^2 + |d_L|^2 + \tfrac{1}{3}|u_R|^2 + \tfrac{1}{3}|d_R|^2 + \text{corrections}$$
$$= \tfrac{1}{2} - \sin^2 \theta_W + (\tfrac{20}{27}) \sin^4 \theta_W + \text{corrections} \quad (8.123)$$

$$R_{\bar{\nu}} = \sigma_{\text{NC}}^{\bar{\nu} N}/\sigma_{\text{CC}}^{\bar{\nu} N} = |u_R|^2 + |d_R|^2 + \tfrac{1}{3}|u_L|^2 + \tfrac{1}{3}|d_L|^2 + \text{corrections}$$
$$= \tfrac{1}{2} - \sin^2 \theta_W + (\tfrac{20}{9}) \sin^4 \theta_W + \text{corrections} \quad (8.124)$$

where the corrections refer to the effects of the $q\bar{q}$ sea and of scale breaking.

Measurements of $R_{\nu,\bar{\nu}}$ are subject to various systematic errors that must be corrected. In the counter experiments, there are three main sources: CC events mistakenly identified as NC events because of failure to observe the final-state muon; events induced by ν_e in K_{e3} decays that always appear as NC events because of the short electron range; and contamination by ν backgrounds in $\bar{\nu}$ beams. Bubble chamber measurements are also subject to neutron-induced backgrounds.

Measurements of $R_{\bar{\nu}}$ and R_ν are summarized in Figure 8.19, and $R_{\nu,\bar{\nu}}$ can be combined with the CC ratio

$$r = \sigma_{\text{CC}}^{\nu N}/\sigma_{\text{CC}}^{\bar{\nu} N} \quad (8.125)$$

to yield two different combinations of coupling constants:

$$|u_L|^2 + |d_L|^2 = \frac{R_\nu - r^2 R_{\bar\nu}}{1 - r^2} + \text{corrections} \tag{8.126}$$

$$|u_R|^2 + |d_R|^2 = \frac{r(R_\nu - R_{\bar\nu})}{1 - r^2} + \text{corrections} \tag{8.127}$$

These equations define annuli in the u_L–d_L and u_R–d_R planes, as shown in Figure 8.20. In these data, corrections must also be applied, since r has a slight energy dependence and some of the targets were not pure isoscalars.

(c) *Deep inelastic scattering from proton and neutron targets.* It is clearly impossible to determine u_L, d_L, u_R, and d_R from the results of Figure 8.20 alone. Additional constraints are provided by experiments carried out in bubble chambers filled with hydrogen or heavier liquids. In hydrogen, the ratios measured are

$$R_\nu^p = \sigma_{NC}^{\nu p}/\sigma_{CC}^{\nu p} \quad \text{and} \quad R_{\bar\nu}^p = \sigma_{NC}^{\bar\nu p}/\sigma_{CC}^{\bar\nu p} \tag{8.128}$$

The theoretical values of $R_{\nu,\bar\nu}^p$ are obtained by integrating $\partial^2\sigma/\partial x\,\partial y$. If we neglect the sea quarks and integrate over x, the distribution functions in the proton are $\bar Q = 0$, $Q_u \simeq 2Qd$. By integrating over y, we obtain

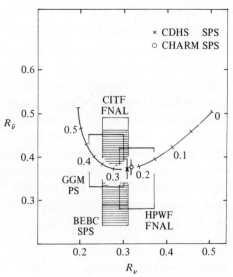

Figure 8.19. Ratio of charged current to neutral-current cross sections for ν and $\bar\nu$. The numbers on the curve refer to $\sin^2\theta_W$.

8.7 Tests of the standard model

$$R_\nu^p \simeq 2|u_L|^2 + |d_L|^2 + \tfrac{2}{3}|u_R|^2 + \tfrac{1}{3}|d_R|^2$$
$$= \tfrac{2}{3} - \tfrac{10}{9}\sin^2\theta_W + \tfrac{39}{27}\sin^4\theta_W + \text{corrections} \tag{8.129}$$

$$R_{\bar\nu}^p \simeq \tfrac{2}{3}|u_L|^2 + \tfrac{1}{3}|d_L|^2 + 2|u_R|^2 + |d_R|^2$$
$$= \tfrac{2}{3} - \tfrac{10}{9}\sin^2\theta_W + \tfrac{7}{9}\sin^4\theta_W + \text{corrections} \tag{8.130}$$

In these experiments the main background is caused by neutrons produced in the shielding around the bubble chamber. This can be reduced by accepting only those events where the total energy of the hadronic shower is large and then requiring a large transverse momentum for the final-state hadrons. Specific models are then employed to determine the effect of the data cuts on $R_{\nu,\bar\nu}^p$, and these contribute to systematic errors. The results of experiments utilizing the BEBC apparatus at CERN and the Fermilab 15′ bubble chamber are listed in Table 8.2.

Neutrino–neutron NC scattering cross sections are much more difficult to obtain because there are no neutron targets. Instead, one determines the ratios

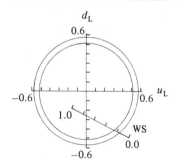

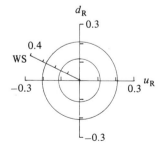

Figure 8.20. Values of coupling constants allowed by deep inelastic νN data on isoscalar target. Standard model predictions lie along the lines labeled WS for different values of $\sin^2\theta_W$ as shown.

Table 8.2. *Measurements of* R_ν *and* $R_{\bar\nu}$

Detector/collaboration	Target	R_ν [a]	$R_{\bar\nu}$ [a]	Reference
Isoscalar				
HPWF	CH_2	0.30 ± 0.04	0.33 ± 0.09	Wanderer et al. 78
CITF	Fe	0.28 ± 0.03	0.35 ± 0.11	Merritt et al. 78
BEBC	H_2–Ne	0.32 ± 0.03	0.39 ± 0.07	Deden et al. 79
CDHS	Fe	$0.307 \pm 0.008(\pm 0.003)$	$0.373 \pm 0.025(\pm 0.014)$	Geweniger 79
CHARM	$CaCO_3$	$0.30 \pm 0.021(\pm 0.006)$	$0.39 \pm 0.024(\pm 0.014)$	Mess 79
R_ν^p				
BEBC	H_2	0.52 ± 0.04	—	Bleitschau et al. 79
15' FNAL	H_2	0.48 ± 0.17	—	Harris 77
15' FNAL	H_2	—	0.42 ± 0.13	Derrick et al. 78
$R_\nu^{n/p}$				
15' FNAL	D_2	1.22 ± 0.35	—	Marriner et al. 77
15' FNAL	D_2	—	1.06 ± 0.20	Bell et al. 79

[a] Numbers in parentheses refer to additional systematic uncertainties.

8.7 Tests of the standard model

$$R_\nu^{n/p} = \sigma_{NC}^{\nu n}/\sigma_{CC}^{\nu p} \quad \text{and} \quad R_{\bar\nu}^{n/p} = \sigma_{NC}^{\bar\nu n}/\sigma_{CC}^{\bar\nu p} \qquad (8.131)$$

by performing experiments on isoscalar targets in bubble chambers. One must determine whether a particular event was caused by a neutrino striking a proton or a neutron in the nucleus. This is very difficult because many hadrons can be produced in a single collision and may come from the interacting quark or the spectator quarks (valence and/or sea) or may even be nuclear fragments.

In order to understand how such problems are solved, we must digress briefly on the general problem of neutrino-induced final states. Here, we wish to find the cross section for production of a particular hadron final state with energy fraction z of the total hadronic energy $z = E_{\text{hadron}}/(m_N + \nu) = E_{\text{hadron}}/(m_N + \omega y)$. (Sometimes z is defined differently: $z = E_{\text{hadron}}/\nu$ or $z = |p_{\text{hadron}}|/|p_{\text{quark}}|$. All of these are equivalent in the high energy limit.) According to the simple quark–parton model, the hadron production cross section is the product of the cross section for hitting a given type of quark times the probability that that quark will decay into the desired hadron summed over all quarks in the target. For example, in charged-current νN scattering where, if the sea is neglected, only d quarks participate, we have

$$\partial^2\sigma/\partial x\, \partial z = (\partial \sigma_{CC}^{\nu N}/\partial x) D_u^h(z, x) \qquad (8.132)$$

where D_u^h is the "fragmentation function" for a scattered u quark to become a hadron h plus anything. In this naive picture, where scattering is due to a single neutrino–quark interaction, it is reasonable to assume that D_u^h should depend only on the energy of the scattered quark, not its previous history. Therefore,

$$D_u^h(z, x) \to D_u^h(z) \qquad (8.133)$$

This hypothesis is called *factorization*. (For further details, see Feynman 72.)

Some of the same features that cause scale breaking will also affect D_u^h, so that, more generally, the cross section may be written as:

$$\frac{\partial^2\sigma}{\partial x\, \partial z} = \frac{G_F^2 m_N \omega}{\pi} F(x, q^2) D_u^h(z, q^2) \qquad (8.134)$$

In any real experiment, one observes a distribution $N^h(z, q^2)$ of hadrons of type h with energy fraction z and derives the fragmentation function through the relation

$$D(z, q^2) = dN^h(z, q^2)/dz \qquad (8.135)$$

where N is the total hadron multiplicity. However, this approach has an obvious flaw: The model developed here assumes that all hadrons in the final state come from reactions with a single scattered quark, but in reality, there must also be hadrons that emerge from the target quarks that were not scattered. The experimental problem, then, consists in separating the scattered quark "current fragments" from the spectator "target fragments" in a given reaction, as illustrated in Figure 8.21, in which the W boson transmits all of its energy ν to a single quark, which then transfers some of it to other quarks through lower-energy gluon exchanges in final-state interactions. Thus the fragments with high z tend to be associated with the quark that was originally scattered in the neutrino collision. An empirical rule is that current fragments should have $z \gtrsim 0.2$.

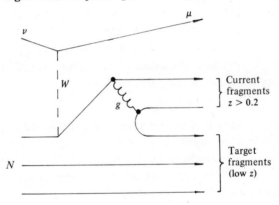

Figure 8.21. Separating current fragments from target fragments.

Once the fragmentation functions are obtained, various predictions can be tested, such as factorization and q^2 dependence. One may also compare specific models of $D(z)$, as in the Field–Feynman model (e.g., Field and Feynman 77) that predicts phenomena such as hadron jets. For the most part, however, it has been possible to test only qualitative predictions, owing to problems of statistics, resolution, and especially contamination from the target fragmentation region.

Most of these problems are not of direct interest to us; however, one result will be needed for our discussion of neutral currents. The fragmentation functions $D_u^{\pi^{\pm}}(z)$, which are the probabilities for a u quark with momentum fraction z to decay into $\pi^{\pm}$, can be obtained in charged-current ν–N scattering, where the scattered quark is almost

8.7 Tests of the standard model

always a u. The experimental results from both $\nu\text{-}N$ and $e\text{-}p$ scattering are shown in Figure 8.22, along with the predictions of the Field–Feynman model in the current fragmentation region. The most striking conclusion from the data is a confirmation of the naive quark model prediction that u quarks decay preferentially to π^+.

We now return to the problem of determining whether the neutrino struck a neutron or proton in the nucleus. From the discussion just given, we know that it is possible to isolate the current fragments from the rest of the target by restricting consideration to those particles with high values of z. If one assumes that the average charge of the fastest

Figure 8.22. Quark fragmentation functions $D_u^{\pi^\pm}(z)$ for $z > 0.2$, where the solid curve represents the results of the Field–Feynman model. (From Lubatti, 79. Reprinted with permission.)

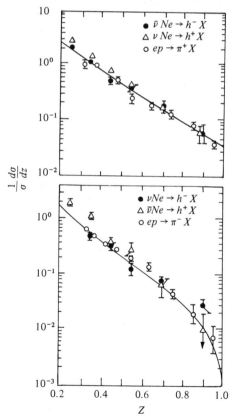

330   8 High-energy neutrino-nucleon collisions

current fragment is equal to the charge of the interacting quark, then neutrons and protons can be separated statistically.[6]

Hadrons with high enough z to qualify as current fragments can be identified in bubble chamber photographs and classified as to charge. After suitable corrections for unseen neutrals and for charged-current muons wrongly identified as hadrons, one thus obtains $R_\nu^{n/p}$ and $R_{\bar\nu}^{n/p}$. These results further constrain the coupling constants, as shown in Figure 8.23.

Figure 8.23. Allowed values of coupling constants consistent with the results of deep inelastic scattering from isoscalar, p, and n targets. On the θ_L–θ_R diagram, × indicates the couplings for $\sin^2 \theta_W = 0.233$. (From Kim et al. 81. Reprinted with permission.)

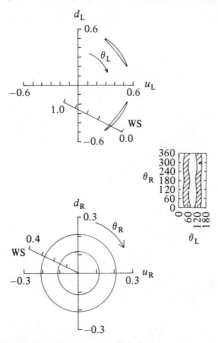

---

[6] This actually depends on the nature of the $q\bar q$ sea. If it consists mainly of $u\bar u$ and $d\bar d$ pairs with few $s\bar s$ pairs, then a scattered u quark has roughly equal probability to pick up a $\bar u$ or $\bar d$ and the average charge is $+\frac{1}{2}$ equal numbers of π^+ and π^0). If the sea is $SU(3)_f$ symmetric, with equal concentrations of $u\bar u$, $d\bar d$, and $s\bar s$, then the hadrons (π^0, π^+, K^+) retain the average charge of the u quark. Charged-current dimuon data reveal that the $SU(3)_f$ symmetry condition is approximately valid and that the assumption in the text is valid also.

8.7.2 Semi-inclusive pion production

The reactions:

$$\nu N \to \nu \pi^{\pm} x \tag{8.136}$$

$$\bar{\nu} N \to \bar{\nu} \pi^{\pm} x \tag{8.137}$$

offer a means, in principle, to determine all four coupling constants, but this is difficult in practice because of the uncertainties associated with hadronization in the current fragmentation region. The method is as follows: one borrows the results of CC scattering to describe the process of hadron formation after the quark is scattered. For example, from Figure 8.22 one obtains the following empirical relation for the pion fragmentation functions:

$$D_u^{\pi^+}(z)/D_u^{\pi^-}(z) = D_d^{\pi^-}(z)/D_{d\pi}^{\pi^+}(z) \simeq 3 \quad \text{at} \quad z \simeq 0.6 \tag{8.138}$$

In addition, we assume that the charge-symmetry equations

$$D_u^{\pi^{\pm}}(z) = D_d^{\pi^{\mp}}(z) \tag{8.139}$$

are true for all values of z.

These results are used to interpret the ratio of π^+ to π^- production in NC scattering in terms of the relative coupling strength between neutrinos and u or d quarks. The following relations are obtained (Hung 77):

$$R_\nu^{+/-} = \left(\frac{\pi^+}{\pi^-}\right)_{\nu \to \nu} = \frac{(|u_L|^2 + \tfrac{1}{3}|u_R|^2)D_u^{\pi^+} + (|d_L|^2 + \tfrac{1}{3}|d_R|^2)D_u^{\pi^-}}{(|u_L|^2 + \tfrac{1}{3}|u_R|^2)D_u^{\pi^-} + (|d_L|^2 + \tfrac{1}{3}|d_R|^2)D_u^{\pi^+}} \tag{8.140}$$

$$R_{\bar{\nu}}^{+/-} = \left(\frac{\pi^+}{\pi^-}\right)_{\bar{\nu} \to \bar{\nu}} = \frac{(|u_R|^2 + \tfrac{1}{3}|u_L|^2)D_u^{\pi^+} + (|d_R|^2 + \tfrac{1}{3}|d_L|^2)D_u^{\pi^-}}{(|u_R|^2 + \tfrac{1}{3}|u_L|^2)D_u^{\pi^-} + (|d_R|^2 + \tfrac{1}{3}|d_L|^2)D_u^{\pi^+}} \tag{8.141}$$

These equations must be modified to include the effects of antiquarks and to correct for the lower limit that must be imposed on z and y to ensure that most of the pions counted are from the current rather than the target fragmentation region. The experimental results, from the Gargamelle bubble chamber at CERN, are (Kluttig et al. 77)

$$R_\nu^{+/-} = 0.77 \pm 0.14 \tag{8.142}$$

$$R_{\bar{\nu}}^{+/-} = 1.64 \pm 0.36 \tag{8.143}$$

which can be combined with the isoscalar inclusive data to yield limits on the coupling constants, as shown in Figure 8.24.

When the results of inclusive isoscalar, inclusive n and p, and semi-inclusive scattering are combined, there remains a 16-fold ambiguity in the values of u_L, u_R, d_L, and d_R or, alternatively, α, β, γ, and δ.

332   8 High-energy neutrino-nucleon collisions

This is simply because deep inelastic scattering is incoherent and, therefore, can reveal only the absolute value of the coupling, with no sign. However, since $|d_R|^2$ is very close to zero, there are, in effect, only eight different solutions, four of which differ from the other four only by simultaneous sign reversal of all four constants.[7] The eight solutions are represented graphically in Figure 8.25 (Hung and Sakurai 77,81).

8.7.3 Exclusive processes

These ambiguities are resolved by data from exclusive reactions, where coherent effects lead to intereference terms in the scattering cross section. These include production of baryon resonances $\nu N \to \nu N^*$, single pion production, elastic νp scattering $\nu p \to \nu p$, and the low-energy dissociation reaction $\bar{\nu}_e D \to \bar{\nu}_e np$.

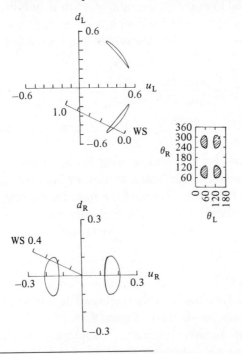

Figure 8.24. Allowed values of coupling constants consistent with the results of deep inelastic and semi-inclusive pion scattering.

---

7 The overall sign is observable; see Chapter 9 and Hung and Sakurai (81).

8.7 Tests of the standard model

(a) $\nu N \to \nu \Delta(1236)$. The baryon resonance $\Delta^+(1236)$ is a member of the $I = \frac{3}{2}$ isomultiplet in the $J^P = \frac{3}{2}^+$ decouplet of flavor $SU(3)_f$. Therefore, the reaction $\nu p \to \nu \Delta^+$ must involve an isovector interaction, denoted by solutions $A(A')$ or $B(B')$ in Figure 8.25. The resonance has now been observed in several experiments, which indicates that solutions $C(C')$ and $D(D')$ are unlikely.

(b) *Exclusive pion production.* Here we consider the reactions
$$\begin{array}{ll} \nu p \to \nu n \pi^+, & \nu p \to \nu p \pi^0 \\ \nu n \to \nu n \pi^0, & \nu n \to \nu p \pi^- \end{array} \qquad (8.144)$$

A large amount of redundancy exists, which, in principle, can place strong constraints on the couplings. However, theoretical uncertainties and nuclear corrections limit the accuracy of quantitative predictions. Even so, several experiments have been performed that distinguish among solutions A, B, C, and D.

The ratios

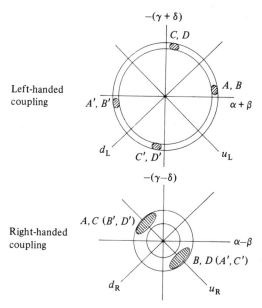

Figure 8.25. Allowed solutions for neutral νN couplings, for inclusive isoscalar, inclusive n and p, and semi-inclusive data. A or A': isovector axial vector dominant; $|\beta|$ large. B or B': isovector vector dominant; $|\alpha|$ large. C or C': isoscalar axial vector dominant; $|\delta|$ large. D or D': isoscalar vector dominant; $|\gamma|$ large.

334   8 High-energy neutrino-nucleon collisions

$$\frac{\nu N \to \nu N \pi^0}{\nu N \to \nu N' \pi^-} \quad \text{and} \quad \frac{\bar{\nu} N \to \bar{\nu} N \pi^0}{\bar{\nu} N \to \bar{\nu} N' \pi^-} \tag{8.145}$$

have been measured in freon (Bertrand-Coremans et al. 76) yielding results that differ from the isoscalar predictions by more than two standard deviations.

The cross sections for the four reactions in (8.144) yield simple predictions for an isoscalar interaction given by Clebsch–Gordan coefficients $\sigma(p\pi^0):\sigma(n\pi^+):\sigma(n\pi^0):\sigma(p\pi^-) = 1:2:1:2$. The experiment of Krenz et al. (78) excludes these ratios to a confidence level of 10^{-4}. Ratios of NC to CC cross sections can also be compared, giving results that favor solutions $A(A')$ or $B(B')$.

Finally, we note that although solutions $A(A')$ and $B(B')$ are predominantly isovector, there is some isoscalar part. Therefore we expect isovector–isoscalar interference, which leads to nonzero values for $\sigma(p\pi^0) - \sigma(n\pi^0)$. These may have been observed, but the statistical precision is still not good (Krenz et al. 78).

(c) νp *elastic scattering*. The results of νp elastic NC scattering are summarized in (8.48) to (8.50). Since g_A is large, solution $B(B')$ is ruled out.

(d) $\bar{\nu}_e D \to \bar{\nu}_e np$. This reaction is carried out with reactor $\bar{\nu}_e$ and occurs near threshold. It involves a transition from the $^3S_1(I=0)$ bound state to the $^1S_0(I=1)$ continuum n–p state. Thus only coefficient β in (8.23) contributes. The experimental results (Pasierb et al. 79) show that

Figure 8.26. All neutrino–hadron data are combined to yield severe constraints on all four coupling constants.

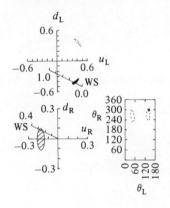

$$|\beta| = 0.9 \pm 0.1 \tag{8.146}$$

which should be compared (Hung and Sakurai 81) with

$$|\beta| = \begin{cases} 0.92 \pm 0.14 & \text{for solutions } A(A') \\ 0.58 \pm 0.14 & \text{for solutions } B(B') \end{cases} \tag{8.147}$$

8.7.4 Conclusions

Figure 8.26 shows that when all neutrino–hadron data are combined, the coupling constants are severely constrained, up to an overall sign. They yield very strong support for the standard model with $\sin^2 \theta_W = 0.23$. Further information on the neutral-current couplings is provided from the results of ν_μ–e and $\bar{\nu}_\mu$–e scattering experiments and e^+e^- annihilation (see Chapter 3) and from observations of parity violation in the electron–nucleon interaction, which we discuss in the next chapter. An overall summary is given in Chapter 12.

Problems

8.1 Show that it is appropriate to neglect the term in $g_3(q^2)$ in (8.5) in the case where the final lepton is highly relativistic.

8.2 Derive (8.17) from (8.5).

8.3 Derive (8.37).

8.4 Show that (8.66) follows from (8.65), assuming electromagnetic current conservation.

8.5 Prove that in (8.75) the form of $\mathcal{A}$ is given by (8.76).

9

The parity-nonconserving eN and NN interactions

9.1 Introduction

An electron and a nucleon may interact not only by photon exchange but also by Z^0 exchange (Figure 9.1). Thus the total scattering amplitude consists of two parts:

$$A = A_{EM} + A_{wk}$$

and the cross section for $e-N$ scattering is proportional to

$$|A|^2 = |A_{EM} + A_{wk}|^2$$
$$= |A_{EM}|^2 \left(1 + \frac{A_{wk}A_{EM}^* + A_{wk}^*A_{EM}}{|A_{EM}|^2} + \frac{|A_{wk}|^2}{|A_{EM}|^2}\right) \quad (9.1)$$

At ordinary laboratory energies, $|A_{EM}| \gg |A_{wk}|$, and the third term on the right-hand side of (9.1) is quite negligible. Therefore, to detect the presence of A_{wk} we must make use of *weak–electromagnetic interference,* as embodied in the second term of (9.1).

The amplitude A_{wk} contains both scalar and pseudoscalar parts:

$$A_{wk} = A_{wkS} + A_{wkP} \quad (9.2)$$

Although the analog of A_{wkS} in purely leptonic weak interactions is observable, for example in the charge asymmetry in the angular distribution of muons in $e^+e^- \to \mu^+\mu^-$ (see Chapter 3), there do not appear to be any observable effects attributable to A_{wkS} in the case of the electron–nucleon interaction.[1] However, the pseudoscalar A_{wkP} can be detected, because its sign depends on whether the experimental coor-

---

[1] A_{wkS} yields a small contribution to the hyperfine structure splitting $\Delta\nu$ in atomic hydrogen (approximately 3 parts in 10^8). Although the splitting has been measured to a precision of better than 1 part in 10^{12}, the theoretical uncertainty is about 1 part in 10^6, so A_{wkS} cannot be isolated here.

9.1 Introduction

dinate system is right (R) or left (L) handed. Therefore, when cross sections for otherwise identical processes of opposite handedness are compared, one obtains from (9.1) the result:

$$\delta \equiv \frac{\sigma_R - \sigma_L}{\sigma_R + \sigma_L}$$
$$= \frac{|A_R|^2 - |A_L|^2}{|A_R|^2 + |A_L|^2} \simeq \frac{A_{\text{wkP}} A^*_{\text{EM}} + A_{\text{EM}} A^*_{\text{wkP}}}{|A_{\text{EM}}|^2} \quad (9.3)$$

This asymmetry is of order G_F and is observable in high-energy deep-inelastic scattering of polarized electrons by nucleons and in low-energy atomic spectroscopy. The results of such experiments impose stringent and complementary constraints on the neutral weak coupling parameters and provide strong support for the standard model.

Figure 9.1. (a) Photon exchange; (b) Z^0 exchange.

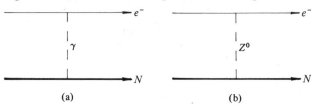

In the polarized-electron scattering experiments, the handedness is defined by the longitudinal polarization of the electron beam. We estimate the asymmetry δ very crudely from Figure 9.1, leaving details for the following section. Since $A_{\text{EM}} \simeq 4\pi\alpha/q^2$ and $A_{\text{wkP}} \simeq G_F$, we have, from (9.3),

$$\delta \simeq G_F q^2/4\pi\alpha \simeq q^2/m_Z^2 \quad (9.4)$$

Thus at $|q| \simeq 1$ GeV/c, δ is approximately 10^{-4}. (This estimate was originally given by Zel'dovich (59), long before the discovery of neutral currents, the invention of polarized high-energy electron beams, or creation of the standard model.) According to the more detailed analysis of Section 9.2, the electron scatters from individual quark–partons. The weak and electromagnetic contributions from each quark are coherent and they interfere, although, at sufficiently high energy, the sum over all quarks is incoherent. The asymmetry predicted on the basis of the standard model, and actually obtained in the experiment performed at SLAC, is still of order 10^{-4}.

In low-energy atomic physics, photon exchange is described by the ordinary Coulomb Hamiltonian H_0, which commutes with the parity operator and possesses energy eigenstates $|\psi_n\rangle$, $|\chi_m\rangle$ of even or odd parity. The Z^0 exchange gives rise to an additional term H' in the atomic Hamiltonian, which possesses both scalar and pseudoscalar parts:

$$H = H_0 + H' = H_0 + H_S + H_P \tag{9.5}$$

The Hamiltonian now no longer commutes with the parity operator, and a given energy eigenstate contains small admixtures of states of opposite parity:

$$|\psi^0\rangle \rightarrow |\psi\rangle = |\psi^0\rangle + \sum_m \frac{|\chi_m^0\rangle\langle\chi_m^0|H_P|\psi^0\rangle}{E(\psi^0) - E(\chi_m^0)} \tag{9.6}$$

In all atomic physics parity-violation experiments performed so far, [e.g., Stark interference in thallium (Tl) and cesium (Cs), optical rotation in bismuth (Bi)], one observes an optical transition between a pair of states $|\psi_i\rangle$, $|\psi_f\rangle$ of the same nominal parity (a magnetic dipole or $M1$ transition). The existence of parity mixing in one or both states, as expressed by (9.6), implies that the transition amplitude contains a parity-violating electric dipole ($E1$) component, in addition to the normal $M1$ component. The interference of these amplitudes, or the interference of the parity-violating $E1$ amplitude with an $E1$ amplitude induced by Stark effect, gives rise to characteristic parity-violating effects. An example is circular dichroism, in which the cross section for photon absorption in the transition i → f depends on the photon helicity in circularly polarized light.

Asymmetries in such experiments would at first seem to be hopelessly small, because of the tiny momentum transfer imparted to a bound electron, where characteristic energies are of order 1 to 10 eV. However, there exist several important enhancement effects in the case of heavy atoms ($Z \simeq 80$), which lead to observed parity-violating asymmetries δ of order 10^{-5} (Stark-interference experiments in thallium and cesium) and 10^{-7} (optical rotation experiments in bismuth). Possibilities for observation of parity violation also exist in atomic hydrogen, where the very small Lamb-shift separation (1,058 MHz) between the $2^2S_{1/2}$ and $2^2P_{1/2}$ states enhances parity violation.

In this chapter we also consider the parity-violating weak nuclear force, which coexists, by virtue of W and/or Z exchange, with the strong and electromagnetic interactions between nucleons. Unfortunately, the complications of nuclear structure make it very difficult to formulate quantitative and precise theoretical predictions here, and one cannot obtain the same sort of constraints on neutral weak coupling parameters that are achieved in the case of eN interaction. However, in one favorable case, it is at least possible to distinguish between the effects of W and Z exchange and to show that experimental results are in reasonable qualitative agreement with the standard model.

9.2 Deep inelastic scattering of polarized electrons

9.2.1 Theoretical analysis

We now consider the helicity-dependent asymmetry in the scattering of high-energy polarized electrons by unpolarized target nucleons, observed in the SLAC experiment (Prescott et al. 78,79). The theoretical analysis has been carried out by a number of authors (see, e.g., Cahn and Gilman 78; also the review by Commins and Bucksbaum 80). Such an analysis is a more refined treatment of the same simple ideas presented in the previous section and is based on the quark–parton model, introduced in Chapter 8. As usual, we assume three valence quarks and a sea of $q\bar{q}$ pairs. Electron and quark masses are neglected, and in this limit $\frac{1}{2}(1 \pm \gamma_5)$ are $\pm$ helicity projection operators, respectively, with vector (γ_μ) and axial vector ($\gamma_\mu \gamma_5$) interactions preserving helicity. Otherwise, the following analysis is model-independent. As we shall see, the experimental results enable us to distinguish unambiguously among alternative models.

The photon couples to the quark or electron through the vector current with a strength given by its charge Q_f^γ ($f = u, d$, or e):

$$Q_e^\gamma = -1, \qquad Q_u^\gamma = \tfrac{2}{3}, \qquad Q_d^\gamma = -\tfrac{1}{3} \tag{9.7}$$

Using $\gamma_m = \tfrac{1}{2}\gamma_\mu(1 + \gamma_5) + \tfrac{1}{2}\gamma_\mu(1 - \gamma_5)$, we may define left- and right-handed charges:

$$Q_{Lf}^\gamma = Q_{Rf}^\gamma = Q_f^\gamma \tag{9.8}$$

Similarly, Z^0 couples to left- and right-handed fermions with strengths Q_{Lf}^Z and Q_{Rf}^Z, respectively; however, these are, in general, different. In any $SU(2)$ model such as (2.103), they are:

$$Q_{Lf}^Z = (L_3 - Q_f^\gamma \sin^2 \theta_W)/\sin \theta_W \cos \theta_W \tag{9.9}$$

$$Q_{Rf}^Z = (R_3 - Q_f^\gamma \sin^2 \theta_W)/\sin \theta_W \cos \theta_W \tag{9.10}$$

and in the standard model, $L_3 = \tfrac{1}{2}$ for up quark, $L_3 = -\tfrac{1}{2}$ for down quark and electron, and $R_3 = 0$ for all fermions. The Z^0-fermion Dirac vertex has the form:

$$\tfrac{1}{2} Q_R{}^Z \gamma_\mu (1 + \gamma_5) + \tfrac{1}{2} Q_L{}^Z \gamma_\mu (1 - \gamma_5) \tag{9.11}$$

Now we consider electron–quark scattering in the laboratory frame and recall from Chapter 8 that if electron and quark have the same initial helicity, the scattering cross section is independent of the inelasticity variable y, whereas if the helicities are opposite, the differential cross section varies as $(1 - y)^2$. Since for each quark, the electromagnetic and weak interactions contribute coherently, we thus obtain the following:

RH e^- on RH quark of type i:

$$d\sigma \propto \left| \frac{Q_{Re}{}^\gamma Q_{Ri}{}^\gamma}{q^2} + \frac{Q_{Re}{}^Z Q_{Ri}{}^Z}{q^2 - m_Z^2} \right|^2 \tag{9.12}$$

RH e^- on LH quark of type i:

$$d\sigma \propto \left| \frac{Q_{Re}{}^\gamma Q_{Li}{}^\gamma}{q^2} + \frac{Q_{Re}{}^Z Q_{Li}{}^Z}{q^2 - m_Z^2} \right|^2 (1 - y)^2 \tag{9.13}$$

LH e^- on LH quark of type i:

$$d\sigma \propto \left| \frac{Q_{Le}{}^\gamma Q_{Li}{}^\gamma}{q^2} + \frac{Q_{Le}{}^Z Q_{Li}{}^Z}{q^2 - m_Z^2} \right|^2 \tag{9.14}$$

LH e^- on RH quark of type i:

$$d\sigma \propto \left| \frac{Q_{Le}{}^\gamma Q_{Ri}{}^\gamma}{q^2} + \frac{Q_{Le}{}^Z Q_{Ri}{}^Z}{q^2 - m_Z^2} \right|^2 (1 - y)^2 \tag{9.15}$$

The asymmetry for longitudinally polarized e^-:

$$\Delta = (\sigma_R - \sigma_L)/(\sigma_R + \sigma_L) \tag{9.16}$$

is now calculated simply by multiplying each of the cross sections (9.12)–(9.15) by the quark distribution $Q_i(x)$, $i = u$ or d (where x is the Bjorken scaling variable defined in Chapter 8) and then adding the four cross sections. This incoherent sum is appropriate since, for the high energies actually used, the de Broglie wavelength of the electron is much smaller than the nucleon radius.

Assuming the charges given by (9.7), (9.9), and (9.10), choosing an isoscalar target (deuterium), and confining our attention to $x > 0.2$ so that the sea may be neglected, we arrive at the following result:

$$\frac{\Delta_{e,D}(x, y)}{Q^2} = a_1 + a_2 \left[\frac{1 - (1 - y)^2}{1 + (1 - y)^2} \right] \tag{9.17}$$

where for $SU(2) \times U(1)$ in general:

9.2 Deep inelastic scattering of polarized electrons

$$a_1 = -\frac{G_F}{2\pi\sqrt{2}\alpha} \frac{9}{10} (1 + 2R_3^e) \left(1 - \frac{20}{9} \sin^2\theta_W + \frac{4}{3} R_3^u - \frac{2}{3} R_3^d\right) \tag{9.18}$$

and

$$a_2 = -\frac{G_F}{2\pi\sqrt{2}\alpha} \frac{9}{10} (1 - 4\sin^2\theta_W - 2R_3^e) \left(1 - \frac{4}{3} R_3^u + \frac{2}{3} R_3^d\right) \tag{9.19}$$

In particular, for the standard model, $R_3^e = R_3^u = R_3^d = 0$, and

$$a_1 = -\frac{G_F}{2\pi\sqrt{2}\alpha} \frac{9}{10} \left(1 - \frac{20}{9} \sin^2\theta_W\right) \tag{9.20}$$

$$a_2 = -\frac{G_F}{2\pi\sqrt{2}\alpha} \frac{9}{10} (1 - 4\sin^2\theta_W) \tag{9.21}$$

Figure 9.2 shows the theoretical asymmetry as a function of y for various choices of $R_3^{e,u,d}$ including the predictions of the standard model, labeled WS. Some of the alternative models include a right-handed doublet composed of a neutral lepton E_0 and the electron e^-. The hypothetical E_0 would have to be massive in order to avoid conflict with the results of charged-current experiments (e.g., β decay).

9.2.2 The SLAC polarized-electron experiment

The asymmetry Δ_{eD} was measured using 19.4-GeV polarized electrons incident on a liquid D_2 target (Prescott et al. 78,79). Since

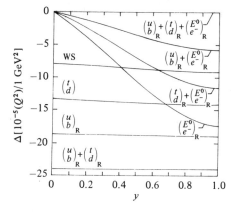

Figure 9.2. Asymmetry for scattering of polarized electrons by deuterons, according to the analysis of Cahn and Gilman (78). Here, $\sin^2\theta_W = 0.23$ is assumed. The curves are for left-handed isodoublets and right-handed isomultiplets as indicated. As explained in the text, E_0 is a hypothetical neutral heavy lepton.

$\Delta \simeq 10^{-4}$ is so small, traditional single-particle counting techniques were not appropriate. Instead, on each 1.5-μsec accelerator pulse, approximately 10^3 electrons were scattered into the detectors, where the total signal was integrated. In this way a statistical precision of 10^{-5} was possible after about 10^7 pulses, or one day of running.

A block diagram of the apparatus is shown in Figure 9.3. The polarized electrons were produced by photoemission from the surface of a GaAs crystal with the aid of circularly polarized light from a laser. The electrons were injected into the linear accelerator, which was then able to deliver about 10^{11} electrons per pulse with an average polarization of $P_0 = \pm 0.37$. After acceleration, the beam was bent through an angle of 24.5° to reach the target. The spins of the highly relativistic electrons underwent a g-2 precession such that the polarization at the target was

$$P = P_0 \cos(E\pi/3.237) \qquad (9.22)$$

(where E is the beam energy in GeV). This provided a source-independent way of reversing the electron helicity, which could be used in addition to reversal of the laser polarization.

Electrons that scattered from the target at 4° entered a spectrometer, which analyzed momentum and accepted a very broad range centered

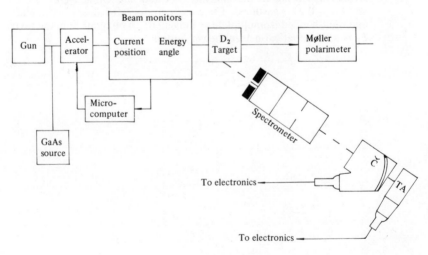

Figure 9.3. Schematic diagram of the SLAC polarized-electron experiment (see text for explanation). The GaAs polarized-electron source could be replaced by the normal SLAC thermionic source for null measurements. The D_2 target was periodically removed to measure the beam helicity with a Møller scattering polarimeter.

9.3 Parity violation in atoms

at 14.5 GeV/c. Since individual particles cannot be discriminated against in a flux-counting experiment, the kinematics were chosen to reduce μ, π, and K backgrounds to a few percent. Two detectors were employed: an atmospheric N_2 Čerenkov counter and a lead-glass shower counter, placed in series. The latter was segmented so that the upper and lower half of the momentum acceptance, representing different average values of y, could be viewed separately.

The results of this remarkable experiment are shown in Figure 9.4. They are clearly consistent with the standard model and give a best fit of $\sin^2 \theta_W = 0.224 \pm 0.020$. The parameters a_1 and a_2 are measured to be:

$$a_1 = -(9.7 \pm 2.6) \times 10^{-5} \tag{9.23}$$

$$a_2 = (4.9 \pm 8.1) \times 10^{-5} \tag{9.24}$$

9.3 Parity violation in atoms

9.3.1 Theoretical analysis

Let us now turn to atomic physics and consider the pseudoscalar Hamiltonian term H_P, which arises from Z^0 exchange between an orbital electron and the atomic nucleus. We assume that the electronic and nucleonic weak neutral currents contain only vector and axial vector components:

$$J_e = V_e + A_e; \quad J_N = V_N + A_N \tag{9.25}$$

Thus we have $H_P = A_e V_N + V_e A_N$. Ignoring momentum-transfer-dependent terms, H_P then becomes $H_P = H_P^{(1)} + H_P^{(2)}$, with

$$H_P^{(1)} \equiv A_e V_N = \frac{G_F}{\sqrt{2}} \sum_i \bar{\psi}_e \gamma_\lambda \gamma_5 \psi_e (C_{1p} \bar{\psi}_{pi} \gamma^\lambda \psi_{pi} + C_{1n} \bar{\psi}_{ni} \gamma^\lambda \psi_{ni}) \tag{9.26}$$

$$H_P^{(2)} \equiv V_e A_N = \frac{G_F}{\sqrt{2}} \sum_i \bar{\psi}_e \gamma_\lambda \psi_e (C_{2p} \bar{\psi}_{pi} \gamma^\lambda \gamma_5 \psi_{pi} + C_{2n} \bar{\psi}_{ni} \gamma^\lambda \gamma_5 \psi_{ni}) \tag{9.27}$$

where the sum is taken over all protons (p) and neutrons (n) in the nucleus. According to the standard model, the coupling coefficients are:

$$C_{1p} = \tfrac{1}{2}(1 - 4 \sin^2 \theta_W) \simeq 0.04 \tag{9.28}$$

$$C_{1n} = -\tfrac{1}{2} \tag{9.29}$$

$$C_{2p} = \tfrac{1}{2} g_A (1 - 4 \sin^2 \theta_W) \simeq 0.05 \tag{9.30}$$

$$C_{2n} = -\tfrac{1}{2} g_A (1 - 4 \sin^2 \theta_W) \simeq -0.05 \tag{9.31}$$

as demonstrated in Section 4.12. Assuming the standard model and

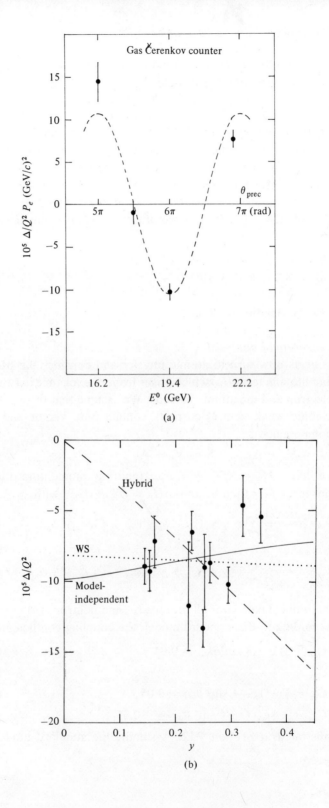

9.3 Parity violation in atoms

performing a nonrelativistic reduction of the nucleonic currents, we obtain from (9.26) the effective electronic weak Hamiltonian corresponding to $A_e V_N$:

$$H_{\text{wk}}^{(1)} = \frac{G_F}{2\sqrt{2}} Q_{\text{wk}} \rho_N(\mathbf{r}) \gamma_5 \tag{9.32}$$

where $Q_W = (1 - 4\sin^2\theta_W)Z - N$ is the "weak charge" of the nucleus and $\rho_N(\mathbf{r})$ the nuclear density, $\mathbf{r}$ being the electron position. The factors of Z and N appear in Q_W because the nucleons contribute *coherently* to the neutral weak $A_e V_N$ coupling. In the limit of a point nucleus and a nonrelativistic electron, (9.32) becomes

$$H_W^{(1)} = \frac{G_F}{4\sqrt{2}} \frac{1}{m_e c} Q_W \{\boldsymbol{\sigma} \cdot \mathbf{p}, \delta^3(\mathbf{r})\} \tag{9.33}$$

where $\boldsymbol{\sigma}$ and $\mathbf{p}$ refer to the electron.

In the case of $H_P^{(2)}$, which corresponds to $V_e A_N$, a nonrelativistic reduction of the axial nucleonic current yields a factor proportional to nucleon spin, which cancels in pairs in the sum over nucleons, leaving at most two unpaired spins for odd–odd nuclei, and otherwise at most one. Thus in the nonrelativistic limit for the electron and for a point nucleus, the standard model gives the effective Hamiltonian:

$$H_W^{(2)} = \frac{G_F}{4\sqrt{2}} \frac{1}{m_e c} (1 - 4\sin^2\theta_W) g_A \boldsymbol{\sigma}_N \cdot \boldsymbol{\sigma} \{\boldsymbol{\sigma} \cdot \mathbf{p}, \delta^3(\mathbf{r})\} \tag{9.34}$$

Since $H_W^{(2)}$ contains no enhancement factor Q_W, its effects are of order $Z^{-1}(1 - 4\sin^2\theta_W)$ relative to those of $H_W^{(1)}$. Thus $H_W^{(2)}$, and also the Hamiltonian describing parity-violating electron–electron interactions in atoms, which has the same order of magnitude as $H_W^{(2)}$, are quite negligible in heavy atoms. Indeed, in the standard model, only C_{1n} has appreciable size, and it is the only coefficient accessible to measurement with present-day heavy-atom experiments.

Matrix elements of $H_W^{(1)}$ are nonzero only for atomic orbitals of opposite parity with nonvanishing value or gradient at the origin ($s_{1/2}$, $p_{1/2}$ orbitals). For a single-valence electron, one finds $\langle \chi | H_{\text{wk}}^{(1)} | \psi \rangle \simeq$

Figure 9.4. (a) Parity-violating asymmetry in the SLAC polarized-electron experiment as a function of beam energy. The dotted curve shows the expected dependence based on relativistic spin precession, Eq. (9.22). (From Prescott et al. 78.) (b) Results of the SLAC experiment, showing y dependence. The dotted line is a best fit to the standard model, and the solid line is a two-parameter fit to the model-independent equation (9.17). (From Prescott et al. 79. Figures reprinted with permission.)

$10^{-19}(e^2/a_0)Z^3K$, where K is a relativistic enhancement factor that grows with Z: $K(Z = 50) \simeq 2.5$; $K(Z = 80) \simeq 10$. The important Z^3 dependence, which greatly enhances parity-violating effects in heavy atoms, was first recognized and analyzed by Bouchiat and Bouchiat (74*ab*,75) and may be understood intuitively as follows. As (9.33) shows, one factor of Z arises from Q_W; another comes from the momentum **p**; and a third arises from the wavefunctions at the origin: $\delta^3(\mathbf{r})$.

As mentioned in Section 9.1, atomic parity-violation experiments involve observation of magnetic dipole transitions between states ψ_i and ψ_f of the same nominal parity. Using (9.6), let us calculate the transition amplitude, taking into account $H_W^{(1)}$ to order G_F. We have:

$$\langle \psi_f | O_{EM} | \psi_i \rangle = \mathcal{M} \pm \mathcal{E}_P \tag{9.35}$$

where

$$\mathcal{M} = \langle \psi_f^0 | M1 | \psi_i^0 \rangle \tag{9.36}$$

is the zeroth-order amplitude and

$$\mathcal{E}_P = \sum_n \left[\frac{\langle \psi_f^0 | E1 | \chi_n^0 \rangle \langle \chi_n^0 | H_{wk}^{(1)} | \psi_i^0 \rangle}{E(\psi_i^0) - E(\chi_n^0)} + \frac{\langle \psi_f^0 | H_{wk}^{(1)} | \chi_n^0 \rangle \langle \chi_n^0 | E1 | \psi_i^0 \rangle}{E(\psi_f^0) - E(\chi_n^0)} \right] \tag{9.37}$$

is the parity-violating electric dipole amplitude. Time-reversal invariance requires that $\mathcal{E}_P$ and $\mathcal{M}$ be relatively imaginary.

As explained in Section 9.1, the $\pm$ sign in (9.35) depends on the handedness of the experimental coordinate system; for example, in a circular-dichroism experiment, it is defined by the helicity of the incoming photons. One may easily demonstrate that the cross section for dichroic absorption in the transition $\psi_i \to \psi_f$ is proportional to

$$|\mathcal{M}|^2 \pm 2 \, \text{Im}(\mathcal{E}_P \cdot \mathcal{M}^*) \tag{9.38}$$

where the $\pm$ sign applies for photon helicity ± 1, respectively. Thus the circular dichroism is

$$\delta \equiv (\sigma_R - \sigma_L)/(\sigma_R + \sigma_L) = 2 \, \text{Im}(\mathcal{E}_P)/\mathcal{M} \tag{9.39}$$

We present a crude numerical estimate of $|\mathcal{E}_P|$. Assuming a typical energy level spacing of $0.05 e^2/a_0$ and typical $E1$ matrix elements, one finds from (9.37) that

$$|\mathcal{E}_P| \simeq 5 \times 10^{-18} e a_0 Z^3 K \simeq 10^{-10} e a_0 \quad \text{for } Z \simeq 80 \tag{9.40}$$

At best, this is a very small amplitude; therefore to obtain a δ of reasonable size, one must choose a transition for which $\mathcal{M}$ is also very small. Particularly favorable cases are the forbidden magnetic dipole transi-

9.3 Parity violation in atoms

tions $6^2S_{1/2} \to 7^2S_{1/2}$ in Cs ($Z = 55$) and $6^2P_{1/2} \to 7^2P_{1/2}$ in Tl ($Z = 81$) (see Figures 9.5 and 9.6). In each case,

$$\mathcal{M} \simeq 10^{-5} e\hbar/2m_e c$$

Moreover, these atoms are described quite well by the single-valence-electron central field approximation, which permits a fairly straightforward and reliable theoretical interpretation. Calculations based on this approximation and employing the standard model with $\sin^2 \theta_W = 0.23$ yield (Neuffer and Commins 77)

$$\delta(\text{Tl}) = 2.1 \times 10^{-3} \tag{9.41}$$

A sophisticated many-body calculation for Cs has been carried out by Das (81) and collaborators. For $\sin^2 \theta_W = 0.23$,

$$\delta(\text{Cs}) = -1.17 \times 10^{-4} \tag{9.42}$$

was obtained, in good agreement with earlier calculations based on the one electron central field approximation.

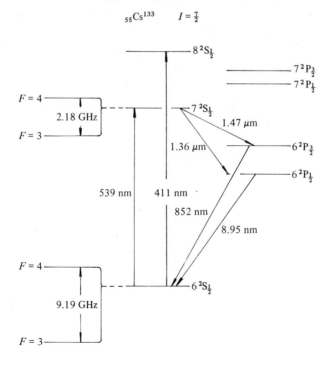

Figure 9.5. Energy levels of Cs ($Z = 55$). Parity violation has been observed in the $6^2S_{1/2} \to 7^2S_{1/2}$ transition at 539 nm.

348   9 *The parity-nonconserving* eN *and* NN *interactions*

9.3.2   Stark interference experiments

In spite of the relatively large values of δ in cesium and thallium, dichroic absorption is very difficult to observe because of small overall transition rates. In actual experiments, one must utilize atomic vapor at rather high densities in order to achieve acceptable signals. Thus collisions and other effects cause weak photon absorption, which is nevertheless much stronger than that expected from $\mathcal{M}$. An effective, if less direct, technique was suggested by Bouchiat and Bouchiat (74a,b;75). Here, one employs an external static electric field and makes use of interference between the resulting Stark amplitude, and $M1$ and parity-violating $E1$ amplitudes. Parity violation gives rise here to a pseudoscalar term in the transition probability proportional to $h(\mathbf{k} \times \mathbf{E} \cdot \mathbf{J})$, where h and $\mathbf{k}$ are the helicity and wave vector of the incoming photon, respectively, $\mathbf{E}$ the external electric field, and $\mathbf{J}$ the atomic angular momentum (see Figure 9.7). The result is an angular mo-

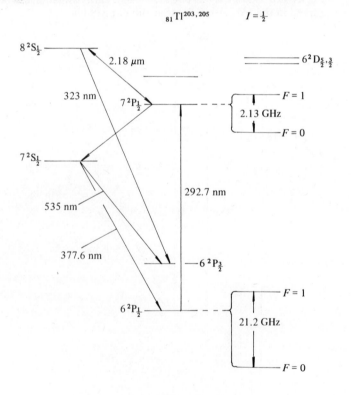

Figure 9.6. Energy levels of Tl ($Z = 81$). Parity violation has been observed in the $6^2P_{1/2} \rightarrow 7^2P_{1/2}$ transition at 293 nm.

9.3 Parity violation in atoms

mentum polarization of the excited state that changes sign with **E**, *h*, and **k**. In a measurement of $\delta(6^2P_{1/2} - 7^2P_{1/2})$ in Tl performed at Berkeley (Conti et al. 79, Bucksbaum et al. 81*ab*), the polarization of the 7*P* state was measured by an optical-pumping technique involving stimulated absorption to the 8*S* state (see Figure 9.6), and observation of decay fluorescence at 323 nm. The result is:

$$\delta = \left(2.8 \, {}^{+1.0}_{-0.9}\right) \times 10^{-3} \tag{9.43}$$

in agreement with the theoretical estimate (9.41). In a somewhat similar experiment on cesium carried out at ENS, Paris, by Bouchiat and co-workers, the polarization of the $7^2S_{1/2}$ state (Figure 9.5) is detected by observation of circular polarization of the decay fluorescence at 1.36 μm. In this case the results obtained are also in agreement with the standard model.

Figure 9.7. Coordinate system **k**–**J**–**E**.

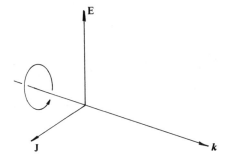

9.3.3 Optical rotation

Optical-rotation experiments have been performed on magnetic dipole transitions in atomic bismuth at the University of Washington (Hollister et al. 81), Oxford (Baird et al. 77), Novosibirsk (Barkov and Zolotorev 79) and Moscow (Bogdanov et al. 80). Each employs a technique suggested by a number of authors (Zel'dovitch 59, Khriplovich 74, Sandars 75, Soreide and Fortson 75) in which a beam of linearly polarized light with frequency close to resonance traverses a cell of length *L* containing atomic vapor of density *N* cm^{-3}. Optical rotation occurs because the linear polarization is a superposition of circular polarization states ($\pm$) that propagate with different indexes of refraction $n_\pm$:

9 The parity-nonconserving eN and NN interactions

$$n_\pm = 1 - \frac{2\pi N}{h} \; [\overline{|\mathcal{M}|^2 \pm 2\,\text{Im}(\mathcal{E}_P \mathcal{M}^*)}] \left\langle \frac{1}{\omega - \omega_0 + (v/c)\omega_0 + \tfrac{1}{2}i\Gamma} \right\rangle \quad (9.44)$$

Here ω is the photon frequency, ω_0 the transition frequency, Γ the natural width of the excited state, v the thermal atomic velocity in the direction of the light beam, and the angle brackets indicate an average over the Doppler width of the line. Also, the bar over the matrix element squared indicates a sum over final and average of initial atomic polarizations. Absorption occurs (by slightly differing amounts for the $\pm$ components) and therefore the light emerging from the cell is elliptically polarized. The absorption coefficients are

$$\alpha_\pm = (2\omega/c)\,\text{Im}(n_\pm) \quad (9.45)$$

The optical rotation angle per unit absorption length is

$$\phi = (\omega l/2c)\,\text{Re}(n_+ - n_-) \quad (9.46)$$

where l is one absorption length. The rotation angle ϕ follows a dispersionlike dependence on $\omega - \omega_0$, and near resonance, one has

$$\phi_{\max} \simeq l\,\text{Im}(\mathcal{E}_P)/\mathcal{M} \quad (9.47)$$

For experiments actually performed on the allowed $M1$ transitions in bismuth ($Z = 83$) (see Figure 9.8), one has:

$6p^3,\ J = \tfrac{3}{2}$ (ground state) $\to 6p^3,\ J = \tfrac{5}{2}$, $\lambda = 648$ nm
$6p^3,\ J = \tfrac{3}{2}$ (ground state) $\to 6p^3,\ J = \tfrac{3}{2}$, $\lambda = 876$ nm

and in each case $\mathcal{M}$ is approximately one electron Bohr magneton. One expects rotation angles of order 10^{-7} rad per absorption length, and such rotations are in fact observed (Barkov and Zolotorev 79, Hollister et al. 81).

Bismuth, unlike thallium or cesium, has three equivalent p electrons outside of closed shells, and jj coupling is dominant. Thus bismuth has a complex structure, and calculations of $\mathcal{E}_P$ are more difficult and somewhat more uncertain than for cesium and thallium. As Figure 9.8 shows, various published estimates disagree by as much as a factor of 2.

The simplest optical rotation experiment would employ a cell containing bismuth vapor between crossed polarizers. However, a rotation of 10^{-7} radian per absorption length is far too small to be observed in this way, since the transmitted intensity is proportional to the square of the rotation angle, for small angles. This difficulty can be overcome by introducing a small but precisely known additional misalignment angle ϕ_F, in which case the total transmitted intensity is

9.3 Parity violation in atoms

$$I = I_0[(\phi_{PNC} + \phi_F + \phi_R)^2 + b] \tag{9.48}$$

where b is an angle-independent background and ϕ_R any residual optical rotation not caused by parity violation. If $\phi_F \gg \phi_{PNC} + \phi_R$, reversal of ϕ_F yields the asymmetry (Baird et al. 77):

$$\frac{I_+ - I_-}{I_+ + I_-} \simeq \frac{2\phi_F(\phi_{PNC} + \phi_R)}{\phi_F^2 + b} \simeq \frac{\phi_{PNC} + \phi_R}{\phi_F} \simeq 10^{-4}$$

where the optimum condition $\phi_F^2 \simeq b$ has been assumed. The residual angle ϕ_R can be eliminated by using the dispersion shape of ϕ_{PNC}.

Observations of parity-violating optical rotations in bismuth have been completed at Seattle (Hollister et al. 81) and Novosibirsk (Barkov and Zolotorev 79), but there is not yet complete agreement on the magnitude of the effect, and a null result has been published by the

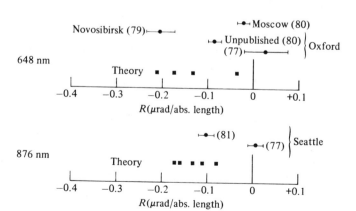

Figure 9.8. Summary of optical rotation experiments and calculations for bismuth ($Z = 83$).

Moscow group (Bogdanov et al. 80). We summarize the situation in bismuth in Figure 9.8.

9.3.4 Parity-violation experiments with atomic hydrogen

Even if theoretical uncertainties associated with parity violation in heavy atoms could be overcome, the experiments would only yield information about the vector-coupling constants C_{1p} and C_{1n} (in particular the latter), associated with $H_{wk}^{(1)}$. The potential $H_{wk}^{(2)}$ may be investigated in experiments utilizing the $2^2S_{1/2}$ metastable state in hydrogen, deuterium, or tritium. Although parity-violation effects in these one-electron atoms are precisely calculable, the experiments are extremely difficult to perform because of apparatus complexity, numerous possibilities for systematic error, and the small size of the expected asymmetries, ($\sim 10^{-7}$). No results have been published, but discussions of the techniques and problems can be found in various original articles (for example, Dunford et al. 78) as well as review articles (Commins and Bucksbaum 80, Commins 81).

9.4 Constraints on neutral-current parameters

The SLAC data and the atomic physics results may be combined to yield useful constraints on neutral weak $e-N$ coupling parameters. For this purpose, we rewrite the weak Hamiltonians $H_p^{(1)}$ and $H_p^{(2)}$ in terms of the nucleon valence quarks: $p = uud$, $n = udd$. Thus we obtain:

$$H_P = \frac{G_F}{2\sqrt{2}} \{\bar{\psi}_e \gamma^\lambda \gamma_5 \psi_e [\tilde{\alpha}(\bar{\psi}_u \gamma_\lambda \psi_u - \bar{\psi}_d \gamma_\lambda \psi_d) + \tilde{\gamma}(\bar{\psi}_u \gamma_\lambda \psi_u + \bar{\psi}_d \gamma_\lambda \psi_d)]$$
$$+ \bar{\psi}_e \gamma^\lambda \psi_e [\tilde{\beta}(\bar{\psi}_u \gamma_\lambda \gamma_5 \psi_u - \bar{\psi}_d \gamma_\lambda \gamma_5 \psi_d) + \tilde{\delta}(\bar{\psi}_u \gamma_\lambda \gamma_5 \psi_u + \bar{\psi}_d \gamma_\lambda \gamma_5 \psi_d)]\}$$
(9.49)

Table 9.1. *Coupling constants for the parity-violating electron–nucleon interactions*

Co-efficient	Atomic coefficient	Space-time	Isospin	Standard model	$\sin^2 \theta_W = 0.23$
$\tilde{\alpha}$	$C_{1n} - C_{1p}$	$A_e V_q$	1	$-1 + 2\sin^2 \theta_W$	-0.54
$\tilde{\beta}$	$\frac{1}{g_A}(C_{2n} - C_{2p})$	$V_e A_q$	1	$-1 + 4\sin^2 \theta_W$	-0.08
$\tilde{\gamma}$	$\frac{1}{3}(C_{1n} + C_{1p})$	$A_e V_q$	0	$\frac{2}{3}\sin^2 \theta_W$	0.15
$\tilde{\delta}$	$\frac{1}{3g_A}(C_{2n} + C_{2p})$	$V_e A_q$	0	0	0

9.4 Constraints on neutral-current parameters

where the coupling constants $\tilde{\alpha}$, $\tilde{\beta}$, $\tilde{\gamma}$, and $\tilde{\delta}$ must be determined by experiment. They represent the various space-time and isospin portions of H_P as summarized in Table 9.1. Hung and Sakurai (81) have shown that the SLAC data may be represented in terms of these couplings as follows:

$$\alpha_1 = \frac{G_F}{\sqrt{2}} \frac{9\tilde{\alpha} + 3\tilde{\gamma}}{5} \tag{9.50}$$

$$\alpha_2 = \frac{G_F}{\sqrt{2}} \frac{9\tilde{\beta} + 3\tilde{\delta}}{5} \tag{9.51}$$

From experimental results (9.23) and (9.24), one obtains:

$$\tilde{\alpha} + \tfrac{1}{3}\tilde{\gamma} = -0.60 \pm 0.16 \tag{9.52}$$

$$\tilde{\beta} + \tfrac{1}{3}\tilde{\delta} = 0.31 \pm 0.51 \tag{9.53}$$

For thallium, an average over the isotopes 203 and 205 yields $Z = 81$ and $N \simeq 123$. Thus one obtains:

$$Q_W(\text{Tl}) = 162 C_{1p} + 246 C_{1n}$$

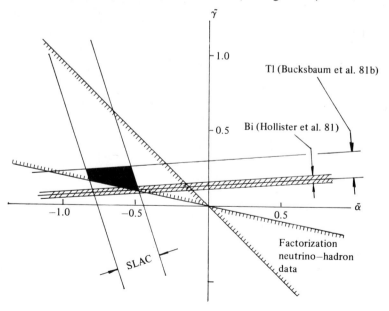

Figure 9.9. Constraints on $\tilde{\alpha}$ and $\tilde{\gamma}$ imposed by the SLAC and Tl, Bi results. The black area represents the overlap of the SLAC result, the factorization constraint, and the Tl (Bucksbaum et al. 81b) or Bi (Hollister et al. 81) results. The result shown for bismuth does not include uncertainties in atomic calculations (see Figure 9.8).

or

$$Q_W(\text{Tl}) = 42\tilde{\alpha} - 627\tilde{\gamma} = -155 \pm 63 \qquad (9.54)$$

where experimental result (9.43) and theoretical prediction (9.41) have been utilized. Note that the SLAC polarized-electron experiment (9.52) and the atomic physics experiments employ almost orthogonal combinations of $\tilde{\alpha}$ and $\tilde{\gamma}$. The two results may be combined to yield allowed values of these coefficients (see Figure 9.9). The results are consistent with the standard model for $\sin^2 \theta_W = 0.23$ (See Hung and Sakurai, 81).

If we assume that a single Z^0 boson is responsible for ν–q and e–q couplings (the "factorization" hypothesis), then we may use the neutrino data to help determine $\tilde{\alpha}$, $\tilde{\beta}$, $\tilde{\gamma}$, and $\tilde{\delta}$. This yields the wedge-shaped region of Figure 9.9. Since it overlaps the allowed region of the $\tilde{\alpha}$-$\tilde{\gamma}$ plane as determined by eN data, independent evidence is provided for factorization.

9.5 Parity violation in nuclear forces

9.5.1 Theory

The simplest contribution to parity violation in the coupling between nucleons is single $W^{\pm}$ or Z^0 exchange (Figures 9.10a,b). These diagrams give rise to an effective pseudoscalar potential between nucleons N_1 and N_2, analogous to the effective potentials (9.33) and (9.34) for the eN neutral weak interaction. For example, the contribution from Figure 9.10a is

$$V_W^P(r) = -\frac{G_F}{2\sqrt{2}} \frac{1}{m_N} \{(1 + \mu^V)(i\boldsymbol{\sigma}_1 \times \boldsymbol{\sigma}_2) \cdot [\mathbf{p}, \delta^3(\mathbf{r})]$$
$$+ (\boldsymbol{\sigma}_1 - \boldsymbol{\sigma}_2) \cdot [\mathbf{p}, \delta^3(\mathbf{r})](\boldsymbol{\tau}_1 \cdot \boldsymbol{\tau}_2 - \tau_1^{(3)} \cdot \tau_2^{(3)})\} \qquad (9.55)$$

where $\mathbf{p} = \tfrac{1}{2}(\mathbf{p}_1 - \mathbf{p}_2)$, $\mathbf{r} = \mathbf{r}_1 - \mathbf{r}_2$, and $\mu^V = \tfrac{1}{2}(g_p - g_n - 2) = 1.85$; $V_W^P(r)$ is nonzero only for $np \to np$ scattering, but an additional term

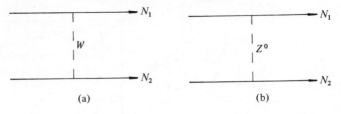

Figure 9.10. The NN interaction by (a) W exchange; (b) Z^0 exchange.

9.5 Parity violation in nuclear forces

$V_Z^P(r)$, corresponding to Z exchange, is of comparable size and contributes to nn or pp scattering.

However, the effect of these zero-range potentials is greatly suppressed because the strong nucleon–nucleon interaction is repulsive at very short distances. The main contribution to the parity-nonconserving interaction must therefore occur by the exchange of light pseudoscalar and vector mesons (Figure 9.11). The relative size of parity-violating and parity-conserving meson-exchange potentials is crudely estimated as

$$V_{\text{PNC}}/V_0 \simeq G_F m_\pi^2 \simeq 2 \times 10^{-7} \tag{9.56}$$

It is very difficult to go beyond this simple estimate because of the complications of nucleon–meson dynamics, but we can make some progress by exploiting various symmetries as follows.

Figure 9.11. Light pseudoscalar and vector meson exchange.

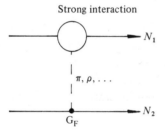

First, we consider the isospin structure of the parity-violating Hamiltonian. The $\Delta S = 0$ portion of the charged hadronic weak current transforms like a vector under isospin rotations, whereas the $|\Delta S| = 1$ current transforms like an isospinor. Also, the neutral hadronic weak current has isoscalar and isovector parts. Since all three currents contribute to the parity-violating weak Hamiltonian, we obtain the following selection rules for the current–current interaction:

$$\Delta S = 0: \quad J_W^{0\dagger} J_W^0 + J_W^0 J_W^{0\dagger} \quad |\Delta I| = 0, 2 \tag{9.57}$$

$$|\Delta S| = 1: \quad J_W^{1\dagger} J_W^1 + J_W^1 J_W^{1\dagger} \quad |\Delta I| = 1 \tag{9.58}$$

$$J_Z^\dagger J_Z + J_Z J_Z^\dagger \quad |\Delta I| = 0, 1, 2 \tag{9.59}$$

Since the strong interaction conserves isospin, these simple conclusions should also apply to the effective Hamiltonian in the presence of strong interactions. However, they only remain valid to the extent to which we can ignore electromagnetic interactions, which are not isospin-invariant. In particular, in heavy nuclei where Coulomb effects are large and isospin is no longer a good quantum number, relations (9.57)–(9.59) cease to be very useful.

Next we consider restrictions imposed by CP invariance, and demonstrate that the π^0 exchange amplitude vanishes, along with all neutral meson exchange amplitudes, in the limit of CP conservation. In addition, charged pion diagrams from W exchange are suppressed by the Cabibbo factor $\sin^2\theta_C$. To prove the first of these statements, we write the most general neutral pseudoscalar $NN\pi^0$ interaction Hamiltonian:

$$H_p^{NN\pi^0} = \frac{G_F}{\sqrt{2}} \int [a\bar\psi_N\psi_N\phi + b\bar\psi_N\gamma^\mu\psi_N\,\partial_\mu\phi + c\bar\psi_N\sigma^{\mu\nu}\psi_N\,\partial_\mu\partial_\nu\phi]\,d^3r$$

(9.60)

where a, b, and c are coefficients and ϕ the π^0 field operator. The latter is even under C (this follows from the fact that $\pi^0 \to \gamma\gamma$ is allowed), which implies that the first term is CP odd. Therefore, coefficient a must be zero. The second term may be transposed by partial integration to $-b\,\partial_\mu(\bar\psi_N\gamma^\mu\psi_N)\phi$, which vanishes from current conservation. This conclusion remains valid if we add momentum-dependent terms such as weak magnetism. Finally, the third term can be transformed by two partial integrations to a form with two derivatives acting on the nucleon spinor; this term then vanishes because of the antisymmetry of $\sigma_{\mu\nu}$.

The argument for charged pions is quite similar. Starting with the most general pseudoscalar Hamiltonian for the charged $NN\pi^\pm$ coupling to pseudoscalar fields $\phi^\pm$, we write

$$\begin{aligned}H_P^{NN\pi^\pm} = \frac{G_F}{\sqrt{2}} \int (&a\bar\psi_N\tau_+\psi_N\phi_- + a'\bar\psi_N\tau_-\psi_N\phi_+ \\&+ b\bar\psi_N\gamma^\lambda\tau_+\psi_N\,\partial_\lambda\phi_- + b'\bar\psi_N\gamma^\lambda\tau_-\psi_N\,\partial_\lambda\phi_+ \\&+ c\bar\psi_N\sigma^{\lambda\nu}\tau_+\psi_N\,\partial_\lambda\,\partial_\nu\phi_- + c'\bar\psi_N\sigma^{\lambda\nu}\tau_-\psi_N\,\partial_\lambda\partial_\nu\phi_+)\,d^3\mathbf{r}\end{aligned}$$ (9.61)

Coefficients b, b', c, and c' vanish as before. As for a and a', we require that together these terms be odd under C so that the Hamiltonian is even under CP. This implies $a + a' = 0$, so that:

$$H_P^{NN\pi^\pm} \propto \frac{G_F}{\sqrt{2}} \int \bar\psi(\tau_+\phi_- - \tau_-\phi_+)\psi\,d^3\mathbf{r} = \frac{G_F}{\sqrt{2}} \int \bar\psi(\boldsymbol{\tau}\times\boldsymbol{\phi})^{(3)}\psi\,d^3\mathbf{r}$$

which obeys $|\Delta I| = 1$. Since the isovector current – current interaction (9.57) permits $|\Delta I| = 2, 0$ only, we conclude that charged-current single-pion exchange is dominated by the isospinor current–current in-

9.5 Parity violation in nuclear forces

teraction (9.58), which is suppressed by the factor $\sin^2\theta_C$. (The charm-changing currents may be neglected because of the large mass of the charmed quark.)

The other $NN\pi$ vertex in a single-pion-exchange diagram is an isospin-conserving strong interaction. Therefore, the effective parity-violating potential may be written as the $|\Delta I| = 1$ operator (Henley 78):

$$V_{PNC}^{\pi}(r) = i\frac{G_{eff}^{\pi}g}{2\sqrt{2}m_N}(\tau_1 \times \tau_2)^{(3)}(\sigma_1 + \sigma_2) \cdot \left[\mathbf{p}, \frac{1}{r}e^{-m_\pi r}\right] \quad (9.62)$$

where $\mathbf{p} = \frac{1}{2}(\mathbf{p}_1 - \mathbf{p}_2)$, $\mathbf{r} = \mathbf{r}_1 - \mathbf{r}_2$, G_{eff}^{π} is the effective dimensionless Fermi constant $G_F m_N^2 \times$ (model-dependent factors), and g the pion–nucleon coupling constant given by $g^2/4\pi = 14.4$. For charged weak currents only, one then has:

$$G_{eff}^{\pi,CC} \simeq G_F \frac{m_N^2 \sin^2\theta_C}{4\pi} \simeq 4 \times 10^{-8} \quad (9.63)$$

The neutral-current contribution does not suffer the $\sin^2\theta_C$ suppression, and it may be shown (Gari and Reid 74) that

$$G_{eff}^{\pi,NC} \simeq 5 \times 10^{-7} \quad (9.64)$$

Since pion exchange with charged weak currents is suppressed, the relative importance of ρ exchange is enhanced. (CP invariance no

Table 9.2. *Phenomenological potential*

$$V_P^{eff} = V_P^{\pi} + \mathbf{s}\cdot\mathbf{v}\bar{U}\frac{(1-\tau_1\cdot\tau_2)}{4} + \mathbf{S}\cdot\mathbf{v}\bar{W}\frac{\tau_1^{(3)} - \tau_2^{(3)}}{2} + \mathbf{s}\cdot\mathbf{v}\bar{V}^{(0)}\frac{3+\tau_1\cdot\tau_2}{4}$$

$$+ \mathbf{s}\cdot\mathbf{v}\bar{V}^{(1)}\frac{\tau_1^{(3)} + \tau_2^{(3)}}{2} + \mathbf{s}\cdot\mathbf{v}\bar{V}^{(2)}\frac{3\tau_1^{(3)}\cdot\tau_2^{(3)} - \tau_1\cdot\tau_2}{2\sqrt{6}}$$

with $\mathbf{v} = \frac{2\pi}{m_N}\{\mathbf{p}, \delta^3(\mathbf{r})\}$, $\mathbf{s} = \frac{1}{2}(\sigma_1 - \sigma_2)$, $\mathbf{S} = \frac{1}{2}(\sigma_1 + \sigma_2)$,

$\mathbf{p} = \frac{1}{2}(\mathbf{p}_1 - \mathbf{p}_2)$, $\mathbf{r} = \mathbf{r}_1 - \mathbf{r}_2$.

Experiments[a]	Parameter measured
^{181}Ta, ^{175}Lu, ^{41}K, ^{19}F	$G_{eff}^{\pi} + \frac{1}{2}m_N^2(2\bar{W} + \bar{U} + 3\bar{V}^{(0)} + \bar{V}^{(1)}$
^{18}F, $np \to D\gamma$[b]	$G_{eff}^{\pi} + 0.1 m_N^2 \bar{W}$
^{16}O, ^{12}C	$\bar{U} + 3\bar{V}^{(0)}$
$np \to D\gamma$[c]	$2\bar{U} - \bar{V}^{(0)} + 2\bar{V}^{(2)}$

[a] For explanation, see text.
[b] Circular polarization.
[c] Asymmetry.
Source: Henley 78.

358   9 *The parity-nonconserving* eN *and* NN *interactions*

longer eliminates the Cabibbo-allowed amplitude because ρ is a vector meson.) The net effect is to replace the δ functions in (9.55) with a ρ-meson Yukawa factor:

$$\delta^3(r) \to m_\rho^2 \frac{1}{4\pi r} e^{-m_\rho r} \tag{9.65}$$

One may then compare with (9.62) to arrive at the estimate

$$G^\rho_{\text{eff}} = G_F m_\rho^2 e^{-2m_\rho r_N}/4\pi \simeq 10^{-6}\text{--}10^{-7} \tag{9.66}$$

where r_N is the hard-core radius. Evidently then, ρ and π exchange are of comparable importance in the parity-violating potential. However, it seems very difficult to draw more detailed theoretical conclusions because of the complications of nuclear dynamics.

It may be shown that all PNC effects in nuclei are describable by means of a purely phenomenological potential containing G^π_{eff} and five additional constants multiplying various isospin operators (see, for example, Table 9.2). Such an approach has the advantage that consistency among experimental results can be established, at least in principle. At present, however, these results are insufficient to place unique constraints on the constants. The data do allow us to conclude that parity-violating effects in heavy nuclei are mainly due to terms proportional to G^π_{eff} and the results are consistent with the neutral-current value (9.64).

9.5.2   *Experiment*

Nuclear-parity-violation experiments make use of the fact that the strong interaction is invariant under both parity and isospin. Thus, if we neglect electromagnetic and weak interactions, an approximation valid for light nuclei, nuclear states are eigenstates of isospin as well as parity. The weak interaction mixes states of opposite parity in the usual way:

$$\psi_n = \psi_n^0 + \sum_m \frac{|\chi_m\rangle\langle\chi|H_p|\psi_n\rangle}{E_n - E_m} \tag{9.67}$$

where the isospin of $|\chi_m\rangle$ differs from that of $|\psi_n\rangle$ by 0, 1, or 2, depending on the $|\Delta I|$ properties of H_P. In the simplest case, only one $|\chi_m\rangle$ gives the dominant contribution:

$$\psi' = \psi + F\chi$$

where

$$F = \langle\chi|H_P|\psi\rangle/(E_\psi - E_\chi) \simeq 10^{-7}\text{--}10^{-6} \tag{9.68}$$

9.5 Parity violation in nuclear forces

In such cases, the isospin structure of H_P can, in principle, be revealed.

(a) *Parity-forbidden transitions.* The simplest application of this general method is to the measurement of a nuclear nonleptonic decay that does not conserve parity. The only observed example is α decay of the 8.87-MeV $J^P = 2^-$, $I = 0$ level of ^{16}O (see Figure 9.12):

$$^{16}\text{O}(J^P = 2^-, I = 0) \rightarrow {}^{12}\text{C}(0^+0) + \alpha(0^+0)$$

The parity of the final state is positive, since both ^{12}C and ^{4}He have positive parity, and angular momentum is conserved. Therefore, the transition can only occur by means of a parity-violating $\Delta I = 0$ perturbation. Mixing in the ground state of ^{12}C and ^{4}He may be neglected, because neither of these nuclei have low-lying levels. The 2^- state of ^{16}O, however, can be admixed with nearby 2^+ levels, some of which are above threshold for α decay themselves and can therefore be measured in subsidiary experiments. Unfortunately, a subthreshold 2^+ level is expected to provide a large contribution also, which interferes destruc-

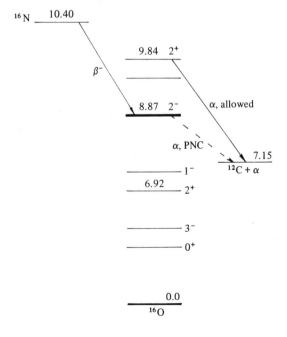

Figure 9.12. ^{16}O α decay. The principal contributions to the parity mixing of the 8.87-MeV 2^- state come from the 6.92-MeV 2^+ state, which is below threshold for α decay, and higher 2^+ states such as that at 9.84 MeV.

tively with the other contributions, and the matrix element cannot be calculated very precisely. Therefore the predicted parity-violating rate is quite uncertain.

The experiment has been done by populating the 2^- state from the β decay of ^{16}N. The PNC rate is expressed as an energy width and is measured to be (Neubeck et al. 74)

$$\Gamma_\alpha^{PNC} = (1.03 \pm 0.28) \times 10^{-10} \text{ eV} \tag{9.69}$$

which should be compared with the standard model prediction Gari 73)

$$\Gamma_\alpha^{PNC} \simeq 1 \times 10^{-10} \text{ eV} \tag{9.70}$$

(b) *Circular polarization and asymmetry of γ rays.* Parity violation results in circular polarization of γ emission from unpolarized nuclei or fore–aft asymmetry in γ emission from polarized nuclei. The transition rate between levels i and f is proportional to

$$I = |\langle \psi_f + \Sigma F_f \chi_f | H_{EM} | \psi_i + \Sigma F_i \chi_i \rangle|^2$$

$$\simeq |\langle \psi_f | H_{EM} | \psi_i \rangle|^2 \times \left\{ 1 + \Sigma F_f \frac{\langle \chi_f | H_{EM} | \psi_i \rangle}{\langle \psi_f | H_{EM} | \psi_i \rangle} + cc \right. \tag{9.71}$$

$$\left. + \Sigma F_i \frac{\langle \psi_f | H_{EM} | \chi_i \rangle}{\langle \psi_f | H_{EM} | \psi_i \rangle} + cc \right\}$$

where factors $F_{i,f}$ are pseudoscalars that reverse sign under coordinate inversion. We write H_{EM} as the sum of electric and magnetic multipole operators:

$$H_{EM} = \sum_L E(L) + \sum_L M(L) \tag{9.72}$$

The pseudoscalar terms in (9.71) then become

$$\sum_{n,L} F_{fn} \frac{\langle \chi_{fn} | E(L)[M(L)] | \psi_i \rangle}{\langle \psi_f | M(L)[E(L)] | \psi_i \rangle} + \sum_{n,L} F_{in} \frac{\langle \psi_f | E(L)[M(L)] | \chi_{in} \rangle}{\langle \psi_f | M(L)[E(L)] | \psi_i \rangle} \tag{9.73}$$

The sum over L is limited by the usual angular momentum selection rules, and generally only one L contributes. For example, if the normal transition proceeds by $E(1)$ γ emission, then parity violation causes an admixture of $M(1)$. The circular polarization is

$$P_\gamma = 2\overline{FR} = 2 \Sigma F_i R_i \tag{9.74}$$

where

$$R_i = \frac{\langle M1 \rangle}{\langle \psi_f | E(1) | \psi_i \rangle} \tag{9.75}$$

9.5 Parity violation in nuclear forces

Similarly the angular distribution of γ rays from nuclei with polarization P is given by

$$I(\theta) = 1 + \Delta_\gamma P \cos\theta \tag{9.76}$$

where

$$\Delta_\gamma = 2\overline{FR}A \tag{9.77}$$

and A is a numerical coefficient of order unity, which depends on the initial and final nuclear spins.

Experiments are chosen to maximize the ratio R_i by selection of cases where the parity-allowed transition is suppressed and the amplitude in the numerator is allowed. Enhancements of 10^2 or more can be obtained, allowing circular polarizations of fore–aft asymmetries of 10^{-4} or more, in some cases.

A transition of particular interest is:

$$^{18}\text{F}^*(0^-, 0) \to {}^{18}\text{F}(1^+, 0) + \gamma(1.081 \text{ MeV}) \tag{9.78}$$

To a very good approximation, parity violation arises solely from mixing of the $(0^-, 0)$ level at 1.081 MeV with the nearby $(0^+, 1)$ level at 1.042 MeV (see Figure 9.13.) Thus the $|\Delta I| = 1$ portion of the weak Hamiltonian is responsible for the effect, and as we have seen, this is dominated by Z^0 exchange. The small energy difference between $(0^-, 0)$ and $(0^+, 1)$ states results in an enhancement of F by a factor of about 100 compared with a "typical" case, and additional enhancement occurs because in

$$R = \frac{\langle 1^+, 0|M1|0^+, 1\rangle}{\langle 1^+, 0|E1|0^-, 0\rangle} \tag{9.79}$$

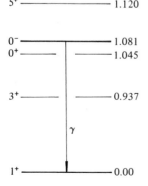

Figure 9.13. Energy levels of ^{18}F. The observed transition is from the 1.081-MeV $(0^-, 0)$ state.

the "allowed" $E1$ amplitude in the denominator is in fact forbidden by isospin selection rules (Gell-Mann and Telegdi 53; see Problem 9.8). The actual suppression of this amplitude is by a factor of about 10^{-5}, which shows that isospin is a good quantum number in ^{18}F and reinforces the conclusion that the $|\Delta I| = 1$ weak interaction is important here.

Since both $M1$ and $E1$ amplitudes have been measured, the main theoretical uncertainty arises from the pseudoscalar matrix element F, which involves complicated nuclear structure calculations. The circular polarization is predicted (Gari et al. 75) to be

$$|P_\gamma| = \begin{cases} |2FR| = 3.6 \times 10^{-4} & \text{(CC only)} \quad (9.80) \\ 3.6 \times 10^{-4} \left(1 + \dfrac{8 \sin^2 \theta_W}{3 \sin^2 \theta_C}\right) = 3.7 \times 10^{-3} \quad (9.81) \end{cases}$$

(NC + CC, standard model)

With such a large effect, one might hope to observe parity-violating neutral weak currents in nuclei quite unambiguously. Unfortunately, however, the experiment is extremely difficult, and only an upper limit has been obtained so far (Barnes et al. 78):

$$P_{\gamma,\text{expt}} = (-0.5 \pm 2.0) \times 10^{-3} \quad (9.82)$$

Excited ^{18}F nuclei are produced in the reaction ^{16}O(^{3}He, p) ^{18}F* with ^{3}He ions incident on a flowing water target. The γ-ray circular polarization is analyzed in a ferromagnetic Compton polarimeter, which has rather poor analyzing power: 2 to 4 percent.

Another case of considerable interest is the fore–aft asymmetry in γ emission from polarized ^{19}F (Figure 9.14):

$$^{19}\text{F}^*(\tfrac{1}{2}^-, \tfrac{1}{2}) \rightarrow {}^{19}\text{F}(\tfrac{1}{2}^+, \tfrac{1}{2}) + \gamma(110 \text{ keV}) \quad (9.83)$$

Here, the only significant parity mixing occurs between the two states participating in the transition. Each contains a small admixture of the other, as follows:

$$|\tfrac{1}{2}^-, \tfrac{1}{2}\rangle = |-\rangle + F|+\rangle \quad (9.84)$$

$$|\tfrac{1}{2}^+, \tfrac{1}{2}\rangle = |+\rangle - F|-\rangle \quad (9.85)$$

Figure 9.14. Low-lying levels of ^{19}F, a case well approximated by a simple two-level system.

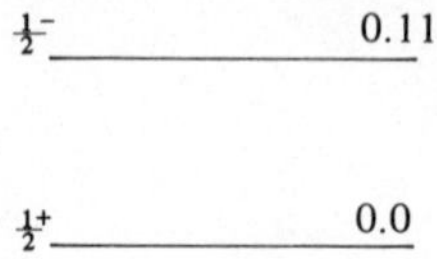

$\tfrac{1}{2}^-$ _____ 0.110 MeV

$\tfrac{1}{2}^+$ _____ 0.0

9.5 Parity violation in nuclear forces

where
$$F = \langle -|H_P|+\rangle/110 \text{ keV} \tag{9.86}$$
and H_P contains $|\Delta I| = 0$ and $|\Delta I| = 1$ portions. The fore–aft asymmetry is
$$\Delta = \frac{2F(\langle +|M1|+\rangle - \langle -|M1|-\rangle)}{\langle +|E1|-\rangle} \tag{9.87}$$

The $M1$ and $E1$ matrix elements have been measured separately, and the latter are quite small, which implies considerable enhancement. However, F must be calculated. In the simplest model for this purpose, $\tfrac{1}{2}^+$ and $\tfrac{1}{2}^-$ states are $2s_{1/2}$ and $1p_{1/2}$ harmonic oscillator hole states in ^{20}Ne, and one can treat the problem in terms of a single particle moving in an effective potential. More sophisticated calculations based on a variety of potentials give values for Δ that differ by as much as a factor of 2. However, this is adequate at least for a qualitative test of the standard model, since the neutral-current Hamiltonian generates a large enhancement in the $|\Delta I| = 1$ case. The asymmetry can be expressed as the sum of $\Delta I = 0$ and $\Delta I = 1$ contributions (Box et al. 76):

$$\Delta = \Delta_0 + \Delta_1 \tag{9.88}$$

$$= \begin{cases} (7.2 - 2.6) \times 10^{-5} = +4.6 \times 10^{-5} & \text{(CC only)} \\ (10.0 - 25.3) \times 10^{-5} = -15 \times 10^{-5} & \text{(NC + CC, standard model, } \sin^2 \theta_W = 0.23\text{)} \end{cases} \tag{9.89}$$

It can be seen that neutral currents reverse the sign of the effect and increase its magnitude substantially. The experimental result (Adelberger et al. 75)

$$\Delta_{\text{expt}} = (-18 \pm 9) \times 10^{-5} \tag{9.90}$$

supports the standard model, at least qualitatively. Figure 9.15 is a sketch of the apparatus.

The reaction
$$n + p \to D(1^+, 0) + \gamma \ (\geq 2.22 \text{ MeV}) \tag{9.91}$$

is the simplest one in which parity violation is observed. There are two observable pseudoscalars: the γ-ray circular polarization P_γ for capture of unpolarized neutrons and the fore–aft asymmetry Δ_γ for polarized neutrons. The effects are much smaller here than in cases described previously because there are no low-lying states of opposite parity. Capture occurs almost entirely from a ^1S continuum state, and therefore the γ ray must be predominantly $M1$. The requirement that the total wavefunction be antisymmetric then specifies the isospin uniquely:

$n + p(^1S, I = 1) \rightarrow D + \gamma(M1)$ (parity-conserving)

(a) $n + p(^1S, I = 1) \rightarrow D + \gamma(E1)$
(b) $n + p(^3S, I = 0) \rightarrow D + \gamma(E1)$ (parity-violating)
(c) $n + p(^3S, I = 0) \rightarrow D + \gamma(M2)$

Each of these reactions occurs via parity mixing with a P state, but because the deuteron has $Z = N = 1$, charge symmetry imposes certain selection rules, as in the case of ^{18}F (see Problem 9.8). In particular, $E1$ transitions are forbidden between states of the same I. Therefore, intermediate (^1P, $I = 0$) states are excluded and only (^3P, $I = 1$) states can contribute to cases (a) and (b). We thus obtain:

(a) $\langle D, \gamma(E1)|H_{EM}|np, {}^1S, I = 1\rangle = F(|\Delta I| = 0)\langle D, \gamma|E1|^3P, 1\rangle$ (9.92)

(b) $\langle D, \gamma(E1)|H_{EM}|np, {}^3S, I = 0\rangle = F(|\Delta I| = 1)\langle D, \gamma|E1|^3P, 0\rangle$ (9.93)

(c) $\langle D, \gamma(M2)|H_{EM}|np, {}^3S, I = 0\rangle = F(|\Delta I| = 0)\langle D, \gamma|M2|^1P, 0\rangle$
$+ F(|\Delta I| = 1)\langle D, \gamma|M2|^2P, 1\rangle$
(9.94)

In the case of circular polarization, a finite effect arises from quantum-mechanical interference, which can only occur if the parity-conserving and parity-violating transitions proceed from the same initial state. Therefore only case (a) [(9.92)] contributes, and solely the $\Delta I = 0$ portion of the weak Hamiltonian is operative. On the other hand, the fore–aft asymmetry requires a triplet initial state, which im-

Figure 9.15. ^{19}F parity-violation experiment. Polarized ^{19}F in the 110-keV state is produced in the reaction ^{22}Ne(p, ^{4}He)^{19}F* in a gas cell by means of polarized protons. Ge(Li) detectors placed fore and aft are used to measure Δ. The proton counters monitor beam motion when the polarization is reversed. (From Adelberger et al. 75. Reprinted with permission.)

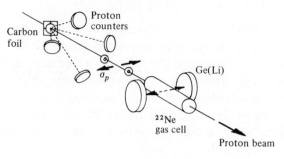

9.5 Parity violation in nuclear forces

plies that (b) and (c) contribute, with the largest amplitude coming from (b), since it is associated with the lowest multipole moment. This implies that $|\Delta I| = 1$ dominates in this case.

P_γ has been measured in an experiment carried out by Lobashov et al. (72a) in Leningrad. Thermal neutrons were captured by protons in an H_2O target placed in the middle of a reactor and the capture radiation was analyzed in a Compton polarimeter. Since the expected asymmetry is very small, a very large number of photon counts was required. Therefore, the photons were not counted one by one but were integrated by the detector. Unfortunately, with this method, one cannot easily discriminate against unwanted backgrounds. One particularly harmful source of systematic error is circularly polarized *bremsstrahlung* arising from left-handed β radiation in the reactor environment. Lobashov et al (72a) found

$$P_{\gamma,\text{expt}} = -(1.30 \pm 0.45) \times 10^{-6} \tag{9.95}$$

a result more than 100 times as large as that anticipated on the basis of the standard model and that disagrees with the ^{16}O α-decay result, also sensitive to H_P ($\Delta I = 0$).

The asymmetry Δ_γ should also be very small, although there is some enhancement from $\Delta I = 1$ neutral-current effects. One predicts:

$$\Delta_\gamma \simeq (4 - 7) \times 10^{-9} \quad \text{(CC only)} \tag{9.96}$$

$$\Delta_\gamma \simeq (6 - 11) \times 10^{-8} \quad \text{(NC + CC, standard model)} \tag{9.97}$$

[(9.97) from Lassey and McKellar 76)]. An upper limit has been obtained at the Grenoble reactor (Cavignac et al. 77). Here, cold neutrons are polarized by reflection from magnetized cobalt mirrors and are guided into a target of liquid parahydrogen, where they are captured without depolarization. The result is

$$\Delta_{\gamma,\text{expt}} = (0.6 \pm 2.1) \times 10^{-7} \tag{9.98}$$

Unfortunately, the precision is insufficient to discern an effect.

Finally, we consider parity-violating neutron spin rotation, a phenomenon first predicted by Michel (64) and only recently observed (Forte et al. 80, Ramsey 81). The principle is essentially the same as for parity-violating optical rotation in atoms (bismuth). In the present case, weak–strong interference causes a slight difference in the phase velocity for neutrons with helicity ± 1 passing through matter. The transverse polarization of a neutron therefore rotates around its momentum by an angle

$$\phi = \frac{n_+ - n_-}{\lambdabar} l \tag{9.99}$$

where l is the distance of travel and $n_\pm$ are the indexes of refraction for neutrons with helicity ± 1. Equation (9.99) may be expressed in terms of the forward-scattering length $a_\pm$ in a solid with atomic density N_a as follows:

$$\phi = \frac{2\pi h}{m_N v} N_a(a_+ - a_-)l = \sqrt{2}\,\frac{\hbar}{m_N c} N_a(G_F m_N^2)Wl \qquad (9.100)$$

where

$$W = ZC_{2n} + [ZC_{np} + (A - Z)C_{nn}]\eta + \alpha(Z, N)$$

C_{2n} is the electron-vector–neutron-axial-vector coupling constant of parity violation in atoms (9.31), C_{np} and C_{nn} the neutron–nucleon parity-violating coupling coefficients, η a nuclear shielding factor, and α the parity violation from nuclear structure effects, such as parity mixing of nuclear excited states and collective effects. For a meaningful test of the standard model, one requires a material in which the term C_{2n} dominates over the other terms or in which these other terms can be described theoretically with some precision. Unfortunately, no such case is known at present; therefore, interpretation of experimental results is difficult.

Table 9.3. *Nuclear parity-violation experiments*

Transition	Values	Reference
$n + p \to D + \gamma$	$\Delta_\gamma = (0.06 \pm 0.2) \times 10^{-6}$	Cavignac et al. 77
	$P_\gamma = (-1.3 \pm 0.5) \times 10^{-6}$	Lobashov et al. 72a
$^{16}\text{O}(2^-) \to {}^{12}\text{C}(0^+)$	$\Gamma_\alpha = (1.03 \pm 0.28) \times 10^{-10}$ eV	Neubeck et al. 74
$^{18}\text{F}(0^- \to 1^+)$	$P_\gamma = (-0.5 \pm 2.0) \times 10^{-3}$	Barnes et al. 78
$^{19}\text{F}(\frac{1}{2}^- \to \frac{1}{2}^+)$	$\Delta_\gamma = (-18 \pm 9) \times 10^{-5}$	Adelberger et al. 75
$^{41}\text{K}(\frac{7}{2}^- \to \frac{3}{2}^+)$	$P_\gamma = (2.0 \pm 0.4) \times 10^{-5}$	Lobashov et al. 72b
$^{114}\text{Cd}(1^+ \to 0^+)$	$\Delta_\gamma = (-4 \pm 1) \times 10^{-4}$	Abov et al. 65
	$P_\gamma = (-6 \pm 1.5) \times 10^{-4}$	Alberi et al. 72
$^{118}\text{Sn}(1^+ \to 0^+)$	$\Delta_\gamma = (4.4 \pm 0.6) \times 10^{-4}$	Benkoula et al. 77
	$\Delta_\gamma = (8.5 \pm 1.5) \times 10^{-4}$	Danilyan et al. 76
$^{175}\text{Lu}(\frac{9}{2}^- \to \frac{7}{2}^+)$	$P_\gamma = (5.5 \pm 0.5) \times 10^{-5}$	Kuphal et al. 74
$^{180}\text{Hf}(8 \to 6)$	$P_\gamma = (-2.4 \pm 0.3) \times 10^{-3}$	Jenschke and Bock 70; Lipson et al. 71
	$\Delta_\gamma = (-1.7 \pm 0.2) \times 10^{-3}$	Krane et al. 72
$^{181}\text{Ta}(\frac{5}{2}^+ \to \frac{3}{2}^+)$	$P_\gamma = (-5.2 \pm 0.5) \times 10^{-6}$	Herb et al. 77
$n + {}^{117}\text{Sn}$ (neutron spin rot.)	$R = (3.7 \pm 0.3) \times 10^{-5}$ rad/cm	Forte et al. 80

A large effect has been observed experimentally in ^{117}Sn:

$$\phi/l = (+36.8 \pm 2.7) \times 10^{-6} \quad \text{rad/cm} \tag{9.101}$$

Let us now summarize the contents of this chapter. In the case of the eN interaction, the results of the SLAC polarized-electron experiment and the atomic physics parity-violation experiments yield values of the parameters $\tilde{\alpha}$ and $\tilde{\gamma}$ in excellent agreement with the standard model. However, improvements in the heavy-atom experiments are still worthwhile, and results from the atomic hydrogen parity-violation experiments would be extremely useful. In the case of the NN interaction, theoretical uncertainties prevent us from drawing unambiguous conclusions from experimental data. However, the results of at least one experiment (in ^{19}F) tend to favor the standard model. Table 9.3 summarizes the various nuclear parity-violation experiments.

Problems

9.1 Estimate the neutral weak eN contribution to the hyperfine splitting of the $1\,^2S_{1/2}$ state of atomic hydrogen.

9.2 Derive an expression for the parity-violating asymmetry in the scattering of polarized electrons from hydrogen, analogous to (9.17)–(9.19) for deuterium. Assume the standard model.

9.3 Fill in the steps leading to result (9.17) with (9.18) and (9.19).

9.4 Consider a forbidden $M1$ transition ($M1$ amplitude $\mathcal{M}$) between a pair of atomic levels, such as the $6\,^2S_{1/2}$–$7\,^2S_{1/2}$ transition in cesium. Parity violation gives rise to an additional electric dipole amplitude $\mathscr{E}_P$, as explained in the text. Let circularly polarized photons with frequency corresponding to the difference in energy between the two levels be incident on atoms in the ground state. Show that the cross section for photon absorption is proportional to (9.38), where the $\pm$ sign applies for photon helicity ± 1, respectively. Prove that $\mathscr{E}_P$ and $\mathcal{M}$ must be relatively imaginary if time-reversal invariance holds.

9.5 Show that (9.32) reduces to (9.33) in the limit of a point nucleus and a nonrelativistic electron.

9.6 If the parity-violating atomic Hamiltonian was not time-reversal-invariant, what observational effects would occur?

9.7 Derive (9.55).

9.8 Consider a light nucleus (for which isospin is a good quantum number) that is self-conjugate ($Z = N$). (An example is ^{18}F.) Prove that the matrix element of the electric dipole operator

$$\sum_{\text{protons}} e\mathbf{r}_i$$

taken between $I = 0$ states, must vanish (Gell-Mann and Telegdi 53). This result is relevant to the question of parity violation in ^{18}F [(9.78)].

10

Lepton mixing, neutrino oscillations, and neutrino mass

10.1 Introduction

In our previous discussions of neutrinos, we relied for the most part on certain "standard" assumptions, as follows:

(i) There exist distinct lepton numbers L_e, L_μ, L_τ; each of which is conserved in all processes. This is based on the results of (a) various "two-neutrino" experiments in which neutrinos generated in $\pi \to \mu$ decay are effective in inducing inverse reactions:

$$\nu_\mu + n \to p + \mu^-$$

but are ineffective in producing electrons:

$$\nu_\mu + n \not\to p + e^-$$

and (b) the very small upper limit on the rate for the transition $\mu \to e\gamma$ (Bowman et al. 79):

$$\frac{\Gamma(\mu \to e\gamma)}{\Gamma(\mu \to e\nu\nu)} \leq 1.9 \times 10^{-10}$$

(ii) Neutrinos have zero mass.

(iii) Neutrinos obey the V–A law. Hence the ν are left-handed, the $\bar{\nu}$ are right-handed, and each is described by the two-component theory.

Although not one of these assumptions is contradicted by any confirmed experimental result, they demand close scrutiny for several reasons. First, neutrino helicity measurements are of limited accuracy (5 to 10 percent precision at best). Second, it can never be demonstrated experimentally that the neutrino mass is precisely zero, and indeed, we shall see that there are indications to the contrary in at least one experiment. Third, there is no compelling theoretical reason to assume zero neutrino mass. We recall that, according to the standard model (Section 2.6), lepton masses come about through a Yukawa-type coupling of the lepton fields to the vacuum expectation value of the Higgs field. In this way, neutrino masses as well as charged lepton masses may be generated, and in the standard model as well as in other theories, the neutrino mass eigenstates and neutrino–lepton-number eigenstates are not necessarily coincident. Thus there may exist neutrino "mixing," which is analogous to quark mixing as described by the Cabibbo angle in the GIM scheme and generalized in the Kobayashi–Maskawa six-quark scheme.

Neutrino mixing gives rise to the hypothetical phenomenon of "neutrino oscillations" (Maki et al. 62, Pontecorvo 68, Bilenky and Pontecorvo 78) in which, for example, a neutrino born in eigenstate ν_e finds itself at some later time in a mixture of states ν_e, ν_μ, and/or ν_τ. In such oscillations, separate lepton numbers L_e, L_μ, and L_τ are not conserved, but the sum $L = L_e + L_\mu + L_\tau$ (total lepton number) is conserved ($\Delta L = 0$).

However, according to the modern point of view, baryon number B and lepton number L do not enjoy the same fundamental status as electric charge; whereas it seems that the latter is strictly conserved in all processes, it is widely thought that at some level baryon and lepton conservation break down, and processes occur in which these conservation laws are violated. This view arises in various grand unified theories, in which an attempt is made to tie together the electroweak and strong interactions. For example, in one such theory, the $SU(5)$ model of Georgi and Glashow (74), quarks and leptons are not truly distinct but are merely different members of a single primitive supermultiplet. Also, there exist, in addition to the 12 intermediate bosons discussed so far (the photon, W^+, W^-, Z^0, and the 8 gluons), certain supermassive intermediate bosons $X^\pm$ and $Y^\pm$ with charges $\pm\frac{4}{3}$ and $\pm\frac{1}{3}$, respectively, and masses of order 10^{15} GeV/c^2. The X and Y couple to quarks *and* leptons, with the result, for example, that the proton can decay to $\pi^0 e^+$ and other final states. The transition probabilities for such processes are exceedingly small, but they are not thought to be beyond the range of experimental observation (the proton lifetime in such models is predicted to be of order 10^{31} yr). Thus neither baryon number nor lepton number is conserved.

Indeed, it is possible to imagine theories in which $\nu \to \bar{\nu}$ oscillations occur, for which $|\Delta L| = 2$. Both $\Delta L = 0$ and $|\Delta L| = 2$ type oscillations

10.2 $\Delta L = 0$ neutrino oscillations

require that at least one type of neutrino possess mass and that the masses of at least two types be unequal.

The questions of lepton mixing, neutrino oscillations, and neutrino mass have interesting and perhaps very important ramifications in astrophysics and cosmology. These topics will be considered in the following chapter. Here, we shall restrict ourselves to a development of the underlying theory and a discussion of laboratory experiments. Many of the topics considered here are discussed in the review by Primakoff and Rosen (81).

10.2 $\Delta L = 0$ neutrino oscillations

Let us recall the discussion of lepton mass in the standard model, Section 2.6. By confining ourselves to the first two lepton generations only and assuming that e^- and μ^- mass eigenstates are the same as the weak interaction eigenstates, we obtain (2.114) for the lepton-mass portion of the Lagrangian:

$$\mathcal{L}_s = \eta[g_{11}\bar{e}e + g_{22}\bar{\mu}\mu] \\ + \eta[h_{11}\bar{\nu}_e\nu_e + h_{22}\bar{\nu}_\mu\nu_\mu + h_{12}(\bar{\nu}_\mu\nu_e + \bar{\nu}_e\nu_\mu)] \tag{2.114'}$$

We rewrite this equation by employing the notation:

$$l = \begin{pmatrix} e \\ \mu \end{pmatrix}, \quad \nu_l = \begin{pmatrix} \nu_e \\ \nu_\mu \end{pmatrix}$$

$$M_l = \begin{pmatrix} \eta g_{11} & 0 \\ 0 & \eta g_{22} \end{pmatrix} \equiv \begin{pmatrix} m_e & 0 \\ 0 & m_\mu \end{pmatrix}$$

$$M_\nu = \begin{pmatrix} \eta h_{11} & \eta h_{12} \\ \eta h_{12} & \eta h_{22} \end{pmatrix} \equiv \begin{pmatrix} m_{\nu_e} & m_{\nu_e \nu_\mu} \\ m_{\nu_e \nu_\mu} & m_{\nu_\mu} \end{pmatrix}$$

as follows:

$$\mathcal{L}_s = \bar{l}M_l l + \bar{\nu}_l M_\nu \nu_l \tag{10.1}$$

We wish to investigate the consequences of the fact that M_ν has off-diagonal elements.

Since M_ν is symmetric, a new basis $\nu_1 \nu_2$ exists with

$$\begin{aligned} \nu_e &= \nu_1 \cos\phi + \nu_2 \sin\phi \\ \nu_\mu &= -\nu_1 \sin\phi + \nu_2 \cos\phi \end{aligned} \tag{10.2}$$

in which $M_\nu \to M'_\nu$ is diagonal:

$$M'_\nu = \begin{pmatrix} m_1 & 0 \\ 0 & m_2 \end{pmatrix} \tag{10.3}$$

The states ν_1 and ν_2 have definite masses m_1 and m_2, respectively. One has

$$m_{\nu_e} = m_1 \cos^2 \phi + m_2 \sin^2 \phi \tag{10.4}$$
$$m_{\nu_\mu} = m_1 \sin^2 \phi + m_2 \cos^2 \phi \tag{10.5}$$

and

$$m_{\nu_e \nu_\mu} = (m_2 - m_1) \sin \phi \cos \phi \tag{10.6}$$

Let us now consider a neutrino formed in state $|\nu_e\rangle$ at time $t = 0$:

$$|\nu_e(0)\rangle = |\nu_1(0)\rangle \cos \phi + |\nu_2(0)\rangle \sin \phi \tag{10.7}$$

Since $|\nu_1\rangle$ and $|\nu_2\rangle$ have the time evolution:

$$|\nu_1(t)\rangle = e^{-iE_1 t}|\nu_1(0)\rangle, \qquad |\nu_2(t)\rangle = e^{-iE_2 t}|\nu_2(0)\rangle$$

where $E_i = (p^2 + m_i^2)^{1/2}$, (10.7) evolves to

$$\begin{aligned} |\nu_e(t)\rangle &= e^{-iE_1 t}|\nu_1(0)\rangle \cos \phi + e^{-iE_2 t}|\nu_2(0)\rangle \sin \phi \\ &= [e^{-iE_1 t} \cos^2 \phi + e^{-iE_2 t} \sin^2 \phi]|\nu_e(0)\rangle \\ &\quad + \cos \phi \sin \phi \, [e^{-iE_2 t} - e^{-iE_1 t}]|\nu_\mu(0)\rangle \end{aligned} \tag{10.8}$$

The probability that the neutrino, originally in state $|\nu_e\rangle$, is in state $|\nu_\mu\rangle$ at time t is then

$$\begin{aligned} P(\nu_e \to \nu_\mu) &= |\langle \nu_\mu | \nu_e(t)\rangle|^2 = \sin^2 \phi |e^{-iE_2 t} - e^{-iE_1 t}|^2 \\ &= \tfrac{1}{2} \sin^2 2\phi \, [1 - \cos(E_2 - E_1)t] \\ &= \tfrac{1}{2} \sin^2 2\phi \left(1 - \cos \frac{m_2^2 - m_1^2}{2p} \frac{c^3}{\hbar} t \right) \end{aligned} \tag{10.9}$$

where we have written the $\hbar$ and c factors explicitly in the last line and it is assumed that $p \gg m_{1,2}$.

Expression (10.9) describes the simplest type of neutrino flavor oscillation, in which the amplitude depends on the mixing angle ϕ (and is a maximum for $\phi = \pi/4$). The oscillation frequency is proportional to $m_2^2 - m_1^2$ and is also inversely proportional to p, a consequence of time dilation for relativistic neutrinos. We may define an oscillation length l by the equation:

$$\frac{m_2^2 - m_1^2}{2p} \frac{c^3}{\hbar} t = \frac{2\pi x}{l}$$

where x is the distance downstream. This gives

$$l = \frac{4\pi \hbar p}{(m_2^2 - m_1^2)c^2} = \frac{2.5 p \, (\text{MeV}/c)}{(m_2^2 - m_1^2)(\text{eV}/c^2)^2} \text{ meters} \tag{10.10}$$

Next, we consider the corresponding oscillations of antineutrinos ($\bar{\nu}_e \leftrightarrow \bar{\nu}_\mu$). CPT invariance implies $P(\bar{\nu}_\mu \to \bar{\nu}_e) = P(\nu_e \to \nu_\mu)$. Also, if

$$\begin{pmatrix} \nu_e \\ \nu_\mu \end{pmatrix} = T \begin{pmatrix} \nu_1 \\ \nu_2 \end{pmatrix}, \qquad \text{where } T = \begin{pmatrix} \cos \phi & \sin \phi \\ -\sin \phi & \cos \phi \end{pmatrix}$$

10.2 $\Delta L = 0$ neutrino oscillations

then

$$\begin{pmatrix} \bar{\nu}_e \\ \bar{\nu}_\mu \end{pmatrix} = T^* \begin{pmatrix} \bar{\nu}_1 \\ \bar{\nu}_2 \end{pmatrix}$$

However, $T^* = T$; thus CP invariance is automatic, and $P(\bar{\nu}_e \to \bar{\nu}_\mu) = P(\nu_e \to \nu_\mu)$.

Flavor oscillations such as those described may also be obtained by a suitable modification of the CC interaction Lagrangian. In the standard model this is

$$\mathcal{L} = -\frac{g}{2\sqrt{2}} \bar{\nu}_l \gamma_\lambda (1 - \gamma_5) l \cdot W^{\lambda+} + hc$$

However, we may apply a rotation to the vector ν_l:

$$\nu_l \to U\nu_l = \begin{pmatrix} \cos\theta & \sin\theta \\ -\sin\theta & \cos\theta \end{pmatrix} \nu_l \quad (10.11)$$

which allows for the possibility of flavor-changing charged-current weak interactions. In muon decay, for example, the charged current is now:

$$j_\lambda = \bar{\nu}_\mu \gamma_\lambda (1 - \gamma_5)\mu \cos\theta - \bar{\nu}_e \gamma_\lambda (1 - \gamma_5)\mu \sin\theta \quad (10.12)$$
$$\text{(allowed)} \qquad\qquad \text{(forbidden)}$$

which is analogous to the Cabibbo-rotated charged hadronic weak current:

$$J_\lambda = \bar{S}\gamma_\lambda(1 - \gamma_5)C \cos\theta_C - \bar{D}\gamma_\lambda(1 - \gamma_5)C \sin\theta_C \quad (10.13)$$

In (10.12) the lepton number change $L_\mu \to L_e$ associated with the second term on the right-hand side occurs at the instant of decay, not in subsequent evolution of the neutrino state vector.

We now include both types of mixing by writing

$$j_\lambda = [\cos\theta\,(\nu_2 \cos\phi - \nu_1 \sin\phi) - \sin\theta\,(\nu_1 \cos\phi + \nu_2 \sin\phi)]\gamma_\lambda(1 - \gamma_5)\mu \quad (10.14)$$

The neutrino state vector then evolves according to the formula:

$$|\nu(t)\rangle = \cos(\theta + \phi)|\nu_2(t)\rangle - \sin(\theta + \phi)|\nu_1(t)\rangle \quad (10.15)$$

The analysis of oscillations may then be repeated as before, except that ϕ is replaced by $\theta + \phi$. The period of oscillation is unchanged, and indeed, if oscillations were observed, it would be impossible to decide whether they were caused by mass mixing (ϕ), Cabibbo rotation (θ), or a combination of the two. We thus have the freedom to define the charged current so that lepton number is conserved in weak decays, that is, so that U is diagonal and $\theta = 0$. This choice is made henceforth.

Let us consider the effect of neutrino mixing on the muon momentum spectrum in $\pi_{\mu 2}$ decay. In the pion rest frame, one has

$$|\mathbf{p}_\mu| = |\mathbf{p}_\nu| = \left[\frac{(m_\pi^2 + m_\mu^2 - m_\nu^2)^2}{4m_\pi^2} - m_\mu^2\right]^{1/2} \quad (10.16)$$

If one could attain sufficient precision, the momentum spectrum would reveal a series of peaks indicative of several two-body decay modes. The positions of the peaks would be a measure of the difference between the neutrino masses, and the heights would be proportional to the quantities $\cos^2(\theta + \phi)$ and $\sin^2(\theta + \phi)$. Of course this is not a practical example: if two neutrino mass eigenstates differed in mass by 1 MeV/c^2, the momentum peaks would be separated by only 14 keV/c compared with 29.79 MeV/c total momentum. A much better test is possible with the electron energy spectrum in the β decay of tritium; here one is sensitive to mass in the range 10 to 50 eV/c^2 (see Section 10.8).

Since (10.12) and (10.13) are so very similar, we may ask why quark oscillations do not occur following the decay of a charmed meson. The reason is that strong interactions perturb the quark state vector from a coherent superposition of d and s into an eigenstate of mass. Thus a K or π meson emerges but not a superposition of the two.

10.3 Neutrino oscillations with three lepton generations

The discussion of Sections 2.6 and 10.2 is readily generalized to three lepton generations (see, e.g., Barger et al. 80). We now have a charged lepton matrix ηg_{ij} and a neutrino matrix ηh_{ij}, each of which contains $3 \times 3 = 9$ elements. A basis may be chosen in which ηg_{ij} is diagonal:

$$\eta g_{ij} = \begin{pmatrix} m_e & & \\ & m_\mu & \\ & & m_\tau \end{pmatrix} \quad (10.17)$$

so that the e, μ, and τ fields correspond to real particles of definite mass. The remaining neutrino matrix $M_\nu = \eta h_{ij}$ has off-diagonal elements. We seek a unitary transformation U that diagonalizes M_ν:

$$M'_\nu = \begin{pmatrix} m_1 & & \\ & m_2 & \\ & & m_3 \end{pmatrix} = U M_\nu U^{-1} \quad (10.18)$$

The restrictions on U are equivalent to those applying to the six-quark matrix (4.70), and one finds that U contains four independent real constants. Since a real 3×3 orthogonal matrix is specified by three real

10.3 Neutrino oscillations with three lepton generations

parameters (angles), there is an additional *CP*-violating phase parameter. With *CP* violation, one no longer has $P(\nu \to \nu') = P(\bar{\nu} \to \bar{\nu}')$.

Matrix U can be written in the Kobayashi–Maskawa form:

$$\begin{pmatrix} \nu_e \\ \nu_\mu \\ \nu_\tau \end{pmatrix} = \begin{pmatrix} c_1 & +s_1 c_3 & +s_1 s_3 \\ -s_1 c_2 & c_1 c_2 c_3 - s_2 s_3 e^{i\delta} & c_1 c_2 s_3 + s_2 c_3 e^{i\delta} \\ -s_1 s_2 & c_1 s_2 c_3 + c_2 s_3 e^{i\delta} & c_1 s_2 s_3 - c_2 c_3 e^{i\delta} \end{pmatrix} \begin{pmatrix} \nu_1 \\ \nu_2 \\ \nu_3 \end{pmatrix} \quad (10.19)$$

where $c_i = \cos\theta_i$ and $s_i = \sin\theta_i$. Needless to say, θ_1, θ_2, θ_3, and δ are not necessarily the same as θ_1, θ_2, θ_3, and δ in (4.70).

Seven independent elements of M_ν may be expressed in terms of the three angles $\theta_{1,2,3}$, the phase δ and the masses m_1, m_2, m_3. The remaining two elements of M_ν are linear combinations of the rest. This is because when ν_e, ν_μ, and ν_τ are expressed in terms of $\nu_{1,2,3}$ we are free to choose which is ν_1 and which is ν_2, but then ν_3 is determined.

An electron–neutrino state produced at $t = 0$ is written as follows:

$$|\nu_e(0)\rangle = c_1|\nu_1\rangle + s_1 c_3 |\nu_2\rangle + s_1 s_3 |\nu_3\rangle \quad (10.20)$$

Thus, for $t > 0$,

$$|\nu_e(t)\rangle = c_1 e^{-iE_1 t/\hbar}|\nu_1\rangle + s_1 c_3 e^{-iE_2 t/\hbar}|\nu_2\rangle + s_1 s_3 e^{-iE_3 t/\hbar}|\nu_3\rangle$$

The probability of finding an electron neutrino at $t > 0$ is

$$|\langle \nu_e | \nu_e(t)\rangle|^2 \equiv P(\nu_e \to \nu_e)(t) = 1 - 2c_1^2 s_1^2 c_3^2 \left[1 - \cos\frac{E_1 - E_2}{\hbar}t\right]$$

$$- 2c_1^2 s_1^2 s_3^2 \left[1 - \cos\frac{E_1 - E_3}{\hbar}t\right]$$

$$- 2s_1^4 s_3^2 c_3^2 \left[1 - \cos\frac{E_2 - E_3}{\hbar}t\right] \quad (10.21)$$

It is convenient to make the substitutions:

$$(E_i - E_j)/\hbar = 2\pi x/\lambda_{ij}$$

in order to express (10.21) in terms of spatial oscillations. We then have, as in (10.10), the wavelengths:

$$\lambda_{ij} = \frac{2\pi \hbar x}{t}\frac{1}{E_i - E_j} \approx \frac{4\pi\hbar|\mathbf{p}|}{c^2|m_i^2 - m_j^2|}$$

$$\approx \frac{2.5|\mathbf{p}|\ (\mathrm{MeV}/c)}{|m_i^2 - m_j^2|\ (\mathrm{eV}/c^2)^2}\ \mathrm{meters} \quad (10.22)$$

Expression (10.21) is quite complicated, but the first oscillation after $t = 0$ is determined by the largest difference $m_i^2 - m_j^2$. In particular, if one mass, say m_1, is much larger than the others, (10.21) is much simplified:

$$P(\nu_e \to \nu_e)(t) \simeq 1 - 2c_1^3 s_1^2 \left[1 - \cos(m_1^2/2p)x\right] \quad (10.23)$$

Generally we may expect some energy spread in the neutrino source. In this case the cosines average to zero for $x \gg L_{ij}$, and $P(\nu_e \to \nu_e)(t)$ reaches an asymptotic limit:

$$P(\nu_e \to \nu_e)(x \to \infty) = 1 - 2c_1^2 s_1^2 c_3^2 - 2c_1^2 s_1^2 s_3^2 - 2s_1^4 s_3^2 c_3^2 \quad (10.24)$$

This has a minimum value of $\frac{1}{3}$ (for $s_1^2 = \frac{2}{3}$, $s_3^2 = \frac{1}{2}$).

The probabilities $P(\nu_e \to \nu_\mu)$ and $P(\nu_e \to \nu_\tau)$ depend on the CP-violating parameter δ. For $x \gg L_{ij}$,

$$P(\nu_e \to \nu_\mu) \to 2c_1^2 s_1^2 c_2^2 + 2s_1^2 s_3^2 c_3^2 (s_2^2 + c_1^2 c_2^2)$$
$$+ 2s_1^2 s_2 s_3 c_1 c_2 c_3 \cos \delta \, (s_3^2 - c_3^2) \quad (10.25)$$

$$P(\nu_e \to \nu_\tau) \to 2c_1^2 s_1^2 s_2^2 + 2s_1^2 s_3^2 c_3^2 (c_2^2 + c_1^2 s_2^2)$$
$$+ 2s_1^2 s_2 s_3 c_1 c_2 c_3 \cos \delta \, (s_3^2 - c_3^2) \quad (10.26)$$

With nonzero δ, there is CP violation; for example $P(\nu_e \to \nu_{\mu,\tau}) \neq P(\bar{\nu}_e \to \bar{\nu}_{\mu,\tau})$. However CPT invariance still ensures $P(\nu_e \to \nu_{\mu,\tau}) = P(\bar{\nu}_{\mu,\tau} \to \bar{\nu}_e)$.

10.4 $|\Delta L| = 2$ oscillations and Majorana neutrinos

According to the conventional viewpoint, neutrinos are Dirac particles, described by the four-component field Ψ_ν:

$$\Psi_\nu = \frac{1}{\sqrt{V}} \sum_{p,s} \left(\frac{m_\nu}{E_\nu}\right)^{1/2} [u_\nu(p,s) b(p,s) e^{ip \cdot x} + v_\nu(p,s) d^\dagger(p,s) e^{ip \cdot x}]$$

$$(10.27)$$

The lepton number $L = L_e + L_\mu + L_\tau + \cdots$ is always conserved in weak processes involving Dirac leptons, since the currents contain only terms that create as many particles as they destroy. However, since neutrinos are electrically neutral and ν and $\bar{\nu}$ satisfy the same Dirac equation, one can take the viewpoint that neutrinos are "Majorana" particles, described by the field

$$\Psi_M = \frac{1}{\sqrt{2}} [\Psi_\nu + \Psi_{\nu C}] \quad (10.28)$$

where $\Psi_{\nu C}$ is the charge-conjugate Dirac field:

$$\Psi_{\nu C} \equiv C \Psi_\nu C^{-1} = i\gamma_2 \gamma_0 \bar{\Psi}_\nu^T = i\gamma_2 \Psi_\nu^* \quad (10.29)$$

in the Pauli–Dirac or Weyl representation. Thus

$$\Psi_M = \frac{1}{\sqrt{2}} \frac{1}{\sqrt{V}} \sum_{p,s} \left(\frac{m_\nu}{E_\nu}\right)^{1/2} \{u_\nu(p,s)[b(p,s) + d^\dagger(p,s)] e^{-ip \cdot x}$$
$$+ v_\nu(p,s)[b^\dagger(p,s) + d(p,s)] e^{ip \cdot x}\} \quad (10.30)$$

10.4 $|\Delta L| = 2$ oscillations and Majorana neutrinos

A charged current containing the Majorana field

$$j_\lambda^M = \overline{\Psi}_l \gamma_\lambda (1 - \gamma_5) \Psi_M \qquad (10.31)$$

thus contains terms that change the lepton number by two units, $|\Delta L| = 2$, such as $b_l^\dagger \bar{u}_l \gamma_\lambda (1 - \gamma_5) d_\nu^\dagger u_\nu$ (see Figure 10.1). Then $|\Delta L| = 2$ oscillations $\nu \leftrightarrow \bar{\nu}$ are possible for Majorana neutrinos and may be understood as follows, where, for simplicity, we consider only one neutrino generation represented by the $\nu - \bar{\nu}$ column vector:

$$\psi = \begin{pmatrix} \nu \\ \nu_c \end{pmatrix} \qquad (10.32)$$

The mass term in the Lagrangian contains

$$M = \begin{pmatrix} m_{11} & m_{12} \\ m_{12}^* & m_{22} \end{pmatrix} \qquad (10.33)$$

and CPT invariance implies $m_{11} = m_{22}$. If m_{12} is real, the mass eigenstates are Majorana neutrino states:

$$\chi_+ = \frac{1}{\sqrt{2}} (\nu + \bar{\nu}) \qquad (10.34)$$

$$\chi_- = \frac{1}{\sqrt{2}} (\nu - \bar{\nu}) \qquad (10.35)$$

There is a small mass difference between χ_+ and χ_-, given by

$$\Delta m / m = 2|m_{12}|/m$$

where $\Delta m = m_+ - m_-$ and $m = \frac{1}{2}(m_+ + m_-)$. Suppose that a neutrino produced by the standard weak interaction at $t = 0$ is in the state

$$|\nu_L\rangle = \frac{1}{\sqrt{2}} [|\chi_+\rangle_L + |\chi_-\rangle_L] \qquad (10.36)$$

At $t > 0$ we have:

$$|\nu_L\rangle = \frac{1}{\sqrt{2}} (e^{-iE_+ t} |\chi_{+L}\rangle + e^{-iE_- t} |\chi_{-L}\rangle) \qquad (10.37)$$

Figure 10.1. Violation of lepton conservation at the $Wl\nu$ vertex by emission of a Majorana neutrino.

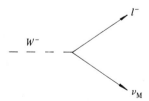

The result is a finite probability for generating a left-handed $\bar{\nu}$ that cannot be absorbed or detected in any way (except by a second $|\Delta L| = 2$ interaction).

However, in general, $|\Delta L| = 2$ processes are inhibited by the vertex factor $\gamma_\lambda(1 - \gamma_5)$ (V–A law). For example, we consider the forbidden $|\Delta L| = 2$ process in which one attempts to employ antineutrinos generated in β decay:

$$n \to p e^- \bar{\nu}_e \qquad (10.38)$$

as "neutrinos" to induce the capture process

$$\nu_e + {}^{37}\text{Cl} \to {}^{37}\text{Ar} + e^- \qquad (10.39)$$

Such an experiment has in fact been completed to demonstrate that $\nu_e \neq \bar{\nu}_e$ (Davis 55). It may be described as a two-step process in which we sum over the spins of the intermediate Majorana neutrino. The matrix element is:

$$\mathcal{M} = \sum_s \langle \text{Ar}, e^{-\prime} | j_\beta{}^M J^\beta | \text{Cl}, \nu_M \rangle \langle p, e^-, \nu_M | j_\alpha{}^M J^\alpha | n \rangle \qquad (10.40)$$

where $J^{\alpha,\beta}$ are the nuclear currents. The leptonic portion of the amplitude is

$$\mathcal{M}_{\alpha\beta}{}^l = \sum_s \langle e^{-\prime} | \bar{\Psi}_e \gamma_\beta (1 - \gamma_5) \Psi_M | \nu_M \rangle \langle \nu_M, e^- | \bar{\Psi}_e \gamma_\alpha (1 - \gamma_5) \Psi_M | 0 \rangle$$
$$\propto \sum_s \bar{u}_{e'} \langle 0 | \gamma_\beta (1 - \gamma_5) \Psi_M | \nu_M \rangle \cdot \bar{u}_e \langle \nu_M | \gamma_\alpha (1 - \gamma_5) \Psi_M | 0 \rangle \qquad (10.41)$$

Since in $j_\alpha{}^M$ the component of Ψ_M that makes a nonzero contribution is $d\dagger v_\nu(p, s) e^{-ip \cdot x}$ and that of $j_\beta{}^M$ is $b u_\nu(p, s) e^{ip \cdot x}$, we have:

$$\mathcal{M}_{\alpha\beta}{}^l = \sum_{p,s} (m_\nu/E_\nu) \bar{u}_{e'} \gamma_\beta (1 - \gamma_5) u_\nu [\bar{u}_e \gamma_\alpha (1 - \gamma_5) v_\nu]$$

The factor in brackets is a spin scalar and may be transposed:

$$\mathcal{M}_{\alpha\beta}{}^l = \sum_{p,s} (m_\nu/E_\nu) [\bar{u}_{e'} \gamma_\beta (1 - \gamma_5) u_\nu] [\tilde{v}_\nu (1 - \gamma_5) \tilde{\gamma}_\alpha \tilde{\gamma}_0 \tilde{u}_e\dagger]$$

which becomes

$$\mathcal{M}_{\alpha\beta}{}^l = -i \sum_{p,s} (m_\nu/E_\nu) \bar{u}_{e'} \gamma_\beta (1 - \gamma_5) u_\nu \bar{u}_\nu (1 - \gamma_5) \gamma_2 \gamma_0 \gamma_\alpha \gamma_0 u_e^* \qquad (10.42)$$

in the Pauli–Dirac representation. However,

$$\sum_s u_\nu \bar{u}_\nu = (\not{p}_\nu + m_\nu)/2m_\nu$$

and in (10.42) the term in $\not{p}_\nu$ vanishes, since $(1 - \gamma_5) \not{p}_\nu (1 - \gamma_5) = (1 - \gamma_5)(1 + \gamma_5) \not{p}_\nu = 0$. The remaining term is of order m_ν/E_ν, relative to the amplitude for a lepton-conserving process. Thus, the helicity of the neutrino, which is a consequence of the V–A law, inhibits

10.4 $|\Delta L| = 2$ oscillations and Majorana neutrinos

$|\Delta L| = 2$ transitions, even though lepton conservation is not an explicit feature of the Majorana formulation. For massless neutrinos we can in fact define a new lepton number L', such that

$$L' = \begin{cases} +1 & \text{for } e^-, \; \tfrac{1}{2}(1-\gamma_5)\nu_M \\ -1 & \text{for } e^+, \; \tfrac{1}{2}(1+\gamma_5)\nu_M \end{cases} \tag{10.43}$$

Massless Majorana neutrino interactions conserve L' just as Dirac neutrino interactions conserve L, and the theories have identical consequences in the limit $m_\nu \to 0$.

Experimental bounds on $\bar{\nu}_e$ mass may be used to estimate the maximum $|\Delta L| = 2$ rate in β decay. Since we know that $m(\bar{\nu}_e) < 60$ eV/c^2 (see Section 10.8), it follows that

$$\text{V--A:} \quad \left.\frac{\Gamma(|\Delta L|=2)}{\Gamma(\Delta L=0)}\right|_{\beta\text{decay}} \lesssim \left(\frac{m}{E_\beta}\right)^2 \simeq \left(\frac{60 \text{ eV}}{1 \text{ MeV}}\right)^2 \simeq 4 \times 10^{-9} \tag{10.44}$$

Larger lepton number violations are possible in Majorana theories with both left- and right-handed currents, however. We rewrite j_λ^M of (10.31) as follows:

$$j_\lambda^M = \frac{1}{(1+|\eta|^2)^{1/2}} \bar{\Psi}_e \gamma_\lambda [(1-\gamma_5) + \eta(1+\gamma_5)]\Psi_M + hc \tag{10.45}$$

where, for simplicity, η is chosen real (no CP violation). Repeating the previous discussion leading to (10.42), we now find:

$$\mathcal{M}^{\text{lept}} \simeq \frac{m_\nu}{E_\nu} \bar{u}'_e[(1-\gamma_5) + \eta(1+\gamma_5)]\frac{\slashed{p}+m_\nu}{2m_\nu}$$

$$\times [(1-\gamma_5) + \eta(1+\gamma_5)]u_e^* \frac{1}{1+\eta^2} \tag{10.46}$$

$$\simeq \frac{\eta}{1+\eta^2}\frac{m_\nu}{E_\nu}\frac{p_\nu}{m_\nu} \simeq \eta$$

Thus the $|\Delta L| = 2$ amplitude is now first order in η and independent of neutrino mass.

Experimental limits on η are obtained from measurements of electron helicity in nuclear β decay or from recoil experiments (Goldhaber et al. 58). One finds

$$\eta \lesssim 0.1 \tag{10.47}$$

which implies that lepton number violations as large as $\eta^2 = 0.01$ are possible. However, the results of double-β-decay experiments impose much more stringent limits, as we shall see later.

While $\Delta L = 0$ oscillations arise quite naturally in the standard model, it is not so easy to construct a theory that yields $|\Delta L| = 2$ oscillations. In certain "grand unified models," there exist $|\Delta L| \neq 0$ couplings between leptons and new Higgs fields that give rise to mass. However, there seems no reason at this time to prefer such models over other "grand unified schemes" in which the neutrino masses are zero.

10.5 Neutrino refractive index

Although we have thus far confined ourselves to a discussion of neutrino oscillations in vacuum, for most real situations neutrinos pass through matter (neutron shielding in reactors, the sun in the case of solar neutrinos, and so on). We wish to consider the possible effects of matter on oscillations (Wolfenstein 78). For simplicity we assume only two neutrino flavors, ν_e and ν_μ, with mass eigenstates ν_1 an ν_2 related by

$$\nu_e = \nu_1 \cos \phi + \nu_2 \sin \phi$$
$$\nu_\mu = -\nu_1 \sin \phi + \nu_2 \cos \phi \tag{10.48}$$

If the neutrino wavefunction at $t = 0$, $z = 0$ is known:

$$\psi = \begin{pmatrix} \psi_1 \\ \psi_2 \end{pmatrix} = \begin{pmatrix} A_0 \\ B_0 \end{pmatrix} = A_0 |\nu_e\rangle + B_0 |\nu_\mu\rangle$$

then at finite z and t it is

$$\psi_i = e^{ikn_{ij}z} e^{-i\epsilon_{ij}t} \psi_j \tag{10.49}$$

where ϵ_{ij} and n_{ij} are 2×2 matrices and i and j run over flavors e and μ. The matrix ϵ_{ij} contains off-diagonal terms because of mass mixing, which give rise to vacuum oscillations, and have already been considered. That is:

$$\epsilon_{ee} = E_1 \cos^2 \phi + E_2 \sin^2 \phi$$
$$\epsilon_{\mu\mu} = E_1 \sin^2 \phi + E_2 \cos^2 \phi$$
$$\epsilon_{e\mu} = \epsilon_{\mu e} = \cos \phi \sin \phi (E_2 - E_1)$$

where $E_i = (m_i^2 + |\mathbf{p}_i|^2)^{1/2}$, $i = 1, 2$, are the energies of $|\nu_1\rangle$ and $|\nu_2\rangle$. The index-of-refraction matrix n_{ij} is given by

$$n_{ij} = \delta_{ij} + (2\pi N/k^2) f_{ij}(0) \tag{10.50}$$

where $f_{ij}(0)$ is the forward-scattering amplitude and N the number density of scattering centers. In first approximation, f_{ij} is diagonal, since there is no weak scattering process, charged or neutral, that causes $\nu_e \leftrightarrow \nu_\mu$ or $\bar{\nu}_e \leftrightarrow \bar{\nu}_\mu$. However, $f_{ee} \neq f_{\mu\mu}$, because ν_e can scatter from electrons in matter via charged *and* neutral weak interaction, whereas ν_μ can only do so via the neutral weak interaction. Furthermore, the

10.5 Neutrino refractive index

imaginary part of f is negligible, because the absorption coefficient for neutrinos (ν_e or ν_μ) is so small:

$$k \times \frac{2\pi N}{k^2} \operatorname{Im} f(0) = \frac{N\sigma_T}{2} \simeq 10^{-23} \quad \text{cm}^{-1}$$

Thus n_{ij} is a real diagonal matrix:

$$n_{ij} = \begin{pmatrix} 1+X & 0 \\ 0 & 1+X \end{pmatrix} + \begin{pmatrix} \frac{2\pi N_e}{k^2} f_e^{CC}(0) & 0 \\ 0 & 0 \end{pmatrix} \tag{10.51}$$

where $X = (2\pi N/k^2)f^{NC}(0)$ is the same for ν_e and ν_μ. The first matrix on the RHS of (10.51) results merely in an unobservable overall phase shift in the wavefunction, and it may be disregarded. However, the second term does alter the neutrino basis, even though it does not cause oscillations by itself. It is easy to show that

$$\tan 2\phi \to \tan 2\phi' = \tan 2\phi \left(1 - \frac{l}{l_0} \sec 2\phi\right)^{-1} \tag{10.52}$$

where l is the vacuum oscillation length (10.10) and

$$l_0 = \frac{2\pi}{G_F N_e} = \frac{2.5 \times 10^9}{\rho_e(g/\text{cm}^3)} \text{ cm} \tag{10.53}$$

The oscillation length in matter then turns out to be

$$l'(k) = l(k) \left[1 + \left(\frac{l(k)}{l_0}\right)^2 - 2\cos 2\phi \frac{l(k)}{l_0}\right]^{-1/2} \tag{10.54}$$

Matter becomes important only if $l \gtrsim 10^3$ km, and then only if more than 10^9 g/cm² of column density is traversed. The net effect is to suppress oscillations.

In some theories other than the standard model, matter can have a much greater effect on neutrino oscillations. For example, if the weak neutral current contains cross terms such as $\nu_\mu \gamma_\lambda (1 - \gamma_5)\nu_e$, non-diagonal terms appear in n_{ij} that produce oscillations even for massless neutrinos (Wolfenstein 78). In principle, matter-induced oscillations of this type could be distinguished from vacuum oscillations by differences in energy dependence. In the vacuum case, time dilation causes

$$l_v \propto 1/k \tag{10.55}$$

whereas for matter,

$$l_m \propto \frac{2\pi}{k(n-1)} = \frac{1}{NG_F} \tag{10.56}$$

which is independent of energy.

10.6 Neutrino oscillation experiments

We now discuss attempts to observe neutrino oscillations that have been made by employing reactors and accelerators. In view of (10.10) or its generalization (10.22), the quantity of importance in a given observation is L/E, where L is a characteristic length of the experiment (in meters) and E the neutrino energy (in MeV). In various observations with laboratory neutrinos, L/E ranges from 10^{-2} to 20 m/MeV. Such observations are thus sensitive to masses roughly in the range 1 to 10 eV/c^2.

10.6.1 Reactor experiments

Nuclear reactors can produce intense fluxes of nearly pure $\bar{\nu}_e$ with energies of several millions of electron-volts. Unfortunately the $\bar{\nu}$ energy spectrum is very difficult to calculate because it depends on the relative proportions of various fission products and their descendants, neutron capture products, reactor operating history, and so forth. Nevertheless, spectra have been calculated (see, e.g., Davis et al. 79) and compared with $\bar{\nu}$ flux measurements, in search of oscillations in the range $1 \lesssim L/E \lesssim 20$ m/MeV (Silverman 81).

Some evidence for oscillations was in fact claimed by Reines (78) and co-workers (80) in experiments carried out at the Savannah River reactor. In one observation, the $\bar{\nu}$ flux above the 1.8-MeV threshold was measured by detection of n and e^+ in the reaction $\bar{\nu} + p \rightarrow n + e^+$. In another, the rate for the CC reaction $\bar{\nu} + D \rightarrow e^+ + n + n$ was compared with that for the NC reaction $\bar{\nu} + D \rightarrow n + p + \bar{\nu}$. Although the CC reaction can only be induced by $\bar{\nu}_e$, the NC reaction can occur with any antineutrino flavor. It was claimed that the results indicate $m_i^2 - m_j^2 \simeq 1$–2 (eV/c^2)^2. However, these results are inconclusive because of poor statistics, smearing over the rather large reactor core size, and uncertainties in the $\bar{\nu}$ spectrum. A similar experiment carried out at the Grenoble reactor, which has a much smaller core size, showed no evidence for oscillations in $\bar{\nu} + p \rightarrow n + e^+$ (Boehm et al. 80).

In principle the dependence on calculated $\bar{\nu}$ spectra may be eliminated and deviations from the inverse square law revealed by simultaneous measurements of neutrino flux at two distances. Such experiments are in progress but no results have been obtained so far.

10.6.2 Beam dump experiments

Neutrinos are produced in weak decays when a high-energy proton beam is "dumped" onto a target. For low average target den-

10.6 Neutrino oscillation experiments

sity, the decaying particles that result are mostly pions and kaons, with some contribution from Λ and Σ hyperons. As the density is increased, these longer-lived particles are reabsorbed before they can decay, and the residual "prompt" neutrinos come mostly from charmed mesons. The latter are expected to produce equal numbers of ν_e and ν_μ, which can then be identified by charged-current events in detectors placed at large distances from the target. After correction for differences in efficiencies and subtraction of the nonprompt neutrino component, any disparity in the cross sections for $\nu_e \to e$ versus $\nu_\mu \to \mu$ may be evidence for oscillations. In practice, it is impossible to eliminate the nonprompt background, which is richer in ν_μ than ν_e. Instead, data are taken at two or more target densities ρ and then extrapolated to $1/\rho \to 0$. A group at CERN using the Big European Bubble Chamber (BEBC) reported

$$N(\nu_e + \bar{\nu}_e)/N(\nu_\mu + \bar{\nu}_\mu) = 0.48 \begin{cases} +0.24 \\ -0.16 \end{cases} \tag{10.57}$$

and a second experiment employing the CHARM detector (placed behind BEBC in the same beam) presented the confirmatory result (Jonker et al. 80b):

$$N(\nu_e + \bar{\nu}_e)/N(\nu_\mu + \bar{\nu}_\mu) = 0.48 \pm 0.12 \tag{10.58}$$

Since the average neutrino beam energy is about 100 GeV, these results could be taken as evidence for oscillations in the range $L/E \simeq 10^{-2}$ m/MeV. However, these results are contradicted by more recent experiments carried out at Fermilab (Reeder 82), where a ratio of unity (with large uncertainty) is obtained.

10.6.3 Accelerator experiments

High-energy neutrino beams of fairly definite energy and composition from π and K decay have been used to search for $\nu_\mu \to \nu_e$, $\nu_e \to \nu_e$, and $\nu_\mu \to \nu_\tau$ oscillations at Fermilab and CERN, in the range $L/E \simeq 0.04$ to 0.01 m/MeV. The results are presented in Table 10.1.

The only consistent interpretation of these data is that no neutrino oscillations exist for the ranges covered and $m_i^2 - m_j^2 < 50$ (eV/c^2)^2 for all neutrino types i and j. Lower-energy neutrinos (0 to 50 MeV) are produced from the decays of stopped muons and pions at "meson factories." Results from the Los Alamos LAMPF facility are:

$$P(\bar{\nu}_\mu \to \bar{\nu}_e) < 0.065 \quad (90\% \text{ C.L.}) \tag{10.59}$$

$$P(\nu_e \to \nu_e) = 1.1 \pm 0.4 \tag{10.60}$$

with $L/E \simeq 0.3$ m/MeV (Willis 80,81), consistent with no oscillations.

Table 10.1. *Accelerator neutrino oscillations*

Oscillation	Confidence level	Reference	Location
$P(\bar{\nu}_\mu \to \bar{\nu}_e)/P(\nu_\mu \to \nu_\mu) < 1.4 \times 10^{-3}$	90%	Blietschau et al. 78	CERN/Gargamelle
$P(\nu_\mu \to \nu_e)/P(\nu_\mu \to \nu_\mu) < 1.3 \times 10^{-3}$	90%	Blietschau et al. 78	CERN/Gargamelle
$P(\nu_e \to \nu_e)/P(\nu_\mu \to \nu_\mu) = 0.92 \pm 0.21$	—	Blietschau et al. 78	CERN/Gargamelle
$P(\nu_e \to \nu_e) = 1.04 \pm 0.15$	—	Deden et al. 81	CERN/BEBC
$\quad\quad\quad\quad\,\, = 1.21 \pm 0.19$	—	Enriquez 81	CERN/BEBC
$|m_{\nu_\mu}{}^2 - m_{\nu_\tau}{}^2| < 3.0 \text{ eV}^2$	90%	Ushida et al. 81	Fermilab

10.7 Double β decay

10.6.4 Electron-capture neutrinos

Finally, a neutrino oscillation experiment has been suggested in which a beam of monoenergetic neutrinos is formed from electron capture in ^{65}Zn. In this case, the smearing of L/E would be caused only by the finite source or detector. This experiment is of special interest because of indications from tritium β decay that $m_{\bar{\nu}_e} \simeq 30$ eV/c^2 and therefore that neutrino oscillations (if they are visible at all) might be in the $L/E = 0.1\text{-}1.0$ m/MeV range (see Section 10.8).

10.7 Double β decay

10.7.1 Double-β-decay rates

Nuclei exist for which ordinary β decay is energetically forbidden or highly suppressed by conservation of angular momentum but

Table 10.2. *Possible $\beta^-\beta^-$ transitions*

Transition	A	Z	Isotopic abundance (%)	Transition energy (MeV)	Intermediate transition energy $(A,Z) - (A,Z+1)$ (MeV)
Ca–Ti	46	20	0.0033	0.985	−1.382
Ca–Ti	48	20	0.185	4.267	+0.289
Zn–Ge	70	30	0.62	1.008	−0.653
Ge–Se	76	32	7.67	2.045	−0.923
Se–Kr	80	34	49.82	0.138	−1.871
Se–Kr	82	34	9.19	3.003	−0.089
Kr–Sr	86	36	17.37	1.240	−0.054
Zr–Mo	94	40	2.80	1.230	−0.921
Zr–Mo	96	40	17.40	3.364	+0.215
Mo–Ru	100	42	9.62	3.034	−0.335
Ru–Pd	104	44	18.5	1.321	−1.145
Pd–Cd	110	46	12.7	2.004	−0.868
Cd–Sn	114	48	28.86	0.547	−1.439
Cd–Sn	116	48	7.58	2.811	−0.517
Sn–Te	122	50	4.71	0.349	−1.622
Sn–Te	124	50	5.98	2.263	−0.653
Te–Xe	128	52	31.79	0.872	−1.268
Te–Xe	130	52	34.49	2.543	−0.407
Xe–Ba	134	54	10.44	0.731	−1.328
Xe–Ba	136	54	8.87	2.718	−0.112
Ce–Nd	142	58	11.07	1.379	−0.777
Nd–Sm	148	60	5.71	1.936	−0.514
Nd–Sm	150	60	5.60	3.390	−0.036
Sm–Gd	154	62	22.61	1.260	−0.718
Gd–Dy	160	64	21.75	1.782	−0.029
U–Pu	238	92	99.275	1.173	−0.117

which are unstable for *double β decay*, a second-order process. For $\beta^-\beta^-$ emission, there are only 26 possible transitions, each of the form $J^P = 0^+ \to 0^+$ between even–even nuclei (see Table 10.2 and the review by Bryman and Picciotto 78). There also exist a number of cases in which double K capture can occur, and a few in which K capture $+ \beta^+$ emission, or even $\beta^+\beta^+$ emission, are possible, but for a variety of reasons, these are even more difficult to observe than is $\beta^-\beta^-$.

Lepton conservation requires that the two decay electrons be accompanied by two antineutrinos, as shown in Figure 10.2a. This process (so-called $\beta\beta_2$) is allowed but greatly inhibited by the phase-space restrictions of the four-lepton final state and because it is of second order in G_F. Thus typical lifetimes for $\beta\beta_2$ are very long: of order 10^{20-21} yr.

Figure 10.2. Double β decay. (a) $\beta\beta_2$; (b) $\beta\beta_0$.

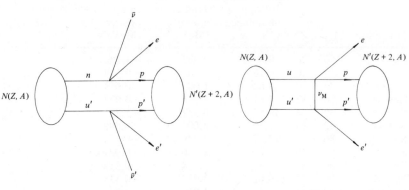

Neutrinoless double β decay ($\beta\beta_0$) is also possible if lepton conservation fails. Here a virtual Majorana neutrino appears in the intermediate state (Figure 10.2b) and one has $|\Delta L| = 2$.

In $\beta\beta_0$ decays, the phase-space factor is not so restrictive as for $\beta\beta_2$. Thus, if there were no inhibition from lepton conservation, $\beta\beta_0$ lifetimes could be as short as 10^{14-15} yr. Of course, the rate is suppressed by the helicity barrier and vanishes if $m_\nu = 0$ or if $\eta = 0$, even for massive Majorana neutrinos. Table 10.3 summarizes the results of approximate calculations of the half-lives for double β decay and related processes.

Table 10.3. *Approximate theoretical half-lives for double β decay and related processes*

Process	Half-life (yr)
$\beta\beta_2$	$\tau_{1/2}(\beta^\pm\beta^\pm) \simeq 2 \times 10^{20\pm2} \left[\frac{\langle E_i - E_m - \frac{1}{2}(\epsilon+2)\rangle}{10}\right]^2 \left(\frac{60}{Z}\right)^2 [1 - \exp(\pm 2\pi\alpha Z)]^2 \left(\frac{8}{\epsilon}\right)^{10}$
	$\tau_{1/2}(\beta^+, \kappa\,\text{capt}) \simeq 3 \times 10^{22\pm2} \left[\frac{\langle E_i - E_m - \frac{1}{2}\epsilon\rangle}{10}\right]^2 \left(\frac{60}{Z}\right)^4 \exp[2\pi\alpha(Z-60)]\left(\frac{8}{\epsilon}\right)^7$
	$\tau_{1/2}(\kappa\kappa\,\text{capt}) \simeq 3 \times 10^{15\pm2} \left[\frac{\langle E_i - E_m - \frac{1}{2}(\epsilon-2)\rangle}{10}\right]^2 \left(\frac{60}{Z}\right)^6 \left(\frac{8}{\epsilon}\right)^5$
$\beta\beta_0$	$\tau_{1/2}(\beta^\pm\beta^\pm) \simeq 10^{15\pm2}\eta^2 \left(\frac{60}{Z}\right)^2 [1 - \exp(\pm 2\pi\alpha Z)]^2 \left(\frac{A}{150}\right)^{2/3} \left(\frac{8}{\epsilon}\right)^6$
	$\tau_{1/2}(\beta^+, \kappa\,\text{capt}) \simeq 5 \times 10^{18\pm2}\eta^2 \left(\frac{60}{Z}\right)^4 \exp[2\pi\alpha(Z-60)]\left(\frac{A}{150}\right)^{2/3} \left(\frac{8}{\epsilon}\right)^4$
	$\tau_{1/2}(\kappa\kappa\,\text{capt}) \simeq 5 \times 10^{23\pm2}\eta^2 \left(\frac{60}{Z}\right)^8 \left(\frac{A}{150}\right)^{2/3} \left(\frac{8}{\epsilon}\right)^5$

Note: ϵ = energy release in units $m_e c^2$; E_i, E_m = initial, intermediate nuclear energies. These rates are calculated by assuming that the two electrons are emitted by two separate neutrons in the nucleus (see, e.g., Rosen and Primakoff 65; Primakoff and Rosen 59, 69, 72). It is possible that nuclear states contain admixtures of baryon resonances of a few percent. The $\Delta(1232) I = \frac{3}{2}, J = \frac{3}{2}$ resonance is particularly significant, since it could, in principle, emit and reabsorb a virtual neutrino with simultaneous emission of two electrons: $\Delta^- \rightarrow p + 2e^-$, $n \rightarrow \Delta^{++} 2e^-$ (see, e.g., Smith et al. 73, Primakoff and Rosen 81).

10.7.2 Double-β-decay experiments

Double β decay was first observed in tellurium and selenium geological formations with the aid of noble gas mass spectroscopy. When the Te or Se nucleus decays, an atom of Xe or Kr is formed, which remains trapped in the rock for billions of years. The gas is removed by heating the rock and is then analyzed in a sensitive mass spectrometer to search for anomalously large abundances of those isotopes that are the products of double β decay. The lifetime can be determined if the age of the rock is measured separately (which is accomplished by applying the same technique in the case of ^{40}K–^{40}Ar single β decay) and it can be ascertained that no gas has leaked out of the rock over the aeons.

The tellurium transitions are particularly interesting since decays of isotopes 128 and 130 may be compared; in the ratio of rates, various possible systematic errors associated with preparation and history of the sample may be assumed to cancel. If one assumes that the nuclear matrix elements are roughly the same for both transitions, the lifetime ratio is predicted to be:

$$\tau(128)/\tau(130) = \begin{cases} 10^{2.4} & \text{for } \beta\beta_0, \quad \eta = 1 \quad (10.61) \\ 10^{3.8} & \text{for } \beta\beta_2 \quad (10.62) \end{cases}$$

Earlier, measurements were made by Takaoka and Ogata (66) and by Srinivasan et al. (73). Perhaps the most clear-cut observations were made by Hennecke et al. (75), who found

$$\tau(128)/\tau(130) = 10^{3.20 \pm 0.01} \tag{10.63}$$

This result may indicate that the assumption of equality for the nuclear matrix elements of the two isotopes is invalid. Or, it may be the only evidence that exists for breakdown of lepton conservation. It implies that the total decay rate is due to a combination of $\beta\beta_2$ and $\beta\beta_0$, with the latter channel dominant (branching ratio 70 percent). The value of η thus arrived at is

$$\eta = (4.3 \pm 0.1) \times 10^{-5} \tag{10.64}$$

In view of the uncertainties associated with geological experiments, however, one must regard this result with caution and skepticism.*

Attempts to observe double β decay in the laboratory have been hampered by the extremely low counting rate, which makes even the most minute background important. Moe and Lowenthal (80) at-

---

* *Note added in proof:* Recent measurements by Kirsten and co-workers (unpublished) reveal no evidence for failure of lepton conservation. They find $\eta \leq 2.4 \times 10^{-5}$ (95% C.L.).

10.7 Double β decay

tempted to observe the decay electrons from ^{82}Se → ^{82}Kr in a cloud chamber. The apparatus consisted of 12 thin sandwiches of aluminized Mylar with 97 percent enriched ^{82}Se deposited on the inner surfaces, for a total ^{82}Se mass of 13.75 g. The sandwiches were suspended inside the cloud chamber, edge-on to a camera that recorded the electron tracks. A magnetic field along the viewing axis was employed to analyze the electron momenta, and a sophisticated trigger-and-veto scheme reduced the background rate to one count every several minutes. A total of 20 events were found in which two electrons emerged from the same spot on the selenium source. In 16 additional events, two electrons were accompanied by an α particle, which is characteristic of the decays of uranium- or thorium-series impurities, especially ^{214}Bi. An analysis of this type of background and others, such as Møller scattering of single β electrons or Compton scattering of γ rays, could not account for the two-electron events. Their energies were distributed in a way that is consistent with the $\beta\beta_2$ spectrum of ^{82}Se. However, the lifetime derived in this way is much shorter than the one established by geological methods:

$$\tau_{1/2} = \begin{cases} (1.0 \pm 0.4) \times 10^{19} \text{ yr} & \text{by direct detection} \\ 2.76 \times 10^{20} \text{ yr} & \text{by geological methods} \end{cases} \quad \begin{matrix}(10.65)\\(10.66)\end{matrix}$$

where the value of (10.66) is from Srinivasan et al. (73). No satisfactory explanation has been found for this discrepancy.

Cleveland (75) performed a direct detection experiment on ^{82}Se that was sensitive mainly to $\beta\beta_0$ events and found:

$$\tau_{1/2}^{\beta\beta_0} > 3.1 \times 10^{21} \text{ yr} \quad (10.67)$$

This experiment and a similar one performed on ^{48}Ca (Bardin et al. 70) made use of the fact that the spectrum of summed energies of the

Table 10.4. *Double-β-decay experimental results*

Decay	$\tau_{1/2}$ (yr)	Double β	Reference
^{48}Ca → ^{48}Ti	$>10^{21.3}$	$\beta\beta_0$	Bardin et al. 70
	$>10^{19.56}$	$\beta\beta_2$	
^{76}Ge → ^{76}Se	$>10^{21.7}$	$\beta\beta_0$	Fiorini et al. 73
^{82}Se → ^{82}Kr	$>10^{21.49}$	$\beta\beta_0$	Cleveland 75
^{82}Se → ^{82}Kr	$(1.0 \pm 0.4) \times 10^{19}$ (total)	$\beta\beta_0 + \beta\beta_2$	Moe and Lowenthal 80
^{82}Se → ^{82}Kr	$10^{20.42\pm.14}$	$\beta\beta_0 + \beta\beta_2$	Srinivasan et al. 73
^{130}Te → ^{130}Xe	$10^{21.34\pm0.12}$	$\beta\beta_0 + \beta\beta_2$	Kirsten et al. 68
^{128}Te → ^{128}Xe ^{130}Te → ^{130}Xe	ratio $= 10^{3.20\pm.01}$	$\beta\beta_0 + \beta\beta_2$	Hennecke et al. 75

double electrons has a single peak nearly equal to T_0 for neutrinoless emission but a broad spectrum for two-neutrino emission. By limiting the accepted $2e$ events to a narrow range around T_0, the background rejection is significantly improved. A similar technique has been used to measure the neutrinoless decay rate of ^{130}Te by direct detection (Zdesenko 80). No decays were found, but the upper limit was $\tau_{1/2}^{\beta\beta_0} > 1.2 \times 10^{21}$ yr (68 percent confidence level). When combined with the geologically determined lifetime, this implies a branching ratio for $\beta\beta_0$ decay of

$$R \leq 83\% \tag{10.68}$$

consistent with (10.63).

Double-β-decay experimental results are summarized in Table 10.4.

10.8 Direct measurements of neutrino mass

The electron energy spectrum in nuclear β decay is affected by finite $\bar{\nu}_e$ mass. We recall from Chapter 5 that the differential transition probability for allowed β decay, averaged over nuclear spin and integrated over e^- and $\bar{\nu}_e$ directions, is

$$dW = \frac{4G_F^2 \cos^2 \theta_1}{2\pi^3} \xi F(Z, E) p_e^2 \, dp_e \, p_\nu^2 \, dp_\nu \delta(\Delta - E - E_\nu) \tag{10.69}$$

Figure 10.3. Kurie plot showing deviation from straight line for nonzero neutrino mass.

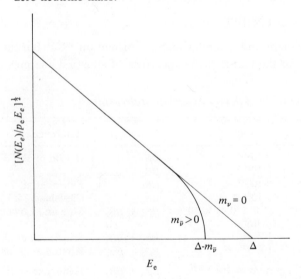

10.8 Direct measurements of neutrino mass

Including the possibility of finite neutrino mass $m_\nu = (E_\nu^2 - p_\nu^2)^{1/2}$, writing $p_\nu^2\, dp_\nu = p_\nu E_\nu\, dE_\nu$, and integrating over E_ν, we obtain the electron energy spectrum:

$$N(E) = \frac{G_F^2}{2\pi^2} \cos^2\theta_1\, \xi F(Z, E) p_e E [(\Delta - E)^2 - m_\nu^2]^{1/2}(\Delta - E) \quad (10.70)$$

However, the final atom may be in an excited state i (the probability for which is reliably calculated by means of the sudden approximation). Since each such state i is associated with a different energy Δ_i, (10.70) must be replaced by

$$N(E) = \frac{G_F^2}{(2\pi^3)} \cos^2\theta_1\, \xi F(Z, E) p_e E \sum_i W_i\{[(\Delta_i - E)^2 - m_\nu^2]^{1/2}(\Delta_i - e)\}$$

$$(10.71)$$

where $W_i = |\langle \psi_{\text{atom}}^f | \psi_{\text{atom}}^i \rangle|^2$. If the sum is dominated by a single term and $m_\nu = 0$, a plot of $[N(E)/pE]^{1/2}$ versus E (Kurie plot) is a straight line with negative slope, which intercepts the E axis at Δ (Figure 10.3). A finite mass causes the line to curve downward near the endpoint and to intersect the axis with infinite negative slope. The existence of other final atomic states usually causes the line to curve the other way, obscuring the effect of finite mass. The situation is even more complicated if the neutrino is not in a mass eigenstate. In this case, the sum in (10.71) must be replaced by a double sum:

$$\sum_i \sum_{j=1}^{3} W_i |u_{e_j}|^2 [(\Delta_i - E)^2 - m_j^2]^{1/2}(\Delta_i - E) \quad (10.72)$$

where u_{e_j} is the matrix element $\langle \nu_e | \nu_j \rangle$ and the $|\nu_j\rangle$ are states of definite mass. The resulting Kurie plot would exhibit kinks from several superimposed end points. $N(E)$ would be modified further by V + A currents. It may be shown that the spectrum $N(E)$ of (10.71) must in this case be multiplied by the factor

Figure 10.4. Tritium β decay. $T_0 = \Delta - m_e c^2 = 18.6$ keV is the maximum electron kinetic energy.

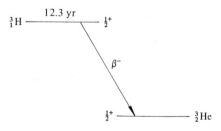

$$f = \left[1 + \frac{|\eta|^2 m_e m_\nu}{EE_\nu}\right]\left[\frac{1}{1 + |\eta|^2}\right] \tag{10.73}$$

that reaches its maximum value

$$f_{\max} \simeq 1 - \frac{|\eta|^2}{1 + m_\nu/E_{\max}}$$

at $E = E_{\max}$. However, since $|\eta|^2 \leq 10^{-2}$, this correction is quite negligible.

The β decay $^3\text{H} \rightarrow {}^3\text{He} + e^- + \bar{\nu}_e$ (see Figure 10.4) is most widely employed in searches for (anti) neutrino mass, for the following reasons: The endpoint energy is low, the transition is superallowed (and the β-decay theory very reliable), the atomic wavefunctions of ^3H and $^3\text{He}^+$ are known, and the lowest excited state of He^+ has very large energy (40.8 eV). For a number of years, the best experimental limit on

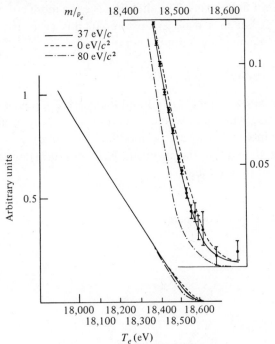

Figure 10.5. Calculated Kurie plots for tritium decay near the endpoint and experimental data from ITEP, Moscow. Here, T_e is the electron kinetic energy. (From Lubimov et al. 80. Reprinted with permission.)

$m_{\bar{\nu}_e}$ was obtained from tritium decay, in a careful experiment done by Bergkvist (72), with the result:

$$m_{\bar{\nu}_e} < 60 \quad \text{eV}/c^2 \tag{10.74}$$

More recently an experiment carried out at ITEP (Moscow) yielded results suggesting a finite mass (Lubimov et al. 80):

$$14 \leq m_\nu \leq 46 \quad \text{eV}/c^2 \quad (99\% \text{ C.L.}) \tag{10.75}$$

In this experiment, the source was a thin layer of valine ($C_5H_{11}NO_{26}$) containing 18 percent ^{3}H. Electrons emitted from the sample traversed a 720° magnetic spectrometer. The resulting Kurie plot is shown in Figure 10.5. The data were fit to a theoretical β spectrum (10.71) folded with an experimental resolution function that depended on energy. Two final states were employed: the ground state (70 percent) and an effective excited state (30 percent) at 43 eV. By means of Monte Carlo simulations, it was demonstrated that the data are inconsistent with $m_{\bar{\nu}} = 0$. The results are quite insensitive to various theoretical assumptions about the valine molecule.

The ITEP experiment is being continued, and several other experiments are underway to confirm or disprove the result. These include tritium experiments (Simpson 81) as well as efforts to observe the inner *bremsstrahlung* spectrum from low-energy orbital electron capture in ^{193}Pt or ^{163}Ho.

10.9 Neutrino decay; neutrino magnetic moment

Any mixing scheme between neutrinos of different mass must inevitably lead to decay of the heavier species, via the process of Figure 10.6. In the standard model, such decays are suppressed by the GIM mechanism; amplitudes with intermediate e^-, μ^-, and τ^- cancel

Figure 10.6. Neutrino decay process.

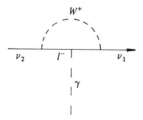

each other. However, this suppression is imperfect because the leptons have different masses, and some decay occurs. The dominant diagram involves an intermediate τ^- lepton, and a straightforward calculation for the case where one neutrino mass eigenstate ν_1 is heavy and the other two ν_2, ν_3 are light yields (de Rujula and Glashow 80):

$$\frac{1}{\tau} \simeq G_F^2 m_1^5 \alpha (1/512\pi^4) \sin^2(2\beta_1) I^2 \tag{10.76}$$

where $\beta = \cos^{-1}\langle \nu_1 | \nu_\tau \rangle$, I is the GIM suppression factor

$$I = (m_\tau/m_W)^2 [\ln(m_W/m_\tau)^2 + \mathcal{O}(1)], \tag{10.77}$$

and m_e and m_μ have been neglected compared with m_τ. If $m_1 = 30$ eV/c^2, this gives a lower limit for the neutrino lifetime:

$$\tau \gtrsim 4 \times 10^{28} \quad \text{yr} \tag{10.78}$$

where the minimum is for $\beta = \pi/4$. Calculations with different assumptions (e.g., more lepton generations or extra neutral heavy leptons) decrease the lifetime, but for most reasonable assumptions, one obtains: $\tau \gtrsim 10^{22}$ yr. An experimental lower limit may be derived from the observed *absence* of $\mu \to e\gamma$, which might be expected to occur as in Figure 10.7. This gives $\tau > 10^{16}$ yr.

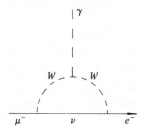

Figure 10.7. Diagram illustrating the decay $\mu \to e\gamma$, which violates lepton conservation and is not observed.

Incidentally, a somewhat analogous diagram describes the weak–electromagnetic transition $\Sigma^+ \to p\gamma$ (Figure 10.8). Here, neutrino mixing is replaced by the usual Cabibbo mixing of quark states. The transition has actually been observed.

Finally, let us mention the electric charge and magnetic moment of the neutrino. From charge conservation and the known charges of p^+, n, and e^-, it follows from neutron β decay that $Q(\nu_e) \leq 10^{-19}e$. An upper limit of 10^{-8} electron Bohr magneton may be placed on the mag-

Figure 10.8. Diagram illustrating $\Sigma^+ \to p\gamma$.

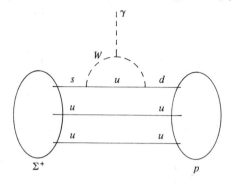

netic moment of ν_μ from observed neutrino–nucleon scattering cross sections. On the theoretical side, the magnetic moment of a massless Dirac neutrino must be zero, whereas the magnetic moment of a Majorana neutrino, massive or massless, must also vanish. However, a massive Dirac neutrino should possess a small magnetic moment from virtual dissociation into $l^\pm W^\mp$ pairs. One finds (Fujikawa and Shrock 80) that, in the standard model,

$$\mu_\nu = 1.85 \times 10^{-27} m_\nu \quad \text{eV/Gauss}$$

where m_ν is the neutrino mass in electron volts per c^2. If $m_\nu = 1\,\text{eV}/c^2$, this gives $\mu_\nu \simeq 10^{-18}$ electron Bohr magneton.

Problems

10.1 Verify (10.21), (10.25), and (10.26).

10.2 Consider β decay of a tritium atom in the $1\,^2S_{1/2}$ state. Use the sudden approximation to calculate the probabilities that the resulting ^{3}He ion is in the $1^2S_{1/2}$, $2^2S_{1/2}$, and $2^2P_{1/2}$ states.

11
Neutrino astrophysics

11.1 Introduction

Astrophysical neutrinos bombard the earth from a variety of sources. Closest at hand are the secondary cosmic-ray decays of kaons, pions, and muons in the atmosphere, which yield neutrinos (chiefly, ν_μ and $\bar\nu_\mu$) over a broad energy spectrum far beyond the range of present or planned accelerators. A small number of events generated by these neutrinos have been observed in deep-mine experiments, during attempts to study weak interactions at high energies and to search for neutrino oscillations. Another goal of the deep-mine detectors and of a planned deep underwater muon and neutrino detector (DUMAND) is to study localized extraterrestrial sources of very-high-energy neutrinos, such as supernova explosions.

The sun is a source of low-energy neutrinos ν_e generated deep in the interior as a by-product of nuclear reactions responsible for the solar luminosity. The results of the first experiment to detect such neutrinos are now available, and they disagree with theoretical estimates based on a standard solar model (although the discrepancy is considerably less significant than was previously thought). Many suggestions for the disagreement have been offered, including the notion that three-flavor neutrino oscillations are responsible. New and more-sensitive solar neutrino detectors are needed to advance our understanding of this problem.

It seems quite impossible to observe neutrinos from other ordinary stars, since the total neutrino flux from the stars divided by that from the sun ought to be roughly equal to the ratio of starlight to sunlight. However, the emission of $\nu\bar\nu$ pairs by a hot dense star in the last stages of thermonuclear evolution is enormous, and it affects this evolution drastically. The final gravitational collapse of a sufficiently massive stellar core is accompanied by an enormous explosion of the outer layers (supernova). At the same time, rapid neutronization occurs in the core:

$$e^- + p \to n + \nu_e$$

yielding a burst of ν_e, followed by a large pulse of thermal neutrinos and antineutrinos (ν_e, ν_μ, $\bar{\nu}_e$, $\bar{\nu}_\mu$, . . .). The time scale of such events is rather sharply defined in theoretical models, as is the neutrino energy (10–20 MeV). Neutrino-burst detectors now operate or are under construction in the United States and the USSR in an effort to observe this effect. They are part of a coordinated program that complements optical searches for supernovas.

The role of neutrinos in cosmology is most speculative. According to the "big bang" model, there should exist a relic sea of neutrinos and antineutrinos, which are an inevitable consequence of reactions in the hot early universe. The neutrino sea is somewhat analogous to the 3K cosmic blackbody photon background. Although cosmological neutrinos are probably unobservable by direct means, their existence may have had a very profound effect on the Hubble expansion, on primordial helium formation, on galaxy formation and stability, and other phenomena.

Neutrino astrophysics is just beginning. Experiments are extraordinarily difficult and time-consuming, and as yet there are hardly any observational results, a state of affairs that encourages the most fanciful theoretical speculations. Still, with all its difficulties, the study of astrophysical neutrinos raises deep and fascinating questions, and we may anticipate extremely interesting developments in the coming decades.

11.2 Neutrinos from the sun

11.2.1 The total flux of solar neutrinos

The sun shines because of energy supplied by exothermic nuclear reactions deep in the interior. The main reactions are those of the proton–proton chain, although a few percent of the energy is contributed by the C–N–O cycle (see Table 11.1). In either case, the net effect is the conversion of four protons into an α particle:

$$2e^- + 4p \to {}^4\text{He} + 2\nu_e + 26.7 \quad \text{MeV} \tag{11.1}$$

where all but about 3 percent of the energy appearing on the right-hand side of (11.1) is delivered in the form of charged-particle and photon energy, which ultimately becomes part of the sun's thermal energy store. The other 3 percent (the exact amount depends on which cycle or chain we consider) is carried off by the neutrinos, which depart from the sun with velocity c and are lost. Therefore, one ν_e is created for each 13 MeV of thermal energy generated. Since the sun is at present quite stable in size and temperature, the rate of thermal-energy generation must be equal to the rate at which the energy is radiated from the surface. We know the solar constant S, the amount of solar flux at the top of the earth's atmosphere:

$$S = 1.37 \times 10^6 \quad \text{erg/cm}^2 \text{ sec}$$

Thus we easily obtain the solar neutrino flux at the earth:

$$\Phi_\nu = S/13 \text{ MeV} = (6 \times 10^{10})\nu_e/\text{cm}^2 \text{ sec} \qquad (11.2)$$

Table 11.1. *Nuclear reactions of importance for solar energy production*

Reaction	Neutrino energy (MeV)		Estimated flux[a] cm^{-2} sec^{-1}
	Average	Maximum	
Proton–proton chain			
$p + p \to D + e^+ + \nu_e$	0.26	0.42	6.0×10^{10}
$p + p + e^- \to D + \nu_e$		1.44[b]	1.5×10^8
$p + D \to {}^3\text{He} + \gamma$			
${}^3\text{He} + {}^3\text{He} \to {}^4\text{He} + 2p$			
or			
${}^3\text{He} + {}^4\text{He} \to {}^7\text{Be} + \gamma$			
${}^7\text{Be} + e^- \to {}^7\text{Li} + \nu_e$		0.86 (90%)[b]	2.7×10^9
${}^7\text{Li} + p \to {}^4\text{He} + {}^4\text{He}$		0.34 (10%)[b]	3.0×10^8
or			
${}^7\text{Be} + p \to {}^8\text{B} + \gamma$			
${}^8\text{B} \to {}^8\text{Be} + e^+ + \nu_e$	7.2	14	3.0×10^6
${}^8\text{Be} \to {}^4\text{He} + {}^4\text{He}$			
C–N–O cycle			
${}^{12}\text{C} + p \to {}^{13}\text{N} + \gamma$			
${}^{13}\text{N} \to {}^{13}\text{C} + e^+ + \nu_e$	0.71	1.19	3.0×10^8
${}^{13}\text{C} + p \to {}^{14}\text{N} + \gamma$			
${}^{14}\text{N} + p \to {}^{15}\text{O} + \gamma$			
${}^{15}\text{O} \to {}^{15}\text{N} + e^+ + \nu_e$	1.00	1.70	2.0×10^8
${}^{15}\text{N} + p \to {}^{12}\text{C} + {}^4\text{He}$			

[a] Bahcall et al. (80).
[b] Monoenergetic neutrinos.

11.2 Neutrinos from the sun

Unfortunately it is very difficult to detect the solar neutrinos. One possibility is the reaction $\nu_e + e^- \rightarrow \nu_e + e^-$, which has no threshold energy and is thus suited for detecting low-energy neutrinos. However, the cross section is small:

$$\sigma(\nu e) = (1.7 \times 10^{-44}) E_\nu \quad \text{cm}^2 \quad (E_\nu \text{ in MeV}) \quad (11.3)$$

and furthermore, this reaction is very difficult to distinguish from Compton scattering, the cross section for which is 19 orders of magnitude larger. Thus electron-scattering solar neutrino detectors do not as yet seem practical, although such possibilities have been considered (Lande 79, Chen 78).

The only known alternative is detection by radiochemical identification of the final nucleus in the endothermic reaction:

$$\nu_e + (Z, N) \rightarrow (Z + 1, N - 1) + e^- \quad (11.4)$$

The requirements for such a detector are that it have a low threshold, that the cross section be reasonably large (corresponding to a "superallowed" β-decay matrix element), and that the target material be cheap and readily available in large quantities, in a chemical form suitable for easy extraction of the final nucleus. The one practical case for which an experiment has been realized is

$$\nu_e + {}^{37}\text{Cl} \rightarrow {}^{37}\text{Ar} + e^- \quad (11.5)$$

a reaction first suggested by Pontecorvo (46) and, independently, by Alvarez (49). The threshold is 0.81 MeV, with a cross section that is particularly large above 6 MeV because of a $T = \frac{3}{2}$, $J^P = \frac{3}{2}^+$ excited state of ^{37}Ar connected by a superallowed matrix element to the ground state of ^{37}Cl (Bahcall 64). Unfortunately, the great bulk of the solar neutrino flux, that from the $pp \rightarrow De^+\nu_e$ reaction, is unobservable with this detector. In fact, although less than 10^{-4} of the total neutrino flux originates in ^{8}B decay, it turns out that these neutrinos, by virtue of their high energy, should account for about 80 percent of the estimated ^{37}Cl capture rate. Now, the total rate of ^{8}B production in the solar interior depends sensitively on the central temperature and composition. Therefore, even though it is simple enough to calculate the total neutrino flux, estimation of the expected ^{37}Cl capture rate requires a precise and detailed solar model.

11.2.2 Theoretical solar models

The equations governing the structure and evolution of the sun relate the pressure $P(r)$, the mass $M(r)$ interior to the distance r from the center, the density $\rho(r)$, the temperature $T(r)$, the energy production by nuclear reactions per gram $\epsilon(r)$, the luminosity $L(r)$, and the opacity $\kappa(r)$ (see, e.g., Clayton 68). Other variables of importance are the fractional abundances of hydrogen, helium, and heavier elements, denoted by X, Y, and Z, respectively, (with $X + Y + Z = 1$). The equation of state $P = P(\rho, T, X, Y, Z)$ is, to a good approximation, that of a perfect gas, but there are significant corrections for electron degeneracy and exchange effects. The opacity is an exceedingly complicated function of density, temperature, and composition, especially of the heavy element content Z. It can be estimated only from a detailed knowledge of atomic transition probabilities for free–free, free–bound, and bound–bound transitions, as well as electron scattering effects in which corrections attributable to many-electron correlations are important. Calculation of ϵ depends on detailed knowledge of cross sections and decay probabilities for the reactions listed in Table 11.1. Of these, the cross section for the primary weak reaction

$$p + p \rightarrow D + e^+ + \nu_e$$

is far too small to be measured in the laboratory. It must be calculated from the theory of β decay (Bethe and Critchfield 38, Bahcall and May 69). The cross sections for the other reactions have been measured, but not without some errors in the past nor without the necessity of extrapolation from laboratory to solar energies. Knowledge of cross sections for the ^{3}He–^{3}He, ^{3}He–^{4}He, and ^{7}Be–p reactions could be improved by modern experimental methods (Rolfs and Trautvetter 78).

To the equations just described, one must add the boundary conditions for mass and luminosity:

$$r = 0: \quad M(0) = 0, \quad L(0) = 0$$
$$r = R: \quad M(R) = M_\odot, \quad L(R) = L_\odot$$

In addition, the pressure near the surface depends on the temperature in a complicated way, which is characteristic of convective stellar envelopes.

In principle, if one is given the mass of a star and its initial composition X, Y, Z, the latter assumed homogeneously distributed at $t = 0$, the start of hydrogen burning, then it is possible to solve the equations of stellar structure to find the radius, the surface luminosity $L(t)$, and the temperature $T(r, t)$, as well as the composition $X(r, t)$, $Y(r, t)$, $Z(r, t)$, at any time $t > 0$. Of course, the composition changes with time as the star evolves and becomes inhomogeneous as the result of nuclear reactions.

In the case of the sun, we know the present mass and present luminosity:

11.2 Neutrinos from the sun

$$M_\odot = 1.989 \pm 0.002 \times 10^{33} \quad \text{g}$$
$$L_\odot = 3.90 \pm 0.04 \times 10^{33} \quad \text{erg/sec}$$

quite accurately, and we also know the time since the start of hydrogen burning:

$$t \simeq 4.7 \times 10^9 \quad \text{yr}$$

reasonably well. However, we do not know the initial composition nor even the present composition of the sun. The fractional abundance of helium Y cannot be estimated from photospheric absorption spectroscopy, since there are no helium lines in the visible spectrum. The heavy-element abundance Z can be obtained from spectroscopy ($Z \simeq 0.015$), but this determination is uncertain.

In light of the foregoing, the procedure for calculating a solar model is as follows. One takes $M_\odot$ and $L_\odot$ as data that the model must satisfy. The photospheric value of Z is taken as representative of Z_i, since Z has not been altered by light nucleus thermonuclear reactions. The opacity is then calculable from knowledge of heavy-element abundances. Trial values of $(Y/X)_i$ are chosen, and various solar models are calculated as a function of Y/X until the correct luminosity is found at $t = 4.7 \times 10^9$ yr. The neutrino flux for each of the contributing reactions in Table 11.1 is also an output of the calculation.

Estimation of the ^{37}Cl capture rate requires not only the neutrino spectrum but also the cross section for absorption of ν_e by ^{37}Cl as a function of energy, a quantity that we have already mentioned. The theoretical results are conveniently expressed in terms of the total conversion rate per ^{37}Cl atom per second:

$$1 \text{ solar neutrino unit (SNU)} = 10^{-36} \text{ capture per target particle per second}$$

Until very recently, the best theoretical estimate was that given by Bahcall et al (80):

$$\text{capture rate}_{\text{theo}} = 7.5 \pm 1.5 \quad \text{SNU} \qquad (11.6a)$$

However, this has been challenged by the estimate of Fillipone (81):

$$\text{capture rate}_{\text{theo}} = 4.8 \pm 1.6 \quad \text{SNU} \qquad (11.6b)$$

The discrepancy between (11.6a) and (11.6b) arises mainly from different values of the cross section for the reaction $^3\text{He} + {}^4\text{He} \rightarrow {}^7\text{Be} + \gamma$.

11.2.3 The solar neutrino experiment

We next turn briefly to the ^{37}Cl experiment carried out by Davis and collaborators (Davis 78). Their apparatus is located 4,850 ft underground in the abandoned Homestake Gold Mine in South Dakota. Such depths are required to reduce the muon cosmic-ray background to a tolerable level. Muons would otherwise produce fast protons, which, in turn, would generate ^{37}Ar via the ^{37}Cl$(p, n)^{37}$Ar reaction. The target is a tank with 390,000 liters (600 tons) of C_2Cl_4 containing the normal isotopic abundance of ^{37}Cl (about 25 percent). If the theoretical estimate (11.6b) were correct, approximately one ^{37}Ar atom would be produced in this target every 2 days from solar neutrinos. However, this value is much larger than the total production rate from all background sources, including radioactivity in the surrounding rock as well as cosmic-ray muon sources. Moreover, as demonstrated in separate calibration experiments, it is possible to extract the minute quantity of ^{37}Ar from C_2Cl_4 with better than 90 percent efficiency. The method is to expose the fluid to solar neutrinos for a month or so (the half-life of ^{37}Ar is 35 days) and then circulate large quantities of ^{4}He gas through the liquid target. The ^{37}Ar is thus flushed out in about 24 hours, then separated from the helium in a series of traps, and pumped into a small proportional counter in which the 2.8-keV Auger electrons accompanying ^{37}Ar electron-capture decay are detected with about 50 percent efficiency. The entire system is calibrated reliably by means of auxiliary experiments.

The result of this remarkable experiment is:

Table 11.2. *Characteristics of solar-neutrino detectors*

Target	Neutrino threshold (MeV)	Cross sections ($\times 10^{-46}$ cm^2)[a]						
		pp	pe^-p	e^- ^{7}Be	^{8}B	^{13}N	^{15}O	^{65}Zn
^{7}Li	0.862	0	600	9.5	3×10^4	41.7	230	225
^{37}Cl	0.814	0	15.6	2.4	1×10^4	1.66	6.61	6.1
^{51}V	0.751	0	11.0	2.2	1×10^4	1.52	4.94	4.3
^{55}Mn	0.231	0.282	4.3	1.7	6×10^3	1.44	2.5	1.8
^{71}Ga	0.236	10.7	157	64	3×10^3	53	92	67
^{81}Br	0.459	0	31.9	10.6	1×10^3	8.6	17.3	13
^{87}Rb	0.115	22.9	182	81	3×10^3	70	112	89
^{115}In	0.120	87.6	640	290	9×10^3	250	400	320
^{205}Tl	0.062	72	340	180	3×10^3	160	230	185

[a] All cross sections are averaged over the appropriate energy spectra for the solar neutrino sources.
Source: Bahcall (78)

11.2 Neutrinos from the sun

$$\text{capture rate}_{\text{expt}} = 2.2 \pm 0.4 \quad \text{SNU} \tag{11.7}$$

The discrepancy between this result and earlier theoretical estimates [e.g., (11.6a)] provoked considerable discussion for a number of years, and many speculative ideas were proposed, including solar neutrino oscillations. The enthusiasm for these speculations has been dampened somewhat by (11.6b), which does not disagree so severely with experiment. In any case it is clear that a better understanding of solar neutrinos requires more detailed experimental results and therefore new and more-sensitive solar neutrino detectors.

11.2.4 Alternative solar neutrino detectors

In addition to ^{37}Cl the isotopes ^{7}Li, ^{51}V, ^{55}Mn, ^{71}Ga, ^{81}Br, ^{87}Rb, ^{115}In, and ^{205}Tl offer attractive possibilities as solar neutrino detectors (Bahcall 78). The calculated neutrino absorption cross sections for these cases are summarized in Table 11.2 for the various solar neutrino transitions and also for ^{65}Zn, which has been proposed as a convenient calibration source:

$$e^- + {}^{65}\text{Zn} \rightarrow {}^{65}\text{Cu} + \nu \quad (0.227 \text{ MeV}, \ 50.75\%)$$
$$e^- + {}^{65}\text{Zn} \rightarrow {}^{65}\text{Cu} + \nu \quad (1.343 \text{ MeV}, \ 47.8\%)$$
$$^{65}\text{Zn} \rightarrow {}^{65}\text{Cu} + e^+ + \nu \quad (0.330 \text{ MeV (max)}, \ 1.45\%)$$

When various practical considerations are taken into account, ^{71}Ga and ^{115}In emerge as most promising. In the case of ^{71}Ga, first discussed by Kuzmin (65), Pomanski (65), and Kuzmin and Zatsepin (66), a prototype $1\frac{1}{2}$-ton detector is being constructed for use at the Homestake site (Kirsten 79), and another detector is under construction in the USSR. The gallium will be in the form of GaCl$_3$ in hydrochloric acid. The expected neutrino reaction by-product is GeCl$_4$, which can be distilled out of the solution at 60°C. The great advantage of the gallium de-

Figure 11.1. Energy levels of ^{115}In and ^{115}Sn.

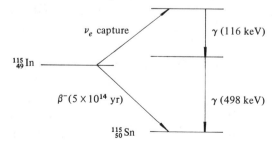

tector is that it is primarily sensitive to solar neutrinos generated in the *pp* reaction.

In Figure 11.1, we show the energy-level scheme associated with ^{115}In (Pfeiffer et al. 78). The cross sections for neutrino capture are very large in this case (see Table 11.2), but there is a very serious problem: ^{115}In is β-unstable and has a decay lifetime of 5×10^{14} yr. Thus a mass of indium sufficient to capture one solar neutrino per day will yield 10^{11} β decays per day! The background can be reduced substantially, however, by detecting the e^- that accompanies neutrino capture in triple coincidence with ^{115}Sn γ rays at 116 and 498 keV. It is then necessary to divide the indium into n cells such that the accidental triple coincidences from β decay in each cell have a smaller rate than solar neutrino capture. One finds that $n = 10^6$ is required. It seems quite possible that ^{71}Ga and/or ^{115}In will be employed successfully as solar neutrino detectors in the coming decade.

11.3 Emission of neutrinos by hot stars

We now turn to a discussion of the various possible mechanisms by which neutrinos are radiated from the hot dense cores of highly evolved stars. It may be helpful if we first sketch briefly the events that lead to the existence of such stars. It is generally believed that stars originate from tenuous clouds of gas that contract, slowly at first, because of self-gravitation. The first stars are thought to have been composed of primordial matter, which was probably hydrogen ($X = 0.75$) and helium ($Y = 0.25$) exclusively. Later generations of stars, including the sun, contain small but significant fractions of heavy elements. These were probably synthesized in earlier, highly evolved stars and then deposited in space by various energy-loss mechanisms, including supernova explosions.

As a young protostar contracts, gravitational energy is converted into kinetic energy of mass motion and then into thermal energy. Thus pressure builds up, which slows the contraction, until ultimately the star reaches a quasi-static configuration (hydrostatic equilibrium). Application of the virial theorem then shows that, if the gas is nonrelativistic and monatomic, half the released gravitational energy must go into thermal energy and half into energy lost by radiation. Thus the temperature of the core continues to rise as the star contracts quasi-statically, until hydrogen-burning reactions (Table 11.1) are ignited at $T \simeq 10^7$ K. At this point, gravitational contraction is halted, the temperature is stabilized, and the rate of energy generation by nuclear

11.3 Emission of neutrinos by hot stars

reactions is just sufficient to account for radiation loss from the surface. The star has thus arrived at its position on the "main sequence" of the Hertzsprung–Russell diagram (Figure 11.2), and there it remains more or less stationary, during the entire period of main-sequence hydrogen burning. This time interval is greater for light stars than for massive ones, and is about 10^{10} yr for the sun.

Once the core hydrogen is exhausted, gravitational contraction resumes, and the core temperature and density once more increase. Meanwhile, hydrogen burning may continue in a spherical shell around the core, while the envelope expands and cools. Thus the star moves relatively rapidly from its main sequence position into the red-giant region (Figure 11.2). Once the core reaches temperatures $T \simeq 10^8$ K (and typical densities $\rho \simeq 10^4$ g/cm³), helium burning commences:

$$3\,^4\text{He} \rightarrow {}^{12}\text{C}^* \rightarrow {}^{12}\text{C} + \gamma \tag{11.8}$$

This reaction supplies sufficient energy to halt gravitational contraction temporarily. Other reactions also begin to occur at this stage, although they are inhibited by high Coulomb barriers:

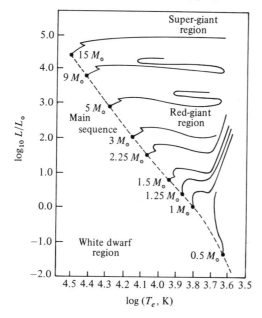

Figure 11.2. Schematic Hertzsprung–Russell diagram. Post-main-sequence evolutionary tracks of typical stars are indicated by solid curves. T_e = effective surface temperature.

$$^4He + {}^{12}C \to {}^{16}O + \gamma \tag{11.9}$$

$$^4He + {}^{16}O \to {}^{20}Ne + \gamma \tag{11.10}$$

$$^4He + {}^{20}Ne \to {}^{24}Mg + \gamma \tag{11.11}$$

$$^4He + {}^{24}Mg \to {}^{28}Si + \gamma \tag{11.12}$$

However, once the core helium is exhausted, gravitational contraction resumes, accompanied as always by heating. What happens next depends on the mass of the star. It can be shown that, for a given central temperature and composition, the central density of a light star is greater than that of a heavy star. If the density is sufficiently great, pressure from electron degeneracy halts the gravitational contraction before further thermonuclear reactions can begin. Thus for light stars, the core may be stabilized at some temperature below 6×10^8 K, with a composition of carbon and/or oxygen. By some mechanisms that are not clearly understood, the envelope of the star may be shed (initiating a planetary nebula), and the remaining core, a white dwarf, cools by radiating away its heat.

If the star is more massive, gravitational contraction continues unabated until at $T = 6$–7×10^8 K, a whole new series of thermonuclear reactions is ignited, involving the burning of carbon and oxygen. At each stage, thermonuclear fuel is exhausted and gravitational contraction sets in and continues either until halted by electron degeneracy pressure or until a temperature is reached where the next thermonuclear fuel can be ignited. However, if the core is sufficiently massive (that is, if its mass exceeds the Chandrasekhar limit), no amount of degeneracy pressure can halt the contraction.

As T approaches 2×10^9 K, carbon and oxygen have burned mainly to ^{28}Si and the thermal radiation has become so intense and energetic that photodisintegration of nuclei begins to proceed very rapidly. Therefore, in this temperature range, the core nuclear abundances rapidly arrive at statistical equilibrium, where the most stable, tightly bound nuclei are produced in profusion. These are in the neighborhood of ^{56}Fe, which has the maximum binding energy per nucleon. At this point, which is reached in the temperature range 3 to 5×10^9 K, no further energy can be extracted from thermonuclear reactions.

Meanwhile, as the temperature has increased past 10^8 K, neutrino-loss mechanisms begin to appear, and their significance mounts rapidly with temperature. The essential and unique feature of neutrinos is that, once they are produced, they have very long mean free paths in stellar material, since they suffer only weak interactions. Thus, they may es-

11.3 Emission of neutrinos by hot stars

cape directly from the stellar core and rob the star of energy that can only be replenished by further gravitational contraction and heating. In the range 10^8 to 10^9 K, the most important of these effects are the photoneutrino process, in which a $\nu\bar{\nu}$ pair replaces the outgoing photon in Compton scattering; and the plasma process, in which a photon in the dense plasma acquires an effective mass and can thus decay via a virtual e^+e^- pair to a $\nu\bar{\nu}$ pair. The latter is of special importance at high densities. At somewhat higher temperatures and moderate densities, the pair process $e^+e^- \to \nu\bar{\nu}$ gains ascendancy over the first two. (For detailed discussions of the photoneutrino and plasma processes, and others as well, see Ruderman 65, Beaudet et al. 67, and Dicus 72. Although the calculations in Ruderman's monograph and in the paper of Beaudet et al. are based on the assumption of charged currents, the results are not so drastically altered by the inclusion of neutral currents, as in the article by Dicus.)

To obtain some feeling for the quantities involved and also because it is important in our discussion of cosmology, we consider the pair process more closely. The transition $e^+e^- \to \nu_e\bar{\nu}_e$ can occur in two ways: by Z^0 exchange and by W exchange, whereas the reactions $e^+e^- \to \nu_\mu\bar{\nu}_\mu$ and $e^+e^- \to \nu_\tau\bar{\nu}_\tau$ proceed only by Z^0 exchange. The relevant diagrams are Figures 11.3a and b, and the amplitudes for $\nu_e\bar{\nu}_e$ emission are:

$$\mathcal{M}_{\text{NC}} = -\frac{G_F}{2\sqrt{2}} \bar{v}_e \gamma_\lambda (1 - 4\sin^2\theta_W - \gamma_5) u_e \bar{u}_\nu \gamma^\lambda (1 - \gamma_5) v_\nu \quad (11.13)$$

$$\mathcal{M}_{\text{CC}} = -\frac{G_F}{\sqrt{2}} \bar{u}_\nu \gamma_\lambda (1 - \gamma_5) u_e \bar{v}_e \gamma^\lambda (1 - \gamma_5) v_\nu \quad (11.14)$$

where the sign of (11.14) is arrived at as in the discussion of Section 3.10. We perform a Fierz rearrangement on (11.14), thereby interchanging u_e and v_ν and introducing a sign change. Then $\mathcal{M}_{\text{NC}}$ and $\mathcal{M}_{\text{CC}}$ may be combined, and we arrive by standard means at the following expression for the cross section σ times the relative velocity v of electron and positron:

$$\sigma v = \frac{G_F^2}{12\pi} \frac{1}{EE'} \{(C_V'^2 + C_A'^2)[m_e^4 + 3m_e^2 p \cdot p' + 2(p \cdot p')^2] + 3(C_V'^2 - C_A'^2)[m_e^4 + m_e^2 p \cdot p']\} \quad (11.15)$$

where E and E' are the energies of e^- and e^+, respectively, and p and p' are their 4-momenta, with $C_V' = \frac{1}{2} + 2\sin^2\theta_W$ and $C_A' = \frac{1}{2}$. In the CM frame, this becomes

$$\sigma v = \frac{G_F^2}{6\pi} m_e^2 [(C_V'^2 + C_A'^2)\left(\frac{4E^2}{m_e^2} - 1\right) + 3(C_V'^2 - C_A'^2)] \quad (11.16)$$

If we also include production of $\nu_\mu \bar{\nu}_\mu$ and $\nu_\tau \bar{\nu}_\tau$ pairs, this expression must be replaced by

$$\sigma v = \frac{G_F^2 m_e^2}{6\pi} [(3C_V'^2 - 2C_V' + 1 + 3C_A'^2)\left(\frac{4E^2}{m_e^2} - 1\right)$$
$$+ 3C_V'^2 - 2C_V' + 1 - 3C_A'^2] \quad (11.17)$$

For $\sin^2 \theta_W \simeq \tfrac{1}{4}$, this becomes

$$\sigma \frac{v}{c} \simeq 1.9 \times 10^{-45} \left(\frac{4E^2}{m_e^2 c^4} - 0.55\right) \text{ cm}^2 \quad (11.18)$$

Figure 11.3. (a) Z^0 exchange; (b) W exchange.

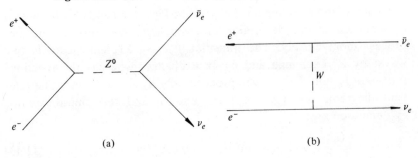

The total energy loss per cubic centimeter per second, Q, arising from the pair process depends not only on σv but also on the available densities of e^- and e^+:

$$Q_{\text{pair}} = \int (E + E') \sigma v \, dn_- \, dn_+ \quad (11.19)$$

In order to calculate the electron and positron densities, we make use of Fermi–Dirac statistics. Let the average number density of protons in the stellar material be $n_0 = N\rho/\mu_e$, where N is Avogadro's number, ρ the mass density, and

$$1/\mu_e = \sum_i X_i Z_i / A_i \quad (11.20)$$

where X_i, Z_i, and A_i are the mass fraction, atomic number, and atomic weight, respectively, of the ith nuclear species. Assuming overall charge neutrality, we have

11.3 Emission of neutrinos by hot stars

$$n_- = n_0 + n_+ = \frac{1}{\pi^2\hbar^3}\int_0^\infty \frac{p^2 dp}{\exp[(E - \mu_-/kT] + 1} \quad (11.21)$$

and

$$n_+ = \frac{1}{\pi^2\hbar^3}\int_0^\infty \frac{p^2 \, dp}{\exp[(E - \mu_+)/kT] + 1} \quad (11.22)$$

where μ_- and μ_+ are the chemical potentials for e^- and e^+, respectively. The physical mechanism whereby electron–positron pairs exist is, of course, pair production from the thermal radiation bath, and at any given temperature, we may safely assume equilibrium between electromagnetic pair production and annihilation:

$$n\gamma \leftrightarrow e^+ + e^- \quad (11.23)$$

For any reaction in equilibrium $A \leftrightarrow B + C$, the chemical potentials satisfy $\mu_A = \mu_B + \mu_C$, and since the chemical potential of thermal radiation is zero, we have $\mu_+ = -\mu_-$. If we ignore the density of nuclear charges, it follows from (11.21) and (11.22) that $\mu_+ = \mu_- = 0$. Then,

$$n_+ \simeq n_- = \frac{1}{\pi^2\hbar^3}\int_0^\infty \frac{p^2 \, dp}{e^x + 1} \quad (11.24)$$

where $x = E/kT = (p^2c^2 + m_e^2c^4)^{1/2}/kT$. For $kT \ll m_ec^2$, we may approximate $(e^x + 1)^{-1}$ by $\exp(-m_ec^2/kT)\exp(-p^2/2m_ekT)$. Thus we obtain:

$$n_+ \simeq n_- \simeq (2m_ekT/\pi\hbar^2)^{3/2} \exp(-m_ec^2/kT) \quad (11.25)$$

For $T \approx 10^9$ K, for example, this yields $n_+ \approx n_- \approx 2 \times 10^{27}$ cm^{-3}. Then, employing (11.18) to obtain $\sigma v \simeq 7 \times 10^{-45}c$ cm^3/sec, and writing $E_+ + E_- \simeq 2m_ec^2$, we have

$$Q_{\text{pair}} \simeq n_+n_-(\sigma v)(E_+ + E_-) \simeq 10^{15} \text{ erg/cm}^3 \text{ sec at } 10^9 \text{ K} \quad (11.26)$$

Although this estimate contains several rough approximations, it is within an order of magnitude of the correct result for the pair process at low and moderate densities and 10^9 K.

The pair energy-loss rate increases very rapidly with temperature. For $kT \gg m_ec^2$, the integral in (11.24) may be approximated by

$$\frac{1}{\pi^2\hbar^3c^3}\int_0^\infty \frac{E^2 \, dE}{\exp(E/kT) + 1} = \frac{(kT)^3}{\pi^2\hbar^3c^3}\int_0^\infty \frac{x^2 \, dx}{1 + e^x}$$
$$\simeq 1.8 \frac{(kT)^3}{\pi^2\hbar^3c^3} \quad (11.27)$$

Therefore, since, in natural units,

$$\sigma v \simeq G_F^2 E^2 \simeq G_F^2(kT)^2 \tag{11.28}$$

for large energies, we have

$$\begin{aligned}Q_{\text{pair}} &\simeq G_F^2 n_+ n_- \sigma v(E_+ + E_-)\\ &\simeq G_F^2(kT)^3(kT)^3(kT)^2(kT) \simeq G_F^2(kT)^9\end{aligned} \tag{11.29}$$

This crude estimate is also quite reliable; a more detailed calculation shows that Q_{pair} increases by about 10 orders of magnitude between 10^9 and 10^{10} K. It is interesting to compare Q_{pair} with ϵ, the total thermal energy stored in electromagnetic radiation per cubic centimeter. We have:

$$Q_{\text{pair}}(10^{10} \text{ K}) \simeq 10^{25} \quad \text{erg/cm}^3 \text{ sec}$$

and

$$\epsilon = aT^4 \simeq 7 \times 10^{25} \quad \text{erg/cm}^3$$

The ratio $\epsilon/Q \approx 10$ sec suggests the sort of time scale one may expect for major changes in stellar structure brought on by neutrino losses at these temperatures.

Several other mechanisms for generating $\nu\bar{\nu}$ pairs may play an important role, but detailed evaluation is difficult and uncertain. Among these is the "Urca" process, in which a β unstable nucleus (Z, N) catalyzes the conversion of electron thermal energy to neutrino pair energy. Here, we consider the β decay:

$$(Z, N) \rightarrow (Z + 1, N - 1)e^-\bar{\nu}_e$$

and denote the amount of kinetic energy available to the leptons as δ. In ordinary matter at low temperatures, the reaction

$$e^- + (Z + 1, N - 1) \rightarrow \nu_e + (Z, N)$$

would not occur, because the electron thermal energy would be insufficient to overcome the threshold δ. However, in a hot stellar core, this is not so and the inverse reaction can proceed (see, e.g., Ruderman 65, Tsuruta and Cameron 65).

11.4 Supernova explosions and neutrino emission

The mechanisms described in the previous section should play an important role in the evolution of blue and red supergiants, in the formation of planetary nebulas, and in the cooling of hot white dwarf stars. However, neutrino processes are expected to have the most spectacular effects in the collapse of massive stellar cores to form neutron stars and even black holes. These dramatic events are presumed to be associated with supernova explosions. To obtain some under-

11.4 Supernova explosions and neutrino emission

standing, we resume our sketchy narrative of the course of events for a highly evolved star of, let us say, 10 solar masses (see, e.g., Freedman et al. 77). There is a dense hot core ($\rho \simeq 10^8$ g/cm^3, $T \simeq 5 \times 10^9$ K) in which thermonuclear reactions have been carried to completion and the material is almost exclusively ^{56}Fe, ^{56}Ni, and/or ^{54}Fe. The core is surrounded by a mantle at somewhat lower temperatures and densities, consisting mainly of ^{28}Si but also of protons, neutrons, α particles, and α-particle nuclei. As we proceed outward from the center, we encounter carbon and oxygen and, at still lower temperatures and densities, a large envelope of helium and hydrogen.

As we have seen, emission of neutrino–antineutrino pairs from the core is enormous, and these losses can no longer be compensated by thermonuclear reactions. The only source of energy is gravitation, so the core contracts, the temperature rises, and the neutrino losses become larger. At a certain point, the iron begins to dissociate:

$$^{56}\text{Fe} \rightarrow 13\alpha + 4n$$

in a reaction that is highly endothermic (it takes about 100 MeV per iron nucleus from the gravitational energy store) but is favored by entropy considerations. This is somewhat analogous to the situation in hydrogen gas at a temperature of about 10^4 K. There, the H$_2$ molecules dissociate to a pair of hydrogen atoms; the reaction costs energy but becomes favored by considerations of statistical equilibrium.

As the iron dissociates to helium and then as the helium dissociates to hydrogen, the core collapses with great rapidity and the density becomes very high. Thus the electron Fermi energy rises to a point where electron capture reactions become energetically favorable:

$$e^- + p \rightarrow n + \nu_e \tag{11.30}$$

The result is neutronization of the core, accompanied by emission of neutrinos ν_e with thermal energies corresponding to temperatures of the order 1 to 2×10^{11} K, that is, 10 to 20 MeV. These neutrinos may carry off as much as 10^{53} ergs of energy. Detailed calculations (Wilson 74, Freedman et al. 77, Mazurek 79, Pethick 79, and Sawyer 79) suggest that this neutrino pulse has a very sharp rise time (1 msec) and that it is followed by the emission of $\nu\bar{\nu}$ pairs accompanying the cooling of the neutron core (Bludman 75). This thermal neutrino pair emission begins a few milliseconds after the initial ν_e burst, peaks after about 30 msec, with a total luminosity of about 10^{53} erg/sec, and decays away in several seconds.

The collapse of the core cannot be halted by any known mechanism until it approaches nuclear densities ($\rho \geq 10^{14}$ g/cm^2), at which point the degeneracy pressure of the neutron gas and the repulsive short-range nucleon–nucleon interaction may balance gravitational crush. In this case the core forms a neutron star, which, if it is spinning, may be observable as a pulsar. However, if the core mass is sufficiently large, even neutron degeneracy and nucleon–nucleon repulsion will be insufficient, and the collapse continues past the Schwarzchild limit, resulting in black-hole formation.

Until now we have assumed that neutrinos emitted from a stellar core proceed outward unhindered. However, the neutrino opacity of the core and mantle may indeed be quite significant, because of several factors. First, the cross sections for neutrino–electron and neutrino–nucleus interactions increase as energy increases, and neutrinos emitted in core collapse are quite energetic (10–20 MeV). Second, neutrino–nucleus elastic scattering by Z^0 exchange is enhanced because the nucleons contribute coherently, provided that the neutrino de Broglie wavelength is large enough to envelop the whole nucleus ($E_\nu \lesssim 40$ MeV). The result is that σ is proportional to $(Z + N)^2$ (Freedman et al. 77). It therefore appears quite possible that neutrinos deposit sufficient energy in the core and mantle to cause a supernova explosion. However, the analysis of such processes is extremely complicated and beyond the scope of our discussions (see, e.g., Pethick 79, Sawyer 79, and Lattimer 81).

11.5 Neutrino-burst detectors

The preceding discussion suggests that supernova explosions and associated neutron star or black-hole formation may be observable from neutrino bursts, as well as by optical means. Let us consider very briefly what the characteristics of a suitable neutrino detector should be (Lande 79). For a source at the galactic center, (3×10^{22} cm distant), a ν_e burst of 10^{53} erg would yield a flux at earth of about 10^{12} ν_e/cm^2 in the 10 to 20-MeV range. These could, in principle, be detected with the existing ^{37}Cl solar neutrino device, although this has an integration time of many days and is thus not very suitable for short bursts.

The thermal $\bar{\nu}$ emission may be observed more easily, however. This requires the reaction

$$\bar{\nu}_e + p \rightarrow n + e^+$$

Since $\sigma = 7.5 \times 10^{-44} E_\nu^2$ cm^2 (E_ν in MeV) and the expected flux is about 10^{11} cm^2 at earth, the expected signal is one secondary e^+ per 6×10^{29} detector protons. Such positrons would have about 10-MeV en-

11.6 Neutrinos of cosmological origin

ergy and would be detectable with a water Čerenkov counter or a liquid scintillation counter. In fact, three detectors are now in operation: a water Čerenkov counter at the Homestake site in South Dakota, and two liquid scintillation detectors in the USSR. Figure 11.4 shows the first of these, a 500-m³ counter hodoscope in the form of a closed hollow box 20 m long, 10 m wide, and 6 m high, which surrounds the solar neutrino detector (see, e.g., Deakyne et al. 78).

The three detectors presently operate in a coordinated search for supernovas with clocks synchronized to better than 100 μsec. This allows for the detection of coincidences in which the relative arrival time of the neutrino burst wavefront at the various detectors provides information as to the location of the source in the sky.

11.6 Neutrinos of cosmological origin

The simplest plausible assumption one can make regarding the large-scale distribution of mass-energy in the universe is that it is

Figure 11.4. The Homestake neutrino detector hodoscope. The sides of the detector are water Čerenkov counters, and the top counters are filled with liquid scintillator. (From Lande 79. Reprinted with permission.)

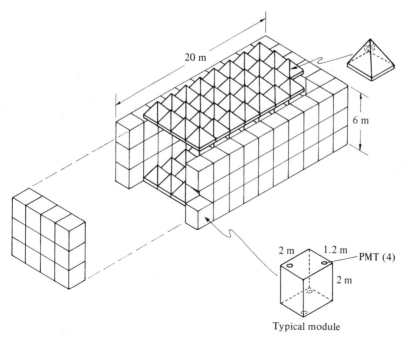

homogeneous and isotropic. This is the *cosmological principle*, which, together with general relativity, forms the basis of the well-known and widely accepted "big bang" model (see, e.g., Schramm and Wagoner 77, Weinberg 77a, and Steigman 79). The origins of the model may be traced to the discovery by A. Friedmann (22) of solutions to Einstein's field equations for a homogeneous isotropic fluid of energy density $\epsilon = \rho c^2$ and pressure p. One solution corresponds to expansion of the fluid with time. Thus, in a sense, Friedmann anticipated the important discovery of the cosmological red shift by the astronomer Edwin Hubble in 1929. As is well known, this phenomenon is interpreted as a Doppler shift of light emitted by distant galaxies, which recede from us with a velocity v, proportional in first approximation to their distance from us:

$$v = H_0 r$$

where H_0 is Hubble's constant.

The homogeneous isotropic universe is described by the Robertson–Walker metric:

$$ds^2 = c^2 dt^2 - R^2(t) \left[\frac{dr^2}{(1 - kr^2)} + r^2(d\theta^2 + \sin^2\theta \, d\phi^2) \right] \quad (11.31)$$

Here, r, θ, and ϕ are co-moving coordinates, t the time measured by a co-moving clock, and the 3-space curvature is related to the constant $k = \pm 1$ or 0. The quantity $R(t)$ is a time-dependent dimensionless scale factor. When the metric of (11.31) is introduced into Einstein's field equations, they reduce to two ordinary differential equations. The first is

$$\tfrac{1}{2}\dot{R}^2 = GM/R + kc^2 + \tfrac{1}{3}\Lambda R^2 \quad (11.32)$$

where $M = \tfrac{4}{3}\pi R^3 \rho$ and Λ is the "cosmological constant." The second equation is

$$d(\rho c^2 V) + p \, dV = 0 \quad (11.33)$$

where V is a definite but arbitrary co-moving volume proportional to R^3.

We may distinguish two extreme cases of special interest. The first is one in which the universe is filled with an ideal gas of nonrelativistic particles. Here the pressure is

$$p = nkT = \rho kT/m \quad (11.34)$$

where m is the mass of a particle. Since the particles are nonrelativistic, $kT \ll mc^2$, and thus $p \ll \rho c^2$. From (11.33) it then follows that

11.6 Neutrinos of cosmological origin

$$\rho_{NR} \propto R^{-3} \tag{11.35}$$

The second extreme case is that of ultrarelativistic particles, for example, photons or zero-mass neutrinos. Here, the pressure is

$$p = \tfrac{1}{3}\rho c^2 \tag{11.36}$$

and (11.33) yields:

$$\rho_{UR} \propto R^{-4} \tag{11.37}$$

At present the average density of nonrelativistic particles (baryons) is known to be greater than 2×10^{-31} g/cm³ from observations of luminous matter in galaxies. An upper limit is obtained from the deceleration parameter. Therefore,

$$2 \times 10^{-31} \text{ g/cm}^3 < \rho_{NR} < 2 \times 10^{-29} \text{ g/cm}^3 \tag{11.38}$$

The ultrarelativistic density is characterized by the cosmic blackbody radiation at $T_0 = 2.9$ K. We have:

$$\rho_\gamma^0 = aT_0^4/c^2 \simeq 5 \times 10^{-35} \text{ g/cm}^3 \tag{11.39}$$

where

$$a = \pi^2 k^4/15\hbar^3 c^3$$

is the Stefan–Boltzmann constant. At present, then, $\rho_{UR} \ll \rho_{NR}$. However, at an earlier time, $R(t)$ was much smaller; thus, taking into account (11.35) and (11.37), we can extrapolate to a very early time when the universe was "radiation dominated" and very hot. [This idea was first suggested by Gamow (46).]

In that regime, (11.32) simplifies considerably, because for small R we may ignore the second and third terms on the right-hand side. Taking into account that $\rho \propto R^{-4}$, we easily integrate (11.32) to obtain

$$\rho = (3/32\pi G)t^{-2} \tag{11.40}$$

If, for the moment, we pretend that there are no other ultrarelativistic particles than photons, then $\rho = \rho_\gamma = aT^4/c^2$ and from (11.40) we obtain

$$T = (3c^2/32\pi Ga)^{1/4} t^{-1/2} \simeq 10^{10} t^{-1/2} \text{ K} \tag{11.41}$$

Of course, this expression is not exactly correct because ρ contains contributions from other ultrarelativistic particles at sufficiently high temperatures. For example, above 3×10^9 K, there are e^+e^- pairs in great numbers and in thermal equilibrium with the photons. The density of each is given by the formula

$$\rho_+ = \rho_- = \frac{g_s}{2\pi^2 \hbar^3 c^3} \int_0^\infty \frac{p^2 \, dp \, pc}{e^{pc/kT} + 1} \tag{11.42}$$

where $g_s = 2$ is the number of possible spin orientations. Thus

$$\rho_+ = \rho_- = \frac{1}{\pi^2\hbar^3c^2}\left(\frac{kT}{c}\right)^4 \int_0^\infty \frac{x^3\,dx}{e^x+1}$$

$$= \frac{7}{8}\frac{1}{c^2}aT^4$$

$$= \tfrac{7}{8}\rho_\gamma \tag{11.43}$$

Nevertheless, (11.41) provides us with a rough guide to the dependence of temperature on time and allows us to follow the evolution qualitatively and to see the significance of the neutrinos.

At $t = 10^{-4}$ sec and $T = 10^{12}$ K (and $kT \simeq 100$ MeV), particles much more massive than 100 MeV/c^2 and with decay lifetimes much less than 10^{-4} sec have ceased to exist. However, pions and nucleons are in thermal equilibrium with muons, e^+e^- pairs, photons, and even with $\nu_e\bar{\nu}_e$, $\nu_\mu\bar{\nu}_\mu$, $\nu_\tau\bar{\nu}_\tau$ pairs. This is because, for $e^+e^- \leftrightarrow \nu\bar{\nu}$, we have, in units $\hbar = c = 1$, $\sigma v \simeq G_F^2 E^2 \simeq G_F^2(kT)^2$ and number densities of order $(kT)^3$ [see (11.27) and (11.28)]. Thus the "time between collisions" for a typical neutrino is

$$t_\nu \simeq (n\sigma v)^{-1} \simeq G_F^{-2}(kT)^{-5}$$

On the other hand, the expansion time scale is set by (11.41):

$$t_{\exp} \simeq 1/\sqrt{G}(kT)^2$$

Thermal equilibrium is maintained as long as $t_\nu \ll t_{\exp}$, and this continues to be true until $T \simeq 10^{10}$ K.

At $t = 10^{-2}$ sec and $T \simeq 10^{11}$ K (and $kT \simeq 10$ MeV), pions and muons have decayed away.

At $t = 1$ sec and $T \simeq 10^{10}$ K (and $kT \simeq 1$ MeV), neutrino pairs have decoupled and the neutrino gases now expand adiabatically. The total density in ultrarelativistic particles is

$$\rho = \tfrac{1}{2}gaT^4/c^2 \tag{11.44}$$

where

$$\frac{g}{2} = \tfrac{1}{2}(g_\gamma + g(e^-) + g(e^+) + g(\nu_e) + g(\bar{\nu}_e) + \cdots)$$
$$= 1 + \tfrac{7}{8} + \tfrac{7}{8} + \tfrac{7}{16} + \tfrac{7}{16} + \cdots \tag{11.45}$$

and in (11.45) we have assumed that the neutrinos are two-component objects, with only one possible spin state ($g_s = 1$). From (11.40), (11.44), and (11.45), it is clear that the relationship between T and R depends on g and thus on the total number of neutrino flavors. This has an important bearing on the neutron/proton number ratio n/p, a

11.6 Neutrinos of cosmological origin

number that "freezes out" in the vicinity of 10^{10} K, as the reactions $n + e^+ \leftrightarrow p + \bar{\nu}_e$, $n + \nu_e \leftrightarrow p + e^-$, and $n \to pe^-\bar{\nu}_e$ fall out of thermal equilibrium. The precise temperature where this occurs depends on g, because the inverse β decay rates depend on baryon density, which is proportional to R^{-3} for given present baryon density. Thus the ratio n/p at "freeze-out" is a rather sensitive function of the number of neutrino flavors.

At $t = 10$ sec and $T \simeq 3 \times 10^9$ K, the e^+e^- pairs annihilate and contribute their energy to the photons. Thus the photon temperature gets a boost and remains somewhat higher than the temperature of the neutrinos. From conservation of entropy, one can show that

$$T_\gamma/T_\nu = (11/4)^{1/3} \tag{11.46}$$

From the known present-day blackbody temperature $T_0 = 2.9$ K, we conclude that there should exist a relic neutrino background at temperature $T \simeq 2.1$ K for each of the neutrino components.

At $t = 100$ sec and $T = 10^9$ K, nuclear fusion commences, with

$$n + p \to D + \gamma \tag{11.47}$$

and subsequent reactions leading to the formation of ^{4}He plus small traces of ^{6}Li and ^{7}Li. This can happen because the time is still early enough so that neutrons have not disappeared by β decay, the temperature is low enough so that deuterons do not immediately suffer photodisintegration, and the temperature is high enough so that nuclear reactions to form ^{4}He are not inhibited excessively by Coulomb barriers. Detailed analysis shows that after the first 20 min, essentially all the neutrons are first imprisoned in deuterium, which is in turn incorporated in ^{4}He. Since there are no stable nuclei with $A = 5$, nucleosynthesis now stops, not to be resumed again for millions of years, until stars are formed.

We now consider the primordial helium mass fraction Y:

$$Y \simeq \frac{2n/p}{n/p + 1} \tag{11.48}$$

Evidently Y depends on the number of neutrino flavors, since it depends on the neutron/proton ratio. It can be shown that three neutrino flavors yield $Y = 0.25$, if the present matter density is greater than 2×10^{-31} g/cm^3 (Yang et al. 79). The predicted value of Y increases by about 0.01 for each additional flavor. Astronomical observations provide the limits

$$0.20 \lesssim Y_{\text{present}} \lesssim 0.29 \tag{11.49}$$

although these observations require a great many assumptions for their interpretation and are subject to considerable uncertainty. In addition, some helium has been made in stars. An estimate of this quantity leads to the conclusion that:

$$0.20 \lesssim Y_{\text{primordial}} \lesssim 0.25 \tag{11.50}$$

This may be taken as evidence that not too many undiscovered neutrino flavors exist; indeed, some authors would claim that no more than one additional neutrino flavor can be accommodated (Schramm 79).

Up to this point, we have considered neutrinos of the conventional type (zero-mass, two-component). We may now ask about constraints from cosmology on the existence of neutrinos with mass. A stable heavy neutral lepton ($m > 1$ MeV/c^2) with ordinary weak interactions would drop out of equilibrium very early in the expansion. The evolution equations then reveal (Lee and Weinberg 77) that the present mass density of such objects is given by the equation

$$\rho \simeq \frac{4.5 \times 10^{-29} \text{ g/cm}^3}{m^2[\text{GeV}/c^2]^2} \tag{11.51}$$

The observational upper limit on ρ [expression (11.38)] is inconsistent with the existence of stable neutral heavy leptons with mass less than 2 GeV/c^2. Dicus et al. (78ab) and Gunn et al. (78) have considered the possibility of unstable massive neutrinos. Here one cannot obtain comparable constraints but can only conclude that the lifetimes of such objects would have to be $< 10^{-6}$ sec.

Finally, we consider the cosmological consequences if the ordinary neutrinos (ν_e, ν_μ, ν_τ) possess small but finite mass. As noted, these neutrinos decoupled at temperatures $\sim 10^{10}$ K, and one can show rather easily that the present neutrino/photon number ratio is 6/11 for each four-component neutrino type. It can then be shown that the mass density is

$$\rho_\nu = 4 \times 10^{-31} m_\nu(\text{eV}/c^2) \quad \text{g/cm}^3 \tag{11.52}$$

where m_ν is the total mass in all neutrino types (Cowsik and McClelland 72):

$$m_\nu = m(\nu_e) + m(\bar{\nu}_e) + m(\nu_\mu) + \cdots$$

From the observational upper limit on ρ [expression (11.38)], we conclude that $m_\nu \lesssim 50$ eV/c^2.

Relic neutrinos with rest mass much in excess of 10^{-4} eV/c^2 would be nonrelativistic and might cluster in halos around galaxies. They thus might add considerably to galactic mass in a way that is not directly observable, and it has been suggested that they could stabilize the dynamics of large disk-shaped galaxies (de Rujula and Glashow 80) and perhaps account for the "missing mass" in clusters of galaxies (Faber and Gallagher 79).

11.7 High-energy neutrino detection

Finally we turn to a short discussion of the various methods for detecting very high energy neutrinos generated in the earth's atmosphere and in remote extraterrestrial sources. For this purpose, several deep-mine muon-detection experiments were undertaken in the 1960s: one at the East Rand Proprietary Gold Mine near Johannesberg, South Africa, at a depth of 8800-m water equivalent (Reines et al. 71); another at the Kolar Gold Field, India, at 7500-m water equivalent (Krishnaswamy et al. 71,76); and a third in Utah (Bergeson et al. 73). In each of the first two detectors, the cosmic-ray muon flux is reduced by about a factor 10^8 or more compared with its value at sea level, and additional discrimination against cosmic-ray muons is obtained by requiring that an acceptable event have a zenith angle θ greater than 50° (note that the cosmic-ray background varies as $\cos^9 \theta$). The Indian and South African experiments succeeded in establishing that the horizontal flux of muon–neutrino-induced muons passing through a plane deep within the earth is 4×10^{-13}/cm² sec sr. This flux is consistent with the assumption that the neutrinos originated from π and K cosmic-ray secondaries and that the muons were produced in the rock surrounding the detector by charged-current reactions with a cross section that rises linearly with energy for $E(\nu) > 1$ GeV. Thus a lower limit to the mass of W was established at 3 GeV/c^2. This limit was extended to 5 GeV/c^2 by the Utah group. There was no evidence for localized extraterrestrial sources of high-energy neutrinos with intensities greater than 10 percent of the cosmic-ray background.

Following these initial attempts, it was recognized that the possibilities exist for a high-energy neutrino detector (DUMAND) with active volume vastly greater than that available in the deep-mine experiments (see, e.g., Roberts 78,81; Stenger 81). The proposed DUMAND is to be realized in the Pacific Ocean near Hawaii, at a depth of 4.5 km. The active volume of water, which serves both as target and as Čerenkov radiator, is to be 1.25×10^8 m³. In this volume will be placed 756 photomultiplier tubes spaced 25 m apart vertically on 36 strings, with a horizontal spacing of 50 m. Such an array can detect muons of energy $E > 100$ GeV with an angular resolution better than 0.5°. Charged-current $\nu_\mu - N$ interactions from outside the array can also be detected, the effective volume for a 2-TeV neutrino being estimated at 4×10^8 m^3.

The DUMAND should possess unique capabilities for high-energy neutrino physics and astronomy. For example, a 10-TeV neutrino beam passing through the center of the earth would be attenuated to 76 percent of its initial intensity if $m_W = \infty$ (Fermi interaction), but to 82 percent if $m_W = 80$ GeV/c^2, as predicted by the standard model. In principle, this difference is detectable with the DUMAND, which also offers unique possibilities for the study of neutrino oscillations. Finally, it should be more sensitive to high-energy neutrinos from supernovas than are deep-mine experiments by several orders of magnitude.

Problems

11.1 Suppose that the neutrino ν_e possesses a magnetic moment of $10^{-9}\mu_0$ (μ_0 = electron Bohr magneton). Estimate crudely the mean free path in solar material for 1-MeV neutrinos. Estimate crudely the heating of the earth by solar neutrinos.

11.2 Derive (11.15), (11.16), and (11.17) from (11.13) and (11.14).

11.3 Derive (11.46).

11.4 Consider a beam of 10-TeV muon neutrinos passing through the center of the earth. Estimate the attenuation of the beam as predicted by the Fermi theory, and by the standard model.

12
Summary and conclusions

Let us now summarize very briefly the state of weak interactions as of 1982. The $SU(2)_L \times U(1)$ standard model accounts for all known electroweak phenomena, at least in principle. The following particles appear in the model.

Leptons
These are in left-handed weak isodoublets and right-handed weak isosinglets:

$$\begin{pmatrix} \nu_e \\ e \end{pmatrix}_L, \begin{pmatrix} \nu_\mu \\ \mu \end{pmatrix}_L, \begin{pmatrix} \nu_\tau \\ \tau \end{pmatrix}_L, \ldots$$

$\nu_{eR}, e_R; \quad \nu_{\mu R}, \mu_R; \quad \nu_{\tau R}, \tau_R, \quad \ldots$

Quarks
These also appear in left-handed weak isodoublets and right-handed weak isosinglets. The quark mass eigenstates are

$$\begin{pmatrix} u \\ d \end{pmatrix}, \begin{pmatrix} c \\ s \end{pmatrix}, \begin{pmatrix} (t) \\ b \end{pmatrix}, \ldots$$

The quark weak eigenstates are related to the quark mass eigenstates by a generalized Cabibbo rotation (the Kobayashi–Maskawa matrix for three generations).

More lepton and quark generations may exist. It is presumed that the numbers of lepton and quark generations are equal. If this is not so, difficulties arise with triangle anomalies. Fragmentary and not very convincing evidence from astrophysics suggests that the number of generations is no more than four.

Intermediate bosons

There are four in the standard model: $W^{\pm}$, Z^0, and γ. Before radiative corrections, one has

$$m_W = \left(\frac{\pi\alpha}{\sqrt{2}\,G_F}\right)^{1/2} \frac{1}{\sin\theta_W}$$

$$m_Z = \frac{m_W}{\cos\theta_W}$$

For $\sin^2\theta_W = 0.23$, this gives $m_W = 78$ GeV/c^2, $m_Z = 89$ GeV/c^2. Radiative corrections alter these values by several percent. The photon mass is zero (experimentally, $m_\gamma \leq 10^{-16}$ eV/c^2).

Higgs boson(s)

In the simplest form of the model, there is one neutral Higgs boson with indeterminate mass m_σ. However, we expect

$$\alpha G_F^{-1/2} \lesssim m_\sigma \lesssim G_F^{-1/2}$$

Electroweak couplings of the leptons and quarks to the gauge bosons are prescribed by (2.91) and (2.103). These Lagrangian terms are constructed to conform to the known results of electrodynamics and low-energy charged-current weak interactions. However, they also contain neutral weak couplings, and the principal predictions of the model are in the domain of neutral weak interactions.

As emphasized in Chapters 3, 8, and 9, the neutral weak couplings in the neutrino-electron, neutrino-quark, electron-quark, and lepton-lepton sectors may be formulated in terms of a model-independent neutral weak Hamiltonian. The experimental results provide very stringent constraints on this Hamiltonian, however, as demonstrated in detailed analyses by Hung and Sakurai (81) and Kim and co-workers (81). In Table 12.1, we employ the notation of Hung and Sakurai as defined in Chapters 3, 8, and 9 to summarize the results. Clearly, excellent agreement is obtained with the standard model, for $\sin^2\theta_W = 0.23$, and we may regard the model as well established, at least for low-energy neutral weak phenomena. However, there is still room for improvement in measurements of various parameters (for example, those that appear in the quark-quark sector and are accessible, in principle, from parity violation in nuclear forces). In Table 12.2, we summarize the various methods for determining $\sin^2\theta_W$.

In the charged-current domain, all results remain consistent with the V-A law (left-handed currents). However, despite heroic ex-

Table 12.1. *Experimental results for neutral weak couplings*

Parameter	Experiment Model-independent	Experiment Factorization-dependent	Standard model	$\sin^2 \theta_W = 0.23$
vq				
α	$\pm(0.589 \pm 0.067)$	0.589 ± 0.067	$1 - 2\sin^2\theta_W$	0.54
β	$\pm(0.937 \pm 0.062)$	0.937 ± 0.062	1	1
γ	$\mp(0.273 \pm 0.081)$	-0.273 ± 0.081	$-\frac{2}{3}\sin^2\theta_W$	-0.153
δ	$\pm(0.101 \pm 0.093)$	0.101 ± 0.093	0	0
ve				
g_V	0.06 ± 0.08	0.043 ± 0.063	$-\frac{1}{2}(1 - 4\sin^2\theta_W)$	-0.04
	or			
	-0.52 ± 0.06			
g_A	-0.52 ± 0.06	-0.545 ± 0.056	$-\frac{1}{2}$	-0.5
	or			
	0.06 ± 0.08			
eq				
$\tilde{\alpha}$	-0.67 ± 0.19	-0.68 ± 0.19	$-(1 - 2\sin^2\theta_W)$	-0.54
$\tilde{\beta}$	—	0.06 ± 0.21	$-(1 - 4\sin^2\theta_W)$	-0.08
$\tilde{\gamma}$	0.22 ± 0.12	0.24 ± 0.10	$\frac{2}{3}\sin^2\theta_W$	0.153
$\tilde{\delta}$	—	0.00 ± 0.02	0	0
eq				
$\tilde{\alpha} + \frac{1}{3}\tilde{\gamma}$	-0.60 ± 0.16	-0.60 ± 0.16	$-(1 - \frac{20}{9}\sin^2\theta_W)$	-0.489
$\tilde{\beta} + \frac{1}{3}\tilde{\delta}$	0.31 ± 0.51	0.06 ± 0.21	$-(1 - 4\sin^2\theta_W)$	-0.08

perimental efforts over the years, the limits on V + A (right-handed currents) from muon decay, β decay, and so forth are none too small. Therefore, one cannot exclude the possibility that electroweak phenomena are described by a "left–right symmetric" model, as outlined in Section 2.12. In such models right-handed currents appear, but the V + A couplings may be made arbitrarily small at low energies by assuming that the mass of the right-handed W boson is sufficiently great. Present experimental limits impose only the modest constraint $m_{WR}/m_{WL} \gtrsim 3$.[1]

In the pure electroweak theory, masses of leptons and quarks arise from Yukawa-type couplings to the vacuum expectation value of the Higgs scalar field. Such a formulation suggests that the neutrino-mass matrix may contain off-diagonal elements and therefore that neutrino oscillations may exist. However, there is no observational evidence for such oscillations so far.

What does the future hold for weak interactions? If history is any guide, the path ahead will take unexpected turns, and phenomena will appear that are undreamed of as this is written. Nevertheless, we can write a short list of important problems that seem accessible, at least in part, to experimental resolution within the foreseeable future.

(a) Do the intermediate bosons $W^{\pm}$ and Z^0 really exist and possess those properties attributed to them in the standard model? This

Table 12.2. *Determinations of* $\sin^2 \theta_W$

Reaction	$\sin^2 \theta_W$
$\nu_\mu + N \rightarrow \nu_\mu + X$	$0.229 \pm 0.09 \ (\pm 0.005)$
$\bar{\nu}_\mu + N \rightarrow \bar{\nu}_\mu + X$	0.230 ± 0.023
$\nu_\mu + p \rightarrow \nu_\mu + p$	0.26 ± 0.06
$\nu_\mu + N \rightarrow \nu_\mu + N + \pi^0$	0.22 ± 0.09
$\bar{\nu}_\mu + N \rightarrow \bar{\nu}_\mu + N + \pi^0$	0.15–0.52
$e^- + D \rightarrow e^- + X$	0.224 ± 0.020
$\nu_\mu + e^- \rightarrow \nu_\mu + e^-$	$0.22 \begin{cases} +0.08 \\ -0.05 \end{cases}$
$\bar{\nu}_\mu + e^- \rightarrow \bar{\nu}_\mu + e^-$	$0.23 \begin{cases} +0.09 \\ -0.23 \end{cases}$
$\bar{\nu}_e + e^- \rightarrow \bar{\nu}_e + e^-$	0.29 ± 0.05

---

[1] See, however, Beall et al. (82), who attempt to derive the constraint $m_{WR} \geq 1.6$ GeV/c^2 from the $K_L^0 - K_S^0$ mass difference.

12 Summary and conclusions

question is only beginning to be answered with $p\bar{p}$ collision experiments, and more detailed investigation must await the construction of large e^+e^- colliding beam machines (LEP at CERN and/or the "linear collider" at SLAC).

(b) Does the t quark exist? What is its mass; and what are the precise values of the Kobayashi–Maskawa matrix elements? To find these values, one must carry out detailed observations and analysis of the weak decays of hadrons containing c, b, and/or t quarks.

(c) Does the neutrino possess mass? Does lepton conservation fail? One needs more sensitive experiments in tritium β decay or alternatives, in neutrino oscillations, in neutrinoless double β decay; and a better limit is required for $\mu \to e\gamma$.

(d) What is the origin of the numerical value of $\sin^2 \theta_W = 0.23$? In one grand unified theory [the $SU(5)$ model of Georgi and Glashow, (74)], the numerical value $\sin^2 \theta_W = 0.20$ arises naturally. This model (among others) predicts also that baryon conservation fails at a certain level and, consequently, that the proton decays with a small but finite probability. However, preliminary negative results on proton decay seem to rule out the $SU(5)$ model, at least in it's simplest form. Further experiments on proton decay may help to distinguish between various grand unified models and thus shed indirect light on the significance of $\sin^2 \theta_W$.

(e) What is the origin of CP violation?

(f) What is the origin of the $|\Delta I| = \frac{1}{2}$ rule?

Answers to these questions probably require a much better understanding of quark–gluon dynamics than exists presently.

Appendix A: Notation and conventions

A.1 **Relativistic notation**

We employ the same relativistic notation as Bjorken and Drell (64). Space-time coordinates are denoted by the contravariant 4-vector

$$x^\mu \equiv (x^0, x^1, x^2, x^3) \equiv (t, x, y, z)$$

The covariant coordinate 4-vector is

$$x_\mu = (x_0, x_1, x_2, x_3) = (t, -x, -y, -z).$$

A covariant 4-vector A_μ and a contravariant 4-vector A^μ are related by

$$A_\mu = g_{\mu\nu} A^\nu$$

where

$$g_{\mu\nu} = g^{\mu\nu} = \begin{pmatrix} 1 & 0 & 0 & 0 \\ 0 & -1 & 0 & 0 \\ 0 & 0 & -1 & 0 \\ 0 & 0 & 0 & -1 \end{pmatrix}$$

The scalar product of two 4-vectors A and B is

$$A \cdot B = A_\mu B^\mu = A^\mu B_\mu = A_0 B_0 - \mathbf{A} \cdot \mathbf{B}$$

The momentum operator in coordinate representation is

$$p^\mu = i \frac{\partial}{\partial x_\mu} = \left(i \frac{\partial}{\partial t}, -i\mathbf{\nabla} \right)$$

$$p_\mu = i \frac{\partial}{\partial x^\mu} = \left(i \frac{\partial}{\partial t}, +i\mathbf{\nabla} \right)$$

In coordinate representation,

$$p^2 = p \cdot p = p^\mu p_\mu = -\frac{\partial}{\partial x^\mu} \frac{\partial}{\partial x_\mu} = -\frac{\partial^2}{\partial t^2} + \nabla^2 = -\Box$$

A.2 The Dirac equation

The Dirac equation for a particle of spin $\frac{1}{2}$ and mass m, in the absence of external fields, is

$$i\gamma^\mu \, \partial\psi/\partial x^\mu - m\psi = 0 \tag{A.1}$$

where ψ is a four-component spinor:

$$\psi = \begin{pmatrix} \psi_1 \\ \psi_2 \\ \psi_3 \\ \psi_4 \end{pmatrix}$$

and the γ^μ's are 4×4 matrices satisfying the commutation relations:

$$\gamma^\mu \gamma^\nu + \gamma^\nu \gamma^\mu = 2g^{\mu\nu} I \tag{A.2}$$

The sixteen 4×4 matrices

$$I, \gamma^\mu, \sigma^{\mu\nu} \equiv \tfrac{1}{2}i(\gamma^\mu\gamma^\nu - \gamma^\nu\gamma^\mu), \gamma^\mu\gamma_5, \gamma_5 \equiv i\gamma^0\gamma^1\gamma^2\gamma^3$$

are linearly independent and form a basis for the vector space of all 4×4 matrices.

The Dirac conjugate spinor $\bar\psi = \psi^\dagger \gamma^0$ satisfies the equation

$$i\frac{\partial\bar\psi}{\partial x^\mu}\gamma^\mu + m\bar\psi = 0 \tag{A.3}$$

Let ψ be a plane-wave solution of the Dirac equation representing a fermion (e.g., an electron) with 3-momentum $\mathbf{p}$ and energy $E = +(|\mathbf{p}|^2 + m^2)^{1/2}$:

$$\psi = \frac{1}{\sqrt{V}} u(p, s) \exp(i\mathbf{p} \cdot \mathbf{x}) \exp(-iEt) \tag{A.4}$$

where V is an arbitrary normalization volume. Then u is a four-component spinor that satisfies

$$(E\gamma^0 - \mathbf{p} \cdot \boldsymbol{\gamma} - m)u = 0$$

or, in covariant notation,

$$(p_\mu \gamma^\mu - m)u \equiv (\not{p} - m)u = 0 \tag{A.5}$$

Similarly, if $\bar u = u^\dagger \gamma^0$,

$$\bar u(\not{p} - m) = 0 \tag{A.6}$$

The corresponding Dirac wavefunction for an antifermion (e.g., a positron) with 3-momentum $\mathbf{p}$ and energy $E = +(|\mathbf{p}|^2 + m^2)^{1/2}$ is:

$$\psi' = \frac{1}{\sqrt{V}} v(p, s)\exp(-i\mathbf{p} \cdot \mathbf{x}) \exp(iEt) \tag{A.7}$$

where

$$(\not{p} + m)v = 0 \tag{A.8}$$

$$\bar{v}(\not{p} + m) = 0 \tag{A.9}$$

The spin-polarization 4-vector s^μ takes the value in the rest frame:

$$s^\mu = (0, \hat{s}) \tag{A.10}$$

where $\hat{s} \cdot \hat{s} = 1$. In general $s \cdot s = -1$, $s \cdot p = 0$.

It is convenient to define the projection operators:

$$u_\alpha(p, s)\bar{u}_\beta(p, s) = \left(\frac{\not{p} + m}{2m} \frac{1 + \gamma_5 \not{s}}{2}\right)_{\alpha\beta} \tag{A.11}$$

$$v_\alpha(p, s)\bar{v}_\beta(p, s) = \left(\frac{\not{p} - m}{2m} \frac{1 + \gamma_5 \not{s}}{2}\right)_{\alpha\beta} \tag{A.12}$$

and

$$[\Lambda_+(p)]_{\alpha\beta} = \sum_{\pm s} u_\alpha(p, s)\bar{u}_\beta(p, s) = \left(\frac{\not{p} + m}{2m}\right)_{\alpha\beta} \tag{A.13}$$

$$[\Lambda_-(p)]_{\alpha\beta} = -\sum_{\pm s} v_\alpha(p, s)\bar{v}_\beta(p, s) = \left(\frac{-\not{p} + m}{2m}\right)_{\alpha\beta} \tag{A.14}$$

In general

$$\bar{u}(p, s)u(p, s) = 1 \tag{A.15}$$

$$\bar{v}(p, s)v(p, s) = -1 \tag{A.16}$$

and

$$\sum_s [u_\alpha(p, s)\bar{u}_\beta(p, s) - v_\alpha(p, s)\bar{v}_\beta(p, s)] = \delta_{\alpha\beta} \tag{A.17}$$

Under homogeneous Lorentz transformations, including reflections,

$\bar{\psi}_1 \psi_2$	transforms as a scalar
$\bar{\psi}_1 \gamma^\mu \psi_2$	transforms as a polar vector
$\bar{\psi}_1 \sigma^{\mu\nu} \psi_2$	transforms as an antisymmetric tensor
$\bar{\psi}_1 \gamma^\mu \gamma_5 \psi_2$	transforms as an axial vector
$\bar{\psi}_1 \gamma_5 \psi_2$	transforms as a pseudoscalar

A.3 Representations of γ matrices

(a) Pauli–Dirac representation.

$$I = \begin{pmatrix} I & 0 \\ 0 & I \end{pmatrix}, \quad \gamma^0 = \begin{pmatrix} I & 0 \\ 0 & -I \end{pmatrix}, \quad \boldsymbol{\gamma} = \begin{pmatrix} 0 & \boldsymbol{\sigma} \\ -\boldsymbol{\sigma} & 0 \end{pmatrix}, \quad \gamma_5 = \begin{pmatrix} 0 & I \\ I & 0 \end{pmatrix}$$

$$\sigma^{ij} = \begin{pmatrix} \sigma_k & 0 \\ 0 & \sigma_k \end{pmatrix}, \quad i, j, k = 1, 2, 3 \text{ (cyclic)}$$

(b) Weyl representation.

$$I = \begin{pmatrix} I & 0 \\ 0 & I \end{pmatrix}, \quad \gamma^0 = \begin{pmatrix} 0 & I \\ I & 0 \end{pmatrix}, \quad \boldsymbol{\gamma} = \begin{pmatrix} 0 & -\boldsymbol{\sigma} \\ \boldsymbol{\sigma} & 0 \end{pmatrix}, \quad \gamma_5 = \begin{pmatrix} I & 0 \\ 0 & -I \end{pmatrix}$$

(c) Independent of representation

$$(\gamma^\mu)^\dagger = \gamma^0 \gamma^\mu \gamma^0$$
$$\gamma^{0\dagger} = \gamma^0$$
$$\boldsymbol{\gamma}^\dagger = -\boldsymbol{\gamma}$$
$$\gamma^\mu \gamma_5 = -\gamma_5 \gamma^\mu$$

(d) In the Pauli–Dirac representation, where $\sim$ indicates transpose,

$$\tilde{\gamma}^0 = \gamma^0, \qquad \gamma^{0*} = \gamma^0$$
$$\tilde{\gamma}^1 = -\gamma^1, \qquad \gamma^{1*} = \gamma^1$$
$$\tilde{\gamma}^2 = \gamma^2, \qquad \gamma^{2*} = -\gamma^2$$
$$\tilde{\gamma}^3 = -\gamma^3, \qquad \gamma^{3*} = \gamma^3$$

A.4 Dirac fields

The Dirac field operator $\Psi(\mathbf{x}, t)$ may be expressed as a superposition of single-particle plane-wave solutions to the Dirac equation:

$$\Psi(\mathbf{x}, t) = \frac{1}{\sqrt{V}} \sum_{p,s} \left(\frac{m}{E}\right)^{1/2} [b(p, s)u(p, s)e^{-ip\cdot x} + d^\dagger(p, s)v(p, s)e^{ip\cdot x}]$$

(A.18)

The Dirac–conjugate field $\overline{\Psi}(\mathbf{x}, t)$ is

$$\overline{\Psi}(\mathbf{x}, t) = \Psi^\dagger(\mathbf{x}, t)\gamma^0$$

(A.19)

$$= \frac{1}{\sqrt{V}} \sum_{p,s} \left(\frac{m}{E}\right)^{1/2} [b^\dagger(p, s)\bar{u}(p, s)e^{ip\cdot x} + d(p, s)\bar{v}(p, s)e^{-ip\cdot x}]$$

The quantities b and d are annihilation operators for fermions and antifermions, respectively, and $b^\dagger$ and $d^\dagger$ are the corresponding creation operators. They satisfy

$$\begin{aligned} \{b^\dagger(p, s), b(p', s')\} &= \delta_{ss'} \delta^3(\mathbf{p}' - \mathbf{p}) \\ \{d^\dagger(p, s), d(p', s')\} &= \delta_{ss'} \delta^3(\mathbf{p}' - \mathbf{p}) \\ \{b^\dagger(p, s), b^\dagger(p', s')\} &= \{b(p, s), b(p', s')\} = 0 \\ \{d^\dagger(p, s), d^\dagger(p', s')\} &= \{d(p, s), d(p', s')\} = 0 \\ \{b(p, s), d^\dagger(p', s')\} &= \{b^\dagger(p, s), d(p', s')\} = 0 \\ \{b(p, s), d(p', s')\} &= \{b^\dagger(p, s), d^\dagger(p', s')\} = 0 \end{aligned}$$

(A.20)

A.5 Parity

In general, the one-particle Dirac equation (A.1) is invariant under the replacement $\mathbf{x} \to -\mathbf{x}$ if we also make the replacement $\psi(\mathbf{x}, t) \to \eta_p \gamma^0 \psi(-\mathbf{x}, t)$, where $|\eta_p| = 1$ to ensure that the transformation is unitary. For the Dirac field we define the parity transformation as

$$P\Psi(\mathbf{x}, t)P^{-1} = \eta_p \gamma^0 \Psi(-\mathbf{x}, t) \tag{A.21}$$

Similarly

$$P\overline{\Psi}(\mathbf{x}, t)P^{-1} = \eta_p^* \overline{\Psi}(-\mathbf{x}, t)\gamma^0 \tag{A.22}$$

Note that

$$\gamma^0 u(\mathbf{p}, s) = u(-\mathbf{p}, s) \tag{A.23}$$

$$\gamma^0 v(\mathbf{p}, s) = -v(-\mathbf{p}, s) \tag{A.24}$$

A.6 Charge conjugation

In the Pauli–Dirac and Weyl representations,

$$\begin{aligned} v(p, s) &= i\gamma^2 \gamma^0 \bar{u}(p, s) = i\gamma^2 u^*(p, s) \\ u(p, s) &= i\gamma^2 \gamma^0 \bar{v}(p, s) = i\gamma^2 v^*(p, s) \end{aligned} \tag{A.25}$$

Under charge conjugation, the Dirac fields transform as follows:

$$\begin{aligned} C\Psi(\mathbf{x}, t)C^{-1} &= \eta_c^* M\overline{\Psi}(\mathbf{x}, t) \\ C\overline{\Psi}(\mathbf{x}, t)C^{-1} &= \eta_c \Psi M \end{aligned} \tag{A.26}$$

where $|\eta_c| = 1$ and $M = i\gamma^2 \gamma^0$.

A.7 Time reversal

The time-reversal properties of a one-particle Dirac wavefunction may be obtained by considering the Dirac equation in an external field:

$$i\gamma^0 \, \partial\psi(t)/\partial t - eV(t)\gamma^0 \psi(t) = -i\boldsymbol{\gamma} \cdot \boldsymbol{\nabla}\psi(t) - e\mathbf{A}(t) \cdot \boldsymbol{\gamma}\psi(t) + m\psi(t) \tag{A.27}$$

where the dependence of V, A, and ψ on $\mathbf{x}$ is understood. From $V(-t) = V(t)$ and $\mathbf{A}(-t) = -\mathbf{A}(t)$ we have, with $t' = -t$,

$$\begin{aligned} -i\gamma^0 \, \partial\psi(-t')/\partial t' &- eV(t')\gamma^0 \psi(-t') \\ &= -i\boldsymbol{\gamma} \cdot \boldsymbol{\nabla}\psi(-t') + e\mathbf{A}(t') \cdot \boldsymbol{\gamma}\psi(-t') + m\psi(-t') \end{aligned} \tag{A.28}$$

Dropping the prime and multiplying on the left by an operator T (which must be antiunitary), with $T\psi(-t) = \psi'(t)$, we have

$$\begin{aligned} -T(i\gamma^0)T^{-1} \, \partial\psi'(t)/\partial t &- eV(t)T\gamma^0 T^{-1}\psi'(t) \\ &= -Ti\boldsymbol{\gamma}T^{-1} \cdot \boldsymbol{\nabla}\psi'(t) + e\mathbf{A}(t) \cdot T\boldsymbol{\gamma}T^{-1}\psi'(t) + m\psi'(t) \end{aligned} \tag{A.29}$$

Also, we choose:
$$Ti\gamma^0 T^{-1} = -i\gamma^0$$
$$T\gamma^0 T^{-1} = \gamma^0$$
$$Ti\boldsymbol{\gamma} T^{-1} = i\boldsymbol{\gamma} \quad (A.30)$$
$$T\boldsymbol{\gamma} T^{-1} = -\boldsymbol{\gamma}$$

recalling that T is antiunitary. Then (A.29) reduces to the original (A.27) with the time-reversed solution $\psi'(\mathbf{x}, t) = T\psi(\mathbf{x}, -t)$. Equations (A.30) define T as follows in the Pauli–Dirac representation: $T = T_0 K$, where T_0 is a unitary matrix and K the complex conjugation operator. From (A.30) we find $T_0 = i\gamma^1 \gamma^3$ and

$$T\psi(\mathbf{x}, t) = \psi'(\mathbf{x}, -t) = i\gamma^1 \gamma^3 \psi^*(\mathbf{x}, t) \quad (A.31)$$

Appendix B: The S-matrix Transition probabilities and cross sections

We consider a physical system in some initial state $|\phi_i(t_0)\rangle$ at a time t_0 in the remote past. We assume there exists a time development operator $U(t, t_0)$, such that $|\phi_i(t)\rangle = U(t, t_0)|\phi_i(t_0)\rangle$ with $t \geq t_0$. We shall be interested in the amplitude for a transition from $|\phi_i(t)\rangle$ to some other state $|\phi_f(t)\rangle$:

$$A_{i \to f} \equiv \lim_{\substack{t \to \infty \\ t_0 \to -\infty}} \langle \phi_f(t)|U(t, t_0)|\phi_i(t_0)\rangle$$

$$\equiv \langle \phi_f(\infty)|S|\phi_i(-\infty)\rangle \quad (B.1)$$

The Hamiltonian of the system may be written as

$$H = H_0 + H_I$$

where H_0 is the zeroth-order Hamiltonian and H_I describes the interaction between particles (or fields). Let H_s, $|\phi_s\rangle$ denote the Hamiltonian and state vector, respectively, in the Schrödinger representation. Then we define the transformation to the *interaction* representation by

$$H = e^{iH_0 t}H_s e^{-iH_0 t} \quad (B.2)$$

$$|\phi\rangle = e^{iH_0 t}|\phi_s\rangle \quad (B.3)$$

It follows that in the interaction representation,

$$i\,\partial_t|\phi\rangle = H_I|\phi\rangle = H_I U(t, t_0)|\phi(t_0)\rangle \quad (B.4)$$

and thus

$$i\,\partial U/\partial t = H_I U(t, t_0) \quad (B.5)$$

A formal solution to (B.5), incorporating the initial condition $U(t_0, t_0) = 1$, is:

The S-matrix

$$U(t, t_0) = 1 - i \int_{t_0}^{t} H_I(t_1) U(t_1, t_0)\, dt_1$$

$$= 1 - i \int_{t_0}^{t} H_I(t) \left[1 - i \int_{t_0}^{t_1} H_I(t_2) U(t_2, t_0)\, dt_2 \right]$$

$$= \cdots \qquad (B.6)$$

$$= 1 - i \int_{t_0}^{t} H_I(t_1)\, dt_1 + \cdots$$

$$+ (-i)^n \int_{t_0}^{t} dt_1 \cdots \int_{t_0}^{t_{n-1}} dt_n\, H_I(t_1) \cdots H_I(t_n) + \cdots$$

Taking the limits $t_0 \to -\infty$, $t \to +\infty$, we obtain

$$S = 1 - i \int_{-\infty}^{\infty} H_I(t_1)\, dt_1 + (-i)^2 \int_{-\infty}^{\infty} dt_1 \int_{-\infty}^{t_1} dt_2\, H_I(t_1) H_I(t_2) + \cdots \qquad (B.7)$$

The integral

$$I = \int_{-\infty}^{\infty} dt_1 \int_{-\infty}^{t_1} dt_2\, H_I(t_1) H_I(t_2)$$

may be rewritten as

$$I = \tfrac{1}{2} \int_{-\infty}^{\infty} dt_1 \int_{-\infty}^{\infty} dt_2\, T\{H_I(t_1), H_I(t_2)\}$$

where

$$T\{H_I(t_1), H_I(t_2)\} = \begin{cases} H_I(t_1) H_I(t_2), & t_1 > t_2 \\ H_I(t_2) H_I(t_1), & t_2 > t_1 \end{cases}$$

is the "time-ordered product." In general,

$$S = 1 + (-i) I_1 + (-i)^2 I_2 + \cdots + (-i)^n I_n + \cdots \qquad (B.8)$$

where

$$I_n = \frac{1}{n!} \int_{-\infty}^{\infty} dt_1 \cdots \int_{-\infty}^{\infty} dt_n\, T\{H_I(t_1), H_I(t_2), \ldots, H_I(t_n)\} \qquad (B.9)$$

The partial amplitudes of (B.1) corresponding to (B.8) are

$$A_{fi}^{(0)} = \delta_{fi} \qquad (B.10)$$

$$A_{fi}^{(1)} = -i \left\langle \phi_f \left| \int_{-\infty}^{\infty} dt\, H_I(t) \right| \phi_i \right\rangle \qquad (B.11)$$

$$A_{fi}^{(2)} = \frac{(-i)^2}{2!} \left\langle \phi_f \left| \int_{-\infty}^{\infty} \int_{-\infty}^{\infty} dt_1\, dt_2\, T\{H_I(t_1), H_I(t_2)\} \right| \phi_i \right\rangle \qquad (B.12)$$

and so forth.

The interaction Hamiltonian H_I can be expressed as an integral over all space of a Hamiltonian density:
$$H_I = \int d^3x \, \mathcal{H}_I(\mathbf{x})$$
Thus
$$A_{fi}^{(1)} = -i \langle \phi_f | \int d^4x \, \mathcal{H}_I(x) | \phi_i \rangle$$
$$A_{fi}^{(2)} = \frac{(-i)^2}{2!} \langle \phi_f | \int d^4x_1 \int d^4x_2 \, T\{\mathcal{H}_I(x_1) \mathcal{H}_I(x_2)\} | \phi_i \rangle$$
and so on.

In all calculations of interest to us, $|\phi_i\rangle$ and $|\phi_f\rangle$ are plane-wave states. The integrals may then be carried out, yielding the result:
$$A_{fi}^{(n)} = (2\pi)^4 \, \delta^4(p_f - p_i) \left(\prod_k \frac{m_k}{E_k V} \prod_l \frac{1}{2\omega_l V} \right)^{1/2} (-i\mathcal{M}_{fi}^{(n)}) \quad (B.13)$$

In (B.13), the factor $(2\pi)^4 \, \delta^4(p_f - p_i)$ expresses energy–momentum conservation. Also $\prod_k$ is a product over all fermions and $\prod_l$ a product over all bosons, where ω_L is the energy of the lth boson. These are simply normalization factors. Physics is contained in the invariant amplitude $\mathcal{M}_{fi}^{(n)}$.

To calculate the transition probability, we compute $|A_{fi}^{(n)}|^2$, using
$$|2\pi^4 \, \delta^4(p_f - p_i)|^2 = 2\pi^4 \, \delta^4(p_f - p_i) \cdot 2\pi^4 \, \delta^4(0) = 2\pi^4 \, \delta^4(p_f - p_i) VT$$
where $T = t - t_0$. The transition probability per unit time is $|A_{fi}^{(n)}|^2/T$:
$$\frac{|A_{fi}^{(n)}|^2}{T} = (2\pi)^4 \, \delta^4(p_f - p_i) V \prod_k \frac{m_k}{E_k V} \prod_l \frac{1}{2\omega_l V} |\mathcal{M}_{fi}^{(n)}|^2 \quad (B.14)$$

However, we really wish to employ the differential transition probability per unit time $d\Gamma$ to a group of final states. This is obtained by multiplying by the phase-space factor
$$\prod_f V \, d^3\mathbf{p}_f/(2\pi)^3$$
where $\prod_f$ is a product over all final particles. Thus, we finally arrive at
$$d\Gamma = (2\pi)^4 \, \delta^4(p_f - p_i) V \prod_k \frac{m_k}{E_k V} \prod_l \frac{1}{2\omega_l V} \prod_f \frac{V \, d^3\mathbf{p}_f}{(2\pi)^3} |\mathcal{M}_{fi}^{(n)}|^2 \quad (B.15)$$

In decay of a single particle to several final particles (e.g., $\mu^- \to e^- \bar{\nu}_e \nu_\mu$), the volume factors V in (B.15) cancel.

Formula (B.15) may also be applied to calculation of the differential cross section for a collision (for example, $A + B \to C + D$, where A, B, C, and D are particles). In the case where these are all fermions, we have

The S-matrix

$$d\Gamma = (2\pi)^4 \, \delta^4(p_C + p_D - p_A - p_B) V \frac{m_A}{VE_A} \frac{m_B}{VE_B} \frac{m_C}{VE_C} \frac{m_D}{VE_D}$$

$$\times \frac{V \, d^3\mathbf{p}_C}{(2\pi)^3} \frac{V \, d^3\mathbf{p}_D}{(2\pi)^3} |\mathcal{M}|^2 \tag{B.16}$$

with obvious modification if any of the particles are bosons. The cross section $d\sigma$ is related to $d\Gamma$ by

$$d\sigma = d\Gamma/j$$

where j is the incident flux. In the rest frame of B, this is just $j = v_A/V$, and we have

$$d\sigma = (2\pi)^4 \, \delta^4(p_C + p_D - p_A - p_B) \frac{m_A m_B}{v_A E_A m_B} \frac{m_C m_D}{E_C E_D} \frac{d^3\mathbf{p}_C}{(2\pi)^3} \frac{d^3\mathbf{p}_D}{(2\pi)^3} |\mathcal{M}|^2 \tag{B.17}$$

However, $v_A E_A m_B = |\mathbf{p}_A| m_B = [(p_A \cdot p_B)^2 - m_A^2 m_B^2]^{1/2}$. Thus:

$$d\sigma = \frac{(2\pi)^4 \, \delta^4(p_C + p_D - p_A - p_B)}{[(p_A \cdot p_B)^2 - m_A^2 m_B^2]^{1/2}} m_A m_B \frac{m_C m_D}{E_C E_D}$$

$$\times \frac{d^3\mathbf{p}_C}{(2\pi)^3} \frac{d^3\mathbf{p}_D}{(2\pi)^3} |\mathcal{M}|^2 \tag{B.18}$$

which is the desired final result.

Appendix C: Dimensional regularization

We illustrate the procedure of dimensional regularization by considering a diagram that would occur in a theory of self-interacting scalar mesons of mass m (Figure C.1). The loop integral is of the form:

$$I_4 = \int \frac{d^4k}{(k^2 - m^2)[(p - k)^2 - m^2]} \tag{C.1}$$

but since $p^2 = m^2$, this becomes

$$I_4 = \int d^4k \frac{1}{k^2 - m^2} \frac{1}{k^2 - 2p \cdot k} \tag{C.2}$$

Let us now assume that k is an "n-vector," where n is not necessarily 4, whereas p remains a 4-vector. Then we can write $k^2 = k_0^2 - \omega^2$, where $\omega^2 = k_1^2 + k_2^2 + \cdots + k_{n-1}^2$. Also $d^4k \to d^nk$, where $d^nk = dk_0 \, [\omega^{n-2} \, d\omega \, d\Omega_{n-1}]$ and $d\Omega_{n-1}$ is a differential solid angle in $n - 1$ spatial dimensions:

$$d\Omega_{n-1} = d\theta_1 \sin \theta_2 \, d\theta_2 \sin^2 \theta_3 \, d\theta_3 \cdots \sin^{n-3} \theta_{n-2} \, d\theta_{n-2} \tag{C.3}$$

Figure C.1. Scalar field with self-interaction.

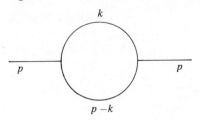

Dimensional regularization

Then, we obtain

$$I_4 \to I_n = \int d^n k \frac{1}{k_0^2 - \omega^2 - m^2} \frac{1}{k_0^2 - \omega^2 - 2mk_0} \tag{C.4}$$

in the rest frame of p_μ, and employing the important formula

$$\frac{\Gamma(x)\Gamma(y)}{\Gamma(x+y)} = 2 \int_0^{\pi/2} \sin^{2x-1}\theta \cos^{2y-1}\theta \, d\theta \tag{C.5}$$

we find

$$I_n = \frac{(2\pi)^{(n-1)/2}}{\Gamma[(n-1)/2]} \int_{-\infty}^{\infty} dk_0 \int_0^{\infty} d\omega \frac{\omega^{n-2}}{(k_0^2 - \omega^2 - m^2)(k_0^2 - 2mk_0 - \omega^2)} \tag{C.6}$$

This integral may now be considered as dependent on the continuous variable n. It diverges for $n \geq 4$ as $\omega \to \infty$ and also for $n \leq 1$ as $\omega \to 0$ but is convergent for $1 < n < 4$. Returning to (C.4), we write:

$$\frac{1}{(k^2 - m^2)(k^2 - 2p \cdot k)} = \int_0^1 \frac{dx}{[k^2 - 2p \cdot kx - m^2(1-x)]^2}$$

which yields

$$I_n = \int_0^1 dx \int \frac{d^n k}{[k^2 - 2p \cdot kx - m^2(1-x)]^2}$$

Then, by means of the substitution $l = k - xp$, we arrive at

$$I_n = \int_0^1 dx \int \frac{d^n l}{[l^2 - m^2(1 - x + x^2)]^2} \tag{C.7}$$

Once more employing (C.3) and (C.5) and carrying out straightforward manipulations, we obtain

$$I_n = i \frac{2\pi^{n/2}}{\Gamma(n/2)} \int_0^{\infty} \frac{l^{n-1} \, dl \, dx}{[l^2 + m^2(1 - x + x^2)]^2}$$

$$= \frac{i 2\pi^{n/2}}{\Gamma(n/2)} \frac{\Gamma(n/2)\Gamma(2 - n/2)}{2\Gamma(2)} \int_0^1 dx \, [m^2(1 - x + x^2)]^{(n/2)-2} \tag{C.8}$$

$$= i\pi^{n/2} \Gamma(2 - n/2) \int_0^1 dx \, [m^2(1 - x + x^2)]^{n/2}$$

Now, since we shall ultimately take the limit as $n \to 4$, in (C.8) we make an expansion about $n = 4$ and find

$$\Gamma(2 - n/2) = \frac{1}{2 - n/2} \Gamma\left(3 - \frac{n}{2}\right) = -\frac{2}{n-4} \Gamma\left(3 - \frac{n}{2}\right)$$

$$= \frac{-2}{n-4} \Gamma(1) + \Gamma'(1) + O(n-4)$$

and thus that

$$I_n \simeq i\pi^2 \left\{ -\frac{2}{n-4} + \Gamma'(1) - \ln \pi - \int_0^1 dx \, \ln[m^2(1-x+x^2)] \right\} \quad (C.9)$$

In addition to evaluating integrals such as I_n, we must also deal with other problems when the number of dimensions is $n \neq 4$. For example, what is the value of $g^{\mu\nu} g_{\mu\nu}$ (normally equal to 4)? How does one define the Dirac matrices and what does their algebra look like? How does one define γ_5? Let us attempt to answer these questions in order.

(a) For consistency, we must define $g^{\mu\nu} g_{\mu\nu} = n$. If we consider the integral

$$J = \int d^n k \, k_\mu / (k^2 - m^2)^2$$

which is zero from covariance, and we perform a shift of the variable of integration $k_\mu \to k_\mu + p_\mu$, it can be shown that we obtain

$$J = \int d^n k \left\{ \frac{k_\mu}{[k^2 - m^2]^2} + \frac{p_\mu}{[k^2 - m^2]^2} - \frac{4k \cdot p k_\mu}{[k^2 - m^2]^3} \right\} + O(p^2)$$

Then if we assume $k_\mu \to 0$, $k_\mu k_\nu \to (1/n) k^2 g_{\mu\nu}$, we obtain

$$J = \int d^n k \left\{ \frac{1}{[k^2 - m^2]^2} - \frac{4k^2/n}{[k^2 - m^2]^3} \right\} p_\mu$$

which can be written as

$$J = i\pi^{n/2} m^{n-4} p_\mu \left[-\frac{2}{n}\left(2 - \frac{n}{2}\right) \Gamma\left(2 - \frac{n}{2}\right) + \frac{2}{n} \Gamma\left(3 - \frac{n}{2}\right) \right] = 0$$

However, if we assume $g_{\mu\nu} g^{\mu\nu} = 4$, we obtain $J \neq 0$, which is an absurdity.

(b) Dirac algebra in n dimensions is then defined as follows:

$$\gamma_\mu \gamma_\nu + \gamma_\nu \gamma_\mu = 2 g_{\mu\nu} I \quad \text{with} \quad g_{\mu\nu} g^{\mu\nu} = n$$

Thus $\gamma^\mu \gamma_\mu = n$ (not 4). Consequently, we find

$$\gamma^\mu \slashed{a} \gamma_\mu = (2-n)\slashed{a}$$
$$\gamma^\mu \slashed{a} \slashed{b} \gamma_\mu = (n-4)\slashed{a}\slashed{b} + 4a \cdot b$$
$$\gamma^\mu \slashed{a} \slashed{b} \slashed{c} \gamma_\mu = -2\slashed{c}\slashed{b}\slashed{a} + (4-n)\slashed{a}\slashed{b}\slashed{c}$$

(c) There is no way to escape all difficulties in defining γ_5. If we define γ_5 such that

$$\{\gamma_5, \gamma^\mu\} = 0, \quad \mu = 0, 1, \ldots, n-1$$
$$\gamma_5^2 = 1$$
$$\text{tr}(\gamma_5 \gamma^\mu \gamma^\nu \gamma^\omega \gamma^\tau) = 4i\epsilon^{\mu\nu\omega\tau} + b(n-4)$$

where b is an arbitrary parameter, then it turns out that there is still an ambiguity, which reflects an ambiguity in the "Adler–Bell–Jackiw triangle anomaly."

Appendix D: Applications of current algebra

D.1 The Adler–Weisberger relation

As an example of the application of the formalism embodied in (4.192)–(4.194), we consider the Adler–Weisberger relation, in which the axial vector coupling constant of neutron β decay is expressed in terms of known π-nucleon scattering cross sections. Our starting point is a particular case of (4.194):

$$[I_+^5, I_-^5] = 2I_3 \tag{D.1}$$

We consider the matrix elements of both sides of (D.1) taken between proton plane-wave states of momentum $\mathbf{q}_1$ and $\mathbf{q}_2$:

$$\langle p(\mathbf{q}_2)|[I_+^5, I_-^5]|p(\mathbf{q}_1)\rangle = \langle p(\mathbf{q}_2)|2I_3|p(\mathbf{q}_1)\rangle$$
$$= \langle p(\mathbf{q}_2)|p(\mathbf{q}_1)\rangle = (2\pi)^3 \,\delta^3(\mathbf{q}_2 - \mathbf{q}_1) \tag{D.2}$$

Now we employ a complete set of states $|\alpha\rangle$ on the left-hand side, with $\Sigma_\alpha |\alpha\rangle\langle\alpha| = 1$:

$$\sum_\alpha \langle p(\mathbf{q}_2)|I_+^5|\alpha\rangle\langle\alpha|I_-^5|p(\mathbf{q}_1)\rangle - \sum_\alpha \langle p(\mathbf{q}_2)|I_-^5|\alpha\rangle\langle\alpha|I_+^5|p(\mathbf{q}_1)\rangle$$
$$= (2\pi)^3 \,\delta^3(\mathbf{q}_2 - \mathbf{q}_1) \tag{D.3}$$

It is convenient to consider separately those states α that correspond on the one hand to a neutron and on the other to all other possible states coupled to a proton by $I_\pm^5$. Thus $\Sigma_\alpha = \Sigma_n + \Sigma_{\alpha \neq n}$. Confining our attention to the neutron states for the moment, we have

$$\sum_n = \sum_{\text{spin}} d^3\mathbf{k}/(2\pi)^3$$

where $\mathbf{k}$ is the neutron 3-momentum. Thus

$$\sum_n = \sum_{\text{spin}} \int d^3k/(2\pi)^3 \int d^3x \int d^3y \, \langle p(\mathbf{q}_2)|g^0_{1+i2}(\mathbf{x},t)|n(\mathbf{k})\rangle$$
$$\times \langle n(\mathbf{k})|g^0_{1-i2}(\mathbf{y},t)|p(\mathbf{q}_1)\rangle \tag{D.4}$$

and where we note that the second sum on the left-hand side of (D.3) is zero. Now,

$$\int d^3x \, \langle p(\mathbf{q}_2)|g^0_{1+i2}(\mathbf{x},t)|n(\mathbf{k})\rangle = -(2\pi)^3 \, \delta^3(\mathbf{q}-\mathbf{k}) C_A \, \bar{u}(\mathbf{q}_2)\gamma^0\gamma_5 u(\mathbf{k}) \tag{D.5}$$

if we neglect momentum-transfer-dependent terms. Thus

$$\sum_n = \sum_{\text{spin}} \int \frac{d^3k}{(2\pi)^3} (2\pi)^3 \, \delta^3(\mathbf{q}_2-\mathbf{k})(2\pi)^3 \, \delta^3(\mathbf{k}-\mathbf{q}_1)$$
$$\times |C_A|^2 \bar{u}(\mathbf{q}_2)\gamma^0\gamma_5 u(\mathbf{k})\bar{u}(\mathbf{k})\gamma_0\gamma_5 u(\mathbf{q}_1) \tag{D.6}$$

However,

$$\sum_{\text{spin}} u(\mathbf{k})\bar{u}(\mathbf{k}) = (\not{k}+m_n)/2m_n$$

Also,

$$\gamma^0\gamma^5(\not{k}+m_n)\gamma^0\gamma_5 u(\mathbf{q}_1) = \gamma^0(\not{k}-m_n)\gamma^0 u(\mathbf{q}_1) \tag{D.7}$$

Furthermore, the delta functions in (D.6) imply $\mathbf{q}_1 = \mathbf{q}_2 = \mathbf{k} \equiv \mathbf{q}$. Thus, (D.7) becomes

$$(\gamma^0 q_0 + \boldsymbol{\gamma}\cdot\mathbf{q} - m_n)u(\mathbf{q}_1) \simeq (\gamma^0 q_0 + \boldsymbol{\gamma}\cdot\mathbf{q} - m_p)u(\mathbf{q}_1)$$
$$= 2(\gamma^0 q_0 - m_p)u(\mathbf{q}_1)$$

where the last step follows from the Dirac equation. Thus, (D.5) becomes

$$\sum_n = (2\pi)^3 \, \delta^3(\mathbf{q}_2-\mathbf{q}_1)|C_A|^2 \bar{u}(\mathbf{q}_2)\frac{(\gamma^0 q_0 - m_p)}{m_p} u(\mathbf{q}_1)$$
$$= (2\pi)^3 \, \delta^3(\mathbf{q}_2-\mathbf{q}_1)|C_A|^2 \left(\frac{q_0^2}{m_p^2}-1\right) \tag{D.8}$$

Next we consider the sum $\Sigma_{\alpha\neq n}$; here the PCAC hypothesis enters. We recall from (4.164) that

$$\partial_\alpha(g_1^\alpha + ig_2^\alpha) = C_A m_p m_\pi^2(\phi_1 + i\phi_2)/g_0 \tag{D.9}$$

In order to apply (D.9) to evaluation of the sum $\Sigma_{\alpha\neq n}$, we consider the Heisenberg equation:

$$[I_+^5, H] = i \, dI_+^5/dt \tag{D.10}$$

where H is the total Hamiltonian. We take the matrix element of (D.10) between $\langle p(\mathbf{q}_2)|$ and $|\alpha\rangle$:

$$\langle p(\mathbf{q}_2)|[I_+^5, H]|\alpha\rangle = i \int \left\langle p(\mathbf{q}_2) \left| \frac{d}{dt}(g_1^0 + ig_2^0) \right| \alpha \right\rangle d^3x \tag{D.11}$$

However, $\int \nabla \cdot (\mathbf{g}_1 + i\mathbf{g}_2) \, d^3x = 0$ by Gauss's theorem, since the "surface" terms may be neglected. Therefore,

$$\int \frac{d}{dt}(g_1{}^0 + ig_2{}^0) \, d^3x = \int \partial_\alpha (g_1{}^\alpha + ig_2{}^\alpha) \, d^3\mathbf{x},$$

and we obtain from (D.11) and (D.9):

$$\langle p(\mathbf{q}_2)|I_+{}^5|\alpha\rangle = \frac{i}{q_{0\alpha} - q_0} \int \langle p(\mathbf{q}_2)|\partial_\alpha(g_1{}^\alpha + ig_2{}^\alpha)|\alpha\rangle \, d^3\mathbf{x}$$
$$= \frac{i}{q_{0\alpha} - q_0} \int \langle p(\mathbf{q}_2)|C_A m_p m_\pi{}^2(\phi_1 + i\phi_2)/g_0|\alpha\rangle \, d^3\mathbf{x} \quad (D.12)$$

Now employing (D.12) and an analogous expression for $\langle \alpha|I_-{}^5|p(\mathbf{q}_1)\rangle$, and so on, we obtain for $\Sigma_{\alpha \neq n}$ in (D.3):

$$\sum_{\alpha \neq n} = \sum_{\alpha \neq n} \frac{1}{(q_0 - q_{0\alpha})^2} \left(\frac{m_\pi{}^2 m_p C_A}{g_0}\right)^2$$
$$\times \int d^3\mathbf{x} \int d^3\mathbf{x}'[\langle p(\mathbf{q}_2)|\phi_1(\mathbf{x}) + i\phi_2(\mathbf{x})|\alpha\rangle\langle\alpha|\phi_1(\mathbf{x}') - i\phi_2(\mathbf{x}')|p(\mathbf{q}_1)\rangle$$
$$- \langle p(\mathbf{q}_2)|\phi_1(\mathbf{x}) - i\phi_2(\mathbf{x})|\alpha\rangle\langle\alpha|\phi_1(\mathbf{x}') + i\phi_2(\mathbf{x}')|p(\mathbf{q}_1)\rangle] \quad (D.13)$$

Expression (D.13) is very similar to sum rules frequently encountered in atomic physics, for example, in the quantum-mechanical treatment of Rayleigh and Raman scattering. It can be seen that the integrand of (D.13) involves matrix elements between proton states and intermediate states $|\alpha\rangle$ consisting of a proton plus pion. It can thus be shown that, in the limit as $q_0{}^2 \to \infty$, (D.13) can be expressed conveniently in terms of the cross sections for $\pi^\pm$-proton scattering. The end result of Adler's calculation, involving numerical integration of pion–nucleon cross sections, is:

$$C_{A,\text{theor}} = 1.24$$

in excellent agreement with the experimental value $C_{A,\text{expt}} = 1.25$.

D.2 The Ademollo–Gatto theorem

We now outline a derivation of the Ademollo–Gatto theorem as it applies to K_{l3} decays. Here we start with a special case of (4.192), namely:

$$[(F_4 + iF_5), (F_4 - iF_5)] = F_3 + \tfrac{3}{2}Y \quad (D.14)$$

We take the matrix element between states $|\pi^+(\mathbf{p}')\rangle$ and $|\pi^+(\mathbf{p})\rangle$ of both sides of (D.14):

$$\langle \pi^+(\mathbf{p}')|[F_4 + iF_5, F_4 - iF_5]|\pi^+)\mathbf{p})\rangle = \langle \pi^+)\mathbf{p}')|F_3 + \tfrac{3}{2}Y|\pi^+(\mathbf{p})\rangle$$
$$= \langle \pi^+(\mathbf{p}')|\pi^+(\mathbf{p})\rangle \quad (D.15)$$

Appendix D

It is convenient to normalize the pion states so that the right-hand side of (D.15) is

$$\langle \pi^+(\mathbf{p}')|\pi^+(\mathbf{p})\rangle = 2p_0(2\pi)^3\,\delta^3(\mathbf{p}' - \mathbf{p}) \tag{D.16}$$

where $p_0 = (|\mathbf{p}|^2 + m_\pi^2)^{1/2}$ is the pion energy. As for the left-hand side of (D.15), we insert $(1/2p_0)\sum|\alpha\rangle\langle\alpha| = 1$ and write

$$\frac{1}{2p_0}\left\{\sum_\alpha \langle \pi^+(\mathbf{p}')|F_4 + iF_5|\alpha\rangle\langle\alpha|F_4 - iF_5|\pi^+(\mathbf{p})\rangle \right.$$

$$\left. - \sum_\alpha \langle p^+(\mathbf{p}')|F_4 - iF_b|\alpha\rangle\langle\alpha|F_4 + iF_b|\pi^+(\mathbf{p})\rangle\right\}$$

As in the discussion of the Adler–Weisberger relation, we break up the sum over α states into two portions: the first corresponding to a $\bar{K}^0$ meson in various momentum states k, and the second, to all other states. In the latter category are particle states not belonging to the pseudoscalar meson $SU(3)_f$ multiplet. Thus:

$$\sum_\alpha = \sum_{\alpha = \bar{K}^0} + \sum_{\alpha \neq \bar{K}^0} \tag{D.17}$$

Now

$$\frac{1}{2p_0}\sum_{\alpha=\bar{K}^0} \rightarrow \frac{1}{(2\pi)^3}\int\frac{d^3k}{2p_0}\langle\pi^+(\mathbf{p}')|F_4 + iF_5|\bar{K}^0(\mathbf{k})\rangle\langle\bar{K}^0(\mathbf{k})|F_4 - iF_5|\pi^+(\mathbf{p})\rangle$$

$$= \frac{1}{(2\pi)^3}\int\frac{d^3k}{2p_0}\int d^3x\, d^3x'\, \langle\pi^+(\mathbf{p}')|j_4^0(\mathbf{x}, t) + ij_5^0(\mathbf{x}, t)|\bar{K}^0(\mathbf{k})\rangle$$

$$\times \langle\bar{K}^0(\mathbf{k})|j_4^0(\mathbf{x}, t) - ij_5^0(\mathbf{x}, t)|\pi^+(\mathbf{p})\rangle$$

$$= \frac{1}{(2\pi)^3}\int\frac{d^3k}{2p_0}\int d^3x\, d^3x'\, [f_+(q'^2)(k_0 + p_0') + f_-(q'^2)(k_0 - p_0')]$$

$$\times e^{i(p_0'-k_0)t}[f_+(q^2)(k_0 + p_0) + f_-(q^2)(k_0 - p_0)]e^{i(k_0-p_0)t}$$

$$\times \exp[i(\mathbf{k} - \mathbf{p})\cdot\mathbf{x}]\exp[i(\mathbf{p} - \mathbf{k})\cdot\mathbf{x}'] \tag{D.18}$$

where $q'^2 = (k - p')^2$ and $q^2 = (k - p)^2$. Now performing the integrations over $\mathbf{x}$ and $\mathbf{x}'$ in (D.18), we obtain

$$\frac{1}{2p_0}\sum_{\alpha=\bar{K}^0} = \frac{1}{(2\pi)^3}\int\left\{\frac{d^3k}{2p_0}(2\pi)^6\,\delta^3(\mathbf{k} - \mathbf{p}')\,\delta^3(\mathbf{p} - \mathbf{k})\right.$$

$$\times [f_+(q_0^2)(k_0 + p_0') + f_-(q_0^2)(k_0 - p_0')]$$

$$\left.\times [f_+(q_0^2)(k_0 + p_0) + f_-(q_0^2)(k_0 - p_0)]\right\}e^{i(p_0'-p_0)t} \tag{D.19}$$

or

$$\frac{1}{2p_0}\sum_{\alpha=\bar{K}^0} = \frac{(2\pi)^3}{2p_0}\delta^3(\mathbf{p}' - \mathbf{p})\,[f_+(q_0^2)(k_0 + p_0) + f_-(q_0^2)(k_0 - p_0)]^2$$

$$\tag{D.20}$$

Since $k_0 = (|\mathbf{p}|^2 + m_K^2)^{1/2}$, $p_0 = (|\mathbf{p}|^2 + m^2)^{1/2}$, and $q_0 = k_0 - p_0$, we obtain in the limit of large p_0:

$$\lim \frac{1}{2p_0} \sum_{\alpha = \bar{K}^0} = (2\pi)^2 2p_0\, \delta^3(\mathbf{p}' - \mathbf{p})\, f_+(0) \quad \text{(D.21)}$$

Thus, comparing (D.21) with (D.16), we see that $f_+(0) = 1$, provided that $\sum_{\alpha \neq \bar{K}^0}$ can in fact be neglected to a good approximation. We therefore turn our attention to the total Hamiltonian $H = H_0 + \lambda H_m$, which we have separated into two portions – an $SU(3)_f$ invariant part H_0 and an $SU(3)_f$ breaking part H_m. Parameter λ must be small ($\lambda \ll 1$), otherwise the hadron mass spectrum would not exhibit any recognizable $SU(3)_f$ symmetry. Consider the matrix element:

$$\langle \pi^+ | F_4 + iF_5 | \alpha \rangle = \frac{\langle \pi^+ | [H, F_4 + iF_5] | \alpha \rangle}{E_\pi - E_\alpha}$$

$$= \lambda \frac{\langle \pi^+ | [H_m, F_4 + iF_5] | \alpha \rangle}{E_\pi - E_\alpha} \quad \text{(D.22)}$$

When we calculate sums over products of the form

$$\langle \pi^+ | F_4 + iF_5 | \alpha \rangle \langle \alpha | F_4 - iF_5 | \pi^+ \rangle \quad \text{(D.23)}$$

only those terms survive for which $p_\alpha = p_\pi$. Thus, $E_\pi - E_\alpha = O(m_\pi - m_\alpha)$ for small momentum. If α represents a particle belonging to the $J^P = 0^-$ octet ($\bar{K}^0$), then $E_\pi - E_\alpha \simeq O(\lambda)$, in which case (D.23) is of order unity from (D.22). However, if α belongs to a different $SU(3)_f$ multiplet, then $E_\pi - E_\alpha$ is of order unity and (D.23) is of order λ^2 from (D.22). Thus

$$\sum_{\alpha \neq \bar{K}^0} \simeq O(\lambda^2) \sum_{\alpha = \bar{K}^0}$$

and the deviation of $f_+(0)$ from unity is of order λ^2.

Appendix E: Physical constants

c = 2.99792458(1.2) × 10^{10} cm/sec
$|e|$ = 4.803242(14) × 10^{-10} esu[1]
$\hbar$ = 1.0545887(57) × 10^{-27} erg sec
m_e = 9.109534(47) × 10^{-28} g = 0.5110034(14) MeV/c^2
m_p = 1836.15152(70)m_e = 938.2796(27) MeV/c^2
χ_e = $\hbar/m_e c$ = 3.8615905(64) × 10^{-11} cm
α^{-1} = $(e^2/\hbar c)^{-1}$ = 137.03604(11)*
G_F = 1.43582(4) × 10^{-49} erg cm^3 = 1.6632(4) × 10^{-5} $(c^2/\text{GeV})^2$
G = 6.6720(4) × 10^{-8} cm^3/g sec^2
k = 1.380662(44) × 10^{-16} erg/K = 8.61735(28) × 10^{-5} eV/K
1 eV = 1.6021892(46) × 10^{-12} erg
1 eV = 11604.50(36) K (from $E = kT$)

* Frequently one writes $\alpha = e^2/4\pi\hbar c$ with e in Heaviside–Lorentz units.

Selected bibliography

Adler, S. L. and R. F. Dashen. *Current Algebras and Applications to Particle Physics.* Benjamin, New York, 1968.

Baldo-Ceolin, M. (ed). *Proceedings of the International School of Physics "Enrico Fermi," 1977: Weak Interactions.* North Holland, Amsterdam, 1979.

Bjorken, J. D. and S. D. Drell. *Relativistic Quantum Mechanics.* McGraw-Hill, New York, 1964.

– *Relativistic Quantum Fields.* McGraw-Hill, New York, 1965.

Commins, E. D. *Weak Interactions.* McGraw-Hill, New York, 1973.

Fritzsch, H. and P. Minkowski, *"Flavordynamics of Quarks and Leptons." Physics Reports 73,* 67, 1981.

Konopinski, E. J. *The Theory of Beta Radioactivity.* Oxford University Press, 1966.

Lee, T. D. and C. S. Wu. "Weak Interactions I, II." *Annual Review of Nuclear Science 15,* 381, 1965; *16,* 471, 1966.

Marshak, R. E., Riazuddin, and C. P. Ryan. *Theory of Weak Interactions in Particle Physics.* Wiley, New York, 1969.

Morita, M. *Beta Decay and Muon Capture.* Benjamin, New York, 1973.

Okun, L. B. *Weak Interactions of Elementary Particles.* Addison-Wesley, Reading, Mass., 1965.

Raymond, P. *Field Theory: A Modern Primer.* Benjamin/Cummings, Menlo Park, Calif., 1981.

Sakurai, J. J. *Advanced Quantum Mechanics.* Addison-Wesley, Reading, Mass., 1967.

Schopper, H. F. *Weak Interactions and Nuclear Beta Decay.* North Holland, Amsterdam, 1966.

Taylor, J. C. *Gauge Theories of Weak Interactions.* Cambridge University Press, 1976.

Wu, C. S. and S. A. Moszkowski. *Beta Decay.* Interscience, New York, 1966.

References

1922
Friedmann, A. *Z. Phys. 10,* 377, 1922.

1927
Dirac, P. A. M. *Proc. R. Soc. London, Sect. A 114,* 243, 710, 1927.

1928
Cox, R. T., C. G. McIlwraith, and B. Kurrelmeyer. *Proc. Natl. Acad. Sci. U.S.A. 14,* 544, 1928.

1929
Heisenberg, W. and W. Pauli. *Z. Phys. 56,* 1, 1929.
Weyl, H. *Z. Phys. 56,* 330, 1929.

1930
Heisenberg, W. and W. Pauli. *Z. Phys. 59,* 169, 1930.

1931
Pauli, W. (private correspondence with O. Hahn and L. Meitner, 1931).

1933
Pauli, W. *Proc. VII Solvay Congress, Brussels* (1933), p. 324. Gauthier-Villars, Paris.

1934
Curie, I. and F. Joliot. *C. R. Acad. Sci. 198,* 254, 1934.
Fermi, E. *Z. Phys. 88,* 161, 1934.

1935
Yukawa, H. *Proc. Phys. Math. Soc. Japan 17,* 48, 1935.

1936
Gamow, G. and E. Teller. *Phys. Rev. 49,* 895, 1936.
Heisenberg, W. *Z. Phys. 101,* 533, 1936.
Neddermeyer, S. H. and C. D. Anderson. *Phys. Rev. 50,* 263, 1936.

1937
Fierz, M. *Z. Phys. 104,* 553, 1937.
Neddermeyer, S. H. and C. D. Anderson. *Phys. Rev. 51,* 884, 1937.

1938
Alvarez, L. *Phys. Rev. 54,* 486, 1938.
Bethe, H. A. and C. L. Critchfield. *Phys. Rev. 54,* 248, 1938.
Heisenberg, W. *Ann. Phys. (Leipzig) 32,* 20, 1938.

1939
Neddermeyer, S. H., and C. D. Anderson. *Rev. Mod. Phys. 11,* 191, 1939.

1946
Conversi, M., E. Pancini, and O. Piccioni. *Phys. Rev. 68,* 232, 1946.
Gamow, G. *Phys. Rev. 70,* 572, 1946.
Pontecorvo, B. National Research Council of Canada Report No. PD.205, 1946.

1947
Conversi, M., E. Pancini, and O. Piccioni. *Phys. Rev. 71,* 209, 1947.
Lattes, C. M., H. Muirhead, G. P. S. Occhialini, and C. F. Powell. *Nature (London) 159,* 694, 1947.
Rochester, G. D. and C. C. Butler. *Nature (London) 160,* 855, 1947.
Wheeler, J. A. *Phys. Rev. 71,* 320, 1947.

1949
Alvarez, L. W. University of California Radiation Laboratory Report No. 328, 1949.
Wheeler, J. A. and S. Tiomno. *Rev. Mod. Phys. 21,* 153, 1949.

1950
Michel, L. *Proc. Phys. Soc., London, Sect. A63,* 514, 1950.

1951
Armenteros, R., K. H. Barker, C. C. Butler, and A. Cachon. *Philos. Mag. 42,* 1113, 1951.

1953
Dalitz, R. *Philos. Mag. 44,* 1068, 1953.
Gell-Mann, M. *Phys. Rev. 92,* 833, 1953.
Gell-Mann, M. and V. L. Telegdi. *Phys. Rev. 91,* 169, 1953.
Nakano, T. and K. Nishijima. *Prog. Theor. Phys. Osaka 10,* 581, 1953.

1954
Dalitz, R. *Phys. Rev. 94,* 1046, 1954.
Fabri, E. *Nuovo Cimento 11,* 479, 1954.
Lüders, G. *Danske Mat. fys. Medd. 28* (5), 1954.
Yang, C. N. and R. Mills. *Phys. Rev. 96,* 191, 1954.

1955
Davis, R. *Phys. Rev. 97,* 766, 1955.
Pauli, W. *Niels Bohr and the Development of Physics,* p. 30. Pergamon Press, Elmsford, N.Y., 1955.

References

1956
Behrends, R. E., R. J. Finkelstein, and A. Sirlin. *Phys. Rev. 101*, 866, 1956.
Lee, T. D. and C. N. Yang. *Phys. Rev. 104*, 254, 1956.
Orear, J., G. Harris, and S. Taylor. *Phys. Rev. 102*, 1676, 1956.

1957
Bincer, A. M. *Phys. Rev. 107*, 1434, 1957.
Bouchiat, C. and L. Michel. *Phys. Rev. 106*, 170, 1957.
Friedman, J. I. and V. L. Telegdi. *Phys. Rev. 105*, 1681, 1957.
Garwin, R., L. Lederman, and M. Weinreich. *Phys. Rev. 105*, 1415, 1957.
Jackson, J. D., S. B. Treiman, and H. W. Wyld, Jr. *Phys. Rev. 106*, 517, 1957;
– *Nucl. Phys. 4*, 206, 1957.
Kinoshita, T., and A. Sirlin. *Phys. Rev. 106*, 110, 1957a;
– *Phys. Rev. 107*, 595, 1957b;
– *Phys. Rev. 107*, 638, 1957c;
– *Phys. Rev. 108*, 844, 1957d.
Landau, L. *Nucl. Phys. 3*, 127, 1957.
Lee, T. D. and C. N. Yang. *Phys. Rev. 105*, 1671, 1957.
Pauli, W. *Nuovo Cimento 6*, 204, 1957.
Pursey, D. L., and S. Kahana. *Nuovo Cimento 6*, 266, 1469, 1957.
Salam, A. *Nuovo Cimento 5*, 299, 1957.
Wu, C. S., E. Ambler, R. W. Hayward, D. D. Hopper, and R. P. Hudson. *Phys. Rev. 105*, 1413, 1957.

1958
Berman, S. M. *Phys. Rev. 112*, 267, 1958.
Bludman, S. *Nuovo Cimento 9*, 433, 1958.
Feynman, R. P. and M. Gell-Mann. *Phys. Rev. 109*, 193, 1958.
Gell-Mann, M. *Phys. Rev. 111*, 362, 1958.
Goldberger, M. L. and S. B. Treiman. *Phys. Rev. 110*, 1178, 1958a.
– *Phys. Rev. 111*, 354, 1958b.
Goldhaber, M., L. Grodzins, and A. W. Sunyar. *Phys. Rev. 109*, 1015, 1958.
Good, M. L. *Phys. Rev. 110*, 550, 1950
Sudarshan, E. C. G., and R. Marshak. *Phys. Rev. 109*, 1860, 1958.

1959
Allen, J. S., R. L. Burman, W. B. Hermannsfeldt, P. Stähelin, and T. H. Braid. *Phys. Rev. 116*, 134, 1959.
Fujii, A. and H. Primakoff. *Nuovo Cimento 12*, 327, 1959.
Kinoshita, T. *Phys. Rev. Lett. 2*, 477, 1959.
Kinoshita, T. and A. Sirlin. *Phys. Rev. 113*, 1652, 1959.
Pais, A. *Phys. Rev. Lett. 3*, 242, 1959.
Primakoff, H. *Rev. Mod. Phys. 31*, 802, 1959.
Primakoff, H. and S. P. Rosen. *Rep. Prog. Phys. 22*, 121, 1959.
Wu, C. S. In *Beiträge zur Physik und Chemie des 20 Jahrkunderts*, ed. O. R. Frisch. Vieweg, Brunswick, 1959.
Zel'dovich, Ya. B. *Zh. Eksp. Teor. Fiz. 36*, 964, 1959 [*Sov. Phys.-JETP 9*, 682, 1959].

1960
Kuznetsov, V. P. *Sov. Phys. JETP 10*, 784, 1960 [*Zh. Eksp. Teor. Fiz. 37*, 1102, 1959].
Plano, R. *Phys. Rev. 119*, 1400, 1960.

1961
Coleman, S. and S. Glashow. *Phys. Rev. Lett. 6*, 423, 1961.
Glashow, S. L. *Nucl. Phys. 22*, 579, 1961.
Goldstone, J. *Nuovo Cimento 19*, 154, 1961.
Kuznetsov, V. P. *Sov. Phys. JETP 12*, 1202, 1961 [*Zh. Eksp. Teor. Fiz. 39*, 1722, 1960].

1962
Berman, S. M. and A. Sirlin. *Ann. Phys. (N. Y.) 20*, 20, 1962.
Kofoed-Hansen, O. and C. J. Christensen. *Handbuch der Physik*, Band 41/2. Springer, Berlin, 1962.
Maki, Z., M. Nakagawa, and S. Sakata. *Progr. Theor. Phys. 28*, 870, 1962.
Messiah, A. *Quantum Mechanics*, Vol. 2, p. 1079. North Holland, Amsterdam/Interscience, New York 1962.

1963
Buhler, A., N. Cabibbo, M. Fidecaro, T. Massam, Th. Muller, M. Schneegano, and A. Zichichi. *Phys. Lett. 7*, 368, 1963.
Cabibbo, N. *Phys. Rev. Lett. 10*, 531, 1963.
Feynman, R. P. *Acta Phys. Pol. 24*, 697, 1963.
Johnson, C. H., F. Pleasanton, and T. A. Carlson. *Phys. Rev. 132*, 1149, 1963.
Lee, Y. K., L. W. Mo, and C. S. Wu. *Phys. Rev. Lett. 10*, 253, 1963.

1964
Ademollo, M. and R. Gatto. *Phys. Lett. 13*, 264, 1964.
Bahcall, J. N. *Phys. Rev. 135*, B137, 1964.
Bjorken, J. D. and S. D. Drell. *Relativistic Quantum Mechanics*. McGraw-Hill, New York, 1964.
Bloom, S., L. A. Dick, L. Feuvrais, G. R. Henry, P. C. Macq, and M. Spighel. *Phys. Lett. 8*, 87, 1964.
Cabibbo, N. *Phys. Rev. Lett. 12*, 62, 1964.
Cabibbo, N. and A. Maksymovicz. *Phys. Lett. 9*, 352, 1964a.
– *Phys. Lett 11*, 360, 1964b.
Christenson, J. H., J. W. Cronin, V. L. Fitch, and R. Turlay. *Phys. Rev. Lett. 13*, 138, 1964.
DeCapua, E., R. Garland, L. Pondrom, and A. Strelzoff. *Phys. Rev. B133*, 1333, 1964.
de Witt, B. *Phys. Rev. Lett. 12*, 742, 1964.
Drell, S. D. and J. D. Walecka. *Ann. Phys. (N.Y.) 28*, 18, 1964.
Duclos, J., J. Heintze, A. DeRujila, and V. Soergel. *Phys. Lett. 9*, 62, 1964.
Englert, F. and R. Brout. *Phys. Rev. Lett. 13*, 321, 1964.
Gell-Mann, M. *Phys. Rev. Lett. 12*, 155, 1964.
Greenberg, O. W. *Phys. Rev. Lett. 13*, 598, 1964.
Guralnik, G. S., C. R. Hagen, and T. W. Kibble. *Phys. Rev. Lett. 13*, 585, 1964.
Gurevich, I. I., et al. *Phys. Lett. 11*, 185, 1964.
Higgs, P. W. *Phys. Lett. 12*, 132, 1964a.
– *Phys. Lett. 13*, 508, 1964b.
Lee, B. W. *Phys. Rev. Lett. 12*, 83, 1964.
Maier, E. J., R. M. Edelstein, and R. T. Siegal. *Phys. Rev. B133*, 663, 1964.
Michel, F. C. *Phys. Rev. B133*, 329, 1964.
Salam, A. and J. C. Ward. *Phys. Lett. 13*, 168, 1964.
Sugawara, H. *Prog. Theor. Phys. 31*, 213, 1964.

References

Wolfenstein, L. *Phys. Rev. Lett. 13*, 562, 1964.
Wu, T. T. and C. N. Yang. *Phys. Rev. Lett. 13*, 380, 1964.

1965

Abov, Y. G., P. A. Kruptchitsky, and Y. A. Oratovskii. [*Yad. Fiz. 1*, 479, 1965]. *Sov. J. Nucl. Phys. 1*, 341, 1965.
Adler, S. L. *Phys. Rev. B140*, 736, 1965.
Bardon, M., P. Norton, J. Peoples, A. M. Sachs, and J. Lee-Franzini. *Phys. Rev. Lett. 14*, 449, 1965. (see also J. Peoples, Columbia University Nevis Cyclotron Report No. 147, 1966).
Cabibbo, N. and A. Maksymovicz. *Phys. Rev. B137*, 438, 1965.
Calaprice, F. P., E. D. Commins, and D. A. Dobson. *Phys. Rev. B137*, 1453, 1965.
Feynman, R. P. and A. R. Hibbs. *Quantum Mechanics and Path Integrals*. McGraw-Hill, New York, 1965.
Fitch, V. L., R. F. Roth, J. S. Russ, and W. Vernon. *Phys. Rev. Lett. 15*, 73, 1965.
Foldy, L. L. and J. D. Walecka. *Phys. Rev. B140*, 1339, 1965.
Kim, C. W., and H. Primakoff. *Phys. Rev. B139*, 1447, 1965a.
– *Phys. Rev. B140*, 566, 1965b.
Kuzmin, V. A. *Phys. Lett. 17*, 27, 1965.
Lee, T. D. *Phys. Rev. B139*, 1415, 1965.
Pomanski, A. A. "On the Possibility of Utilizing ^{71}Ga as a Detector of Solar Neutrinos." Report of the Lebedev Physics Institute, Moscow, USSR, 1965.
Rosen, S. P. and H. Primakoff. In Alpha, Beta, and Gamma Ray Spectroscopy, ed. K. Siegbahn. North Holland, Amsterdam, 1965.
Ruderman, M. *Rep. Prog. Phys. 28*, 411, 1965.
Tsuruta, S. and A. G. W. Cameron. *Can. J. Phys. 43*, 2056, 1965.

1966

Dobrzynski, L., N.-H. Xuong, L. Montanet, M. Tomas, J. Duboc, and R. A. Donald. *Phys. Lett. 22*, 105, 1966.
Higgs, P. W. *Phys. Rev. 145*, 1156, 1966.
Konopinski, E. J. *The Theory of Beta Radioactivity*. Oxford University Press, 1966.
Kuzmin, V. A. and G. T. Zatsepin. *Proceedings of the 9th International Conference on Cosmic Rays, London,* Vol. 2, p. 1023. The Institute of Physics and the Physical Society, London, 1966.
Schopper, H. F. *Weak Interactions and Nuclear Beta Decay*. North Holland, Amsterdam, 1966.
Takaoka, N. and K. Ogata. *Z. Naturforsch. 21a*, 84, 1966.
Weisberger, W. I. *Phys. Rev. 143*, 1302, 1966.
Wu, C. S. and S. A. Moszkowski. *Beta Decay,* Interscience, New York, 1966.

1967

Beaudet, G., V. Petrosian, and E. Salpeter. *Astrophys. J. 150*, 979, 1967.
Callan, C. G., and S. B. Treiman. *Phys. Rev. 162*, 1494, 1967.
de Witt, B. *Phys. Rev. 162*, 1195, 1967.
Fadeev, L. D. and V. N. Popov. *Phys. Lett. B25*, 29, 1967.
Kibble, T. W. *Phys. Rev. 155*, 1554, 1967.
Kistner, O. C. *Phys. Rev. Lett. 19*, 872, 1967.
Matthews, P. I. In *High-Energy Physics,* Vol. I., p. 471–8, ed. E. H. S. Burhop. Academic Press, New York, 1967.
Quaranta, A. A., et al. *Phys. Lett. B25*, 429, 1967.
Sakurai, J. J. *Advanced Quantum Mechanics*. Addison-Wesley, Reading, Mass., 1967.

Schwartz, D. M. *Phys. Rev. 162*, 1306, 1967.
Sherwood, B. *Phys. Rev. 156*, 1475, 1967.
Van Royen, R. and V. F. Weisskopf. *Nuovo Cimento 50*, 617, 1967a.
− *Nuovo Cimento 51*, 583, 1967b.
Weinberg, S. *Phys. Rev. Lett. 19*, 1264, 1967.

1968
Atac, M., B. Chrisman, P. Debrunner, and H. Frauenfelder. *Phys. Rev. Lett. 20*, 691, 1968.
Clayton, D. R. *Principles of Stellar Evolution and Nucleosynthesis*. McGraw-Hill, New York, 1968.
De Pommier, P., et al. *Nucl. Phys. B4*, 189, 1968.
Fryberger, D. *Phys. Rev. 166*, 1379, 1968.
Kirsten, T., O. A. Schaeffer, E. Norton, and R. W. Stoner. *Phys. Rev. Lett. 20*, 1300, 1968.
Mandelstam, S. *Phys. Rev. 175*, 1580, 1968a.
− *Phys. Rev. 175*, 1605, 1968b.
Pontecorvo, B. *Sov. Phys.-JETP 26*, 984, 1968 [*Zh. Eksp. Teor. Fiz. 53*, 1717, 1967].
Salam, A. Nobel Symposium, No. 8, p. 367, ed. N. Svartholm. Almquist and Wiksell, Stockholm, 1968.
Veltman, M. *Nucl. Phys. B7*, 637, 1968.
Weisskopf, M. C., J. P. Carrico, H. Gould, E. Lipworth, and T. S. Stein *Phys. Rev. Lett. 21*, 1645, 1968.

1969
Adler, S. *Phys. Rev. 177*, 2426, 1969.
Bahcall, J. N. and R. N. May. *Astrophys. J. 155*, 501, 1969.
Behrens, H. and J. Jänecke. In *Numerical Tables for Beta Decay and Electron Capture Landolt-Börnstein, New Series Group 1*, Vol. 4, ed.-in-chief K.-H. Hellwege, ed. H. Schopper. Springer, New York, 1969.
Bell, J. S. and R. Jackiw. *Nuovo Cimento 60*, 47, 1969.
Bjorken, J. D. and E. A. Paschos. *Phys. Rev. 185*, 1975, 1969.
Christensen, C. J., V. E. Krohn, and G. R. Ringo. *Phys. Lett. B28*, 411, 1969.
Derenzo, S. *Phys. Rev. 181*, 1854, 1969.
Harrison, G. E., P. G. H. Sandars, and S. J. Wright. *Phys. Rev. Lett. 22*, 1263, 1969.
Primakoff, H. and S. P. Rosen. *Phys. Rev. 184*, 1925, 1969.
Wilson, K. G. *Phys. Rev. 179*, 1499, 1969.

1970
Bardin, R. K., P. J. Gollon, J. D. Ullman, and C. S. Wu. *Nucl. Phys. A158*, 337, 1970.
Callan, C. G. *Phys. Rev. D2*, 1541, 1970.
Christensen, C. J., V. E. Krohn, and G. R. Ringo. *Phys. Rev. C1*, 1693, 1970.
Erozolimskii, B. G., L. N. Bondarenko, Y. A. Mostovoy, B. A. Obinyakov, V. A. Titov, V. P. Zacharova, and A. I. Frank. *Phys. Lett. B33*, 351, 1970.
Fearing, H. W., E. Fischbach, and J. Smith. *Phys. Rev. D2*, 542, 1970.
Glashow, S., J. Iliopoulos, and L. Maiani. *Phys. Rev. D2*, 1285, 1970.
Jenschke, P. and P. Bock. *Phys. Lett. B31*, 65, 1970.
Paul, H. *Nucl. Phys. A154*, 160, 1970.
Rock, S., et al. *Phys. Rev. Lett. 24*, 748, 1970.
Schubert, K. R., B. Wolff, J. C. Chollet, J.-M. Gaillard, M. R. Jane, T. J. Ratcliffe, and J.-P. Repellin. *Phys. Rev. Lett. B-31*, 662, 1970.

References

Symanzik, K. *Commun. Math. Phys. 18*, 227, 1970.
Veltman, M. *Nucl. Phys. B21*, 288, 1970.
Wilkinson, D. H., and D. E. Alburger. *Phys. Rev. Lett. 24*, 1134, 1970.

1971

Adler, S. *Lectures on Elementary Particles and Quantum Field Theory, Brandeis 1970*, p. 1, eds. S. Deser, M. Grisaru, and H. Pendleton, MIT Press, Cambridge, Mass., 1971.
Frampton, P. H. and W.-K. Tung. *Phys. Rev. D3*, 1114, 1971.
Krishnaswamy, M. R., et al. *Proc. R. Soc. London 323*, 489, 1971.
Lipson, E. D., F. Boehm, and J. C. Vanderleeden. *Phys. Lett. B35*, 307, 1971.
Pais, A. *Ann. Phys. (N.Y.) 63*, 361, 1971.
Reines, F., et al. *Phys. Rev. D4*, 80, 1971.
Thacker, H. and J. J. Sakurai. Phys. Lett. B36, 103, 1971.
't Hooft, G. *Nucl. Phys. B33*, 173, 1971a.
– *Nucl. Phys. B35*, 167, 1971b.
Tsai, Y. S. Phys. Rev. D4, 2821, 1971.

1972

Alberi, J. L., I. Schroder, and R. Wilson. *Phys. Rev. Lett. 29*, 518, 1972.
Bardeen, W. A., R. Gastmans, and B. Lautrup. *Nucl. Phys. B46*, 319, 1972.
Bergkvist, K. E. *Nucl. Phys. B39*, 317, 1972a.
– *Nucl. Phys. B39*, 371, 1972b.
Bouchiat, C., J. Iliopoulos, and P. Meyer. *Phys. Lett. B38*, 519, 1972.
Chounet, L. M., J. M. Gaillard, and M. K. Gaillard. *Phys. Rep. 4*, 201, 1972.
Cowsik, R. and M. McClelland. *Phys. Rev. Lett. 29*, 669, 1972.
Devanathan, V., R. Parthasardthy, and P. R. Subramanian. *Ann. Phys. (N.Y.) 73*, 291, 1972.
Dicus, D. *Phys. Rev. D6*, 941, 1972.
Feynman, R. P. *Parton–Hadron Interactions*, Benjamin, New York, 1972.
Fujikawa, K., B. W. Lee, and A. I. Sanda. *Phys. Rev. D6*, 2923, 1972.
Krane, K. S., C. E. Olsen, and W. A. Steyert *Phys. Rev. C5*, 1663, 1972.
Lee, B. W. *Phys. Rev. D5*, 823, 1972.
Lee, B. W. and R. Zinn-Justin. *Phys. Rev. D5*, 3121, 1972a.
– *Phys. Rev. D5*, 3137, 1972b.
– *Phys. Rev. D5*, 3155, 1972c.
Lobashov, V. M., et al. *Nucl. Phys. A197*, 241, 1972a.
Lobashov, V. M., V. A. Nazarenko, N. A. Bozovoi, L. M. Smotritskii, G. I. Kharkevich, and V. A. Knyaz'kov. *Soc. J. Nucl. Phys. 15*, 632, 1972b [*Yad. Fiz. 15*, 1142, 1972].
Primakoff, H. and S. P. Rosen. *Phys. Rev. D5*, 1784, 1972.
Rajasekaren, G. *Phys. Rev. D6*, 3032, 1972.
't Hooft, G. and M. Veltman. *Nucl. Phys. B44*, 189, 1972a.
– *Nucl. Phys. B50*, 318, 1972b.
Weinberg, S. *Phys. Rev. 5*, 1412, 1972.

1973

Abers, E. S. and B. W. Lee. *Phys. Rep. 9C(1)*, 1, 1973.
Bergeson, H. E., G. L. Cassiday, and M. B. Hendricks. *Phys. Rev. Lett. 31*, 66, 1973.
Coleman, S. and E. Weinberg. *Phys. Rev. D7*, 1888, 1973.
Cornwall, J. M., D. N. Levin, and G. Tiktopoulos. *Phys. Rev. Lett. 30*, 1268, 1973.

Eichten, T., et al. *Phys. Lett. B46*, 274, 1973.
Fiorini, E., A. Pullia, G. Bertolini, F. Capellani, and G. Restelli. *Nuovo Cimento A13*, 747, 1973.
Gari, M. *Phys. Rep. 6C*, 319, 1973.
Kobayashi, M. and T. Maskawa. *Prog. Theor. Phys. Japan 49*, 652, 1973.
Langacker, P. and H. Pagels. *Phys. Rev. Lett. 30*, 630, 1973.
Lee, T. D. *Phys. Rev. D8*, 1226, 1973.
Llewellyn-Smith, C. *Phys. Lett. B46*, 233, 1973.
Mann, W. A., et al. *Phys. Rev. Lett. 31*, 844, 1973.
Morita, M. *Beta Decay and Muon Capture*, Benjamin, New York, 1973.
Paschos, E. A., and L. Wolfenstein. *Phys. Rev. D7*, 91, 1973.
Smith, D., C. Picciotto, and D. Bryman. *Phys. Lett. B46*, 157, 1973.
Srinivasan, B. et al. *Econ. Geol. 68*, 252, 1973.
Yokoo, Y., S. Suzuki, and M. Morita. *Prog. Theor. Phys. Japan. 50*, 1894, 1973.

1974
Altarelli, G. *Riv. Nuovo Cimento 4*, 335, 1974.
Altarelli, G. and L. Maiani. *Phys. Lett. B52*, 351, 1974.
Bouchiat, M. A. and C. Bouchiat. *Phys. Lett. B48*, 111, 1974a.
– *J. Phys. (Paris) 35*, 899, 1974b.
Calaprice, F. P., E. D. Commins, and D. C. Girvin. *Phys. Rev. D9*, 519, 1974.
Erozolimskii, B. G., Y. A. Mostovoy, V. I. Fedunin, and A. I. Frank. *JETP Lett. 20*, 345, 1974 [*Pis'ma Zh. Eksp. Teor. Fiz. 20*, 745, 1974].
Gaillard, M. K. and B. W. Lee. *Phys. Rev. Lett. 33*, 108, 1974a.
– *Phys. Rev. D10*, 897, 1974b.
Gari, M. and J. H. Reid. *Phys. Lett. B53*, 237, 1974.
Georgi, H. and S. L. Glashow. *Phys. Rev. Lett. 33*, 438, 1974.
Geweniger, C., et al. *Phys. Lett. B52*, 108, 1974.
Gjesdal, S., et al. *Phys. Lett. B52*, 113, 1974.
Gourdin, H. *Phys. Rep. 1C*, 30, 1974.
Holstein, B. R. *Rev. Mod. Phys. 46*, 789, 1974.
Jane, M. R. et al. *Phys. Lett. B48*, 260, 1974.
Khriplovich, I. B. *Pis'ma Zh. Eksp. Teor. Fiz. 20*, 686, 1974 [*JETP Lett. 20*, 315, 1974].
Kuphal, E., P. Dewes, and E. Kankeleit. *Nucl. Phys. A234*, 308, 1974.
Neubeck, K., H. Schober, and H. Waffler. *Phys. Rev. C10*, 370, 1974.
Pati, J. C. and A Salam. *Phys. Rev. D10*, 275, 1974.
Politzer, H. D. *Phys. Rep. 14C*, 129, 1974.
Salam, A. and J. Strathdee. *Nature 252*, 569, 1974.
Sirlin, A. *Nucl. Phys. B71*, 29, 1974.
Wilson, J. R. *Phys. Rev. Lett. 32*, 849, 1974.

1975
Adelberger, E. G., H. E. Swanson, M. Cooper, J. W. Tape, and T. A. Trainor. *Phys. Rev. Lett. 34*, 402, 1975.
Bludman, S. A. *Ann. N.Y. Acad. Sci. 262*, 181, 1975.
Bouchiat, M. A. and C. Bouchiat. *J. Phys. (Paris) 36*, 493, 1975.
Braun, H., et al. *Nucl. Phys. B89*, 210, 1975.
Bryman, D. and C. Picciotto. *Phys. Rev. D11*, 1337, 1975.
Calaprice, F. P., S. J. Freedman, W. C. Mead, and H. C. Vantine. *Phys. Rev. Lett. 35*, 1566, 1975.
Cleveland, B. T. *Phys. Rev. Lett. 35*, 737, 1975.

de Rujula, A., H. Georgi, and S. L. Glashow. *Phys. Rev. Lett. 35,* 69, 1975.
Ellis, J., M. K. Gaillard, and D. Nanopoulos. *Nucl. Phys. B100,* 313, 1975.
Gaillard, M. K., B. W. Lee, and J. L. Rosner. *Rev. Mod. Phys. 47,* 277, 1975.
Gari, M., J. B. McGrory, and R. Offermann. *Phys. Lett. B55,* 277, 1975.
Hennecke, E. W., O. K. Manuel, and D. D. Sabu. *Phys. Rev. C11,* 1378, 1975.
Krohn, V. E. and G. R. Ringo. *Phys. Lett. B55,* 175, 1975.
Lichtenberg, D. B. *Lett. Nuovo Cimento 13,* 346, 1975.
Mohapatra, R. N., J. Pati, and L. Wolfenstein. *Phys. Rev. D11,* 3319, 1975.
Pais, A. and S. B. Treiman. *Phys. Rev. D12,* 2744, 1975.
Perl, M. L., et al. *Phys. Rev. Lett. 35,* 1489, 1975.
Raman, S., T. A. Walkiewicz, and H. Behrens. *At. Data Nucl. Data Tables 16,* 451, 1975.
Salam, A. and J. Strathdee. *Nucl. Phys. B90,* 203, 1975.
Sandars, P. G. H. *Atomic Physics,* Vol. 4, p. 27, Plenum, New York, 1975.
Soreide, D. C. and E. N. Fortson. *Bull. Am. Phys. Soc. 20,* 491, 1975.
Sugimoto, K., I. Tanihata, and J. Göring. *Phys. Rev. Lett. 34,* 1533, 1975.
Wilczek, F., A. Zee, R. L. Kingsley, and S. B. Treiman. *Phys. Rev. D12,* 2768, 1975.

1976
Bertrand-Coremans, G., et al. *Phys. Lett. B61,* 207, 1976.
Bjorken, J. D. Proceedings of the Summer Institute on Particle Physics. Stanford Linear Accelerator Center, Stanford University, Stanford, Calif. 1976.
Blietschau, J., et al. *Nucl. Phys. B114,* 189, 1976.
Box, M. A., A. J. Gubric, and B. H. J. McKellar. *Nucl. Phys. A271,* 412, 1976.
Calaprice, F. P. and B. Holstein. *Nucl. Phys. A273,* 301, 1976.
Danilyan, G. V., V. V. Novitskii, V. S. Pavlov, S. P. Borovlev, B. D. Vodennikov, and V. P. Dronyaev. *JETP Lett. 24,* 344, 1976 [*Pis'ma Zh. Eksp. Teor. Fiz. 24,* 380, 1976].
Ellis, J., M. K. Gaillard, and D. V. Nanopoulos. *Nucl. Phys. B106,* 292, 1976a.
– *Nucl. Phys. B109,* 213, 1976b.
Erozolimskii, B. G., A. I. Frank, Y. A. Mostovoy and S. S. Arzumanov. *JETP Lett. 23,* 663, 1976 [*Pis'ma Zh. Eksp. Teor. Fiz. 23,* 720, 1976].
Fritzsch, H. and P. Minkowski. *Nucl. Phys. B103,* 61, 1976.
Halprin, A., B. W. Lee, and P. Sorba. *Phys. Rev. D14,* 2343, 1976.
Kingsley, R. L., F. Wilczek, and A. Zee. *Phys. Lett. B61,* 259, 1976.
Kleinknecht, K. *Annu. Rev. Nucl. Sci. 26,* 1, 1976.
Koks, F. W. J. and J. Van Klinken. *Nucl. Phys. A272,* 61, 1976.
Krishnaswamy, M. R., M. G. K. Menon, V. S. Narasimham, N. Ito, S. Kawakami, and S. Miyake. *Proceedings of the International Neutrino Conference, Aachen,* Eds. H. Faissner, H. Reithler, and P. Zerwas, p. 197 Vieweg und Sohn, Braunschweig, 1976.
Langacker, P. *Phys. Rev. D14,* 2340, 1976.
Lassey, K. R. and B. H. J. McKellar. *Nucl. Phys. A260,* 413, 1976.
Linde, A. *JETP Lett. 23,* 64, 1976 [*Pis'ma Zh. Eksp. Teor. Fiz. 23,* 73, 1976].
Morita, M., M. Nishimura, A. Shimizu, H. Ohtsubo, and K. Kubodera. *Prog. Theor. Phys. (Japan), Suppl. 60,* 1, 1976.
Reines, F., H. Gurr, and H. Sobel. *Phys. Rev. Lett. 37,* 315, 1976.
Sikivie, P. *Phys. Lett. B65,* 141, 1976.
Steinberg, R. I., P. Liand, B. Vignon, and V. W. Hughes. *Phys. Rev. D13,* 2469, 1976.
Weinberg, S. *Phys. Rev. Lett. 36,* 294, 1976a.
– *Phys. Rev. Lett. 37,* 657, 1976b.

1977

Alles, W., Ch. Boyer, and A. J. Buras. *Nucl. Phys. B119*, 125, 1977.
Bailey, J., et al. *Phys. Lett. B68*, 191, 1977.
Baird, P. E. G., M. W. S. M. Brimicomb, R. G. Hunt, G. J. Roberts, P. G. H. Sandars, and D. N. Stacey. *Phys. Rev. Lett. 39*, 798, 1977.
Baltrusaitis, R. M. and F. P. Calaprice. *Phys. Rev. Lett. 38*, 464, 1977.
Bambynek, W., et al. *Rev. Mod. Phys. 49*, 77, 1977.
Barish, B., et al. *Phys. Rev. Lett. 39*, 1595, 1977.
Bég, M. A., R. V. Budny, R. Mohapatra, and A. Sirlin. *Phys. Rev. Lett. 38*, 1252, 1977.
Benkoula, H., J. C. Cavignac, J. L. Charvet, D. H. Koang, B. Vignon, and R. Wilson. *Phys. Lett. B71*, 287, 1977.
Casperson, D. E., et al. *Phys. Rev. Lett. 38*, 956, 1977.
Cavignac, J. F., B. Vignon, and R. Wilson. *Phys. Lett. B67*, 148, 1977.
Chinowsky, W. *Annu. Rev. Nucl. Sci. 27*, 393, 1977.
Field, R. D. and R. P. Feynman. *Phys. Rev. D15*, 2590, 1977.
Freedman, D. Z., D. N. Schramm, and D. L. Tubbs. *Annu. Rev. Nucl. Sci. 27*, 167, 1977.
Gerber, G., D. Newman, A. Rich, and E. Sweetman. *Phys. Rev. D15*, 1189, 1977.
Goldhaber, G. Presented at the Conference on Weak Interaction Physics, Bloomington, Ind., 1977 (unpublished).
Harris, F. A., et al. *Phys. Rev. Lett. 39*, 437, 1977.
Herb, S. W., et al. *Phys. Rev. Lett. 39*, 252, 1977.
Holder, M. et al. *Phys. Lett. B69*, 377, 1977.
Holstein, B. R. and S. B. Treiman. *Phys. Rev. D16*, 2369, 1977.
Hung, P. Q. *Phys. Lett B69*, 216, 1977.
Hung, P. Q. and J. J. Sakurai. *Phys. Lett. B72*, 208, 1977.
Hwang, W.-Y. P., and H. Primakoff. *Phys. Rev. C16*, 397, 1977.
Kaina, W., V. Soergel, H. Thies, and W. Trost. *Phys. Lett. B70*, 411, 1977.
Kluttig, H., J. G. Morbin, and W. Van Doninck. *Phys. Lett. B71*, 446, 1977.
Kubodera, K. and A. Arima. *Prog. Theor. Phys. Japan 57*, 1599, 1977.
Langacker, P. *Phys. Rev. D15*, 2386, 1977.
Lee, B. W., C. Quigg, and H. Thacker. *Phys. Rev. Lett. 38*, 883, 1977.
Lee, B. W., and S. Weinberg. *Phys. Rev. Lett. 39*, 165, 1977.
Marriner, J., et al. Lawrence Berkeley Lab Report No. LBL-6438, 1977.
Mohapatra, R. N. and D. P. Sidhu. *Phys. Rev. Lett. 38*, 667, 1977.
Neuffer, D. V. and E. D. Commins. *Phys. Rev. A16*, 844, 1977.
Possoz, A., et al. *Phys. Lett. B70*, 265, 1977.
Schramm, D. N. and R. V. Wagoner. *Annu. Rev. Nucl. Sci. 27*, 37, 1977.
Shifman, M. A., A. I. Vainshtein, and V. I. Zakharov. *Nucl. Phys. B120*, 316, 1977.
Towner, I. S., J. C. Hardy, and M. Harvey. *Nucl. Phys. A284*, 269, 1977.
Van Dyck, R., P. Schwinberg, and H. Dehmelt. *Phys. Rev. Lett. 38*, 310, 1977.
Veltman, M. *Phys. Lett. B70*, 253, 1977.
Weinberg, S. *The First Three Minutes,* Basic Books, New York, 1977a.
– *Trans. N.Y. Acad. Sci., Ser 2, 38*, 185, 1977b.
Wilczek, F. *Phys. Rev. Lett. 39*, 1309, 1977.
Wilkinson, D. H. *Phys. Lett. B67*, 13, 1977.
Wu, C. S., Y. K. Lee, and L. W. Mo. *Phys. Rev. Lett. 39*, 72, 1977.

1978

Anselm, A. A. and D. I. Dyakonov. *Nucl. Phys. B145*, 271, 1978.

References

Appelquist, T., R. M. Barnett, and K. Lane. *Annu. Rev. Nucl. Part. Sci.* 28, 387, 1978.
Asratyan, A. E., et al. *Phys. Lett.* B76, 239, 1978.
Bahcall, J. N. *Rev. Mod. Phys.* 50, 881, 1978.
Barnes, C. A., et al. In *Unification of Elementary Forces and Gauge Theories*, p. 235, eds. D. Cline and F. Mills. Harwood, London, 1978.
Bilenky, S. M. and B. Pontecorvo. *Phys. Rep.* C41, 225, 1978.
Blietschau, J., et al. *Nucl. Phys.* B133, 205, 1978.
Brändl, H., et al. *Phys. Rev. Lett.* 40, 306, 1978a.
– *Phys. Rev. Lett.* 41, 299, 1978b.
Bryman, D. and C. Picciotto. *Rev. Mod. Phys.* 50, 11, 1978.
Cabibbo, N. and L. Maiani. *Phys. Lett.* B73, 418, 1978a.
– *Phys. Lett.* B76, 663, 1978b.
– *Phys. Lett.* B79, 109, 1978c.
Cahn, R. N. and F. J. Gilman. *Phys. Rev.* D17, 1313, 1978.
Camani, M. et al. *Phys. Lett.* B77, 326, 1978.
Chen, H. H. Presented at the Informal Conference on Status and Future of Solar Neutrino Research, Brookhaven National Laboratory, Upton, N.Y., 1978 (unpublished).
Davis, R., J. C. Evans, and B. Cleveland. Presented at Neutrino – 1978, Conference Proceedings, Purdue University, Lafayette, Ind. 1978 (unpublished).
Deakyne, M., et al. Presented at Neutrino – 1978, Conference Proceedings, Purdue University, Lafayette, Indiana, 1978 (unpublished).
Derrick, M., et al. *Phys. Rev.* D18, 7, 1978.
Deshpande, N. G. and E. Ma. *Phys. Rev.* D18, 2574, 1978.
Dicus, D. A., E. N. Kolb, and V. Teplitz. *Astrophys, J.* 221, 327, 1978a.
Dicus, D. A., E. N. Kolb, V. Teplitz, and R. V. Wagoner. *Phys. Rev.* D17, 1529, 1978b.
Dunford, R. W., R. R. Lewis, and W. L. Williams. *Phys. Rev.* A18, 2421, 1978.
Faissner, H., et al. *Phys. Rev. Lett.* 41, 213, 1978.
Fakirov, D. and B. Stech. *Nucl. Phys.* B133, 315, 1978.
Finjord, J. *Phys. Lett.* B76, 116, 1978.
Gaemers, K. and G. Gounaris. *Phys. Lett.* B77, 379, 1978.
Gilman, F. J. and D. H. Miller. *Phys. Rev.* D17, 1846, 1978.
Greenberg, O. W. *Annu. Rev. Nucl. Part. Sci.* 28, 327, 1978.
Gunn, J. E., B. W. Lee, I. Lerche, D. N. Schramm, and G. Steigman. *Astrophys. J.* 223, 1015, 1978.
Hagberg, E., J. C. Hardy, B. Jonson, S. Mattson, and P. Tidemand-Petersson. *Nucl. Phys.* A313, 276, 1978.
Heller, K., et al. *Phys. Rev. Lett.* 41, 607, 1978.
Henley, E. M. *J. Phys. Soc. Japan Suppl.* 44, 812, 1978.
Hung, P. Q. *Phys. Rev.* D17, 1893, 1978.
Kawamota, N. and A. Sanda. *Phys. Lett.* B76, 446, 1978.
Krenz, W. et al. *Nucl. Phys.* B135, 45, 1978.
Lebrun, P., et al. *Phys. Rev. Lett.* 40, 302, 1978.
Marciano, W. and H. Pagels. *Phys. Rep.* 36, 137, 1978.
Merritt, F. S., et al. *Phys. Rev.* D17, 2199, 1978.
Pfeiffer, L., A. P. Mills, R. S. Raghavan, and E. A. Chandross. *Phys. Rev. Lett.* 41, 63, 1978.
Pham, T. N., C. Roiesnel, and T. N. Truong. *Phys. Lett.* B78, 623, 1978.
Prescott, C. Y., et al. *Phys. Lett.* B77, 347, 1978.
Reines, F. In *Proceedings of the Ben Lee Conference on Parity Nonconservation,*

Weak Neutral Currents, and Gauge Theories, Batavia, Illinois, 1977, p. 103. eds. D. Cline and F Mills. Harwood, London, 1978.
Roberts, A. Presented at the DUMAND Summer Workshop, 1978, Scripps Institute of Oceanography, San Diego 1978 (unpublished).
Rolfs, C. and H. Trautvetter. *Annu. Rev. Nucl. Part. Sci. 28,* 115, 1978.
Shrock, R. E. and L. L. Wang. *Phys. Rev. Lett. 41,* 1692, 1978.
Sirlin, A. *Rev. Mod. Phys. 50,* 573, 1978.
Stratowa, C., R. Dobrozemsky, and P. Weinzierl. *Phys. Rev. D18,* 3970, 1978.
Towner, I. S. and J. C. Hardy. *Phys. Lett. B73,* 20, 1978.
Tsai, Y. S. Stanford Linear Accelerator Center SLAC-PUB-2405, 1978.
Vainshtein, A. I., V. I. Zakharov, and M. A. Shifman. *Sov. Phys.-JETP 45,* 670, 1978.
Wanderer, P., et al. *Phys. Rev. D17,* 1679, 1978.
Wolfenstein, L. *Phys. Rev. D17,* 2369, 1978.

1979
Abrams, G. S. et al. *Phys. Rev. Lett. 43,* 1555, 1979.
Bacino, W., et al. *Phys. Rev. Lett. 42,* 749, 1979.
Bailey, J. et al. *Nucl. Phys. B150,* 1, 1979.
Barish, B., et al. Cal. Tech. Preprint No. CALT 68-734, 1979a.
Barish, S. J. et al. *Phys. Rev. D19,* 2521, 1979b
Barkov, L. M. and M. S. Zolotorev Phys. Lett. B85, 308, 1979.
Bell, J., et al. *Phys. Rev. D19,* 1, 1979.
Benvenuti, A., et al. *Phys. Rev. Lett. 42,* 1317, 1979.
Bleitschau, J., et al. *Phys. Lett. B88,* 381, 1979.
Bowman, J. D., et al. *Phys. Rev. Lett. 42,* 556, 1979.
Chanowitz, M., M. Furman, and I. Hinchliffe. *Nucl. Phys. B159,* 225, 1979.
Christenson, J. H., J. H. Goldman, E. Hummel, S. D. Roth, T. W. L. Sanford, and J. Sculli. *Phys. Rev. Lett. 43,* 1209, 1979a.
– *Phys. Rev. Lett. 43,* 1212, 1979b.
Ciampolillo, S., et al. *Phys. Lett. B84,* 281, 1979.
Colley, D. C., et al. *Z. Phys. C2,* 187, 1979.
Conti, R., P. Bucksbaum, S. Chu, E. Commins, and L. Hunter. *Phys. Rev. Lett. 42,* 343, 1979.
Daum, M., G. H. Eaton, R. Frosch, H. Hirschmann, J. McCulloch, R. C. Minehart, and E. Steiner. *Phys. Rev. D20,* 2692, 1979.
Davis, B. R., P. Vogel, F. M. Mann, and K. E. Schenter. *Phys. Rev. C19,* 2259, 1979.
Deden, H., et al. *Nucl. Phys. B149,* 1, 1979.
de Groot, J. G. H., et al. *Z. Phys. C1,* 143, 1979.
Donoghue, J. and L. Li. *Phys. Rev. D19,* 945, 1979.
Ellis, J. and M. K. Gaillard *Nucl. Phys. B150,* 141, 1979.
Entenberg, A., et al. *Phys. Rev. Lett. 42,* 1198, 1979.
Faber, S. M. and J. S. Gallagher. *Annu. Rev. Astron. Astrophys. 17,* 135, 1979.
Farley, F. J. M. and E. Picasso. *Annu. Rev. Nucl. Part. Sci. 29,* 243, 1979.
Geweniger, C. In *Proceedings of the International Conference on Neutrino Physics and Astrophysics, Bergen, Norway, 1979,* Vol. 2, p. 392, eds. A. Haatuft and C. Jarlskog. University of Bergen, Bergen, 1979.
Gilman, F. and M. Wise. *Phys. Lett. B83,* 83, 1979.
Harari, H. *Phys. Lett. B86,* 83, 1979.
Hwang, W.-Y. P. *Phys. Rev. C20,* 805, 1979a.
– *Phys. Rev. C20,* 814, 1979b.

Kirkby, J. In *Proceedings of the 1979 International Symposium Lepton and Photon Interactions at High Energies*, eds. T. B. W. Kirk and H. D. I. Abarbanel., p. 107. FNAL, Batavia, Ill., 1979.

Kirsten, T. In *Proceedings of the International Conference on Neutrino Physics and Astrophysics, Bergen, Norway, 1979*, Vol. 2, p. 452, eds. A. Haatuft and C. Jarlskog, University of Bergen, 1979.

Lach, J. and L. Pondrom. *Annu. Rev. Nucl. Part. Sci. 29,* 203, 1979.

Lande, K. *Ann. Rev. Nucl. Part. Sci. 29,* 395, 1979.

Lubatti, H. J. In *Proceedings of the International Conference on Neutrino Physics and Astrophysics, Bergen, Norway, 1979,* Vol. 2, p. 543, 1979, eds. A. Haatuft and C. Jarlskog, University of Bergen, 1979.

Masuda, Y., T. Minamisono, Y. Nojiri, and K. Sugimoto. *Phys. Rev. Lett. 43,* 1083, 1979.

Mazurek, T. J. In *Proceedings of the International Conference on Neutrino Physics and Astrophysics, Bergen, Norway, 1979,* Vol. 2, p. 438, eds. A. Haatuft and C. Jarlskog, University of Bergen, 1979.

Mess, K. In *Proceedings of the International Conference on Neutrino Physics and Astrophysics, Bergen, Norway, 1979,* Vol. 2 p. 371, eds. A. Haatuft and C. Jarlskog, University of Bergen, 1979.

Pasierb, E., H. S. Gurr, J. Lathrop, F. Reines, and H. W. Sobel. *Phys. Rev. Lett. 43,* 96, 1979.

Peterman, A. *Phys. Rep. 53,* 157, 1979.

Pethick, C. J. In *Proceedings of the International Conference on Neutrino Physics and Astrophysics, Bergen, Norway, 1979,* Vol. 2, p. 78, eds. A. Haatuft and C. Jarlskog, University of Bergen, 1979.

Poggio, E. C. and H. J. Schnitzer. *Phys. Rev. D20,* 1175, 1979.

Prescott, C. Y., et al. *Phys. Lett. B84,* 524, 1979.

Quigg, C. and J. L. Rosner. *Phys. Rep. 56,* 161, 1979.

Reithler, H., et al. *Phys. Bl. (Germany) 35,* 630, 1979.

Sawyer, R. F. In *Proceedings of the International Conference on Neutrino Physics and Astrophysics, Bergen, Norway, 1979,* Vol. 2, p. 429, eds. a. Haatuft and C. Jarlskog, University of Bergen, 1979.

Schramm, D. In *Proceedings of the International Conference on Neutrino Physics and Astrophysics, Bergen, Norway, 1979,* Vol. 2, eds. A. Haatuft and C. Jarlskog, Univ. of Bergen, 1979.

Shabalin, E. P. *Sov. J. Nucl. Phys. 28,* 75, 1979 [*Yad. Fiz. 28,* 151, 1978].

Shrock, R. E. and S. B. Treiman. *Phys. Rev. D19, 2148, 1979.*

Shupe, M. A. *Phys. Lett. B86,* 87, 1979.

Smith, J. In *Proceedings of the International Conference on Neutrino Physics and Astrophysics, Bergen, Norway, 1979,* Vol. 2, p. 101, eds. A. Haatuft and C. Jarlskog, Univ. of Bergen, 1979.

Steigman, G. *Annu. Rev. Nucl. Part Sci. 29,* 313, 1979.

Suzuki, M. *Nucl. Phys. B145,* 420, 1979.

Willutzki, H. J. In *Proceedings of the International Conference on Neutrino Physics and Astrophysics, Bergen, Norway, 1979,* Vol. 2, p. 92, eds. A. Haatuft and C. Jarlskog, University of Bergen, 1979.

Yang, J., D. N. Schramm, G. Steigman, and R. T. Rood, *Astrophys. J. 227,* 697, 1979.

1980

Allison, W., et al. *Phys. Lett. B93,* 509, 1980.

Bahcall, J. N., et al. *Phys. Rev. Lett. 45,* 945, 1980.

Barger, V., K. Whisnant, and R. J. N. Phillips. *Phys. Rev. D22*, 1636, 1980.
Bogdanov, Yu. V., I. I. Sobel'man, V. N. Sorokin, and I. I. Struk. *JETP Lett. 31*, 522, 1980 [*Pis'ma Zh. Eksp. Teor. Fiz. 31*, 234, 1980].
Boehm, F., et al. *Phys. Lett. B92*, 310, 1980.
Brandelik, P., et al. *Phys. Lett. B92*, 199, 1980.
Buras, A. J. *Rev. Mod. Phys. 52*, 199, 1980.
Commins, E. D. and P. H. Bucksbaum. *Annu. Rev. Nucl. Part. Sci. 30*, 1, 1980.
de Rujula, A. and S. L. Glashow. *Phys. Rev. Lett. 45*, 942, 1980.
Donoghue, J. F. and B. R. Holstein. *Phys. Rev. D21*, 1334, 1980.
Egger, J. *Nucl. Phys. A335*, 91, 1980.
Forte, M., B. R. Heckel, N. F. Ramsey, K. Green, G. L. Greene, J. Byrne, and J. Pendlebury. *Phys. Rev. Lett. 45*, 2088, 1980.
Fujikawa, K. and R. Shrock. *Phys. Rev. Lett. 45*, 963, 1980.
Glashow, S. L. *Rev. Mod. Phys. 52*, 539, 1980.
Goldhaber, G. and J. E. Wiss. *Annu. Rev. Nucl. Part Sci. 30*, 337, 1980.
Greco, M., G. Pancheri-Srivastava, and T. Srivastava. *Nucl. Phys. B171*, 118, 1980.
Guberina, B. and R. D. Peccei. *Nucl. Phys. B163*, 289, 1980.
Hagelin, J. S. and M. B. Wise. Harvard Preprint No. HUTP-80/A070, 1980.
Heisterberg, R. H., et al. *Phys. Rev. Lett. 44*, 635, 1980.
Hitlin, D. In *Proceedings of the Summer Institute on Particle Physics, SLAC, 1980*, Pub. 234, p. 67, ed. A. Mosher. Stanford University press, 1980.
Jonker, M., et al. *Phys. Lett. B93*, 203, 1980a.
– *Phys. Lett. B96*, 435, 1980b.
Kelly, R. L., et al. *Rev. Mod. Phys.* 52 (2), Part 2, 1980.
Krasemann, H. *Phys. Lett. B96*, 397, 1980.
Lubimov, V. A., E. G. Novikov, V. Z. Nozik, E. F. Tretyakov, and V. S. Kosik. *Phys. Lett. B94*, 266, 1980.
Moe, M. K. and D. D. Lowenthal. *Phys. Rev. C22*, 2186, 1980.
Nanopoulos, D. V., A. Yildiz, and P. Cox, *Ann. Phys. (N.Y.) 127*, 126, 1980.
Oka, M. and K. Kubodera. *Phys. Lett. B90*, 45, 1980.
Perl, M. *Annu. Rev. Nucl. Part Sci. 30*, 299, 1980.
Quigg, C. *Z. Phys. C4*, 55, 1980.
Reines, F., W. H. Sobel, and E. Pasierb. *Phys. Rev. Lett. 45*, 1307, 1980.
Salam, A. *Rev. Mod. Phys. 52*, 525, 1980.
Sciulli, F. In *Proceedings of Summer Institute on Particle Physics SLAC, 1980*, Pub. 234, p. 29, ed. A. Mosher, Stanford University Press, 1980.
Shabalin, E. P. *Sov. J. Nucl. Phys. 32*, 228, 1980 [*Yad. Fiz. 32*, 443, 1980].
Terazawa, H. *Phys. Rev. D22*, 184, 1980.
Weinberg, S. *Rev. Mod. Phys. 52*, 515, 1980.
Willis, S. E., et al. *Phys. Rev. Lett. 44*, 522, 1980.
Wise, J., et al. *Phys. Lett B91*, 165, 1980.
Zdesenko, Yu. G. *JETP Lett. 32*, 65, 1980 [*Pis'ma Zh. Eksp. Teor. Fiz. 32*, 62, 1980].

1981

Adler, S. L. *Phys. Rev. D23*, 2905, 1981.
Altarev, I. S., et al. *Phys. Lett. B102*, 13, 1981.
Bander, M. *Phys. Rep. 75* (4), 205, 1981.
Barber, D. P., et al. *Phys. Rev. Lett. 46*, 1663, 1981.
Bartel, W., et al. *Phys. Lett. B99*, 281, 1981.
Bebek, C., et al. *Phys. Rev. Lett. 46*, 84, 1981.

References

Bingham, H. H. In *Proceedings of the 1981 APS Conference on Particles and Fields*, University of California, Santa Cruz, 1981 (unpublished).
Bucksbaum, P. H., E. D. Commins, and L. R. Hunter. *Phys. Rev. Lett. 46*, 640, 1981a.
— *Phys. Rev. D24*, 1134, 1981b.
Campbell, M. K., et al. *Phys. Rev. Lett. 47*, 1032, 1981.
Chadwick, K., et al. *Phys. Rev. Lett. 46*, 88, 1981.
Commins, E. D., In *Atomic Physics*, Vol. 7, p. 121, eds. D. Kleppner and F. M. Pipkin. Plenum, New York, 1981.
Cox, P. T., et al. *Phys. Rev. Lett. 46*, 877, 1981.
Das, T. P. (private communication, 1981).
Deden, H., et al. *Phys. Lett. B98*, 310, 1981.
Erriquez, O., et al. *Phys. Lett. B102*, 73, 1981.
Fillipone, B. W. In *Proceedings of the 1981 International Conference on Neutrino Physics and Astrophysics Maui, Hawaii*, Vol. 1, p. 19, ed. R. Cence. University of Hawaii, Honolulu, 1981.
Fritzsch, H. and G. Mandelbaum. *Phys. Lett. B 102*, 319, 1981.
Hollister, J. H., G. R. Apperson, L. L. Lewis, T. P. Emmons, T. G. Vold, and E. N. Fortson. *Phys. Rev. Lett. 46*, 643, 1981.
Hung, P. Q. and J. J. Sakurai. *Annu. Rev. Nucl. Part. Sci. 31*, 375, 1981.
Kim, T. E., P. Langacker, M. Levine, and H. H. Williams. *Rev. Mod. Phys. 53*, 211, 1981.
La Rue, G. S., J. D. Phillips, and W. M. Fairbank. *Phys. Rev. Lett. 46*, 967, 1981.
Lattimer, J. M. *Annu. Rev. Nucl. Part. Sci. 31*, 337, 1981.
Lipkin, H. J. *Phys. Rev. D24*, 1437, 1981a.
— *Phys. Rev. Lett. 46*, 1307, 1981b.
Matsuda, M., M. Nakagawa, and S. Ogawa. *Prog. Theor. Phys. Japan 65*(1), 397, 1981.
Mueller, A. H. *Phys. Rep. 73*(4), 237, 1981.
Némethy, P. et al. *Phys. Rev. D23*, 262, 1981.
Primakoff, H. and S. P. Rosen. *Annu. Rev. Nucl. Part. Sci. 31*, 145, 1981.
Ramsey, N. F. In *Atomic Physics*, Vol. 7, eds. D. Kleppner and F. M. Pipkin. Plenum, New York, 1981.
Roberts, A. In *Proceedings of the 1981 International Conference on Neutrino Physics and Astrophysics, Maui, Hawaii*, Vol. 2, p. 240, ed. R. Cence. University of Hawaii, Honolulu, 1981.
Reya, E. *Phys. Rep. 69*(3), 195, 1981.
Sakurai, J. J. In *Proceedings of the 1981 International Conference on Neutrino Physics and Astrophysics, Maui, Hawaii*, ed. R. Cence. University of Hawaii, Honolulu, 1981.
Shizuya, K. *Phys. Lett. B105*, 406, 1981.
Silverman, D. *Phys. Rev. Lett. 46*, 467, 1981.
Simpson, J. J. *Phys. Rev. D23*, 649, 1981.
Spencer, L. J., et al. *Phys. Rev. Lett. 47*, 771, 1981.
Stenger, V. J. In *Proceedings of the 1981 International Conference on Neutrino Physics and Astrophysics, Maui, Hawaii*, Vol. 2, p. 233, ed. R. Cence. University of Hawaii, Honolulu, 1981.
Suzuki, M. *Nucl. Phys. B177*, 413, 1981.
Tadić, D. and J. Trampetić. *Phys. Rev. D23*, 144, 1981.
Trilling, G. *Phys. Rep. 75*(2), 57, 1981.
Ushida, N., et al. *Phys. Rev. Lett. 47*, 1694, 1981.

Wilkinson, C., et al. *Phys. Rev. Lett. 46,* 803, 1981.
Wise, J., et al. *Phys. Lett. B98,* 123, 1981.

1982
Adeva, B. et al. *Phys. Rev. Lett. 48,* 1701, 1982.
Beall, G., M. Bander, and A. Soni. *Phys. Rev. Lett. 48,* 848, 1982.
Blocker, C. A. et al. *Phys. Rev. Lett. 48,* 1586, 1982.
Cline, D. B., C. Rubbia, and S. van der Meer. *Sci. Am. 246*(3), 48, 1982.
Deshpande, N. G., G. Eilan, and W. L. Spence. *Phys. Lett. B108,* 42, 1982.
Feldman, G. J. et al. *Phys. Rev. Lett. 48,* 66, 1982.
Reeder, D. Fermi National Accelerator Report, May 1982, ed. F. T. Cole, R. Donaldson, and L. Voivodich, Fermilab, Batavia, Illinois.
Roesch, L. Ph., et al. *Am. J. Phys.,* in press, 1982.
Ushida, N. et al. *Phys. Rev. Lett. 48,* 844, 1982.
Vella, E., et al. *Phys. Rev. Lett. 48,* 1515, 1982.

Index

Abers, E., 67
Ademollo–Gatto theorem, 206, 220, 441–3
Adler, S., 82–3, 176, 441
Adler–Bell–Jackiw anomaly, 82–3, 421, 438
Adler–Weisberger relation, 439–41, 442
$\alpha_s(Q^2)$, effective strong coupling in QCD, 229, 314
anticharm production, 317
^{37}Ar, and solar neutrino detector, 399
Argonne Laboratory, 298
asymmetry of γ emission and parity violation in nuclear forces, 360–5
asymptotic freedom, 6, 131, 228, 229
atomic physics
  parity violation coefficients, 343, 352–4
  weak–electromagnetic interference and parity nonconservation in, 29, 35, 172–3, 343, 345–54
axial vector coupling, charged-current weak interactions; see Adler–Weisberger relation, β decay, muon decay, PCAC, V–A law
axial vector coupling, neutral weak interactions, 57, 172–3, 300–1, 320–35
axion(s) and CP violation, 283

B mesons, 31, 149, 204, 241–3
  B^0–$\bar{B}^0$ and CP violation, 272, 282, 283–7
  B meson decays, gluon radiative corrections in, 242
BEBC (Big European Bubble Chamber), 294, 322, 325, 383
baryon(s), 1, 2, 3, 139, 141, 143, 149
  charmed, 149, 204

$J^P = \frac{1}{2}^+$ octet, 141–2, 143
$J^P = \frac{3}{2}^+$ decuplet, 141–2
magnetic moments, 143, 145–7, 225
number conservation, 3, 24, 138, 370
semileptonic transitions, 13, 27, 157; see also hyperon, semileptonic decays.
Becquerel, H., 7
β decay, 1, 7–13, 18, 24, 27, 86, 152, 178–96
  $A = 12$ and CVC, second-class currents, 192–6
  allowed, energy spectrum, 391
  allowed, Fermi integral, 185
  allowed, ft values, 184–6
  allowed, selection rules, 9, 180–1
  allowed, transition probability, 182–90
  allowed, Z^0 radiative correction, 186
  amplitude, two-component reduction, 179
  and antineutrino mass measurement, 390–3
  ^{35}Ar $\to$ ^{35}Cl $e^+ \nu_e$, 189, 190
  and Cabibbo's hypothesis, 13
  circular polarization of γ rays following, 11, 187–8
  Coulomb correction factor, 181–2
  coupling constants, 9, 10, 11, 12, 191–2
  coupling constant, axial, 11, 186, 188, 190
  coupling constant, vector, 11, 13, 166, 185
  discovery of β^+ decay, 8
  double, see double β decay
  e–ν angular correlations, 11, 186–8
  e,ν asymmetries in decay of polarized nuclei, 11, 86, 188–90

464 Index

β decay (cont.)
  effects of "recoil order" in, 192–6
  forbidden transition amplitudes in, 181, 193
  Gamow–Teller matrix element, 181
  impulse approximation, 178
  inverse, 28
  meson exchanges in, 178
  mirror-pair ft values and second-class currents, 195
  $n \to pe^-\bar{\nu}_e$, 8, 11, 12, 161–2, 169–71, 178, 186, 189, 190
  $^{19}\text{Ne} \to {}^{19}\text{F}\, e^+\, \nu_e$, 189, 190
  polarization of $e^\pm$ in, 11, 182–4
  radiative corrections in Fermi theory, 18
  and second-class currents, 181, 192, 195–6
  of tritium, 374, 385, 392–3
  weak magnetism and CVC, 181, 189, 192–6
  $0^+ \to 0^+$ nuclei, 11, 185
big bang model, 397, 414
bismuth, parity violation in, 338, 349–52
Bjorken, J., 88, 304
black hole, formation of, 412
bosons, unphysical scalar, 72, 73, 76, 77, 78; see also Goldstone bosons, Higgs boson
Bouchiat, C., 346
Bouchiat, M., 346, 349
^{81}Br, solar neutrino detector, 402
bremsstrahlung, inner, 393
  in muon decay, 101
Brookhaven National Laboratory, 148, 221, 302
  Brookhaven AGS, 260, 302

C invariance; see charge conjugation, invariance
CDHS neutrino detector, 293, 319, 322
CERN, European High Energy Physics Laboratory, 302, 325, 331, 383
  CERN SPS $p\bar{p}$ collider, 121
CESR (Cornell Electron Storage Ring), 242
CHARM neutrino detector, 293, 309, 323, 383
C–N–O cycle of solar burning, 397, 398
CP invariance, 206, 211, 213, 214, 356
CP violation, 29, 163, 206, 213, 244–88, 425
  ϵ, 256, 257, 270–1
  ϵ', 259, 271
  electromagnetic interaction, 272, 274–5
  η^{+-} parameter, 259–62

η^{00} parameter, 259–62
  and extra Higgs bosons, 281
  and the Kobayashi–Maskawa model, 22, 152, 163, 204, 277–281
  in left–right symmetric models, 282
  millistrong interaction, 272–3
  milliweak interaction, 272, 275–6
  and neutrino oscillations, 375
  summary of experimental data, K^0–$\bar{K}^0$ system, 272
  and superweak interaction, 272, 276–7
CPT invariance, 15, 102, 131–2, 163–5, 205, 250, 255, 257, 258, 259, 269–70
  test in muon g − 2 experiment, 102
CVC (conserved vector current) hypothesis, 27, 58, 132, 165–8, 169, 181, 189, 192–6, 200, 206, 296, 299, 308
Cabibbo, N., 13
  Cabibbo angle(s), 13, 20–2, 150, 151, 152, 185, 189, 206, 322, 370
  hypothesis 13, 16, 21, 27, 148, 206, 218–23, 234, 300
  factor, rotation(s), 30, 35, 62, 85, 156, 158, 285, 305, 356, 421
  rotation, neutrino, 373
Callan–Gross relation, 310
Callan–Symanzik equations, 229, 230
cascade (Ξ) hyperons, 145
Čerenkov radiator, 419, 413
cesium, parity violation in, 338, 347–9, 350
Chadwick, J., 7
Chandrasekhar limit, 406
charge asymmetry in $e^+e^- \to \mu^+\mu^-$, 25, 115, 336
charge conjugation (C), 13, 14, 15, 30, 163
  and Dirac equation, summary, 430
  invariance, in strong interaction, 273–4
charge conservation, 24
charge retention ordering, 99
charged current ν–N elastic scattering cross section, 297
  deep inelastic ν–N interactions, 303–16
charged/neutral leptonic reactions, 26
charged weak interactions, 4
charm, 19–22, 147
  charm threshold, 232
  charmed-meson decays, 232–41
  charmed-meson decays, $D^+ \to l\nu X$, 233–4
  charmed-meson decays, gluon-radiative corrections, 233, 235–6

Index

charmed-meson decays, spectator approximation, 233-5
charmed-meson lifetimes, 235
charmed-particle production in neutrino beams, 317-20
charmed-quark mass, 148, 254
charmonium, 149, 232, 254
Christenson, J. H., 30
circular dichroism and parity violation, 338, 346
circular polarization of γ rays and parity violation in nuclear forces, 360-4
^{37}Cl, solar neutrino detector, 399, 402
closure, and muon capture theory, 197
Coleman, S., 145
color, 5, 6, 143-4, 301
 and τ-hadronic decays, 112
 and cancelation of triangle anomalies, 83
 and $Z^0 \to q\bar{q}$, 116
Compton scattering, 83, 84, 389, 399
confinement, quark, 3, 6
conserved vector current; see CVC hypothesis
constituent quark model, 131-47, 156
cosmic-ray background in neutrino detectors, 419
cosmic-ray decays, 396
cosmological constant, 414
cosmological principle, 414
cosmological red shift, 414
coupling constant, 19, 24
 axial vector, and Adler-Weisberger relation, 439-41
 strong, in QCD, 229, 314
 weak (Fermi), see G_F
covariant derivative, 40, 41, 49
 of lepton fields, 55
 of quark fields, 57
 $SU(3)_L$, 228
Cox, R. T., 11
Cronin, J., 30, 245
current algebra, 175-6, 298, 439-43
current
 axial vector-isovector, 53
 charged weak hadronic, 12, 21, 26, 58, 166, 206
 charged weak leptonic, 12, 52, 58
 charged weak quark, 57, 58, 150, 152
 electronmagnetic, 7, 58, 165-6
 fragments, 328-30
 first-class, and G parity, 168-9
 neutral weak hadronic, 21, 58
 neutral weak leptonic, 57, 58
 second-class, and G parity; see second-class currents
 vector-isovector, 42, 58

D meson(s), 3, 31, 148, 149, 153, 156, 232-41
D^0-$\bar{D}^0$ and CP violation, 272, 282, 283-7
DUMAND, 396, 419
Δ^+ (1236), 333
$\Delta C = -\Delta Q$ rule, 173
$\Delta S = \Delta Q$ rule, 173-74, 205
$|\Delta I| = \frac{1}{2}$ rule, 30, 174-5, 205, 210, 211, 213, 214, 215, 226, 227, 259, 425
Dalitz, R., 159-61
Dalitz diagrams, 159-61, 208
Davis, R., 378, 401
deceleration parameter, 415
deep inelastic scattering
 ν-N charged current, 28, 303-16
 e-N, 306-7
 neutral-current ν-N, 29, 320-30
 neutral current from proton and neutron targets, 324-30
 polarized electron-nucleon and parity nonconservation, 29, 35, 337, 339-43
dichromatic neutrino beam, 290
dilepton events, in high-energy, ν-N collisions, 28
dimuon events and charm production, 318-20
dimuons, same sign, 320
dimensional regularization, 37, 84-5, 436-8
Dirac, P. A. M., 7, 15, 18, 36, 38, 39, 41, 53, 85, 98, 144, 146, 163, 165, 168, 340, 427-8, 429, 438, 440
 equation, summary, 427-8
 fields, summary, 429
 matrices and dimensional regularization, 438
 neutrinos, 379
 particle, 297
divergence, degree of, 78
divergences, in Fermi theories, 17
divergents, primitive, 80, 81
double β decay, 24, 178, 385-90, 425
 experiments, 388-90
 neutrinoless, 386
 rates for, 385
double K capture, 386-7
Dyson, F. J., 36, 67

East Rand Proprietary Gold Mine, 419
elastic neutrino-nucleon scattering, 29
electromagnetic interaction and CP violation, 272, 274-5
electron $e^\pm$, 2
 e capture, double, 386-7

electron $e^{\pm}$ (cont.)
 e capture, monoenergetic neutrinos from, 385
 e capture, orbital, 8, 27, 178
 e electric dipole moment, 274
 e–e interaction, weak–electromagnetic interference in, 345
 $e^+e^- \to e^+e^-$, 25, 127
 $e^+e^- \to \mu^+\mu^-$, 25, 113–15, 127, 129
 $e^+e^- \to \nu_e\bar{\nu}_e$, 26, 407
 $e^+e^- \to \tau^+\tau^-$, 25, 108–11
 $e^+e^- \to W^+W^-$, 118–19
 $e^+e^- \to Z^0$, 25, 117–18
 e–N electromagnetic deep inelastic scattering, 306–7
 e–N interaction, weak-electromagnetic interference in, 29, 35, 337, 339–43
 e–N parity non conserving interaction, 336–54, 367
 e–p elastic scattering, 296
 g-factor anomaly of, 102
electroweak interaction, 1, 5, 228, 421–5
ϵ CP-violation parameter, 256, 257, 270–1
ϵ' CP-violation parameter, 259, 271
η^{+-} CP-violation parameter, 259–62
η^{00} CP-violation parameter, 259–62
Euler, L., 44
 Euler angles, 152
 Euler–Lagrange field equations, 38, 39, 55
exclusive neutral ν–N processes, 321, 332

F meson(s), 3, 31, 149, 233–41
 $F^+ \to \tau^+\nu_\tau$, 26, 156
$^{18}F^* \to {}^{18}F + \gamma$ parity-violating transition, 361–2
$^{19}F^* \to {}^{19}F + \gamma$ parity-violating transition, 362–3
ft values, allowed β decay, 184–6
 ft values of β-decay mirror pairs, 195
factorization, inclusive scattering, 327
factorization hypothesis and neutral weak coupling parameters, 354
Fadeev, L. D., 36, 68
^{56}Fe and supernovae, 411
Fermi, E., 7–9, 12, 36
 Fermi coupling constant; see G_F
 Fermi–Dirac statistics, 408
 Fermi integral and β decay, 185
 Fermi interaction, 9
 Fermi matrix element in β decay, 180
 theory of β decay, 7–9, 18, 81, 180–1
 theory of β decay, difficulties of, 16–18, 81
 Fermi–Yang model, 134

Fermilab (FNAL), 383
 Fermilab 15' bubble chamber, 325
 Fermilab Tevatron $p\bar{p}$ collider, 121
fermion masses, 35, 52
Feynman, R. P., 12, 13, 16, 18, 19, 35, 36, 58, 67, 78, 93
 Feynman rules, 35, 36, 37, 56, 71–8, 122
field, non-Abelian, 35, 36, 43, 44, 68
Field–Feynman model, 328, 329
Fierz transformation, 99, 123, 252, 407
 Fierz interference coefficients, 191–2
Fitch, V., 30, 245, 265
form factor(s)
 baryon semileptonic decay, 157, 168
 deep inelastic ν–N scattering, 308
 hyperon nonleptonic decays, 164–5
 induced pseudoscalar, 170–2, 197, 199
 isoscalar and isovector electric and magnetic, 165–6, 296–9
 K_{l3} decay, 156, 204–6, 207–8
 neutral weak hadronic current, 172–3, 300
 $\pi^+ \to \pi^0 e^+\nu_e$, 167
fractional momentum variable ξ, 304
fractional momentum variable x, 304
fragmentation functions, 328–30

G parity, 141, 168–9
G_F weak (Fermi) coupling constant, 8, 24, 104, 185, 189
 precise determination and numerical value, 104
^{71}Ga solar neutrino detector, 402
γ matrices, summary, 428–9
Gamow, G., 8
 Gamow–Teller matrix element, in β decay, 181
gap method and $K_L - K_S$ mass difference, 265–8
Garwin, R., 10, 104
Gargamelle, 302, 331
gauge
 Landau, 71
 R, 72, 74
 R_ξ, 36, 37, 71–8
 't Hooft–Feynman, 72, 74
 unitary, 50, 72, 73, 74
gauge-fields, isovector, in Yang–Mills theory, 42
gauge invariance
 and fermion–fermion–Higgs coupling, 58–62
 global, 33
 local, 1, 5, 33
 and path integrals, 68–71
 $SU(2)$, and Yang–Mills field, 41–4
 $U(1)$, and electrodynamics, 37–41

Index

gauge quanta, 33, 34, 35
gauge transformation, 33, 68, 70
  chiral $SU(2)$, 53
  $SU(2)$, 33, 41–4
  $SU(2) \times U(1)$, 34, 49
  $U(1)$, 37–41
Gell-Mann, M., 12, 13, 16, 18, 19, 29, 35, 56, 58, 93, 193
generating functional, 66, 67
Georgi–Glashow model, 370, 425
ghost fields, 36, 37, 68, 70, 71, 72, 73, 78
GIM model, 19–22, 57, 62, 85, 150, 151, 152, 251, 253, 299, 370, 393
Glashow, S., 1, 20, 21, 57, 145, 147, 150
gluon(s), 2, 5, 6, 30, 44, 131, 204
  radiative corrections, in charmed-meson decays, 233, 235–6
  radiative corrections, in hyperon nonleptonic decays, 232
Goldberger–Treiman relation, 171–2, 198, 200
Goldstone bosons, 34, 48–9, 50, 72, 73
grand unified theory, 6, 370
gravity, 6
Green's function
  n-point, 66
  and path integral method, 63–5, 68
  in standard model, 72
Grenoble reactor, 382
Gross–Llewellyn-Smith sum rule, 311

H, atomic, parity violation in, 338, 352
^{3}H → ^{3}He $e^- \bar{\nu}_e$, 374, 385, 392–3
hadron(s), 2, 3, 26, 131
hadronic current, neutral weak, 300
haplon, 3
Harvard–Brookhaven–Pennsylvania collaboration, 302
^{3}He ion, 395
Heisenberg, W., 7
Hertzsprung–Russell diagram, 405
helicity
  of $e^\pm$ in β decay, 11, 182–4
  of ν_e, 11, 14, 186–8
  of ν_μ, 200–3
Higgs boson(s), 2, 5, 35, 49, 52, 58–62, 72–5, 78, 86–8, 281, 370, 380, 422
  and CP violation, 281
  mass limits, 87
Higgs' phenomenon, 72, 86
^{163}Ho and neutrino mass, 393
Homestake Gold Mine, 402, 403, 413
Hubble's constant, 414
Huygens' principle, 64
hydrostatic equilibrium and stellar structure, 404
hypercharge, 138
  weak, 54, 57, 83

hyperon(s), 9, 142
  magnetic, moments, 225
  nonleptonic decays, 157–9, 223–7
  nonleptonic decays, $\alpha, \beta, \gamma, \phi$ parameters, 224–5
  nonleptonic decays, final-state interaction, 225
  nonleptonic decays and gluon-radiative correction, 232
  nonleptonic decays, polarization of final baryon, 225
  semileptonic decays, 27, 152, 164, 185, 215–23
  semileptonic decays, asymmetry coefficients for decay of polarized baryons, 217, 218
  semileptonic decays, lepton–neutrino angular correlation, 217, 221–2
  semileptonic decays, and neutral weak hadronic form factors, 300
  semileptonic decays, polarization of final baryon, 217, 218
  semileptonic decays, transition probabilities, 215–8, 221–2

ITEP Laboratory, 393
Iliopoulos, J., 20, 21, 57, 147, 150
impulse approximation and β decay, 178
^{115}In, solar neutrino detector, 402–4
inclusive ν–N scattering, 320–30
inelasticity variable y, 305
infrared slavery, 131
infinite-momentum frame, 304
inner *bremsstrahlung*, 393
interference, weak–electromagnetic in eN interaction, 336–54
intermediate vector boson
  charged; see $W^\pm$
  heavy, $X^\pm, Y^\pm$, 370
  hypothesis, naive, 18–19
  neutral, see Z^0
inverse muon decay; see muon decay, inverse
isoscalar targets, results for inclusive NC ν–N scattering, 322

^{40}K → ^{40}Ar dating, and double β-decay, 388
K-capture, double, 386–7
K meson ($KK^\pm, K^0, \bar{K}^0$), 2, 3, 9, 10, 13
K nonleptonic decays, 209–15, 244–88
K^0-$\bar{K}^0$
  particle mixtures, 29
  decay matrix, 249–50, 254–55
  mass matrix, 249–54
$K_L - K_S$ mass difference, 251–4, 262–8
  gap method of determination, 265–8
  vacuum insertion approximation, 252

Index

$K_L^0 \to \mu^+\mu^-$, 20, 21
$K_L^0 \to \pi^+\pi^-$, $\pi^0\pi^0$, 30, 244–88
$K_L^0 \to \pi\pi$ transition amplitudes, 258–9
K_{l2} decay, 13, 20, 26, 155
  K_{l2} decay constant f_K, 155, 253
K_{l3} decays, 27, 152, 156, 164, 204–9
  and CP violation, 268–9
$K_{\mu 3}$ decay, muon polarization in, 206
$K_S^0 \to \pi^+\pi^-$, $\pi^0\pi^0$, 30, 159–61, 244–88
  $K_S^0 \to \pi\pi$ transition amplitudes, 258–9
Kobayashi, M., 21
Kobayashi–Maskawa model, 21, 152–53, 277–81, 370, 421, 425
  and CP violation, 277–81
Kolar Gold Field, 419
Konopinski, E., 178, 198
Kurie plot, 391

LAMPF accelerator, 383
LEBC, Little European Bubble Chamber, 295
LEP (CERN), 425
Lagrangian density
  β decay (Fermi), 8
  complex isodoublet scalar field, 47, 51, 52
  current–current (Feynman–Gell-Mann), 12
  Dirac electron theory, 38–41
  minimal, of QED, 7, 39–41
  standard model, 35, 37, 47, 51, 52, 55, 58, 62–63, 299, 371
Λ^0 hyperon, 2, 13, 145
Λ_c^+ baryon, 3, 149, 240–1
  decay modes, 240–1
  mass, lifetime, 241
Landau, L. D., 30, 71
Langacker, P., 169
Lederman, L., 104
Lee, B. W., 37, 67
Lee–Sugawara relation, 227
Lee, T. D., 10, 12
left-handed lepton fields, 53, 421
left–right symmetric theories, 85–6, 189, 424
Legendre transformation, 72
lepton(s) 1, 2, 5, 52, 58, 421
  coupling to scalar field, 58–62
  doublets, 2, 52, 108
  generation(s), see lepton doublets
  masses in standard model, 61–2
  mixing, 371
  neutral heavy, 394, 418
  number conservation, 2, 11, 22, 24, 98, 113, 128, 369, 425
  number conservation, and τ decay, 113
  number conservation, and inverse muon decay, 128
  number, nonconservation of separate, 370
leptonic weak currents, see current(s)
leptonic weak interactions, 22–26, 92–130
^{7}Li, solar neutrino detector, 402
logarithmic scaling violations, 316
longitudinal structure function, 310
Lorentz invariance, proper, and restrictions on weak amplitudes, 8, 98, 131, 156–61, 191
Los Alamos Laboratory, 383

m_R: mass of right-handed W boson, 86, 424
Maiani, L., 20, 21, 57, 147, 150
main sequence, stellar evolution, 405
Majorana
  field, 377
  neutrinos, massless, 379
  neutrinos, oscillations of, 376–80
Marshak, R., 12
Maskawa, T., 21, 370
mass mixing, neutrino, 373
meson(s), 1, 2, 3, 139, 143
  factory, 383
  pseudoscalar, 140, 141, 144
  vector, 140, 141, 148
Michel parameters; see muon decay, τ decay
millistrong interaction and CP violation, 272, 273
milliweak interaction and CP violation, 272, 275–6
Mills, R., 33; see also Yang–Mills field
mirror reflection, see parity
^{55}Mn, solar neutrino detector, 402
Møller scattering, 389
Morita, M., 178, 198
muon $\mu^\pm$, 1, 2, 9
$\mu^- \to e^-e^+e^-$, 22, 369
$\mu \to e\gamma$, 22, 369, 394, 425
muon capture, 1, 9, 27, 178, 197–203
  effective Hamiltonian for, 199
  impulse approximation for, 197
  in ^{12}C, 197, 200–203
  in ^{12}C and muon helicity, 200–2
  induced pseudoscalar form factor for, 197
  inverse, 28
  weak magnetism and CVC in, 197
muon decay, 1, 9, 10, 12, 13, 18, 19, 22, 56, 86, 92–108, 197
  δ parameter, 98–101, 107
  $e^\pm$ asymmetries in polarized, 97, 106–7
  $e^\pm$ helicities, 97, 99, 104–5
  Michel parameters, 98–107
  η parameter, 98–101, 106

Index

inverse, 22, 92, 108, 121, 127–9
polarization parameter ξP, 86, 98–101, 106–7
ρ parameter, 86, 98–101, 105–6
radiative corrections, 18, 100, 101
summary of observed parameters, 107–8
transition probability formulas, 93, 97, 98, 100, 101
vector coupling constant, 13
muon $g - 2$ experiments, 101–4
muon, g-factor anomaly, 77, 101–4
muon magnetic moment, 101, 104
muon mass, 104
muon parity, 22
muonium, 104

Nachtman variable, 316
neutral
  current ν–N elastic scattering, 299–303
  current ν–N reactions, as tests of the standard model, 320–35
  heavy leptons, 394, 418
  leptonic reactions, 24, 35
  ν–N collisions, 29, 35, 320–35
  weak couplings, summary, 422–3
  weak current, and neutrino refractive index, 381
  weak hadronic current, 300
  weak Lagrangian density for ν–q interactions, 299
$\bar{\nu}_e + D \to \bar{\nu}_e + n + p$, 29, 321, 333, 334
$\nu_e e^- \to \nu_e e^-$, 26, 121
$\bar{\nu}_e e^- \to \bar{\nu}_e e^-$, 26, 121
$\nu_\mu e^- \to \nu_\mu e^-$, 24, 121
$\bar{\nu}_\mu e^- \to \bar{\nu}_\mu e^-$, 24, 121
$\nu N \to \nu \Delta(1236)$
ν_p elastic scattering, 334
neutrino astrophysics, 396–420
neutrino beams, charmed-particle production in, 317–20
neutrino-burst detector, 412
neutrino charge, and dimensional regularization, 84, 85, 394–5
neutrinos, cosmological, 413–9
neutrino coupling
  to gauge fields, 52–6
  to scalar field, 58–62
  to Z^0, 75
neutrino decay, 393–5
neutrino, definition, 2
neutrino detectors, 292–5
neutrino, dichromatic beam, 290
neutrino–electron scattering, 92, 121–129, 399
  in Fermi theory, 16–7
neutrino, emission in hot stars, 290, 404–10
neutrino, energy spectrum in high-energy beams, 290
neutrinos, experimental methods for production at high energy, 289–92
neutrino flavors, 2, 25
neutrino helicity
  $h(\nu_\mu)$ and muon capture in ^{12}C, 200–3
  in μ decay, 97
  $h(\nu_e)$, 11, 14, 186–8
neutrino, highband beam, 292
neutrino, high-energy beams, 383
neutrino, hypothesis of Pauli, 7
neutrino magnetic moment, 393–5, 420
neutrino mass, 16, 24, 35, 52, 61, 62, 107, 113, 371, 425
  direct measurement of $m_{\bar{\nu}_e}$, 390–3
  limit on m_{ν_μ}, 107
  limit on m_{ν_τ}, 133
neutrino mixing, 370
neutrino, ν_τ, 24
neutrino–nucleon deep inelastic charged-current interactions, 303–16
neutrino–nucleon scattering, 28, 29, 35, 166, 295–9, 303–16, 320–35
neutrino–nucleon neutral current reactions, as tests of the standard model, 320–35
neutrino oscillations, 24, 35, 61, 62, 370, 425
neutrino oscillations
  CP violation in, 375
  $\Delta L \approx 0$, 371–4
  experiments, 382–5
  flavor, 372–4
  of Majorana neutrinos, 376–80
  $\nu \to \bar{\nu}$, 370
  three lepton generations, 374–6
neutrinos, prompt, 385
neutrino, refractive index of matter for, 380–1
neutrinos, relic from big bang, 397
neutrino, solar flux, 398
neutrino, two-component theory, 15, 16
neutrino, wideband beam, 291
neutron, 2, 3, 8, 11, 12, 145
  electric dipole moment, 274, 281
  magnetic moment and CVC hypothesis, 166
  $n + p \to D + \gamma$ parity violating transition, 363–5
  $n \to p e^- \bar{\nu}_e$, 8, 11, 12, 161–2, 169, 189, 190
  spin rotation and parity violation, 365–7
^{56}Ni, 411
nonleptonic weak transitions, 26, 29

nuclear force, weak, 31, 338–9, 354–67
nucleon–nucleon, parity nonconserving interaction, 31, 338–9, 354–67

$^{16}O \to {}^{12}C + \alpha$ parity-violating transition, 359
octet dominance, 211–12, 227, 228
Ω^- hyperon, 3, 142
Ω^- nonleptonic decays and $|\Delta I| = \frac{1}{2}$ rule, 227, 232
optical rotation and parity violation, 338, 349–52

PCAC (partially conserved axial current hypothesis), 169–72, 176, 217, 440
parity (P), 8, 13–15, 30, 141, 163, 274
and Dirac equation, summary, 430
parity violation, 10–12, 13–15, 35, 53, 98, 115, 187, 189
eN interaction, 336–54
NN interaction, 354–67
in NN interaction, asymmetry in γ emission, 360–5
in NN interaction, circular polarization of γ rays, 360–4
parton, 28
Paschos–Wolfenstein relation, 323
path integral method, 36, 63–71
in quantum mechanics, 63–5
in field theory, 65–7
Pauli, W., 7
Pauli–Dirac representation, 163, 376, 378, 428, 429
Pauli matrices, 133, 255
Pauli moment, 165
Pauli principle and color, 143
penquin diagrams, 231
ϕ^0 meson, 3
ϕ^4 interaction, 47
photon γ 2, 4, 7, 8, 25, 28, 33, 40, 52, 72, 78, 422
photoneutrino process for cooling stars, 407
physical constants, summary, 444
pion $\pi^\pm$, π^0, 2, 3, 9, 10, 12, 131, 134
$\pi^0 \to \gamma\gamma$ and color hypothesis, 143
pion β decay $\pi^+ \to \pi l \nu$, 26, 156, 167–8
$\pi_{e2}/\pi_{\mu 2}$ branching ratio, 26, 154, 155
pion fragmentation, 331
pion leptonic decay $\pi_{\mu 2}$, 13, 14, 15, 26, 153–5
transition probability, 154
effect of neutrino mixing on, 374
π–N scattering cross-sections and Adler–Weisberger relation, 439–41

Planck mass, 7n
plasma process for cooling stars, 407
polarized electron–nucleon scattering and parity nonconservation, 29, 337, 339–43
Popov, V. N., 36, 68
power counting, naive, 80
preon, 3
Primakoff, H., 196, 197, 198
primordial helium mass fraction, 417
proton, 2, 3, 131, 145
decay, 24, 370, 425
electric dipole moment, 274
magnetic moment, and CVC hypothesis, 166
proton–proton solar reaction chain, 397, 398, 399
pseudoscalar coupling, 9, 11, 26, 108
neutral, limit from inclusive ν–N neutral–current scattering, 322
ψ/J meson, 2, 148
^{193}Pt and neutrino mass, 393

Q_W (weak charge), 345
QCD (quantum chromodynamics), 1, 6, 28, 30, 44, 131, 228–32, 314
anomalous dimensions, 230–2
gluon-radiative corrections in weak decays, 230, 232, 233, 235–6
scale-breaking corrections, 310
QED (quantum electrodynamics), 4, 5, 7, 24, 35, 36, 37–41, 68–70, 78, 80, 81, 82, 83, 102
quadrilinear gauge couplings, and Yang–Mills field, 44, 75, 76
quark(s), 1, 2, 3, 4, 5, 6, 20, 21, 28, 35, 52, 57, 58, 62, 83, 88, 112, 116, 120, 131, 137–76, 421
quark–antiquark ocean, 3, 28
colors, 5, 83, 112, 116, 133–44
confinement; see confinement, quark
distribution functions, 311, 316
flavors, 2
fractional charge, experimental evidence from ν–N scattering, 311
generations, 57
quark–gluon dynamics, 425
mass, constituent, 145–7
mass, current, 145–7
model, color symmetric $SU(6)$, 145–7, 223, 301
quark–parton model, 303, 309, 311
quark–parton model and polarized electron–nucleon scattering, 339–41
valence, 3, 28, 139
quarkonium, 241

Index

R gauge, 72, 74
R_ξ gauge, 36, 37, 71–8
radiation-dominated universe, 415
radiochemical neutrino detector, 399
^{87}Rb, solar neutrino detector, 402
reactor experiments, neutrino oscillations, 382
red giant, 405
refractive index of matter for neutrinos, 380–1
regeneration and K^0–$\bar{K}^0$ mixing, 247–9, 262–5
regularization, 81, 84, 169
relativistic notation, summary, 426
renormalization, 5, 17, 18, 37, 73, 78–85
renormalization group methods in QCD, 228
right-handed weak isosinglets, 421
rishon, 3
Robertson–Walker metric, 414
Rosenbluth formula, 296

S-matrix, 432–435
SLAC (Stanford Linear Accelerator Laboratory), 148, 309
  SLAC polarized-electron experiment, 337, 339, 341–3, 352–4
SLC (SLAC linear collider), 425
SPEAR (Stanford Positron Electron Asymmetric Ring), 108, 241
$SU(2)$ algebra, 41, 44, 133–4
$SU(3)$ algebra, 44, 134–42
  irreducible representatives, 136–42, 211–12
  structure constants, 135–6, 228
  weight diagrams, 137–42
$SU(3)_c$, 143–4, 146, 228
$SU(3)_f$, 131, 137–42, 143, 144–5, 147, 148, 206, 211–12, 218–21, 236–40, 300
  and current algebra, 175–6
  axial currents in, 145, 211–12, 218–21
  vector currents in, 144, 206, 211–12, 218–21
$SU(4)_f$, 131, 148, 236–40
  irreducible representations, 148–9
$SU(5)$ grand unified model, 370, 425
$SU(6)_f$, 142–7
  color-symmetric quark model, 301
$SU(8)_f$, 149
$SU(n)$, 132
Salam, A., 1, 35, 74, 189
Sakurai, J. J., 153
Savannah River Reactor, 383
scalar coupling, 8, 9, 11, 22, 27, 98, 108

neutral, limit from inclusive ν–N neutral-current scattering, 322
  in K_{l3} decays, 206, 207
scalar field equations, 39, 47
scale-breaking QCD corrections, 310
scaling, 28, 309
  violations, 312, 314–6, 322
Schopper, H., 178
Schwarzchild limit, 412
^{82}Se → ^{82}Kr double β decay, 388–9
second-class currents, 168–9, 181, 192, 195–6, 296
  $A = 12$ β decay, 192–6
  β-decay mirror-pair ft values, 195
second-order weak process, double β decay, 386
semi-inclusive ν–N scattering, 29, 321
semi-inclusive pion production, 331
semileptonic weak transitions, 26
semiweak coupling constant, 19, 24
sequential leptons, 24, 108
^{28}Si and stellar evolution, 411
$\Sigma^\pm$, Σ^0 hyperon(s), 3, 13, 145
$\Sigma^\pm \to p\gamma$, 394
six-quark model; see Kobayashi–Maskawa model
solar
  age, since the start of hydrogen burning, 401
  constant S, 398
  luminosity, 400–1
  mass, 400–1
  models, 399–401
  neutrinos, 397–9
  neutrino flux, 398
  neutrino unit (SNU), 401
spatial inversion, see parity
spectator quarks, 327
spontaneous symmetry breaking, 5, 34, 44–52, 85, 86
  and elastic instability in classical mechanics, 44–6
standard model, 1, 5, 22, 24, 29, 31, 33–91, 92, 127–9, 421–5
  determination of parameters by leptonic interactions, 127–8
  and e–N interaction, 339, 341, 343, 345, 354
  predictions for ν–q interactions, 299
  tests with neutral ν–N reactions, 320–35
Stark interference and parity violation, 338, 348–9
stars, neutrino emission by hot, 404–10
strangeness, 13, 20, 28, 29, 138, 244
strangeness-changing neutral weak interactions, 20–2

472 Index

strangeness oscillations, 1, 30, 204, 247–9
Strathdee, J., 189
structure functions, 306
Sudarshan, E., 12
supernovas, coordinated search for, 413
supernova explosion, 396, 410–12
superweak interaction and CP violation, 272, 276–7

T invariance, see time-reversal invariance
't Hooft, G., 5, 37, 84
target fragments, 328–30
τ lepton ($\tau^\pm$), 1, 2, 24, 92, 108–13
  τ decay, 1, 22, 24, 92, 110, 112–13, 155–6
  τ decay, hadronic, 112, 155–6
  τ decay, leptonic, 22, 24, 112
  τ decay, lifetime, 112
  τ decay, ρ parameter, 112
  τ production cross section, 110–11
τ–θ puzzle, 10
Te → Xe double β decay, 388–90
Teller, E., 8
tensor couplings, 9, 11, 22, 27, 98, 108
  in K_{l3} decays, 206–7
thallium, parity violation in, 338, 347–9, 350, 353, 354
time reversal (T) invariance, 11, 15, 99, 131–2, 161–5, 205, 225, 251, 255, 257, 258, 273, 296, 430–1
  and triple correlations, 161–2, 189–90, 209, 275–6
time reversal (T) transformation, 15
^{205}Te, solar neutrino detector, 402
toponium, 88
triangle anomaly, 82–83, 421, 438
trilinear gauge couplings and Yang–Mills field, 44, 75, 76, 84, 229
trimuon events, 320
Turlay, R., 301
two-neutrino experiment, 369

URCA process, 410
unitarity
  and $e^+e^- \to W^+W^-$, 119
  failure of in Fermi and naive IVB theories, 17, 19
  failure of, in triangle diagrams, 83
unitary symmetry and the quark model, 132–50
unphysical scalar bosons, 72, 76, 77, 78
Υ (upsilon) resonances, 149, 241–2

^{51}V, solar neutrino detector, 402
V–A law, 12–15, 22, 24, 35, 92, 93, 97, 98, 101, 106–8, 112, 113, 127–9, 178, 191–2, 200, 201, 202, 204, 299, 369, 422
V + A currents, 85, 191, 391, 424
vacuum-insertion approximation,
  $K_L - K_S$ mass difference, 252
vacuum polarization, 228, 229
valence quark(s); see quark(s), valence
valine, 393
vector (V) coupling, see β decay, muon decay
Veltman, M., 37, 84, 87

$W^\pm$, 2, 4, 5, 18–19, 20, 22, 24, 28, 31, 33, 35, 50, 51, 52, 55–9, 72–6, 84, 85–6, 88, 93, 117, 118–21, 129, 422
  decay rate, 117
  magnetic moment, 76
  mass, 24, 56, 57
  mass, limit m_R for right-handed boson, 86, 424
  production in e^+e^- collisions, 92, 118–19
  production in $p\bar{p}$ collisions, 120–1
  propagator, 18, 56, 73, 74
Ward, J., 1
weak–electromagnetic interference, definition, 25
weak hypercharge, 60, 61
weak magnetism, 166–7, 181, 189, 192–6
Weinberg, S., 1, 35, 74, 146
Weinberg angle, 50, 56, 57, 323, 422–4
Weisberger, W., 176
Weyl, H., 15
  Weyl representation, 15, 376, 429
white dwarf, 406
Wick, G., 36
Wigner–Eckhart theorem, 214, 226
Wu, C. S., 10, 178

x, fractional momentum scaling variable, 304

y, inelasticity variable in deep inelastic scattering, 305, 340
y distribution for charm production, 319
Yang, C. N., 10, 12, 33
Yang–Mills field, 33, 36, 37, 41–4, 52, 68–71, 78, 81, 83, 88, 118, 146, 228, 282
  $SU(3)_c$, 228
Young diagrams, 139

Yukawa, H., 28, 35, 59, 358
Yukawa coupling, 59–62, 375

Z^0 boson, 2, 4, 5, 20, 24, 25, 26, 29, 31, 33, 35, 50–2, 55–9, 72–4, 88, 101, 110, 113–18, 120, 121, 129, 186, 422
   decay, 25, 115–17
   mass, 24, 56, 57
   production in e^+e^- collisions, 25, 92, 117–18
   production in $p\bar{p}$ collisions, 120–1
   propagator, 56, 73, 74
Zinn–Justin, R., 37
^{65}Zn, monoenergetic neutrino source, 385
Zweig's rule, 280

PHYSICS LIBRARY

RETURN TO: PHYSICS LIBRARY
351 LeConte Hall   510-642-3122

LOAN PERIOD 1 **1-MONTH**	2	3
4	5	6

ALL BOOKS MAY BE RECALLED AFTER 7 DAYS.
Renewable by telephone.

DUE AS STAMPED BELOW.

MAR 2 9 2004 MAY 1 5 2005		DEC 1 4 2008
OCT 0 5 2009 JUL -2 2010 SEP 2 7 2010		
DEC 1 5 2010		

FORM NO. DD 22   UNIVERSITY OF CALIFORNIA, BERKELEY
500   4-03   Berkeley, California 94720–6000